Railway Systems Engineering: Security Concerns

Railway Systems Engineering: Security Concerns

Edited by **Marshall Roy**

Clanrye International

New Jersey

Published by Clanrye International,
55 Van Reypen Street,
Jersey City, NJ 07306, USA
www.clanryeinternational.com

Railway Systems Engineering: Security Concerns
Edited by Marshall Roy

International Standard Book Number: 978-1-63240-438-1 (Hardback)

Printed in China.

Contents

Preface

In my initial years as a student, I used to run to the library at every possible instance to grab a book and learn something new. Books were my primary source of knowledge and I would not have come such a long way without all that I learnt from them. Thus, when I was approached to edit this book; I became understandably nostalgic. It was an absolute honor to be considered worthy of guiding the current generation as well as those to come. I put all my knowledge and hard work into making this book most beneficial for its readers.

This book is an important source of information on railway systems engineering. In railway applications, performance studies are important for increasing the durability of railway systems. One of the main objectives is to validate whether the working conditions are reliable and safe. This task not only requires analysis of the whole traction chain, but also needs authentication that the railway infrastructure is efficiently working. Therefore, many tests for detecting any dysfunctions on their proper operation have been developed. This book discusses the various basics of railway traction, providing suggestions on exploratory approaches to spot any dysfunctions as well as reliability and safety issues. This book aims to serve as a valuable reference for students, educators, scientists, faculty members, researchers, and engineers.

I wish to thank my publisher for supporting me at every step. I would also like to thank all the authors who have contributed their researches in this book. I hope this book will be a valuable contribution to the progress of the field.

Editor

Part 1

Introduction to Railway Traction

1

New Advances in Analysis and Design of Railway Track System

J. Sadeghi
Iran University of Science and Technology, Iran

1. Introduction

A ballasted railway track system comprises several components among which steel rails, rail fasteners, timber, steel or concrete sleepers, granular ballast, sub-ballast, and subgrade materials are the main parts. Railway track systems are constructed to provide a smooth and safe running surface for passenger or freight trains. They are designed to sustain the force of lateral, longitudinal and vertical loads imposed on the track structure.

Due to the wide range of mechanical characteristics of the track complements, as well as the components' complex interaction, there is a lack of a comprehensive and precise understanding of track mechanical behaviour, particularly when the nonlinear and dynamic properties of the track are considered [1]. Several important criteria have been defined in the conventional track design methods to ensure that the weight of the load is safely transferred to the ground. These criteria include limits on rail and rail fastener stresses, rail deflections, sleeper stresses, contact pressure between sleepers and the ballast, and the pressure transferred to the supporting layers underneath the track [1].

The recent extensive increases in axle loads, speed and traffic volume, along with the need to improve passenger comfort and reduce track life cycle costs, have caused the subject of track design optimization to arise. Furthermore, complementary decision support systems require a more precise analytical and mechanistic approach to meet the design needs of modern railway track systems. These aspects highlight the necessity of a thorough review and revision of the current railway track designs.

A thorough review of conventional track design method was made by this author elsewhere [1]. This chapter (in section 2) presents a summery of the track design review adapted directly from the author work in [1]. The recent researches in the field of railway track analysis and design are also discussed. Addressing the limitations of the current codes of practice, some methods for improvement of the current track design methods are presented. These methods consider new research findings in the following areas;

- application of dynamic impact factors,
- accurate loading pattern for the sleepers,
- incorporation of nonlinear characteristics of the rail support system, and
- incorporation of plastic deflection of the track support system.

2. Conventional methods in analysis and design of railway tracks

A wide range of railway design codes including AREMA Manual, Euro Codes, UIC Leaflets, Australian Standards (AS), South African Railway Codes, and Indian Standards are available worldwide. The design methods in all of the available codes follow the same approach although they might have some minor differences. Due to great variety of structural elements used in the track system, the railway standards consider each track component as a single structural unit and suggest each complement be designed independently. Such an approach, subsequently, includes the interaction between track components by defining suitable boundary conditions and load transfer patterns [1]. Furthermore, since the dynamic response characteristics of the track are not sufficiently well understood to form the basis of a rational design method, current practices greatly rely on relating the observed dynamic response to an equivalent static response; this is carried out by making use of various load factors [1]. In the conventional methods, railway tracks are designed utilizing load bearing approach to ensure that the concentrated loads of the wheels are transferred to the track formation while the strength of the components are not exceeded [1]. Several important criteria are defined to secure this objective. They are detailed under.

2.1 Rail

Rail, as the most important track element, should be able to securely sustain wheel loads in the vertical, lateral, and longitudinal directions and subsequently transfer them to the underlying supports. Rail is a track element which is in direct contact with the rolling stock and therefore, it is very important to ensure its proper functioning [1].

The most important criteria used in the conventional design procedure are presented in Figure 1 [1]. As it is illustrated in this figure, rail design criteria are mainly divided into two categories: structural strength and serviceability. Structural strength criteria include wheel-rail contact stresses and rail bending stresses. Having satisfied the structural strength criteria, the serviceability requirements should be met to ensure the rail proper structural and operational performance. Moreover, it is important that track design engineers should have a deep understanding of the track operating conditions in order to make appropriate assumptions in the design process [1].

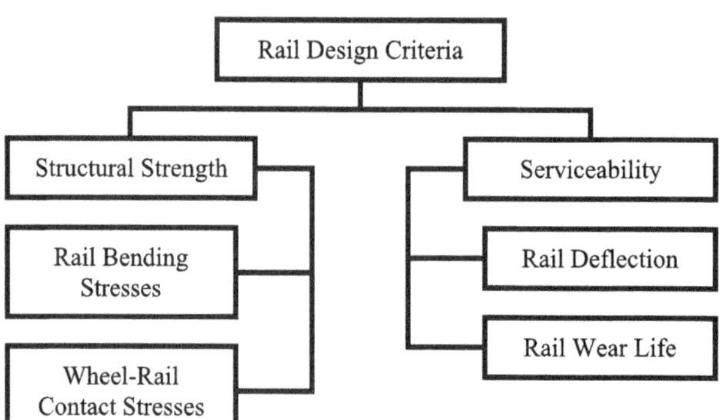

Fig. 1. Recommended rail design criteria [1]

Current practices in the calculation of rail bending moments and vertical deflections are mainly based on the theory of "beam on elastic foundation model" [1]. This model was proposed for the first time by Winkler in 1867 and thereafter developed by Zimmerman in 1888 [2]. The basic assumption in the Winkler model is that the deflection of the rail at any point is proportional to the supporting pressure under the rail (Figure 2) [1].

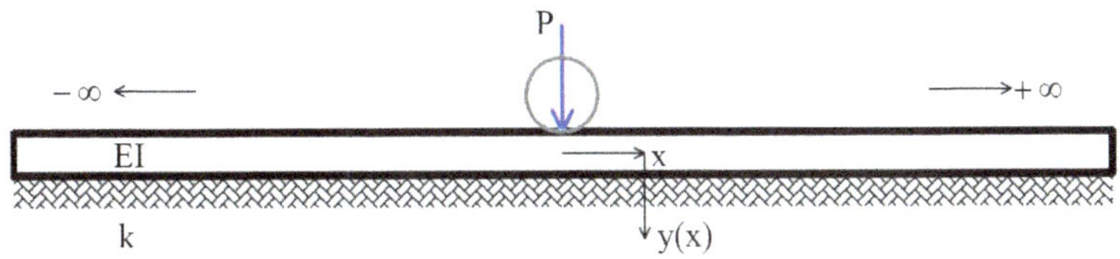

Fig. 2. Beam on elastic foundation model [1]

The corresponding equations for the calculation of rail bending moment and rail deflection are as follows [1].

$$y(x) = \frac{P\beta e^{-\beta x}}{2u}(\cos\beta x + \sin\beta x) \tag{1}$$

$$M(x) = \frac{P}{4\beta}e^{-\beta x}(\cos\beta x - \sin\beta x) \tag{2}$$

where, y(x) and M(x) are the vertical deflection and the bending moment of the rail at the distance "x" from the load point, respectively. Parameter β is defined by the following equation:

$$\beta = \left(\frac{u}{4EI}\right)^{0.25} \tag{3}$$

Winkler model is basically developed for a continuously supported beam on an elastic foundation. This approach neglects some conditions of railway tracks [1]. First, the assumption of continuous support under the rail does not reflect the effects of actual discrete support provided by cross sleepers. Second, this model does not include various track supporting layers (i.e. ballast, sub-ballast, and subgrade) and simply uses the Bernoulli-Euler beam theory to calculate rail deflections and bending moments. Third, it is assumed that the sleepers do not resist against rail bending despite their rotational stiffness [1]. Due to the above limitations, some researchers have questioned the reliability of the Winkler model [1].

Design procedure for a specific rail section always starts with the calculation of the design wheel load. This load is defined as the product of static wheel load and a corrective factor known as dynamic impact factor (DIF) to compensate for dynamic impact of wheel loads irregularities [1]. Taking into account various parameters which affect the magnitude of dynamic impact factor, several relationships for the estimation of this parameter have been proposed [1]. The main DIF suggestions are presented in Table 1.

Developer	Equation
AREMA	$\varphi = 1 + 5.21\frac{V}{D}$
Eisenmann	$\varphi = 1 + \delta.\eta.t$
ORE	$\varphi = 1 + \alpha' + \beta' + \gamma'$
BR	$\varphi = \frac{8.784(\alpha_1 + \alpha_2)\, V}{P_s}\left[\frac{D_j\, P_u}{g}\right]^{1/2}$
India	$\varphi = 1 + \frac{V}{58.14u^{0.5}}$
South Africa	$\varphi = 1 + 4.92\frac{V}{D}$
Clarke	$\varphi = 1 + \frac{19.65V}{D\, u^{1/2}}$
WMMTA	$\varphi = (1 + 3086 * 10^{-5} V^2)^{0.67}$
Sadeghi	$\varphi = 1.098 + 8\times10^{-4} V + 10^{-6} V^2$

Table 1. Dynamic impact factor [1-2]

The magnitude of vertical rail deflection calculated using Equation (1) is greatly dependent upon track modulus [1]. Track modulus is defined as the load which causes unit vertical deflection in unit length of the rail. For typical tracks with light to medium rails, AREMA [6] recommends a value of 13.8 MPa [1].

The rail bending stress is usually calculated at the center of the rail base assuming the pure bending conditions are applicable [1]. The bending stress at the lower edge of the rail head also may be critical if the vehicles impose high guiding forces between wheel flange and rail head. Having calculated the magnitude of the rail bending stress, comparisons should be made between this stress and the allowable limit. AREMA [6] has recommended a practical methodology for calculation of rail bending stress based upon fatigue consideration and through the determination of several safety factors. According to this method, the allowable bending stress is defined as [1,7]:

$$\sigma_{all} = \frac{\sigma_y - \sigma_t}{(1 + A)(1 + B)(1 + C)(1 + D)} \tag{4}$$

where, σ_y is the yield stress of rail steel and σ_t is the longitudinal stress due to temperature changes. σ_t can be calculated using the following equation [1]:

$$\sigma_t = E.\alpha\Delta t \tag{5}$$

The parameters A, B, C, and D in Equation (4) are safety factors to account for rail lateral bending, track condition, rail wear and corrosion, and unbalanced super-elevation of track, respectively [1]. Some of the recommended values for the safety factors are presented in table 2 [1].

Wheel-rail contact stresses mainly include rolling and shear stresses [1]. The magnitude of these stresses is greatly dependent upon the geometry of ellipsoidal wheel-rail contact patch. Many investigations have been carried out to develop reliable formulations for the calculation of these stresses. The most applicable formulas are those suggested by Eisenmann [8]. He conducted an analysis of rolling and shear stress levels in which a simplifying assumption of uniform distribution over the wheel-rail contact area was made. In this analysis wheel and rail profiles were also represented by a cylinder and a plane, respectively (Figure 3) [1].

Safety Factor	Researcher		
	Hay	Clarke	Magee
A	15%	15%	20%
B	25%	25%	25%
C	10%	10%	35%
D	15%-20%	25%	15%

Table 2. Values of rail bending stress safety factors [7]

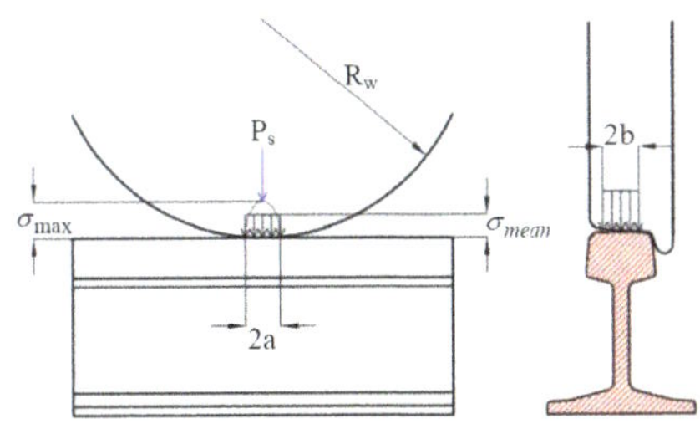

Fig. 3. Uniform distribution of wheel-rail contact stress [5]

Based on Hertz's theory, Eisenmann [8] suggested the following formula for the calculation of the mean value of the rolling contact stress [1]:

$$\sigma_{mean} = \frac{P_s \times 10^3}{2a \times 2b} \tag{6}$$

where, 2b (mm) is the width of wheel-rail contact area. Eisenmann [8] adopted the value of 12 (i.e., 2b=12 mm). The contact length (2a) is also calculated by the following formula [1]:

$$2a = 3.04 \times \left[\frac{P_s . R_w \times 10^3}{2b.E} \right]^{0.5} \tag{7}$$

The values of wheel loads transferred to the rail head through the contact area often exceed the yield limit of the contacting materials [1]. In this situation, the resulting surface plastic deformations jointed with wear processes acts to flatten out the contact area. Therefore, the contact surface can be approximated by a rectangle of the length of 2a and the breadth of w based upon the assumption of contact between a plane (rail) and a cylinder (wheel). For such a condition, Smith and Liu [9] suggested the following formula for the calculation of the contact length [1]:

$$2a = 3.19 \times \left[\frac{P_s.(1-\nu^2).R_w \times 10^3}{w.E} \right]^{0.5} \quad (8)$$

Considering required fatigue strength for rail steel, Eisenmann [8] proposed the limit value for mean rolling contact stress as a percentage of the ultimate tensile strength of rail steel. Based on this assumption, subsequent criterion was suggested [1]:

$$\sigma_{all(roll)} = 0.5\sigma_{ult} \quad (9)$$

Shear stress distribution is chiefly occurs in the rail head area and is in a close relationship with the magnitudes of normal principal stresses [1]. Eisenmann [8] observed that the values of major and minor stresses do not follow the same reduction patterns with increasing depth from the rail head surface. Such a discrepancy results in the appearance of a maximum value of shear stress at a depth corresponding to half of the contact length [2]. Maximum shear stress value is simply interrelated to mean rolling contact stress values and is given by the following equation [1]:

$$\tau_{\max} = 0.3\sigma_{mean} \rightarrow \tau_{\max} = 410\sqrt{\frac{P_s}{R_w}} \quad (10)$$

As indicated earlier, the magnitudes of shear and rolling contact stresses are interrelated [1]. Using the theory of shear strain energy applied for the condition in which the two principal stresses are compressive, the subsequent criterion for the shear stress limit can be obtained [1]:

$$\tau_{all} = \frac{1}{\sqrt{3}}\sigma_{mean} \rightarrow \tau_{all} = 0.3\sigma_{ult} \quad (11)$$

Criteria related to the performance of the rails under operating conditions mainly include rail vertical deflection and rail wear life [2]. These criteria are presented hereunder.

AREMA [6] has proposed a limiting range for the magnitudes of vertical rail deflections. According to this recommendation, vertical rail deflections should be kept within the range of 3.175 to 6.35 millimetres [1]. Lundgren and his colleagues [10] has incorporated this recommendation and proposed a diagram presented in Figure 4 which presents the limit values of vertical rail deflection. This diagram is based upon the capability of the track to carry out its design task [1].

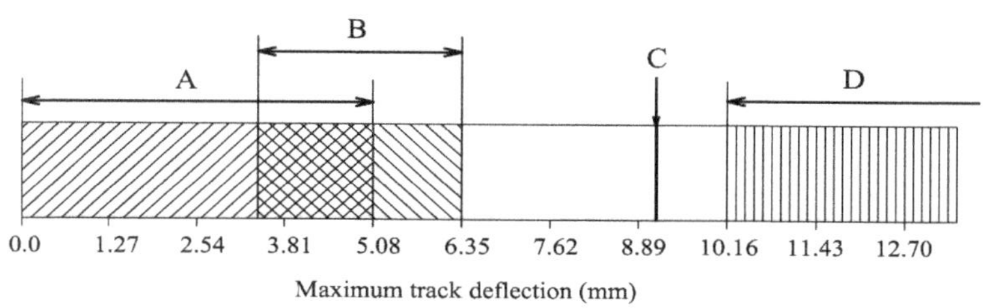

Fig. 4. Track deflection criteria for serviceability [2-10]

Domains indicated in Figure 4 are described as follows [1,2].

A: Deflection range for track which will last indefinitely.

B: Normal maximum desirable deflection for heavy track to give requisite combination of flexibility and stiffness.

C: Limit of desirable deflection for track of light construction (with rails weigh < 50 kg/m)

D: Weak or poorly maintained track which will deteriorate quickly.

The other serviceability criterion is the rail wear life [1]. Although many investigations have been carried out to develop a rational method for estimation of this parameter, the results at the best are still empirical and have no theoretical support [1].

The University of Illinois has conducted a research on some U.S. railway tracks in order to investigate the rail wear rate and consequently develop a method for the estimation of rail wear life [1]. The following formula is suggested by the Illinois University for the estimation of annual rail head area wear (mm^2/year) [11]:

$$W_a = W_t.(1 + K_w.D_c).D_A \tag{12}$$

Having estimated W_a and considering the maximum rail head limit (θ_A), the rail wear life could be calculated from the following formula [1]:

$$T_y = \frac{\theta_A}{W_a} \tag{13}$$

Danzig and his colleagues [12] from the AREMA association also carried out extensive investigations to find a proper formulation for the estimation of rail wear life. Based on the results obtained, they suggested the following equation which represents the rail wear life in terms of MGT passed over a specific time period [1]:

$$T = \frac{1.839 K_C.K_G.K_R.R_{Wt}.(1.102 D_A)^{1.565}}{\sum_i^n \left[\frac{1.102 D_i}{K_{V_i}.K_{A_i}.K_{S_i}} \right]} \tag{14}$$

Each railway industry has established its own allowable rail wear limits [1]. For instance, the envelope of maximum rail wear values for different rail sections set by Canadian National Railway is presented in Figure 5. Using the diagrams outlined in this figure, maximum allowable rail head height and width losses can be determined [1].

The acceptable rail wears usually range from 20 to 50 percent of the rail head area [1]. The weight of unit length of the rail, the amount of MGT passed over the track during its service life, and the train speed are the most important parameters which determine the proper values of allowable rail wear limits to be chosen. The more weight of unit length of the rail, the greater amount of rail head area reduction would be allowed. On the other hand, the greater amount of MGT and higher values of train speed call for more limited rail head area reduction [1].

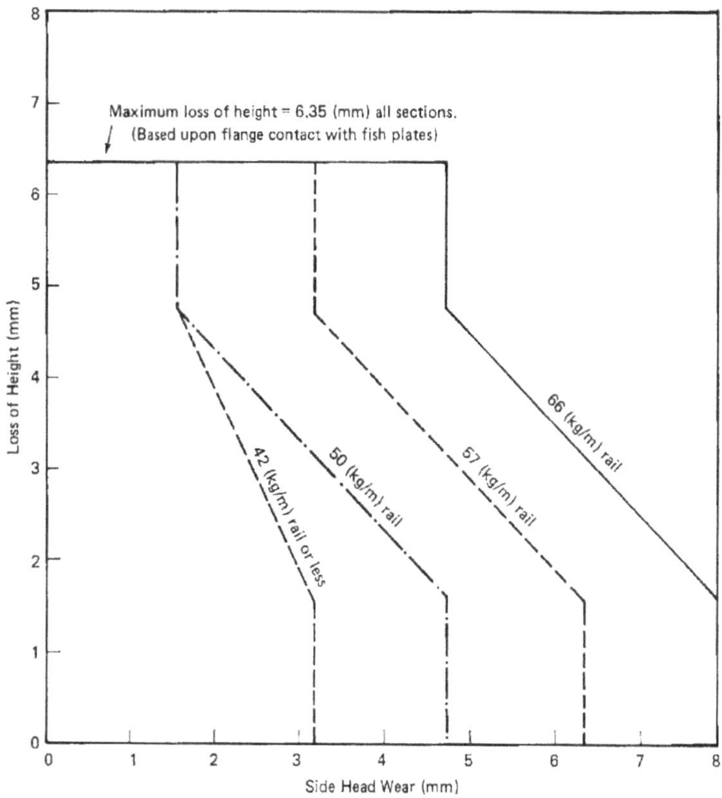

Note: Values used in drawing the envelope of rail wear limits are:

Loss of Height (mm)	Side Head Wear (mm)			
	66 (kg/m)	57 (kg/m)	50 (kg/m)	≤42 (kg/m)
1.57	7.92	6.35	4.75	3.18
3.18	6.35	4.75	3.18	2.39
4.75	4.75	3.18	1.57	1.57

Fig. 5. Envelope of rail wear limits for loss of rail head height and width [2]

2.2 Sleeper

Sleepers play important roles in railway track system [1]. The primary function of the sleepers is to transfer the vertical, lateral and longitudinal rail seat loads to the ballast, sub-ballast and subgrade layers. They also serve to maintain track gauge and alignment by providing a stable support for the rail fasteners [1-2].

The vertical-loads cause bending moments in the sleeper which their amounts are dependent upon the degree and quality of ballast layer compaction underneath the sleeper [1]. The performance of a sleeper to withstand lateral and longitudinal loading is relied on the sleeper's size, shape, surface geometry, weight, and spacing [2].

Current practices regarding the analysis and design of sleepers comprise three steps. These are [1]: 1) estimation of vertical rail seat load, 2) assuming a stress distribution pattern under the sleeper, and 3) applying vertical static equilibrium to a structural model of the sleeper.

Vertical wheel load is transferred through the rail and distributed on certain numbers of sleepers due to rail continuity. This is usually referred to as vertical rail seat load [1]. The exact magnitude of the load applied to each rail seat depends upon several parameters including the rail weight, the sleeper spacing, the track modulus per rail, the amount of play between the rail and sleeper, and the amount of play between the sleeper and ballast [2]. Based on these considerations, various relations are proposed for the amount of rail-seat loads. They are summarized in Table 3 [1].

Most of the expressions are simplified by reducing the number of influencing factors [1]. For instance, AREMA [13] recommends diagrams in which the percentage of wheel load transferred to the sleeper is only dependent on the sleeper spacing, track modulus, and the type of sleepers (Figure 6) [1].

Developer	Formula
Talbot [2]	$q_r = S.u.y_{\max}.F_1$
ORE [14]	$q_r = \overline{\varepsilon}.C_1.P$ *
UIC [15] (Concrete Sleepers)	$q_r = \frac{P_s}{2}(1+\gamma_p \times \gamma_v)\times \gamma_d \times \gamma_r$
Austraila [16] (Concrete Sleepers)	$q_r = j \times P_s \times \frac{D.F.}{100}$
Austraila [17] (Steel Sleepers)	$q_r = 0.5 \times F_2 \times P \times S \times \beta$
Sadeghi [18]	$q_r = 0.474P \times (1.27S + 0.238)$

* $\overline{\varepsilon}$ is defined as the ratio of $\overline{q}_r / \overline{P}_s$ in which $\overline{q}_r$ and $\overline{P}_s$ are mean values of rail seat load and static wheel load, respectively. C_1 is a coefficient usually equals to 1.35.

Table 3. Relations for the calculation of rail seat load [1].

The exact contact pressure distribution between the sleeper and the ballast and its variation with time are highly important in the structural design of sleepers [1]. When track is freshly tamped the contact area between the sleeper and the ballast occurs below each rail seat. After the tracks have been in service the contact pressure distribution between the sleeper and the ballast tends towards a uniform pressure distribution [1]. This condition is associated with a gap between the sleeper and the ballast surface below the rail seat [2]. The most accepted contact pressure distribution patterns between sleeper and ballast are presented in Tables 4 and 5 [1].

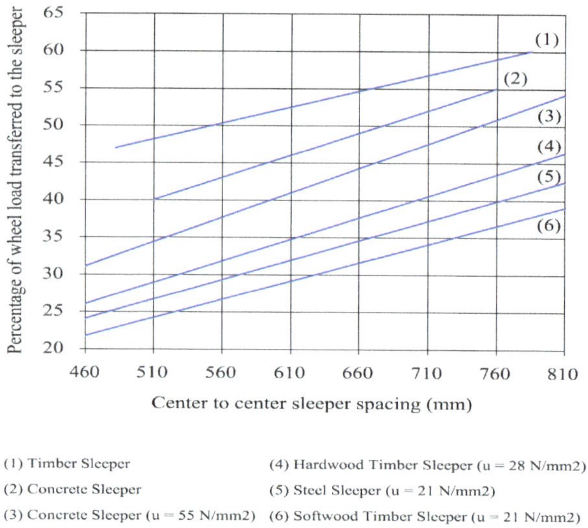

Fig. 6. Estimation of rail seat load [13]

As indicated in Table 6, it is usually presumed that the uniform pressure under the sleeper distribute in certain portions of the sleeper length (area) [1]. This length (area) is referred to as "Effective Length (Area)" and commonly shown with "L (A_e)" in the literature. This assumption is made to facilitate the procedure of design calculations. The static equilibrium in vertical direction is then applied to acquire the magnitude of contact pressure under the sleeper. A factor of safety is also included to account for pressure variations in the sleeper support. Therefore, the average contact pressure between the sleeper and the ballast P_a (kPa) can be obtained by [1]:

$$P_a = \left(\frac{q_r}{B.L}\right).F_3 \qquad (15)$$

Having determined sleeper loading pattern, the sleeper can be analyzed. For this purpose several structural model have been proposed in the literature. Table 6 indicates the sleeper structural model used in the current practice [1].

Pressure Distribution	Remarks
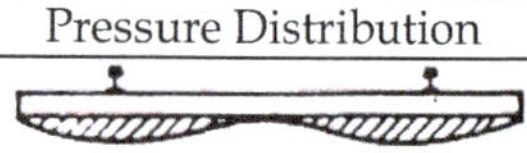	Laboratory test
	Principal bearing on rails
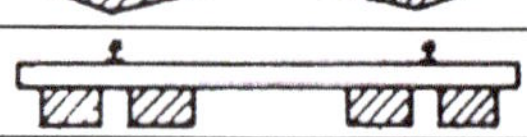	Tamped either side of rail
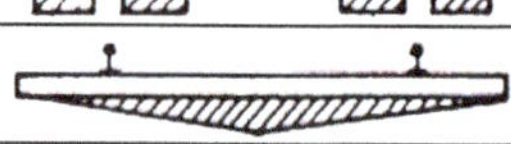	Maximum intensity in middle
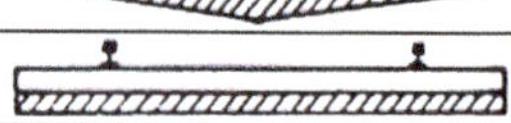	Uniform pressure

Table 4. Some contact pressure distribution patterns [16]

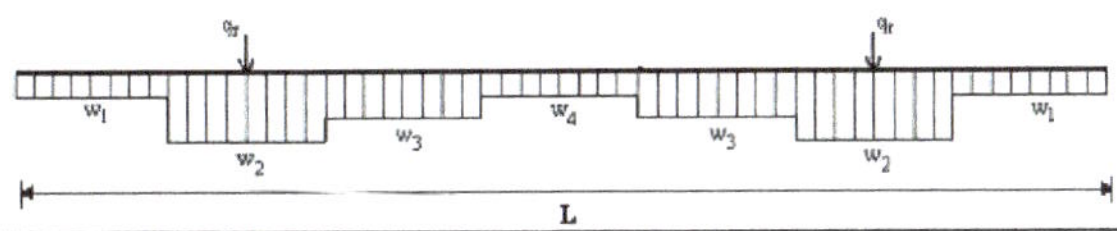

Pressure distribution pattern beneath sleeper	
After Tamping	$w_1 = 1.267q_r / L$ $w_2 = 2.957q_r / L$ $w_3 = 1.967q_r / L$ $w_4 = 1.447q_r / L$
After Accumulative Loading	$w_1 = 1.596q_r / L$ $w_2 = 2.436q_r / L$ $w_3 = 1.974q_r / L$ $w_4 = 1.687q_r / L$

Table 5. In-track sleeper loading pattern [18]

It should be noted that there are some differences between the sleeper structural models (i.e. sleeper loading patterns) suggested for the calculation of bending moments for each type of sleepers. It is indicated in Table 7 [1].

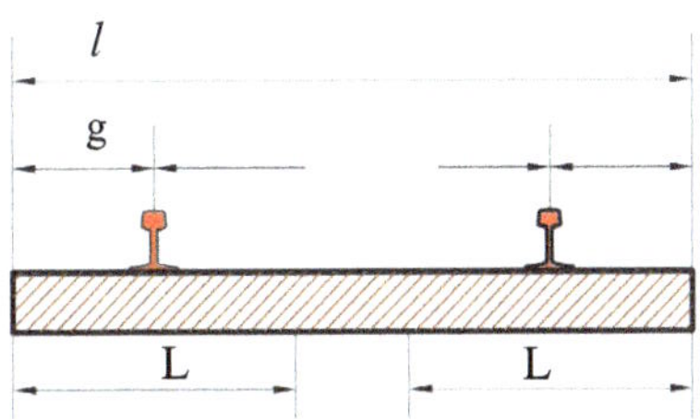

Developer	Description
AREMA [13]	Ae: Two third of sleeper area at its bottom surface
UIC [14-15]	$A_e = 6000$ cm² for $l = 2.5$ m $A_e = 7000$ cm² for $l = 2.6$ m
Australia [16-17]	$L = (l - g)$ (1) $L = 0.9 \times (l - g)$ (2)
Schramm [2]	$L = \frac{l-g}{2}$
Clarke [2]	$L = (l - g)\left(1 - \frac{(l-g)}{125t^{0.75}}\right)$ (3)
Clarke (simplified) [2]	$L = \frac{l}{3}$

1. For bending moment calculation at rail seat
2. For bending moment calculation at sleeper center
3. the parameter "t" is the sleeper height

Table 6. Effective length (area) of sleeper support at rail seat [1].

Sleeper Type	Developer	Rail Seat Moment		Center Moment	
		M_r^+ (kN.m)	M_r^- (kN.m)	M_c^+ (kN.m)	M_c^- (kN.m)
Timber	Battelle [2]	$q_r\left(\frac{l-g}{2}\right)$ (1)	***	***	$q_r\left(\frac{g}{2}\right)$
	Schramm [2]	$q_r\left(\frac{l-g-n}{8}\right)$	***	***	***
	Raymond [2]	***	***	***	$q_r\left(\frac{2g-l}{4}\right)$
Steel	Australian Standard [16]	$q_r\left(\frac{l-g}{8}\right)$	***	$0.05\times q_r\times(l-g)$	$q_r\left(\frac{2g-l}{4}\right)$
Concrete	UIC [15]	$\gamma_i.q_r.\frac{\lambda}{2}$ (2)	$0.5M_r^+$	$1.2M_{dr}+\times\frac{I_c}{I_r}$	$0.7M_c^+$
	Australian Standard [16-17]	$q_r\left(\frac{l-g}{8}\right)$	Max {$0.67M_r^+$, 14}	$0.05\times q_r\times(l-g)$	$q_r\left(\frac{2g-l}{4}\right)$

1. Less conservative and more realistic formula is also suggested by Battelle as $q_r\left(\frac{l-g}{8}\right)$
2. The effective lever arm can be obtained from $\lambda = \frac{L_p - e}{2}$

Table 7. Comparison of methods for calculation of sleepers bending moment [1]

From analyses of the sleeper, the bending moments are calculated [1]. The obtained bending moments are then compared with the sleeper ultimate bending capacities. AREMA [13] has developed a practical method for the estimation of ultimate bending capacities of sleepers. In this method, based on sleeper length, and type of bending moment (i.e. positive or negative), limit values are determined [1].

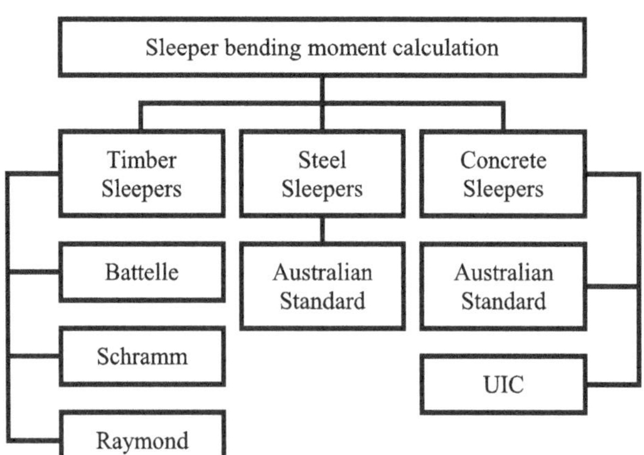

Fig. 7. Calculation of sleeper bending moment [1]

2.3 Rail fastener

Rail fasteners also known as fastening systems are used in railway track structure to fasten the rails to the sleepers and to protect the rail from inadmissible vertical, lateral, and longitudinal movements [1]. Moreover, these components serve as tools for gauge restraining, wheel load impact attenuation, increasing track elasticity, etc. There are various types of rail fasteners which mainly are classified from two points of view as presented in Figure 8 [1].

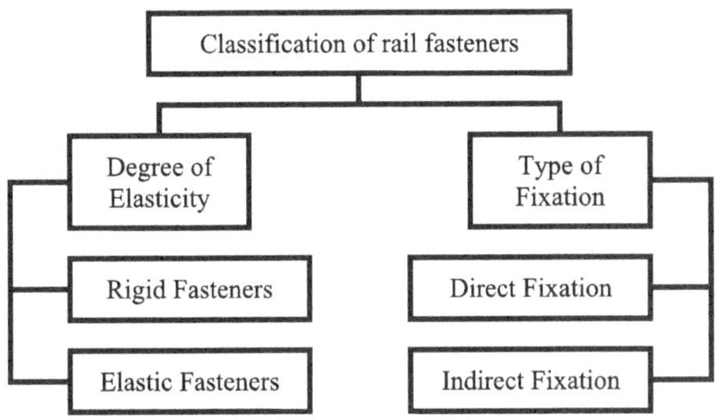

Fig. 8. Classification of rail fasteners [1]

Qualification Test	Design Code		
	AREMA	AS 1085.19	EN 13146
Uplift restraint	✓	✓	✓
Longitudinal restraint	✓	✓	✓
Repeated load	✓	✓	×
Torsional restraint	✓	×	✓
Lateral restraint	×	✓	×
Clip spring rate	×	✓	×
Fatigue strength	×	✓	×
Impact attenuation	×	✓	✓

Table 8. Comparison of design codes for rail fastener qualification tests [1]

Despite the important roles that rail fasteners play in track system, there is no practical and sufficient approaches available in the literature related to the analysis and design of rail fasteners [1]. Design criteria available in all railway standards related to fasteners design are limited to those dealt with laboratory qualification tests. AREMA manual [20-21], Australian Standard [22], and European Standards [23-27] are the most important design codes which include such criteria. A comparison of these standards is presented in Table 8 [1].

2.4 Ballast and sub-ballast layers

Ballast and sub-ballast layers are composed of granular materials and used in track structure mainly to sustain the loads transferred from the sleeper [1]. Other important functions of these layers include [1]: 1) to reduce the stress intensity transferred to the subgrade layer to the allowable level, 2) to absorb impact, noise and vibration induced by the wheels, 3) to restrict the track settlement, 4) to facilitate track maintenance operations, particularly those related to the correction of track geometry defects, and 5) to provide adequate drainage for the track structure. The sub-ballast layer is used as a separation layer between ballast and subgrade layers [1].

The sizes of aggregates in the ballast and sub-ballast layers are proposed in the majority of the railway standards or codes of practice. Theoretical, semi-empirical, and empirical methods are used in order to determine the depth of these layers [1].

Theoretical determination of minimum required depth based on Boussinesq elastic theory (applied to a uniform rectangular loaded area) (Figure 9) is performed using the numerical solution of the following equation [1]:

$$x\,\sigma_z = \frac{3P_a}{2\pi} \int_{\varepsilon=-A}^{\varepsilon=+A} \int_{\eta=-B}^{\eta=+B} \frac{d\varepsilon . d\eta}{\left\{(x-\varepsilon)^2 + (y-\eta)^2 + z^2\right\}^{5/2}} . \tag{16}$$

It should be noted that, ballast and sub-ballast layers are assumed as a single homogenous and isotropic layer in Boussinesq elastic theory [1]. Although such an assumption seems to be insufficiently accurate, ORE [28-29] have indicated the validity of Boussinesq elastic theory based on the available field tests results (Figure 10). As it is apparent from Figure 10, the results obtained from Boussinesq method reasonably remain within the envelope obtained from experimental investigations [1].

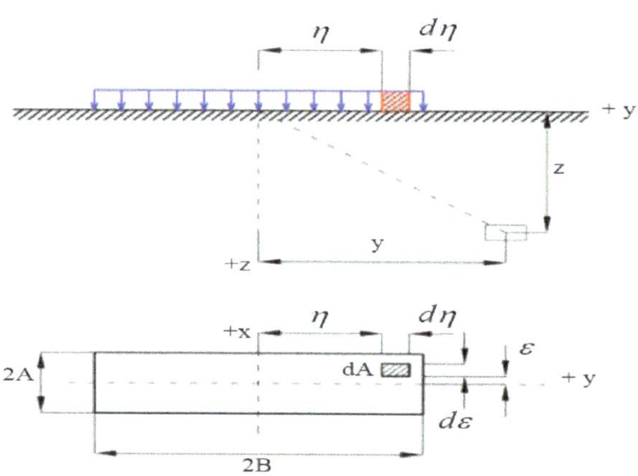

Fig. 9. Application of Boussinesq's elementary single vertical concentrated load over a uniformly loaded rectangular bearing area [2]

Simplified semi-empirical methods are also employed in the analysis of the ballast [1]. It is assumed that the load is distributed vertically with a load spread with a slope of 1 vertical to 1 horizontal or a slope of 2 vertical to 1 horizontal. It is also assumed that the stress distribution is uniform at any given depth below the surface [1]. Only the average vertical pressure at any depth can be calculated by this method while the Boussinesq method leads to the maximum vertical pressure at a depth below the loaded area [1]. A comparison of the vertical stress distribution calculated for both 1:1 and 2:1 load spread assumptions with the theoretical Boussinesq solution is presented in Figure 11 [2]. It is clearly apparent that the assumed 2:1 load spread distribution of vertical pressure more closely approximates the Boussinesq pressure distribution than that of 1:1 load spread distribution [2].

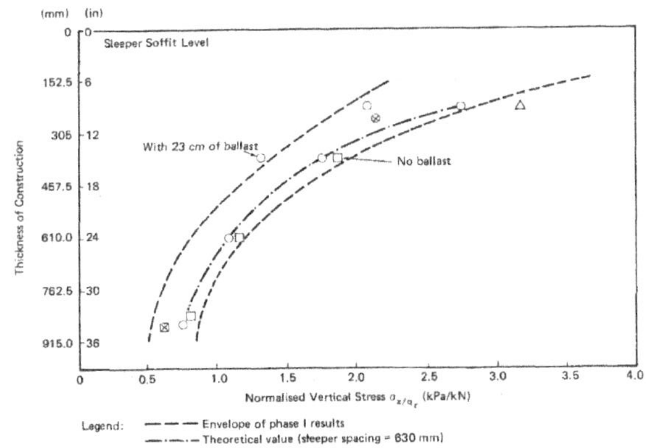

Fig. 10. Comparison of experimental vertical stress distribution with depth and the Boussinesq solution [2-28]

Considering the 2:1 load spread distribution as indicated in Figure 12, the minimum required ballast and sub-ballast depth can be calculated from the following formula [1]:

$$\sigma_z = P_a \frac{B.L}{(B+z)(L+z)} \tag{17}$$

Having determined the allowable subgrade load carrying capacity and substituting it in the equations (16) or (17) the minimum required ballast and sub-ballast depth can be calculated [1].

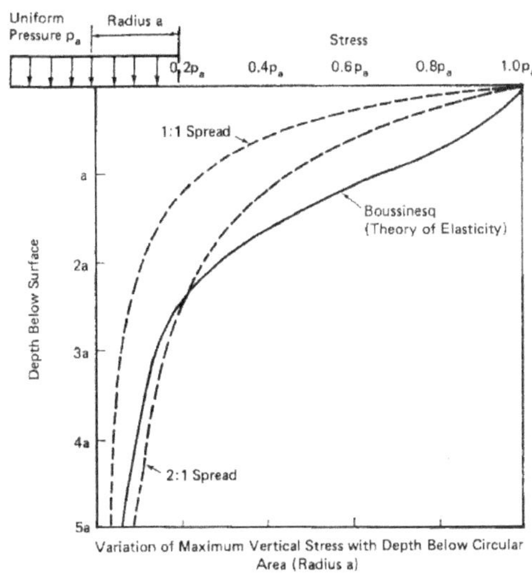

Fig. 11. Comparison of vertical stress distribution under a uniformly loaded circular area based on Boussinesq equations and 1:1 and 2:1 distributions [2]

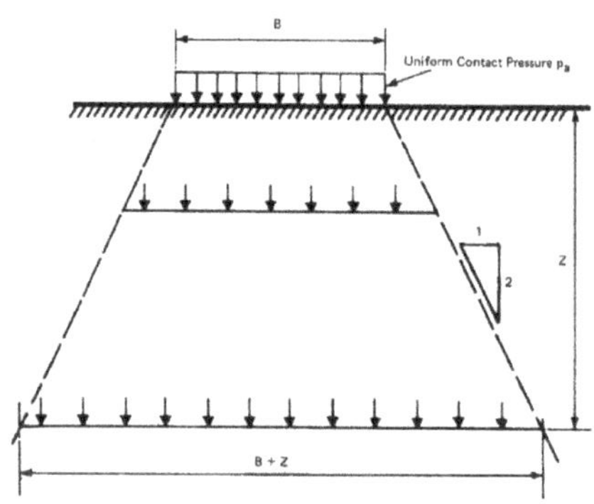

Fig. 12. Recommended semi-empirical pressure distribution in track supporting granular layers [2]

3. New developments in design of railway track system

As discussed in Section 2, the conventional track design methods are based on the Winkler model (a continuously supported beam on an elastic foundation). It includes several unrealistic assumptions, including:

- static point loads for the trains,
- continuous support under the rail,
- linear characteristics for the track support system, and
- linear uniform pressure distribution under the sleeper.

Although several factors have been taken into account to compensate for the errors caused by these assumptions, there are sometimes large discrepancies between the results obtained from conventional methods and those obtained in railway fields. This indicates that the effectiveness of the conventional track design is questionable from two aspects: cost efficiency and accuracy.

In the last decade, some improvements have been made in providing a better understanding of railway track systems. The new developments can be used to improve the current track design approach. They are detailed as follows.

3.1 Dynamic impact factor (DIF)

As indicated in the last section, various formulations have been suggested for the calculation of the track dynamic impact factor. Rail deflection and rail bending stresses have been the main criteria in the development of the majority of the available dynamic impact factor. It means that the current DIF can be only justifiable for the design of rails [29].

The mostly used suggestions for calculation of DIF for the design of rails are compared in Figure 13. It is apparent from this figure that train speed can considerably influence the values of dynamic impact factor. Sadeghi and his colleagues conducted a thorough field investigation into the effect of various track and rolling stock parameters on track dynamic impact factor [30]. They indicated that the suggestion made by AREMA and South African standards are rather accurate for track with lower speed while for the high speed transit, implementation of the suggestion made by ORE (UIC) is more appropriate [30].

Since AREMA recommendations are basically in accordance with heavy haul operations in which trains have low speeds, it would not be practical to use the AREMA formula for the railway high-speed operations. Therefore, the use of the AREMA recommendation (Table 1) for dynamic impact factor is suggested for heavy haul railway with the train speed less than 80 km/h. On the other hand, the use of mathematical expression proposed by ORE (indicated in Table 1) is more justifiable for the calculation of dynamic impact factor in high speed railway tracks [31].

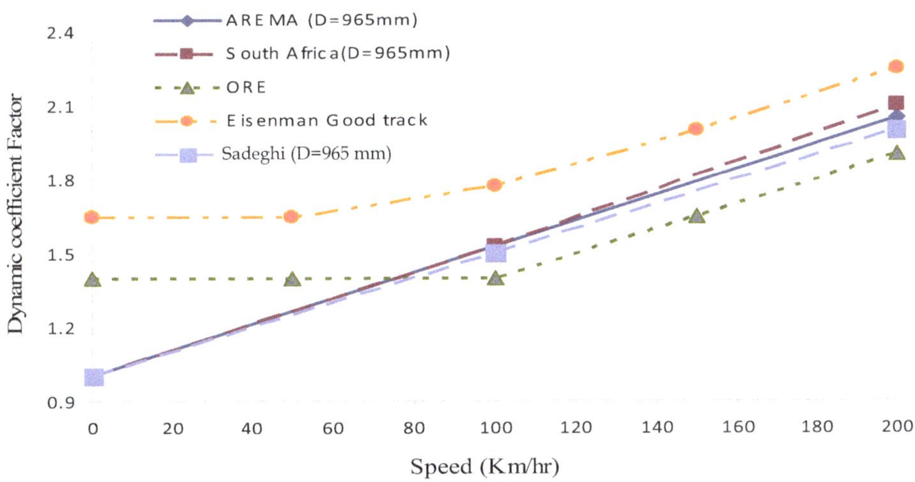

Fig. 13. Comparison of dynamic impact factor for design of rails [1]

The effects of the dynamic characteristics of the load on the sleeper and the ballast (Rail-seat load and ballast-sleeper contact pressure) have been recently investigated by Sadeghi and his colleagues [31]. They indicated that the dynamic characteristics of the load have more impact on the amount of ballast-sleeper pressure at the points which experience less pressure. That is, the less pressure, the more influence due to the load dynamics. Since with an increase in load speed the pressure in the middle and cantilever parts of the sleeper considerably increases, we are led to believe that the load distribution under the sleeper will become increasingly uniform with increasing load speeds [31]. This means that the assumption of a uniform pressure distribution under the sleeper for train speeds of more than 120 km/h is justifiable [31].

The impacts of the dynamic properties of the wheel load on the increase of the loads between ballast and sleepers experimentally investigated by Sadeghi and his colleagues are presented in Figure 14. They have obtained this figure from dividing the results obtained from the load cells (installed under the sleepers) when trains were moving with the speed of 20 to 120 km/h by the static load cell values (i.e., speed is zero). The vertical axis in this figure is the dynamic coefficient factor or dynamic impact factor. Since the DIF calculated in figure 14 is based on the effects of dynamic load on sleeper loading patter, implementation of this figure in the design of sleepers and ballast is more justifiable and lead to a better accuracy [31].

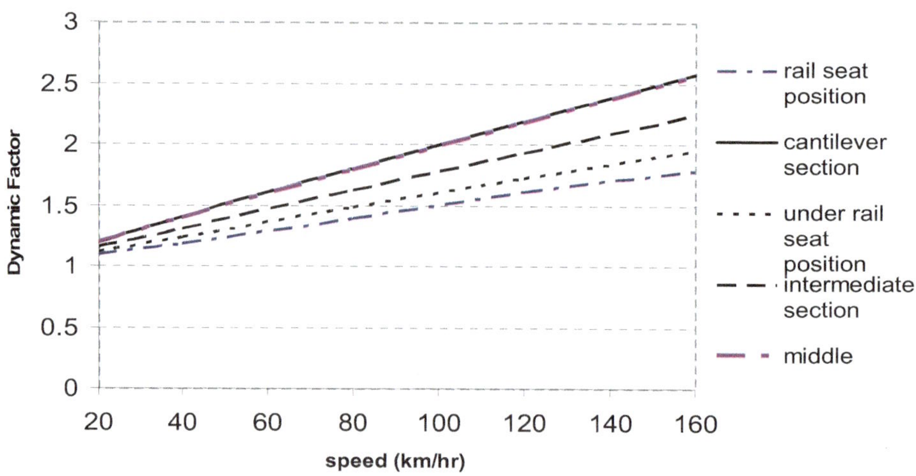

Fig. 14. DIF versus speed for calculation of pressure in various sections of sleepers [31]

3.2 Continuous support under the rail

Because of the limitations of the theory of beam on elastic foundation model, discussed in the last section, several attempts have been made to develop new modeling techniques to simulate railway track behavior. First, a model of beam on discrete support has been developed and more recently analyzed using a practical energy approach [3]. Second, Pasternak foundation and double beam models are also introduced [4] to take into account the interaction between track supporting layers and multi-layer nature of track support, respectively. Kerr [5] has reported the result of a research in which he showed that the effect of the rotational stiffness of the sleepers in calculation of rail deflection and bending moment are negligible.

Sadeghi and his colleagues investigated the importance of considering the discontinuity of the rail supports by comparing the results from the Winkler model (WM) and the discrete support model (DSM) [32]. They have shown that the maximum differences in the results from each method are at the most 2.5% and 11% for rail deflections and rail bending moments, respectively [32]. The difference between rail deflections decreases as the track stiffness increases but the difference in bending moments increases as the track stiffness increases [32]. A comparison of the results confirms that the assumption of continuous support under the rail (Winkler model) slightly increases rail bending moments and negligibly decreases rail deflections [32]. These negligible differences between the results obtained from the two methods along with the simplicity and practicality of the conventional method indicate that the assumption of continuous support under the rail is justifiable [32].

3.3 Non-linearity of track support system

The results obtained in the literature illustrate large differences between analyses of the track with and without consideration of nonlinearity of the track support system [2, 37], indicating a need for incorporation of the nonlinear properties and plastic deformations of the track support system. This is more apparent where the long term behavior of the track is concerned. The major differences between results obtained from the conventional method and the nonlinear track analyses come from the consideration of track plastic deformations caused by accumulative loadings [33-34]. Note that the impact made by the consideration of plastic deformation is influenced by the amount of wheel loads. This means that there are two key influencing factors to be considered in the calculation of track design parameters: the amount of accumulative loadings and the amount of wheel loads [33-34].

There are two ways to include the nonlinear properties of the rail support system in the design procedure [32]: first, by using a new approach which includes nonlinear analysis of the track; or second by applying appropriate correction factors to the conventional method. Because the conventional track design approach is simple and user-friendly, the second method seems more practical. For this purpose, two correction factors have been developed in the literature for rail deflection and rail bending moment to compensate for the assumption of linear properties for the track support system [32].

Sadeghi and his colleagues have obtained the ratios of rail deflections and rail bending moments obtained from track nonlinear analysis to those from linear analysis for different track conditions [32]. They have plotted these ratios against Mega Gross Tons passed (MGTP) and against wheel loads. Using a curve-fitting method, they draw mathematical correlations (expressions) between ratios of nonlinear to linear track responses and wheel loads. Using the results obtained, they have proposed correction factors for the rail deflection and rail bending moment as follows [32]:

$$f_{D-T} = (10^{-4}P^2 - 0.03P + 2.85) + (3\times10^{-5}P^2 - 8\times10^{-3}P + 0.62)T \quad P \geq 20\ kN \tag{18}$$

$$f_{D-C} = (10^{-4}P^2 - 0.02P + 2.45) + (10^{-5}P^2 - 3\times10^{-3}P + 0.28)T \quad P \geq 20\ kN \tag{19}$$

$$f_{M-T} = (-0.001P + 1.126) + (-3\times10^{-4}P + 0.08)T \quad P \geq 20\ kN \tag{20}$$

$$f_{M-C} = (-0.001P + 1.09) + (-3\times10^{-4}P + 0.04)T \quad P \geq 20\ kN \tag{21}$$

where, f_{D-C} and f_{M-C} are correction factors for rail deflection and rail bending moments in concrete sleeper tracks respectively, f_{D-T} and f_{M-T} are correction factors for rail deflection and rail bending moments in timber sleeper tracks, respectively, P is the wheel loads (kN), and T is the amount of accumulative loadings in MGT. The rail bending moment and rail deflection calculated by the conventional method should be multiplied by these correction factors to compensate for the unrealistic assumption of linear properties for the track supports. The use of these correction factors would improve the accuracy of the current track analysis method, particularly when the track has undergone a large amount of accumulative loading [32].

3.4 Sleeper loading pattern

As indicated in the previous section, a uniform contact pressure distribution is assumed between ballast and sleeper in the current practice. In other words, the current design methods do not clearly include the effect of ballast material degradation on the ballast-sleeper pressure distribution pattern [35]. This means that the sleeper design approach awaits further improvements by the incorporation of long term effects of the loads and in turn, consideration of the ballast degradation.

In 2008, Sadeghi and his colleagues conducted a thorough field investigation into the relationship between rail seat loads, static wheel loads and train speeds [36]. Using the results obtained from the field, they proposed the following relationships for the calculation of rail seat load as a function of static wheel loads, train speeds, and sleeper spacing for the three types of sleepers. This reflects the concept of dynamic impact factor which can be used for the design of sleepers [36].

$$q_r = (-3 \times 10^{-6} V^2 + 0.0018V + 0.4274)(1.27S + 0.238)P \text{ Timber sleepers} \quad (22)$$

$$q_r = (-3 \times 10^{-6} V^2 + 0.0017V + 0.4659)(1.27S + 0.238)P \text{ Steel sleepers} \quad (23)$$

$$q_r = (-3 \times 10^{-6} V^2 + 0.0018V + 0.4936)(1.27S + 0.238)P \text{ Concrete sleepers} \quad (24)$$

Comparisons of the experimental results obtained here and those currently used indicate that the American Code is slightly over-estimated for speeds higher than 80 km/h [36]. This can be due to the US railway trend towards higher axle loads and heavy haul operations which limits the reliability and applicability of American Code for high speed operations. On the other hand, the European suggestion is an under-estimation of the ratio of the rail seat load to the static wheel load for train speeds more than 60 km/h [36]. Results indicate that European suggestion is more reliable for timber sleeper in comparison with the steel or concrete sleepers. The results obtained in the field are more agreeable with that suggested by the Australian standard for steel and concrete sleepers, particularly for train speeds between 60 to 120 km/h [36].

Sadeghi and his colleagues have also experimentally investigated the pressure distribution between ballast and various sleepers [36]. They illustrated that the pressure between the ballast and the sleeper increases from the sleepers' edge and reaches its maximum value under the rail-seat position, then with a decreasing pattern, comes to the minimum value at the center of the sleeper. They showed that the contact pressure occurred under pre-stressed concrete sleepers is more uniform when compared with that under timber and steel sleepers [36]. The most non-uniform shape of the sleeper-ballast contact pressure occurs under the timber sleepers; this could be due to the higher bending rigidity of concrete sleepers in comparison with wood and steel sleepers [36].

In order to make the results practical, Sadeghi and his colleagues have brought the obtained stress distribution beneath the sleeper into a uniform shape [36]. It is indicated in Table 6 which presents uniform pressure distributions beneath the timber, steel and concrete sleepers with a reasonable accuracy. Most of the railway standards suggest the consideration of a uniform pressure distribution beneath all the area (or effective area) of the sleeper. Implementation of the new model (suggested in Table 9) can improve the accuracy of the current sleeper design approach [36]. It should also be mentioned that this new sleeper loading pattern is applicable to the curves as long as there is no cant deficiency or cant excess along the curve length (i.e., when the sleeper symmetric load distribution pattern is maintained [36].).

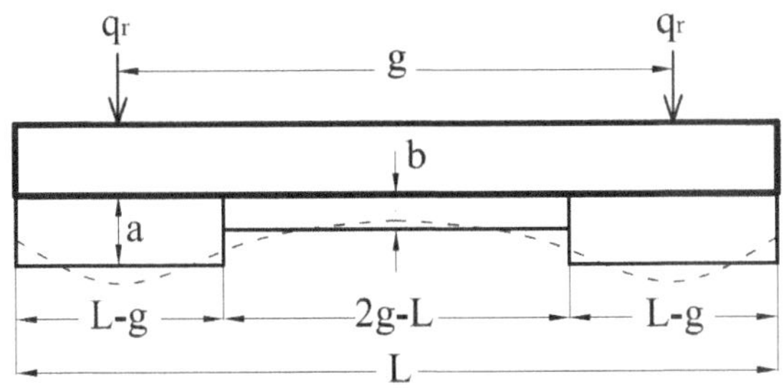

Sleeper Type	Parameter	
	a	b
Timber	$0.753q_r / L$	$0.494q_r / L$
Steel	$0.719q_r / L$	$0.561q_r / L$
Pre-stressed concrete	$0.709q_r / L$	$0.581q_r / L$

Table 9. Load distribution pattern under different sleepers [36]

4. Summary and conclusions

In this chapter, the algorithm and limitations of the current track designed approach were discussed. In the current practice, several assumption are made: (1) each track complement (rail, sleeper, fastening system, and ballast) is design independently; (2) loads are considered static; (3) the theory of beam on elastic foundation is mainly used for the analyses of the track components; and (4) track support system is considered to have a linear mechanical behavior. Although several factors have been taken into account to compensate for the errors caused by these assumptions, there are sometimes large discrepancies between the results obtained from conventional methods and those obtained in railway fields. This indicates that the effectiveness of the conventional track design is questionable.

In the last decade, several experimental and theoretical investigations have been conducted in the field of railway engineering related to the track mechanical behaviors. Incorporation of new research findings into the track design method can lead to substantial improvements in the accuracy and reliability of the conventional methods.

1. The results recently obtained from field investigations indicate that the use of the AREMA suggestion for dynamic impact factor is appropriate for heavy haul railway with the train speed less than 80 km/h while the mathematical expression proposed by ORE (UIC) is more justifiable for the calculation of dynamic impact factor in high speed railway tracks. However, these expressions or formulations have been developed based on the rail design criteria and therefore, their accuracy for the design of sleeper or ballast is questionable. The use of newly obtained dynamic impact factor which has been developed based on the sleeper and ballast design criteria would improve the accuracy of the track design when sleepers and ballast layers are concerned.

2. Results from field investigations and finite element analyses indicate that there are negligible differences between track models with a continuous support and those with discrete supports. Therefore, considering the practicality and simplicity of the conventional method, the assumption of continuous support under the rail seems to be justified.
3. It has been experimentally shown that there is a large impact on the track design parameters due to the nonlinearity of the track support system, indicating the necessity of improving current codes of practice. Using results from track fields, several models have been developed for the nonlinear response function of the track support system for different track loading and material conditions. Using these models and conducting a nonlinear analysis of the track, it was shown that there was a substantial increase in the maximum rail deflection and rail bending moment (the main track design parameters) when considering the nonlinearity of the track support system. In contrast with the current practice of assuming a linear relationship between wheel loads and rail deflection, a cubic relationship between wheel loads and rail deflations has been observed. It has been shown that the major differences between the conventional method and nonlinear track analysis come from the consideration of track plastic deformation caused by accumulative loadings. It has been also illustrated that the impact of the track nonlinearity on the analysis results is influenced by the magnitude of wheel loads. Therefore, plastic deformation and applied wheel loads are the key influencing factors in the nonlinear behavior of track support systems and they are in turn important factors to be considered in the track analysis and design. To compensate for the errors caused by the assumption of a linear response for the track support system, correction factors have been recently developed for rail deflection and rail bending moment, to be multiplied by the rail bending moment and rail deflection. The influences of the nonlinear properties of the track support system on the track design parameters have been the basis for the calculation of the correction factors. They are functions of the magnitude of wheel loads and the amount of accumulative loadings. It has been proved that the incorporation of these factors improves the accuracy and reliability of the conventional track design method.
4. It has been shown that there is a second order relationship between maximum rail seat loads and train speeds. From this point, formulae for the calculation of rail seat loads as a function of wheel loads, train speeds, and sleeper spacing for timber, steel and concrete sleepers have been recently proposed. It has experimentally shown that the pressure distribution pattern between the ballast and the sleeper is not a uniform contact pressure as proposed earlier by American codes (AREMA). The pressure between the ballast and sleeper for all three types of sleepers was found to decrease from the sleepers' edge, reach its maximum value under the rail seat position, and then decrease to a minimum value at the center of the sleeper. This is in close agreement with the sleepers' behaviour proposed by UIC. The pressure distribution beneath the concrete sleepers is more uniform than those of timber or steel sleepers. As the speed and axle loads increase, this pressure distribution becomes less uniform, indicating a need for improvement of the current track design approaches for heavy haul and high speed tracks. The assumption of a uniform pressure distribution between ballast and sleeper in the current sleeper design methods can be replaced with a more realistic uniform load distribution pattern beneath concrete, steel and timber sleepers newly proposed.

5. List of symbols

B	Sleeper breadth, m
c	height of the rail neutral axis from rail base, mm
D	Wheel diameter, mm
D_A	Annual gross tonnage, MGT/year
D_c	Degree of curve, degrees
D_i	Sub-tonnage, MGT/year
D_j	Track stiffness at the joint, kN/mm
D.F.	Load distribution factor
e	Width of the rail seat load distribution along sleeper thickness
E	Rail modulus of elasticity, N/mm^2
F_1	Track support variation safety factor
F_2	Factor accounting for adjacent wheels interactions
F_3	Factor depending on the sleeper type and the standard of track maintenance
g	Gravitational constant, m/s^2
I	Rail moment of inertia, mm^4
I_c	Horizontal moment of inertia of the center of the sleeper cross section, mm^4
I_r	Horizontal moment of inertia of the sleeper cross section at rail seat position, mm^4
j	Load amplification factor
K_{A_i}	Wheel load class factor
K_C	Track curvature and lubrication factor
K_G	Track gradient factor
K_R	Rail factor
K_{S_i}	Service type factor
K_{V_i}	Speed class factor
K_w	Wear factor varying with the degree of curve
l	Sleeper length, m
L_p	Distance between rail seat axles and the end of the sleeper
n	Length of steel rail plate
P	Wheel load, kN
P_s	Static wheel load, kN
P_u	Unsprung weight of the wheel, kN
R_w	Wheel radius, mm
R_{wt}	Rail weight per unit length, kg/m
S	Sleeper spacing, mm
T	Rail wear life, MGT
T_y	Rail wear life, year
u	Track modulus, N/mm^2
V	Train speed, km/h
W_t	Average rail head area wear term, mm^2/MGT

Greek symbols

α	Coefficient of expansion
α'	Speed coefficient
β'	Speed coefficient
δ	Factor related to track condition
Δt	Temperature variation
γ'	Speed coefficient
γ_d	Load distribution factor
γ_i	Dynamic increment of bending moment factor due to sleeper support irregularities
γ_p	Impact attenuation factor of rail fasteners
γ_r	Sleeper support condition factor
γ_v	Speed related amplification factor
η	Speed factor
λ	Effective lever arm, m
ν	Poisson's ratio

6. References

[1] Sadeghi, J. & Barati, P., Evaluation of conventional methods in Analysis and Design of Railway Track System, International Journal of Civil Engineering, Vol. 8, No.1, 2010.

[2] Doyle, N.F., "Railway Track Design: A Review of Current Practice", Occasional Paper No. 35, Bureau of Transport Economics, Commonwealth of Australia, Canberra, 1980.

[3] Sadeghi J., "Fundamentals of Analysis and Design of Railway Ballasted Track", IUST Publication Survive, Tehran, 2010.

[4] Esveld, C., "*Modern Railway Track*", MRT-Productions, the Netherlands, 2001.

[5] Kerr, A.D., "*Fundamentals of Railway Track Engineering*", Simmons-Boardman Books, Inc, 2003.

[6] American Railway Engineering and Maintenance of way Association, Manual for Railway Engineering, Volume 4, Chapter 16, Part 10, "Economics of Railway Engineering and Operations - Construction and Maintenance Operations", 2006.

[7] Robnett, Q.L., Thompson, M.R., Hay, W.W., Tayabji, S.D., Peterson, H.C., Knutson, R.M., and Baugher, R.W., "Technical Data Bases Report", Ballast and Foundation Materials Research Program, FRA Report No. FRA-OR&D-76-138, National Technical Information Service, Springfield, Virginia, USA, 1975.

[8] Eisenmann, J., "Stress Distribution in the Permanent way due to Heavy Axle Loads and High Speeds", AREMA Proceedings, Vol. 71, pp. 24-59, 1970.

[9] Smith, J.O., and Liu, C.K., "Stresses Due to Tangential and Normal Loads on an Elastic Solid with Application to some Contact Stress Problems", Transactions of ASME, Journal of Applied Mechanics, pp. 157-166, 1953

[10] Lundgren, J.R., Martin, G.C., and Hay, W.W., "A Simulation Model of Ballast Support and the Modulus of Track Elasticity", (Masters Thesis), Civil Engineering Studies, Transportation Series, No. 4, University of Illinios, 1970.

[11] Hay, W.W., Schuch, P.M., Frank, M.W., and Milskelsen, M.J., "Evaluation of Rail Sections", 2nd Progress Report, Civil Eng., Transportation Series No. 9, Univ. Illinois Project No. 44-22-20-332, Univ. Illinois, Urbana, 1973.

[12] Danzing J. C., Hay W., and Reinschmidt A., "Procedures for analyzing the economic costs of railway roadway for pricing purposes", Volume 1 and 2, Report No. RPD-11-CM-R, Tops on-line Service Inc, 1976.

[13] American Railway Engineering and Maintenance of Way Association, Manual for Railway Engineering, Volume 1, Chapter 30, Part 1, "Ties - General Considerations", 2006.

[14] Office of Research and Experiments (ORE), Stresses in the Formation, Question D17, "Stresses in the Rails, the Ballast and the Formation Resulting from Traffic Loads", Report D71/RP8/ E, Utrecht, 1968.

[15] International Union of Railways, "Design of Monoblock Concrete Sleepers", UIC CODE, 713 R, 1st Edition, 2004.

[16] Australian Standard, AS 1085.14, Railway track materials, Part 14: "Prestressed Concrete Sleepers", 2002.

[17] Australian Standard, AS 1085.17, Railway Track Materials, Part 17: "Steel Sleepers", 2002.

[18] Sadeghi, J., "Experimental Evaluation of Accuracy of Current Practices in Analysis and Design of Railway Track Sleepers", Canadian Journal of Civil Engineering. Vol. 35, pp. 881-893, 2008.

[19] Zekeri, J. Sadeghi, J., Field Investigation on Load Distribution under Railway Track Sleepers , Journal of Mechanical and Science and Technology, Vol. 21, 1948-1956, 2007.

[20] American Railway Engineering and Maintenance of Way Association, Manual for Railway Engineering, Volume 1, Chapter 30, Part 4, "Ties - Concrete Ties". 2006.

[21] American Railway Engineering and Maintenance of Way Association, Manual for Railway Engineering, Volume 1, Chapter 5, Part 9, "Track - Design Qualification Specifications for Elastic Fasteners on Timber Cross Ties", 2006.

[22] Australian Standard, AS 1085.19, Railway track materials, Part 19: "Resilient fastening, 2009.

[23] European Committee for Standardization, EN 13146-1, Railway Applications - Track, Test Methods for Fastening Systems, Part 1: "Determination of Longitudinal Rail Restraint", 2002.

[24] European Committee for Standardization, EN 13146-2, Railway Applications - Track, Test Methods for Fastening Systems, Part 2: "Determination of Torsional Rail Resistance", 2002.

[25] European Committee for Standardization, EN 13146-3, Railway Applications - Track, Test Methods for Fastening Systems, Part 3: "Determination of Attenuation of Impact Loads", 2002.

[26] European Committee for Standardization, EN 13146-4, Railway Applications - Track, Test Methods for Fastening Systems, Part 4: "Effect of Repeated Loading", 2002.

[27] European Committee for Standardization, EN 13146-7, Railway Applications - Track, Test Methods for Fastening Systems, Part 7: "Determination of Clamping Force", 2002.

[28] Office of Research and Experiments (ORE), Stresses in the Formation, Question D17, "Stresses in the Rails, the Ballast and the Formation Resulting from Traffic Loads", Report D71/RP8/ E, Utrecht, 1968.

[29] Sadeghi, J., Field Investigation on Vibration Behavior of Railway Track Systems, International Journal of Civil Engineering, IUST, 8 (3), pp. 232-241, 2010.

[30] Sadeghi J. and Barati B., Field investigation on railway track dynamic impact factor, Research report N. F/11-1389, School of Railway Engineering IUST, Iran, 2011.

[31] Sadeghi, J., "Field Investigation on Dynamics of Railway Track Pre-stressed Concrete Sleepers", Journal of Advances in Structural Engineering, Vol. 13, No. 1, 2010.

[32] Sadeghi, J. & Barati, P., Improvements of Conventional Methods in Railway Track Analysis and Design, Canadian Journal of Civil Engineering, Vol.37, DOI: 10.1139/L10-010, 2010.

[33] Sadeghi, J. & Barati, P., Investigation on the effect of rail support system on Railway analysis, Journal of Transportation Engineering, Vol. 1, No. 1, 2010.

[34] Sadeghi, J. & H. Askarinejad, Railway Track Non-Linear Model Applying a Modified Plane Strain Technique, Journal of Transportation Engineering, Proceeding of ASCE, J. Transp. Eng. Volume 136, Issue 12, pp. 1068-1074, 2010.

[35] Sadeghi, J., and Yoldashkhan, M., "Investigation on the accuracy of current practices in analysis of railway track sleepers", International Journal of Civil Engineering, Vol. 3, No. 1, pp. 9-15, 2005.

[36] Sadeghi, J., Barati, P., Comparisons of Mechanical Properties of Timber, Steel and Concrete Sleepers, Journal of Structure and Infrastructure Engineering, DOI: 10.1080/ 15732479.2010.507706, 2010.

[37] Costa P. A., R., Calçada R., Cardoso A.S. and Bodare, Influence of soil non-linearity on the dynamic response of high-speed railway tracks, J. of Soil Dynamics and Earthquake Eng. 30(4) (2010) 221-235, 2010

Research on Improving Quality of Electricity Energy in Train's Traction

Huang Yuanliang
Jinan University
China

1. Introduction

To serve as one of the main transportations and carriers of the world nowadays, electrified railway is developing rapidly in its speed and length of line. In order to improve the power supply capability and power quality of traction system, scientific researchers have put forward many methods. Adding shunt capacitor compensator or series capacitor compensator in traction system are both effective methods. They can filter, regulate voltage, raise the utilization rate of the power supplying equipment capacity, and improve power factor.

This section will analyze the performance of power supplying capability increased equipment and the influence of each other, besides, study the strategy of coordinated control.

2. Shunt capacitor compensation of traction power supply system

Shunt capacitor compensation of traction power supply system means to connect a capacitor group and control device in parallel on feeder line or supply arm so as to improve power factor and power quality. Now dynamic capacitor compensation is used primarily.

2.1 Principle of shunt capacitor compensator

The connection and equivalent circuit diagram of shunt capacitor compensator is shown as figure 1, where $\dot{U}_1$ is supply voltage, and γ_1+jW_1 is each phase impedance of power that converting into low-voltage side by internal impedance, line impedance and traction transformer. $\dot{U}_2$ is traction bus bar voltage of traction substation, X_c is capacitor of shunt compensative capacitor banks, X_L is inductance of reactor connected with capacitor group in series, Z is traction load impedance, I_c is static var compensator current of shunt condenser, and I_q is current of traction load. Before installing traction shunt capacitor compensator, current flow through the traction transformer is,

$$I_q = \frac{\dot{U}_1}{Z + \gamma_1 + jX_1}$$

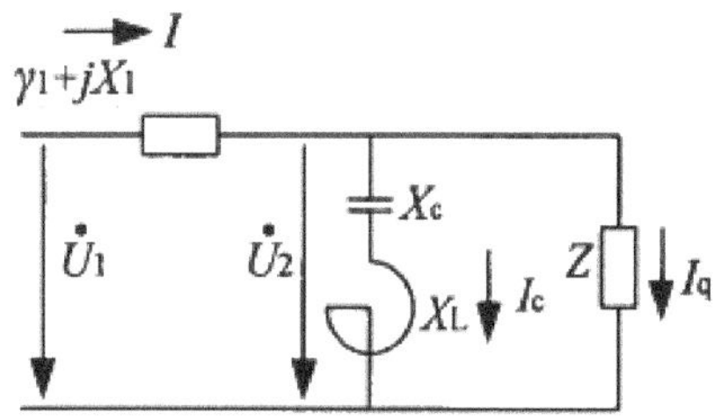

Fig. 1. Chart of traction power supply shunt capacitor compensator

After installing, the current flow through the traction transformer is,

$$I = \frac{\dot{U}_1}{\frac{Z * j(X_C + X_L)}{Z + j(X_C + X_L)} + \gamma_1 + jX_1}$$

As two equations above indicates, after installing, the current flow through the traction transformer I_q reduces to I, active power remains unchanged, and reactive power decreases.

$$P = \dot{U}_1 I_q \cos\Phi_1 = \dot{U}_1 I \cos\Phi_2$$

$$Q_1 = \dot{U}_1 I_q \sin\Phi_1$$

$$Q_2 = \dot{U}_1 I_q \sin\Phi_2 < Q_1$$

It can be seen from 3 equations above, after installing, traction power factor can increase from $\cos\Phi_1$ to $\cos\Phi_2$.

As figure 2 shows,

$$I_C = I_q \sin\Phi_1 - I\sin\Phi_2 = P(\tan\Phi_1 - \tan\Phi_2) / U_1$$

While $I_C = U_1 \div X_C = 2\pi fCU_1$, in order to make power factor increase from $\cos\Phi_1$ to $\cos\Phi_2$, the capacitor it needs to compensate is capacitor C,

$$C = P(\tan\Phi_1 - \tan\Phi_2) \div (2\pi fU_1^2)$$

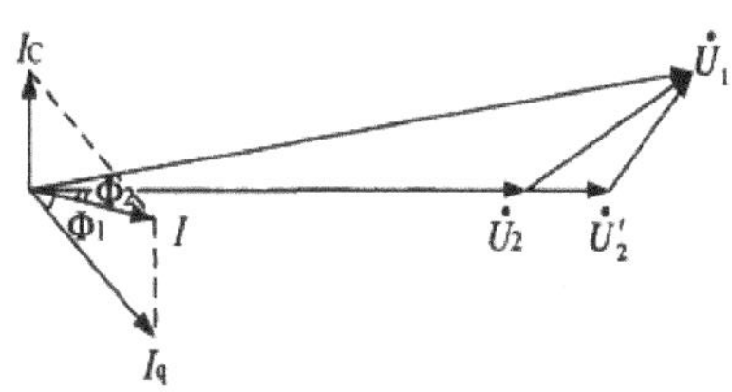

Fig. 2. The column picture of parallel compensation

In Traction Power Supply system, when calculating and defining the capability of shunt capacitor, we should be based on the maximum average of traction load to choose shunt capacitor.

2.2 Dynamic shunt capacitor compensator

Nowadays in order to track changing traction loading timely, provide with rational reactive power compensation, and save the energy cost, dynamic compensator with shunt capacitor is widely used in electric railway. There are 3 commonly used devices.

2.2.1 Static var compensator (SVC)

The immediate purpose of parallel var compensator is to lower reactive current and negative sequence current. Three-phase SVC is an effective mature technology of improving harmonics, negative sequence and voltage fluctuation caused by the load of electric railway. SVC is by means of thyristor controlled reactance (TCR) and the strategy of three-phase balanced control based on "Steinmetz" to make dynamic reactive power compensation, control voltage, and improve negative sequence. This method can not only balance three-phase reactive power, but also balance three-phase active power. Therefore, it is an effective method.

2.2.2 Static synchronous compensator (STATCOM)

In contrast with SVC, STATCOM has its advantages of fast speed, great loading rate adaptation, high work efficiency, and small output harmonic content. Especially, adopting two-phase structure can achieve four-phase control of active and reactive power, provide two supply arms of power substation with dynamic reactive compensation, besides, regulate active flow of two supply arms, so as to dynamically balance the loading.

Since 1980s, the researches on the technology of STATCOM dynamic voltage compensation have become one of the hot topics in the field. Because STATCOM is usually supported by DC voltage provided by capacitor on DC side, it can't provide continues active power. But if change a power supply on DC side, STATCOM, served as voltage source inverter, can exchange energy with system. Connect a single-phase converter on both supply arms of Scott transformer, which are interconnected through intermediate links. The work mode of controlling two transformers makes them be DC power supply of the other one so as to achieve active power facility between both supply arms. When adopting impedance matching balance transformer or Scott wiring transformer, active load of both supply are balanced. Therefore it can eliminate the negative sequence completely. Accordingly, in order to achieve power factor compensation and negative sequence improvement of traction substations, we can install two single-phase STATCOM on traction transformer secondary side to carry on reactive compensation and phase transformations of active power.

2.2.3 Static var generator/ absorber (SVG)

Compared with traditional SVC, the regulation speed of SVG is faster, and the range of operation is wider. Besides, adopting multiplex, multi-level or PWM can significantly reduce the content of harmonics in compensation current. More importantly, because of the smaller size of capacity cell in SVG, the size and cost will reduce greatly.

The basic principle of SVG is connecting self-communicated bridge circuit in parallel on grid or through a reactor, then regulating the phase and amplitude of output voltage of bridge circuit on AC side or controlling the AC current directly to make the absorption and output of reactive current meet the demands. Finally achieve dynamic compensation. Because SVG converts the DC side voltage to AC side output voltage which has the same frequency with the grid through the switching of power semiconductor, just like a voltage-type inverter, when considering fundamental frequency, SVG can be seen as an AC voltage source whose phase and amplitude could both be controlled. It is connected on power grid through AC reactor. The working principle is shown as figure 3.

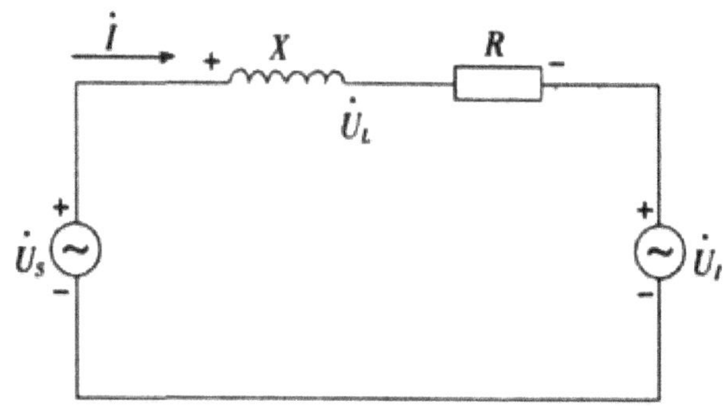

Fig. 3. Single-phase equivalent diagram

The output AC voltage of power grid and SVG are $\dot{U}_s$ and $\dot{U}_I$, and the voltage of reactance X is $\dot{U}_L$. It is the phasor difference of U_S and U_I. The current *I*, flowing through reactance X, absorbed from power grid by SVG, is controlled by voltage of reactance. Hence converting the amplitude of U_I on AC side and the relative phase with U_S can convert the voltage of reactance so as to control the phase and amplitude of current absorbed by SVG, and the nature and size as well. Using appropriate control methods can control the currents running through each branch in three-phase bridge converter circuit individually, and make energy storage capacitor absorb energy from light load, and release energy to heavy load, so as to compensate for reactive power and negative sequence of traction load at the same time. If adopting PWM with appropriate testing method, it still can active filter.

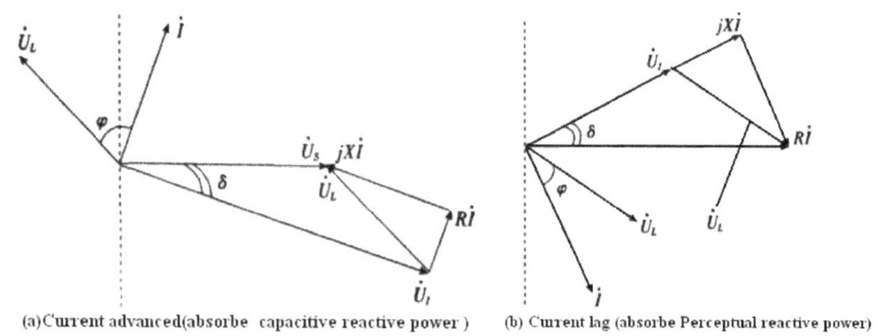

Fig. 4. The original vector graphics of SVG

3. Series compensator of traction power supply system

Series compensation of Traction Power Supply system is a capacitor group and control and protection device connected on feeder line or supply arm. Its function is to raise terminal voltage of supply arm and the power factor, then improve power quality.

3.1 The principle of series compensator of traction power supply system

Connect a capacitor in series on traction substation or power-supply section. The capacitor could offset part of inductance of Traction Power Supply system, and reduce voltage loss. The voltage loss is:

$$\Delta U_C = -IX_C \sin\varphi = -I \sin\varphi \cdot X_C \, .$$

Where φ is the angle between load-terminal current vector and voltage vector.

The voltage loss is negative when load current running through series capacitor. It means output voltage of capacitor is greater than entrance voltage. The effect of voltage improvement has some connection with load current and power factor. The higher the load current, and the lower the power factor, the better the effect. The absolute value of voltage loss of capacitor is proportional to load current. It has automatic regulation to the voltage of supply arms, which is compatible with the drastic changes of the load of electric traction.

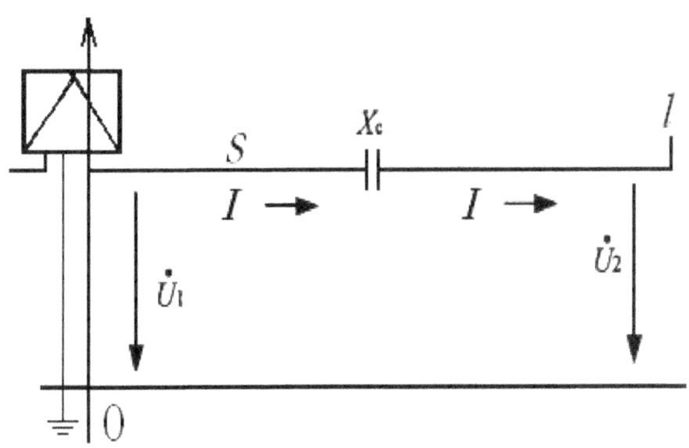

Fig. 5. The graphic of Series Compensation of Power-Supply System of Electric Traction

The installed capacity of the series compensation is decided by its rated current and rated voltage. Series compensation is different from other electric devices for its working voltage is proportional to its current. Therefore, when choosing the content of the series compensation, the higher the rated current, the lower the working voltage. It means the lower the compensative voltage, the worse the compensative effect. That is the compensative effect is in inverse proportion to the installed capability

The working voltage of series capacitor is proportional to its current. When choosing the series capacitor, the higher the rated current, the lower the working voltage. It means the lower the compensation voltage, the worse the compensation effect. The compensation capability is in inverse proportion to the installed capability. So the key point to meet the demands of compensation effect is to choose the rated current of series capacitor.

Series capacitor X_C can be calculated through rated voltage U_{CN}, rated current I_{CN}, and the power factor $\cos\varphi$ of the line. $X_C = \frac{U_{CN}}{I_{CN}}$, that is,

$$\Delta U_C = I_C \bullet \frac{U_{CN}}{I_{CN}} \sin\varphi = I_C \frac{U_{CN}^2}{Q_{CN}} \sin\varphi$$

Obviously, when the rated voltage U_{CN} and the current flows through are the same, the higher the I_{CN} (Q_{CN}), the lower the ΔU_C. ΔU_C is proportion to I_{CN} (Q_{CN}).

In order to meet the demands of compensation effect, the key point is choosing the rated current of series capacitor.

3.2 Thyristor controlled series compensation capacitor

Thyristor controlled series compensation capacitor includes Thyristor Switched Series Capacitor (TSSC) and Thyristor Controlled Series Capacitor (TCSC) commonly.

TSSC is consisted of a series of series capacitor. Each capacitor is connected with a transistor valve, which has a pair of anti-shunt transistors. TSSC adopts discrete step to increase or decrease the connected capacitor so as to control the compensation capacitor.

TCSC is a good way of series compensation, which is consisted of a series capacitor and a thyristor controlled reactor L. Actually, TCSC is also consisted of a protective device. It is shown as figure 2.

When the thyristor is turned off completely, the reactor is at non-conducting state, and TCSC manifests itself as a series capacitor compensator. When the conduction angle of thyristor is increasing gradually, the capability of the reactor of circuit branch and capacitor connected in parallel is increasing as well. When the reactor reaches the size of capacitor, shunt parameter will resonate, and the impedance of TCSC will be infinite. When the conduction angle of thyristor is increasing further, TCSC is manifesting itself as an inductive impedance, and the inductive impedance is decreasing gradually. When the thyristor is conducting completely, the impedance is the least. Therefore, TCSC can provide capacitive or inductive equivalent impedance within a certain range by controlling the conduction angle of thyristor.

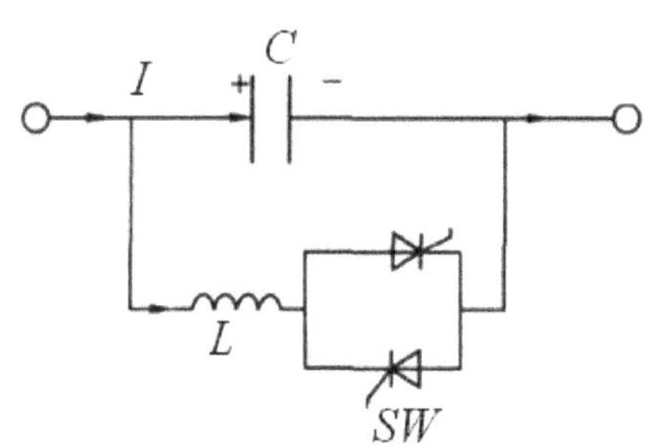

Fig. 6. The structural drawing of TCSC

4. Unified power flow controller (UPFC)

The UPFC consists of two voltage-sourced AC-DC converters connected back-to-back with some means (e.g., a chopper or specially designed1 converters) to permit interchange of power between the two converters.

Research work directed to analyze and build a scaled model of the UPFC has been sponsored by the Western Area Power Administration, the Electric Power Research Institute, and the Westinghouse Science and Technology Center. Refer to figure 7 for details.

Among the unique capabilities of the UPFC is the ability to control the flow of both real and reactive power on its series transmission line simultaneously and independently, as well as the ability to control bus voltage by operation of its shunt (STATCOM) element.

5. The interaction and coordinated control of traction power supplying system capability increased devices

5.1 The interaction between devices

Take the interaction between Thyristor Switched Capacitor (TSC) and series compensation of Traction Power Supply system for example to explain the interaction between ordinary equipments of the system. The circuit of power-supplying capacity increased system of electric traction which has dynamic shunt compensation equipment and series compensation equipment is shown as figure 8. I_q is the current of traction loading.

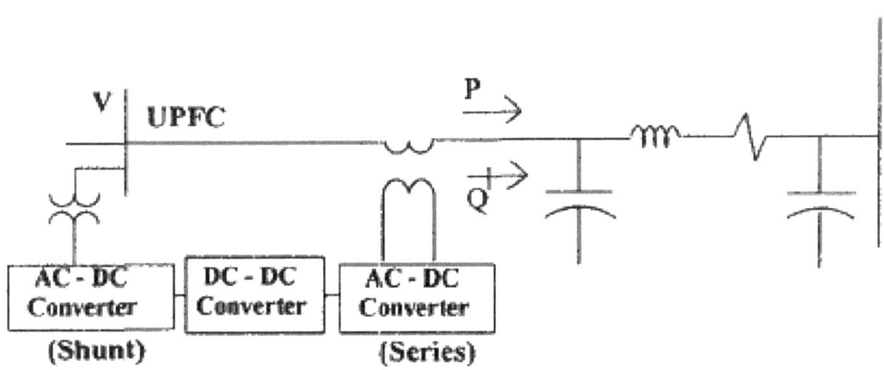

Fig. 7. Unified Power Flow Controller (UPFC), with DC-DC converter between the self-commutated AC-DC converters

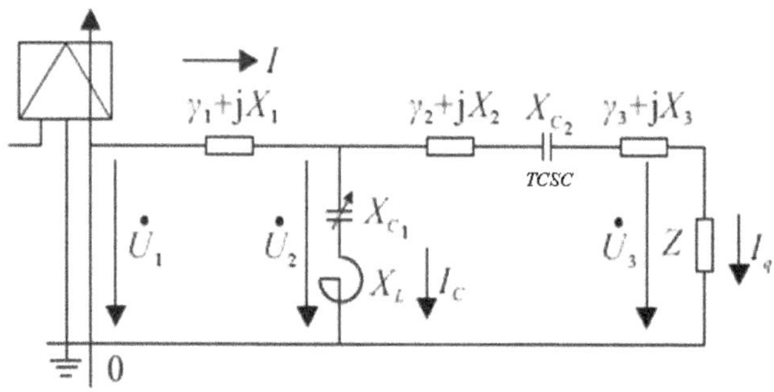

Fig. 8. Figure of power-supplying capacity increased system of electric traction

Let me describe the interactions between them. Assume the voltage of high side of traction substation is $\dot{U}_S$, the voltage of low side is $\dot{U}_1$, while main transformation ratio is k. Other parameters are shown in figure 8. Before installing the dynamic shunt compensation equipment, system current $I=I_q$, and the system can be described as :

$$\dot{U}_1=\dot{U}_3+[(\gamma_1+\gamma_2+\gamma_3)+j(X_1+X_2+X_3-X_{C2})]\dot{I}_q$$

After installing TSC, following changes occur :

a. The power factor changes from $\cos\Phi_1$ to $\cos\Phi_2$.

The power factor increases from $\cos\Phi_1$ to $\cos\Phi_2$ after paralleling a capacitor X_c.

$$X_C = P(\tan\Phi_1-\tan\Phi_2)\div(2\pi f U_1^2)$$

b. $\dot{U}_1$ increases.

Since TCS generates a compensative current contrary to the reactive component, the current flows through γ_1+jX_1 is decreasing, and a voltage rise appears in $\dot{U}_1$:

$$\Delta U_1=(\gamma_1+jX_1)I(\sin\Phi_1-\sin\Phi_2)$$

c. The reactive power absorbed from system by transformer increases.

It is shown as figure 9.

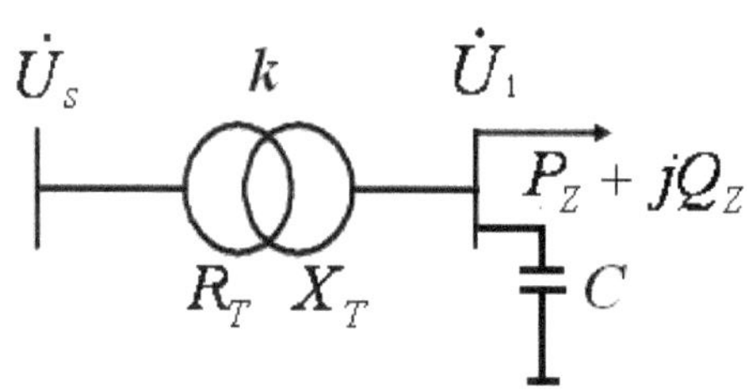

Fig. 9. Circuit of transformer of power-supply system of electric traction

$$U_S - U_1 = (Q_z X_T) \div U_1$$

So

$$Q_z = (U_S - U_1)\, U_1 \div X_T$$

Because $U_S = kU_1$, therefore,

$$Q_z = (k-1)\, U_1^2 \div X_T$$

It can be seen from the equation above: The reactive power absorbed from system by transformer is proportional to its voltage. When the voltage increases from U_1 to ΔU_1, the reactive power absorbed from system by transformer increases as well.

$$\Delta Q_z = (k-1)[(U_1+\Delta U_1)^2 - U_1^2] \div X_T = (k-1)[2U_1 * \Delta U_1 + \Delta U_1^2] \div X_T$$

d. The current flowing through the series compensation decreases, and it leads to decrease of the terminal voltage of supply arms.

The series capacitor X_C can be calculated by the rated voltage of series capacitor U_{CN}, the rated current I_{CN}, and the power factor $\cos\Phi_1$. If the current flowing through the series compensation is I before installing the parallel devices, the compensative voltage is:

$$\Delta U = I \times X_C \sin\Phi 1$$

After installing TSC, if the current flowing through the series compensation is $I+I_C$, because of the contrary direction of the reactive current I_C and I we can get $|I+I_C| < |I|$, so

$$\Delta U' = |I+I_C| \times X_C \sin\Phi_2 < |I| \times X_C \sin\Phi_2 < |I| \times X_C \sin\Phi_1 = \Delta U$$

It means the compensative capacity of the series compensation will decrease after installing TSC, and can't meet the expectations of compensation effect.

e. The active power loss of the system increases.

From the respect of TSC only, the increased active power loss is:

$$\Delta P = X_C \mathrm{tg}\alpha$$

α is the dielectric loss angle of the capacitor.

From the above analysis, if seeking a coordinate control strategy without proceeding from the entire system, each control system will disturb others mutually, and then lead to the difficulty of reaching ideal operating condition.

Adopting TCSC and thyristor controlled parallel capacitor in Traction Power Supply system, and optimally distributing the capacity of TCSC and thyristor controlled parallel capacitor according to the real-time condition of electric traction can not only save power, improve the power quality of Traction Power Supply system, but also protect the equipments, increase the safety of operation of trains. Solving the problem of coordinately control the electric traction reasonably is meaningful for the development of Traction Power Supply system.

5.2 Common analytical methods of interaction

Traction Power Supply is a kind of flexible AC transmission mode. The technology of Flexible AC Transmission System (FACTS) is one of the most attracting directions of nowadays new technology of power system. It becomes an effective method of solving the problems of economic operation, security and stability in power system.

Nowadays in the field of interaction analysis of FACTS, mainly there are two kinds of methods: ①Simulation of nonlinear time-domain. This method can only observe some sever interaction through simulation waveforms, while can't obtain accurate quantitative analysis. ②Traditional modal analysis of characteristic root. This method can study the influence on system from FACTS. Its objects mainly focus on the changes of one or some oscillation modals. However, it can't obtain quantitative analysis of the degree of interaction among controllers.

In the last 30 years, research on interaction has attracted widely focus. There are many documents in the field of analysis of steady-state and dynamic interactive control. Introductions of analytical methods of interaction are shown below.

5.2.1 Analytical method of modal

As a traditional analytical method, analytical method of modal has been widely used in controlled field. It also gets a lot of achievements in the research on the analysis of interaction among power system equipments. The analysis of sensitivity of characteristic root is one of the important achievements of analytical method of modal. It represents the strength of the changes of characteristic root caused by every parameter. The real component stands for damping capacity for modal oscillations. If the real component is positive, it means instable. The imaginary part stands for the frequency of natural oscillation.

Traditional analytical method of modal is built on characteristic root solved of entire state equation (Full-dimension analytical method). When the order of the state equation is 200, QR algorithm which usually considered effective can't get the right characteristic root. In order to overcome the shortcoming of Full-Dimension Analytical Method, reduced dimension analytical technology gets certain research. We get analytical method of modal through the state equation after reducing dimension. The equation has the characteristic root and characteristic vector which need studying. Reference [3] separated the modal, then got a system transfer function that include main modal only. Based on it, put forward multiple indicators of site selection and designation of controller of FACTS. Even though these indicators aren't accurate, the order of the system reduces effectively.

Recent years, with the development of non-linear system theory and application of modern mathematic methods, as a successful mathematical analysis tool and linear modal analysis, Normal Form (NF) is widely used in the analysis of mode interaction, study on stability region of system, and the designation of controller. Reference [4] put forward an analytical technology to reduce the order of state equation of system to second-order equation near working point and revealed the interaction among modals. NF provides new way for site selection and designation of controller. Reference [5] adopted NF to predict the separating phenomenon of oscillation modals in regional system. It revealed the importance of the influence on the stability of system from non-linear factors. Reference [6] and [7] put forward a research method of using NF mathematical tools to analyze the interactions among controllers of FACTS. Linear analysis ignores the influence on system response from non-linear factors, and considers based on NF by preserving second-order equation. Then put forward a non-linear interactive index to evaluate the intensity of interaction among SVC, SVC and TCSC, and UPFC. In the cases of study of multi-machine power system, the effectiveness of NF analytical method was verified through theoretical analysis and time-domain simulation.

5.2.2 Analytical method of relative gain array

Relative Gain Array (RGA), put forward by Bristol in 1996, is an effective method of analyzing the interaction of multi-variable control system. Also, it is a widely used designing tool of controlling system [8]. As for multi-variable control system, RGA is a method of providing optimal combination of control variable and controlled variable by observing the interaction among each input and output variables. It can also provide interaction messages among different control processes. Therefore, RGA is widely used now. Reference [9] studied the possibility of the application of RGA in static analysis of power system. It compared traditional residue, characteristic root and factor analysis method, indentified and studied the problem of site selection of PSS. Reference [10] studied the possibility of designing damping FACTS controller by improving RGA.

5.2.3 Singular value decomposition method

Singular Value Decomposition (SVD) Method is an important method of analyzing the interaction of input and output variables of system [11]. When apply SVD in a frequency range, the main point is, as opposed to original vector, singular vector decreases little, and the interaction of control circuit is small. The advantage of SVD is capable of dealing with time-lag system and non-positive-order system. But when the singular values are too close, SVD will be very sensitive or even unreliable.

5.2.4 NI method

NI method, put forward by Niederlin ski in 1971, is a controller matching method [12]. NI method is widely used in the choice of variable matching of control system just like RGA. Both methods are easy, and they can indicate the degree of coupling of control circuit only depend on object model. The standard of NI method based on variable matching was introduced in reference [13]. In reference [13], it said that NI value is an indicator of global interaction. And it can solve the problem of the choice of variable matching. Reference [14] and [15] put forward the combination of RGA and NI method. But commonly NI indicator is served as a tool of RGA, when analyze the interaction and pairing pattern among controllers. Use RGA to pair, then check if the system stable through NI method.

5.2.5 Analytical method of interaction based on gramian

Gramian is built on dynamic model of state space of system. It can describe the controllability and observability of one stable system, and it suits for continuous or discrete system. Take Gramian of system as quantitative analytical way to describe the signal of input and output as well as the controllability and observability of system [16]. Reference [17] took Gramian as representation of information content, combined with Homology group and information content, put forward an optimum distribution of rotor angle measurement device. Gramian is model based on state space of system, but when a system is represented by four matrixes (A, B, C, D), if feedback matrix of D exists, this method is not applicable.

5.2.6 The standard of Eigen value of Jacobian determinant

The standard of Eigen value of Jacobian Determinant, put forward based on difficulty analysis of calculating inverse matrix of steady state gain matrix G, can be used to guide variables pairing. Best match of variables can make single circuit independent. They are matches that off-diagonal elements of G have least inverse matrix effect on G. For a system that doesn't have interaction, inverse matrix of G is equal to the inverse of a matrix consisted of diagonal elements.

From Reference [18] and [19], we knew that when the inverse of G G_{21} is represented by Jacobian and diagonal matrix, the necessary and sufficient condition of converging G_{21} is that the value of max eigen value of Jacobian matrix must be less than 1. Therefore, the value of eigen value is the least among all possible pairings get from Jacobian matrix. But the difficulty of applying this method is that there are many pairings needed detecting, and the off-diagonal elements of G have significant influence on the process of inversing. So this method doesn't apply to predict pairing.

5.2.7 RHP zero method

RHP zero method can be used to guide variables pairing. Considering pairing different outputs and inputs, system would have different zeros. On some conditions, RHP zero exists. These zeros would make the control performance of closed-loop worse, which we should try to avoid when pairing controller.

From reference [12], we can get that zero would not change with feedback control, but the pole would change with feedback control. As the feedback control gain decreases, the pole of closed-loop would move to the position of the pole of open-loop. And it is going to make closed-loop system unstable possibly. Therefore, the principle of choosing pairing output and input is let closed-loop have least RHP zeros. Especially, we should avoid there are RHP zeros in the frequency region considered.

6. Coordinated control of interaction

Theoretically, the most effective method of solving the problem of interaction among FACTS controllers is to apply multi-variable coordinated control. Because a MIMO system is consisted of a flexible AC transmission system that has many control function (may be many FACTS devices), so a MIMO is the easiest method of solving the problem of interaction among controllers and the stability of closed-loop system. However, because of the complexity of multi-variable control, actually it's difficult to apply it in power system. Usually, we use coordinated control to solve the problem of the interaction among controllers.

6.1 Coordinated control of multi-objective of single FACTS

FACTS has many functions. Generally, controllers are designated aiming for different functions respectively. It makes every control function isolated or even contradicted, so it's time to consider the coordinated control among multiple targets. Reference [21] pointed out that STATCOM which has static parameters couldn't have both satisfying voltage control accuracy and damping control effect. In order to achieve the coordination between two targets, controller based on rules was designated. It means defining the structure and parameters of controller according to the operating condition.

Reference [22] introduced a new control method of multi-objective of FACTS. That is an intelligent control method of FACTS device. This method combined with the advantages of predictive control and inverse system control. The process of the signal of optimal control of FACTS has two sub-processes. First, filter the candidate outputs of FACTS device, and get the best output. Second, inversed calculate the actual control signal of FACTS device according to the best output. In every sub-process, Artificial Nervous Network Technique (ANNT) and fuzzy reasoning are used to solve different problems. Reference [22] took Advenced Static Var Generator (ASVG) as an example to introduce designation steps of intelligent predictive controller based on the method of intelligent predictive. Electromechanical simulation example proved the effectiveness of this new control method.

6.2 The theory of multi-objective evolution

In real world, most optimum problems are relative to multiple targets. Those targets are not alone. They are usually coupling and competing. Each target has different meanings and dimensions. The complexity and competitiveness would make optimization difficult [23].

The optimal solution of single target has clear definition, however, the definition couldn't promote to multiple targets. It's different from the definition of single target that multi-objective problem doesn't have global optimal solution, but it has a collection of global optimal solution. The optimal solution of multi-objective problem is called collection of global optimal solution, and the elements of it incomparable for global target.

Commonly multi-objective problem can be described as[227]:

$$\begin{aligned} &\min/\max\{f_m(x)\},\ m = 1,2,\cdots,m; \\ &\text{subjct to } g_j(x) \geq 0,\ j = 1,2,\cdots,J; \\ &h_k(x) = 0,\ k = 1,2,\cdots,K \end{aligned}$$

Among, decision vector: $x \in R^n$, target vector $f(x) \in R^n$, and $g_i(x)$ and $h_k(x)$ are inequality and restrain condition of equation respectively.

The essence of multi-objective is finding a group of decision vector which is meeting the demand of restraint condition so as to make target vector be the maximum or minimum at feasible region. It's different from optimal problem of single target that it's almost impossible that all target functions are maximum or minimum. Therefore, people put forward the concept of optimal solution of Pareto.

6.3 ZMOEA method

Revolution algorithm based on population can implicit parallel search for multiple solutions in solution space, and it can also improve the efficiency of calculating through similarity of different solution. The combination of revolution algorithm and the concept of optimal

Pareto can produce real revolution algorithm based on the concept of optimal Pareto so as to search for the non-inferiority optimal solutions.

Using revolution algorithm to solve multi-objective problem is like traditional algorithms. Objective function must be scalarized. Fitness, served as the assessed value of individual of next generation, must be scalarized. The process of scalarization should monotonic transformation of coordinate of objective function so as to let the individual endow the best fitness of Pareto. This transformation isn't the only one. It concludes of the preference information of designer when he is assessing individual. Generally, if one scalared fitness is achieving, revolution algorithms use common selection method to progress [24]. There are 3 kinds of revolution algorithms to solve the multi-objective problem [25].

Quantitative method: Multiple targets usually convert to single one so as to optimize them. This kind of method includes weighted method, minimax method, and objective vector method and so on. These methods are usually the same as single target optimum method.

Non-Pareto method based on population: Different objective functions have effect on choice of different individuals of population in turns. The method that the different individuals of population are sorted according to every objective function belongs to non-Pareto method based on population.

Sorting method based on Pareto: Sorting population directly according to Pareto.

6.4 Steps of multi-objective evolution

Steps of Multi-Objective evolution are as follows.

a. Decompress a group which has N chromosomes initially. Coding chromosomes in floating mode.
b. Decoding chromosomes and transferring it to real value of parameter region of controller optimized. Then substitute it into equation unknown. Sorting target values in Pareto way. Then calculating the initial fitness value.
c. Choosing and copying, then producing new population. According to the difference among fitness values, choosing in the way of roulette, and generating new population.
d. Crossover the individuals of new population, and mutate them. Increase the diversity of population. Interleaved mode adopts single-point crossover. And mutation can prevent algorithm from being optimum partially to some extent.
e. Operating Elitism [26], and combining the population produced above with old population. Sorting them, and generating a new population.
f. After evolution of one generation and producing next generation, go back to step B until it meet the termination conditions. Find out some optimum point of Pareto or the some close to it.

Termination conditions can be maxgenterm, or we can choose other indicators to judge restraint condition.

6.5 Multi-objective optimal design of TCSC and SVC

Among evolutionary algorithms, fitness function decides the evolutionary direction of population. In order to transfer coordination problem to multi-objective optimal design problem, we should choose a function consisted a series of performance index of system response. Choose control objectives of TCSC and SVC, it means active power of the line, signal of node voltage supported by SVC, and variance integral of reference value set by controller to be performance index. As mentioned above, the problem of multi-objective optimal design of TCSC and SVC can be described as:

$$\min \begin{cases} F_V(K_P, K_I, K_A) = \int_0^{t_s} |\Delta V_{SVC}|\, dt \\ F_V(K_P, K_I, K_A) = \int_0^{t_s} |\Delta P_{TCSC}|\, dt \end{cases}$$

Constrain condition of optimized parameters is:

$$0 \le K_P \le 30;$$
$$0 \le K_I \le 30;$$
$$0 \le K_{SVC} \le 150.$$

Where K_P is scale factor of TCSC PI controller, K_I is double integral TCSC PI controller, and K_{svc} is the gain factor of SVC voltage regulator.

6.6 The coordinate control case of TCSC and SVC

As figure 7 shows, install TCSC and SVC in system. TCSC is installed on supply arms. The purpose of installing TCSC is to control active power of line by changing the reactance of line. SVC is installed on bus bar in parallel. The purpose of installing SVC is to make the voltage of installation point stable. Adopt multi-objective evolution algorithm to optimize 3 parameters of both controllers of FACTS

Let the population size be 50, let maximum optimum algebra be 50, let crossover frequency be 0.8, and let mutation rate be 0.07. Figure 10 is convergence rate curve of MOEA, and figure 11 is non-optimal Pareto solution. It can be seen from figure 10. The progress ratio of MOEA is relatively high initially. After evolution, the progress ratio gradually tends to zero. It means population is close to Pareto optimal solution. As for Pareto optimal solution described by figure 11, according to the definition of Multi-Objective problem, first, the non-optimum solution is a solution in feasible region. Second, when comparing any solution with other feasible solutions of non-optimal solution, at least one object is better than other feasible solution. Table 1 shows partial Pareto solution.

Group	K_P	K_I	K_{SVC}
1	0.9456774	11.61006	81.08093
2	0.9456834	13.04966	83.08023
3	0.9366024	5.395905	103.3966
4	0.6860377	6.610057	58.17931
5	1.00992	6.395556	49.3839
6	0.4875605	28.76538	112.8552

Table 1. Partial Pareto optimal solution

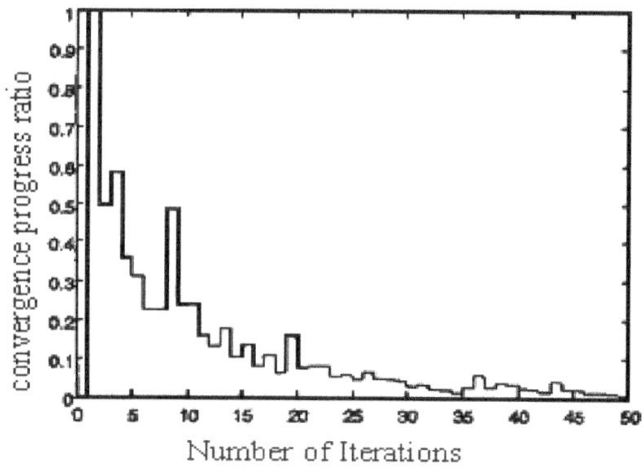

Fig. 10. Progress ratio of multi-objective evolution progress

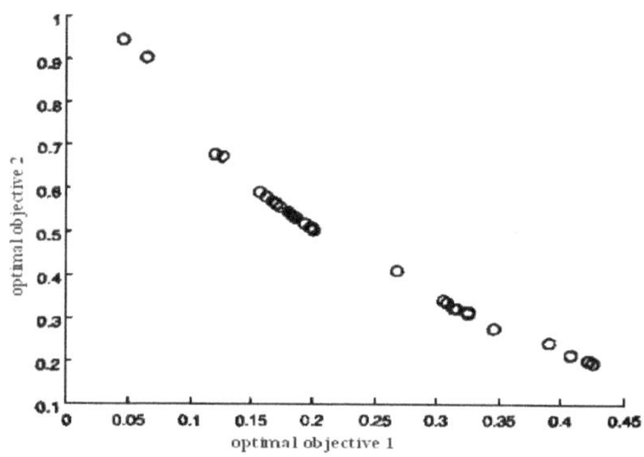

Fig. 11. Pareto-optimal front for two-objective design

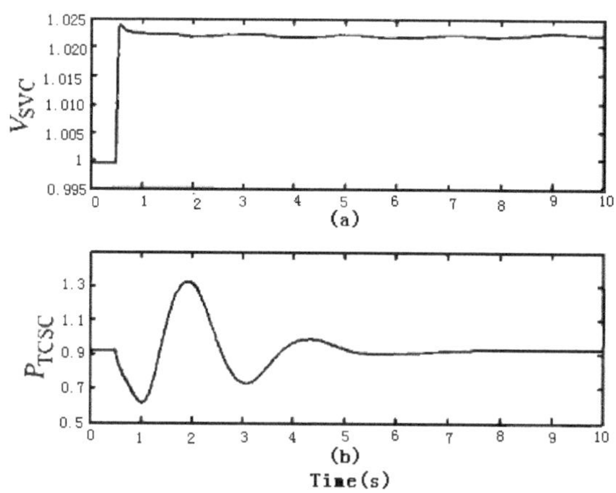

(a) Order step response curve of SVC when TCSC is closed.
(b) Changing curve of an Active power controlled by TCSC when voltage node of SVC steps

Fig. 12. Combined-operation performance of SVC with TCSC

As you can see in figure 12, the coordinate design of TCSC and SVC is successful, and the difference between different groups of parameters of Pareto solution set is: one group of optimal solution has better control effect on TCSC and worse effect on the stability of voltage of SVC. The purpose of Multi-Objective coordinate designation is to analyze which control object is more important according to actual operation when the different objects of SVC and TCSC interact each other, and lead to the worst situation that both objects couldn't be optimum. Then choose a group of optimal solution among Pareto solution to solve the problem of coordinate operation after installing SVC and TCSC.

Now analyze the problem of coordinate control considering the least energy consumption.

The impedance of single-phase traction transformer $R_T + jX_T$ can be calculated by short circuit voltage $U_d\%$, short circuit power loss ΔP_C with reduction to 27.5kV side.

$$R_T = \frac{\Delta P_C U_N^2}{1000 S_N}(\Omega),$$

$$X_T = \frac{U_d\% U_N^2}{100 S_N}(\Omega),$$

U_N=27.5KV, and S_N is the rated capacity of traction transformer. We can get the voltage loss of transformer winding through transformer impedance $R_T + jX_T$

$$\Delta U = (R_T\cos\Phi_2 + X_T\sin\Phi_2)(I_a + I_b),$$

I_a and I_b are currents of both supply arms.

Active power losses of traction transformer are iron loss ΔP_0 and copper loss. The iron loss is no-load loss, which is fixed, and has nothing to do with current flows through load. ΔP_0 can be calculated by copper loss, which is called load loss or changeable loss, and changing with current flows through load. Active power loss of single-phase is:

$$\Delta P = I^2 R_T$$

According to the principle of reasonably utilize the place and the analysis above, assume installing TSC and series compensation which are on 30-meter supply arm on the place at the distance of 8km from transformer, and use l and s to represent the distances of two sections. The high-voltage side and low-voltage side of transformer are 100kV and 27.5kV. Assume the capacitors of SVC and TCSC are X_{C1}, and X_{C2}. X_{C1} is a continuous variable, and X_{C2} is a undetermined variable. The energy consumption function of entire system is:

$$P(X_{C1}, X_{C2}) = (X_{C1} + X_{C2})tg\alpha + \int_0^s I_q[U_1 - \rho I x]dx$$
$$+ \int_s^l I_q[U_1 - \rho I s + I_q X_{C2}\sin\Phi_1 - \rho I_q(x-s)]dx + I^2 R_T$$

$$Q(X_{C1}, X_{C2}) = U_1 I\sin\Phi_2 + U_2 Iq\sin\Phi_2 + (k-1)\, U_1^2 \div X_T - I_q^2 X_{C2}\sin\Phi_2$$

From equation (13) and (14), we can get the objective function of system is:

$$W = \min\{P^2(X_{C1}, X_{C2}) + Q^2(X_{C1}, X_{C2})\}^{0.5}$$

The functions of restrain condition are：

$$U_1 \le 30\text{kV};$$
$$U_3 \ge 20\text{kV};$$
$$\cos\Phi_2 > 0.9。$$

Let's discuss the minimum value of W. Because W and W^2 are the least, in order to discuss easily, we use the minimum value of W^2 to take the place of W, then,

$$U_1 = U_S - (R_T\cos\Phi_2 + X_T\sin\Phi_2)I,$$

$$U_2 = U_1 - (\gamma_1\cos\Phi_2 + X_1\sin\Phi_2)I = U_S - [(R_T+\gamma_1)\cos\Phi_2 + (X_T + X_1)\sin\Phi_2]I \text{。}$$

At the same time, from the equation below,

$$\frac{I_C}{\sin(\Phi_1 - \Phi_2)} = \frac{I_q}{\sin(\pi/2 + \Phi_2)}$$
$$\frac{I}{\sin(\pi/2 - \Phi_1)} = \frac{I_q}{\sin(\pi/2 + \Phi_2)}$$

we can get,

$$I_C = I_q \frac{\sin(\Phi_1 - \Phi_2)}{\cos\Phi_2}$$
$$I = I_q \frac{\cos\Phi_1}{\cos\Phi_2}$$

From $I_C = U_2 \div X_C$ and $I_C = I_q\sin\Phi_1 - I\sin\Phi_2$, we can get,

$$U_2 = (I_q\sin\Phi_1 - I\sin\Phi_2)\times X_C = I_q X_C (\sin\Phi_1 - \text{tg}\Phi_2\cos\Phi_1)$$

$$U_1 = U_S - I_q (R_T + X_T\,\text{tg}\Phi_2)\cos\Phi_1$$

then,

$$U_S - I_q [(R_T+\gamma_1) + (X_T + X_1)\text{tg}\Phi_2]\cos\Phi_1 = I_q X_C (\sin\Phi_1 - \text{tg}\Phi_2\cos\Phi_1)$$

We can get from the equation above,

$$tg\Phi_2 = \frac{U_S - I_q[(R_T + \gamma_1)\cos\Phi_1 + X_C\sin\Phi_1]}{(X_T + X_1 - X_C)I_q\cos\Phi_1}$$

or

$$X_C = \frac{U_S - I_q[(R_T + \gamma_1) + (X_T + X_1)tg\Phi_2]\cos\Phi_1}{I_q(\sin\Phi_1 - tg\Phi_2\cos\Phi_1)}$$

In order to get the best compensation value of capacitor, find partial derivative of W^2,

$$\frac{\partial W^2}{\partial X_{C1}} = 2P(X_{C1}, X_{C2})\{tg\alpha + \int_0^s I_q[\frac{dU_1}{dX_{C1}} - \rho x \frac{dI}{dX_{C1}}]dx + \int_0^s I_q[\frac{dU_1}{dX_{C1}} - \rho(x-s)\frac{dI}{dX_{C1}}]dx + 2IR_T\frac{dI}{dX_{C1}}\}$$

$$+2Q(X_{C1}, X_{C2})\{I_q \cos\Phi_1 tg\Phi_2 \frac{dU_1}{dX_{C1}} + U_1 I_q \cos\Phi_1 \frac{dtg\Phi_2}{dX_{C1}} + I_q \sin\Phi_2 \frac{dU_2}{dX_{C1}}\}$$
$$+(U_2 I_q - I_q^2 X_{C2})\cos\Phi_2 \frac{d\Phi_2}{dX_{C1}}$$

$$\frac{\partial W^2}{\partial X_{C2}} = 2P(X_{C1}, X_{C2})\{tg\alpha + I_q \sin\Phi_1 (l-s)\} - 2Q(X_{C1}, X_{C2}) I_q^2 \sin\Phi_2$$

$$\begin{cases} \dfrac{\partial W^2}{\partial X_{C1}} = 0 \\ \dfrac{\partial W^2}{\partial X_{C2}} = 0 \end{cases}$$

Calculate the equations above, and we can get the best compensation value $\{X_{C1}^*, X_{C2}^*\}$ which cause least energy cost.

In conclusion, the capacity of compensator with parallel capacitor XC1 is decided by U_2, and the capacity of TCSC is decided by X_{C2}. In return, after installing TCSC, the voltage drop caused by it subtracts the voltage drop caused by inductance of contact system. It could be considered that capacitor subtracts inductance. Because the voltage drop of supply arms caused by TCSC and the voltage drop caused by inductance cancel out, the voltage loss of system reduces obviously, and the power factor improves a lot.

Adopting TCSC and optimally distributing the capacity of SVC and TCSC according to the real-time condition of electric traction can not only save power, improve the power quality of Traction Power Supply system, but also protect the equipments, increase the safety of operation of trains. Solving the problem of coordinately control the electric traction reasonably is meaningful for the development of Traction Power Supply system.

7. References

[1] Huang Yuanliang, QIAN Qing-quan. Research on improving quality of electricity energy in train's traction[J], Control and Decision, 2010, 25(10): 1575-1579(in Chinese).

[2] Cao Yi-jia *et al.* Research Progress on Interaction and Coordinated Control Among FACTS Controllers [J], Proceedings of the CSU-EPSA: 2008 20(1): 1~8.(in Chinese)

[3] Larsen E V, Sanchez2Gasca J J , Chow J H. Concepts for design of FACTS controllers to damp power swings[J]. IEEE Transon Power System s, 1995, 10(2): 948~956.

[4] Thapar J, VittalV, Kliemann, *et al.* Application of the normal form of vector fields to predict interarea separation in power system s [J]. IEEE Transon Power Systems, 1997, 12 (2): 844- 850.

[5] Li Yinghui, Zhang Baohui. A new method to determine the transient stability boundary using nonlinear theory [J]. Proceedings of the CSEE, 2000, 20 (1): 41- 44.

[6] Zou Z Y, J iang Q Y, Cao Y J , *et al* . Application of the normal forms to analyze the interact ions among the multi-control channels of UPFC [J]. International Journal of Electrical Power and Energy System s, 2005, 27 (8): 584- 593.

[7] Zou Z Y, J iang Q Y, Cao Y J, *et al*. Normal form analysis of the interact ions among multiple SVC controllers in power system s [J]. IEE Proceedings-Generation, Transmission and Distribution, 2005, 152 (4) : 469- 474.

[8] B risto l E H. O n a new measure of interact ion for multivariable process control [J]. IEEE Trans on Automatic Control, 1966, 11 (1): 133- 134.

[9] M ilanovic J V, Duque A C S. Identification of electromechanical modes and placement of PSS susing relative gain array [J]. IEEE Transon Power Systems, 2004, 19 (1): 410- 417.

[10] Zhang Pengxiang, *et al*. Application of relative gain array method to analyze interact ion of multi-fuctional facts controllers [J]. Proceedings of the CSEE, 2004, 24 (7): 13- 17.

[11] Hamdan A M A. A n invest igat ion of the significance of singular value decomposition in power system dynamics [J]. International Journal of Electrical Power and Energy System s, 1999, 21 (6): 417- 424.

[12] Niederlinski A. A heuristic app roach to the design of linear multivariable interacting control systems [J]. Automatica, 1971, 7(6): 691-701.

[13] Zhu Z X, Jutan A. A new variable pairing criterion based on the Niederlinski index [J]. Chemical Engineering Communications, 1993, 121 (18): 235-250.

[14] Yu C C, L uyben W L. Robustness with respect to integral controllability [J]. Industrial and Engineering Chemistry Research, 1987, 26 (5): 1043-1045.

[15] Zhu Z X, Jutan A. Stability robustness for decentralized control systems [J]. Chemical Engineering Science, 1993, 48 (13) : 2337- 2343.

[16] Conley A, Salgado M E. Gramian based interaction measure [C] || P roceeding of the 39th IEEE Conference on Decision and Control, Sydney, NSW, Australia: 2000.

[17] Chu Xiaodong,Liu Yutian. Optimal placement of rotor angle transducers[J]. Proceedings of the CSEE, 2003, 23 (9): 132- 136.

[18] M ijares G, Co le J D, Naugle N W, *et al*. A new criterion for the pairing of control and manipulated variables [J]. A IChE Journal, 1986, 32(9): 1439-1449.

[19] Schm idt H, Jacobsen E W. Selecting control configurations for performance with independent design [J]. Computers and Chemical Engineering, 2003, 27 (1): 101- 109.

[20] Farsangi M M , Song Y H, Lee K Y. Choice of FACTS device control inputs for damping interarea oscillations [J]. IEEE Trans Power System s, 2004, 19 (2) : 1135- 1143.

[21] Li Chun, Jiang Qirong, Wang Zhonghong. Design of a rule based controller for STA TCOM [J]. Proceedings of the CSEE, 1999, 19 (6): 56- 60.

[22] Lu Qiang, *et al*). Target oriented intelligent predictive evaluating control method for FACTS devices (PART 1) [J]. Power System Techno logy, 1998, 22 (4): 6- 9, 12.

[23] K. Deb. Multi-Objectiveive Optimization using Evolutionary Algorithms. N Y: John Wiley & Sons. I nc.

[24] D. Cvetkovic, I. C. Parmee. Preferences and Their Application in Evolutionary Multi-Objectiveive Optimization. IEEE Transactions on Evolutionary Computation, 2002, 6(1): 42 -57.

[25] C. M. Fonseca, P. J. Fleming, An Overview of Evolutionary Algorithms in Multi objective Optimization. Evolutionary Computation, 1995, 3(1): 1-16

[26] P. Ju, E. Handschin, F. Reyer. Genetic algorithm aided controller design with application to SVC. IEE Proceedings of Generation, Transmission and Distribution, 1996, 143 (3): 2 58-262.

3

Railway Traction

José A. Lozano, Jesús Félez, Juan de Dios Sanz and José M. Mera
Universidad Politécnica de Madrid
Spain

1. Introduction

The railway as a means of transport is a very old idea. At its beginnings, it was mainly utilised in the central European mines with different means of traction being applied. But it did not come into general use until the invention of the steam engine. Since the 18th century it has developed faster and faster until in the 21st century it has become the most efficient means of transport for medium distances thanks to the development of *High Speed*.

The main factors that have driven the enormous development of the railway, as any other means of transport, have been, and continue to be *safety, speed* and *economy*. On top of all this, as every day passes, its *environmental impact* is minimum, if not zero. In the case of the railway, one of the determining factors behind its development was the type of track used, either because of its gauge or the materials used in its construction. At the start, these were made of cast iron, but they turned out to be lacking in safety as they easily broke due to their fragility. Towards the end of the 19th century steel began to be used as it was a less fragile and much stronger material. Nowadays, plate track is a key element in the development of High Speed trains making wooden sleepers a thing of the past. The rubber track mountings which are currently used to support the tracks have led to enormous reductions in vibration and noise both in the track and the rolling stock. Moreover, each country had a different track gauge due to strategic reasons of commerce and defence. Today's global markets leave no other option but to standardise track gauges or failing that, to produce rolling stock that can be adapted to the different gauges quickly and automatically.

The birth of the railway is linked to the birth of the steam engine, while the tremendous development of the railway in the 20th century was linked to the electrification of the railway lines. In addition, the diesel locomotive played a very important role because of its autonomy, particularly on those lines where electrification was unviable. The rivalry between these three types of locomotive was long and hard with each competing to see which was the safest, fastest and cheapest. The first electric locomotives appeared in the last third of the 19th century, while the first Diesel locomotive was built at the beginning of the 20th century (Faure, 2004). Diesel locomotives never reached the speeds of electric locomotives; however, the latter require a greater investment in infrastructure to electrify the line. As we all know, it is electric locomotives that are at the forefront of railway traction, while the Diesel locomotive is kept for some very specific uses. The final decade of the 20th century and the first decade of the 21st century were marked by the enormous rise in High Speed due to the huge leaps forward in electric locomotive technology.

After this very brief historical introduction, in this chapter, we will now focus on the different types of railway traction and their importance in the development of the railway, with particular reference to *safety, speed, economy* and *environmental impact*. The basic aim is to give a detailed description of how the most generally used traction and engine systems work in present-day railways. An analysis of the critical or least efficient points of the way they work will lead us to discover the main criteria for optimising railway traction systems. This could be a starting point for conducting research and seeking innovations to produce ever more effective and efficient traction systems. The final part of this work will suggest and justify some ideas that could be useful for improving the operating efficiency of railway traction systems.

In order to fulfil the objectives described in the above paragraph, a methodology will be applied that is based on the consideration of physical models that represent the behaviour of the railway traction systems under study, taking account of all the processes of transformation, regulation and use of energy and even being able to recover it. Special attention is placed on the elements that can lead to losses of energy and efficiency. Specifically, these models are developed using the Bond-Graph technique. This technique is widely accepted for its capacity to model dynamic systems that embrace various fields of science and technology. Modelling is performed systematically in a way that enables all the dynamic, mechanical, electrical, electromagnetic and thermal phenomena, etc, that may be involved, to be taken into account. Moreover, this technique has more than proven itself to be suited to modelling vehicular and rail systems. By taking the corresponding conceptual models, the Bond-Graph models are generated, and, by means of very mechanical procedures the behaviour equations of the dynamic systems can be found. By then taking these behaviour equations or mathematical models, computer applications can be built to simulate, analyse and validate the behaviour of the developed models.

2. General aspects of railway traction

What really causes the train to accelerate or brake are the adhesion forces that appear in the wheel-rail contact. For these adhesion forces to appear, tractive or braking effort needs to be applied to the wheels. This torque must be generated in a motor or braking system that transforms a certain amount of energy into the required mechanical energy. These three stages form a general summary of the fundamentals of railway traction. All that needs to be done is to develop a simple model that looks at the phenomena produced from an energy point of view, from the moment the energy is captured until it reaches the wheel-rail contact. As a result of all this, the train accelerates or brakes.

2.1 General traction equation, resistance forces

Railway traction is considered to be one of the main problems of longitudinal rail dynamics (Faure, 2004; Iwnicki, 2006). It is seen as a one-dimensional problem located in the longitudinal direction of the track, governed by the Fundamental Law of Dynamics or Newton's Second Law, applied in the longitudinal direction of the train's forward motion:

$$\sum F = M \cdot a \qquad (1)$$

Where the term to the left of the equal sign is the sum of all the forces acting in the longitudinal direction of the train, "M" is the total mass and "a" is the longitudinal acceleration experienced by the train. The sum of forces consists of the tractive or braking effort "F_t" and the passive resistances opposing the forward motion of the train, "ΣF_R", (Fig. 1.-). The tractive or braking effort "F_t", in a final instance, is the resultant of the longitudinal adhesion forces that appear in the wheel-rail contact zones, either when the train's motors make the wheels rotate in the direction of forward motion, or when the braking forces act to stop the wheels rotating (in this case, these forces will obviously be negative or counter to the direction of the train's forward motion).

Fig. 1. Longitudinal train dynamics.

Let us now focus on the passive resistances that are opposing the train's forward motion, "ΣF_R". These basically consist of five types of forces:

a. Rolling resistances of the wheels.
b. Friction between the contacting mechanical elements.
c. Aerodynamic resistances to the train's forward motion.
d. Resistances to the train's forward motion on gradients.
e. Resistances to the train's forward motion on curves.

The phenomena associated with the appearance of the aforementioned resistances to the train's forward motion are widely known (Andrews, 1986 ; Faure, 2004; Coenraad, 2001 ; Iwnicki, 2006). For this reason, we will simply present one of the most common expressions for modern passenger trains:

$$\sum F_R = (266{,}3+27{,}7\cdot V+0{,}05168\cdot V^2)+(r_g+\frac{500}{R})\cdot(L+Q) \tag{2}$$

Where "V" is the train's speed in (Km/h), "r_g" is the inclination of the gradient as a ($^o/_{oo}$), "R" is the radius of the curve in (m), "L" is the weight of the locomotives in (Tm) and "Q" is the towed weight or the weight of the coaches in (Tm).

As can be seen from equation (2), the resistances to forward motion consist of three components: one constant, another that is linearly dependent on speed and another that depends on the speed squared. Moreover, when the train is running on a curve or a gradient, the final term that is not dependent on speed needs to be added.

We will now develop the Bond-Graph model, (Karnoppet alia, 2000), for the longitudinal train dynamics expressed in equation (1). In this model, shown in Figure 2, there are three basic elements:

a. The most important element is the train itself, whose longitudinal motion is represented by the Inertial port "I". The parameter of this port is the train's total mass, "M".
b. The tractive or braking efforts "Ft". In whatever case, this is an element that supplies energy according to a defined force. In Bond-Graphs, these energy sources with a

defined effort are represented by the Source of Effort port "SE". In this case, the parameter of the port is the defined effort.

c. The third element comprises the resistances to forward motion, which are represented by the resistance port "R" in the Bond-Graph. This port will have a variable parameter so that it can satisfy the equation (2).

The three ports comprising the Bond-Graph are brought together in a type "1" node, since all three phenomena are produced at the same speed, which is the speed of the train's forward motion. The inertial port satisfies the following: the input force "e_1" is equal to the first temporal derivative of the quantity of motion "P":

$$e_1 = \frac{dP}{dt} \tag{3}$$

In node "1", given that all the bonds connected to it have the same speed, the following is satisfied: the algebraic sum of the forces is zero. Therefore, the effort "e_1" equals:

$$e_1 = e_2 - e_3 \tag{4}$$

Where effort "e_2" is the tractive effort "F_t", (or braking, if that is the case), and "e_3" is the force of the passive resistances (given by equation (2)).

On the other hand, as we know, the Quantity of Motion "P" is equal to the product of the mass "M" and the speed "V", of the train in this case. By taking all this and entering into equation (3), it may be easily deduced that we will reach an expression that is identical to that shown in equation (1). In fact, it is the Inertial port "I" that resolves the fundamental equation of dynamics given by equation (1).

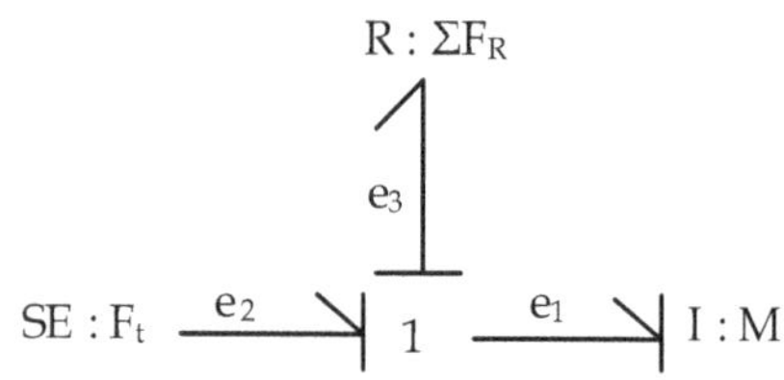

Fig. 2. Bond-Graph diagram of the longitudinal dynamics of the train.

2.2 Tractive efforts, adhesion, power

As already pointed out, we are not attempting to develop a complex model that takes account of all the driving phenomena. The main focus is to study the dynamic and energy phenomena associated with train traction or braking. For the interested reader, there is a wide biography on the subject (Iwnicki, 2006).

Let us now consider the steel wheel shown in Figure 3.-, moving along a longitudinal plane at a speed "V", in contact with a steel rail and to which we apply a traction torque "T". Also coming into play is the gravitational force "Mg", where "M" is the mass suspended on the wheel and "g" is the acceleration due to gravity, the vertical reaction of the rail "N", which balances the vertical forces, and the forces opposing the train's forward motion, "ΣF_R", already mentioned in the previous section. Finally, thanks to the *adhesion* in the wheel-rail contact zone, the tangential adhesion force "F_t" appears, which satisfies the following expression:

$$F_t = \mu \cdot N \tag{5}$$

Where "μ" is the so-called *adhesion coefficient,* whose rolling values for wheel and steel rail is dependent on temperature, humidity, dirt, etc, and particularly on speed. An expression that will enable us to find the value of the adhesion coefficient according to speed is the following:

$$\mu = \frac{\mu_o}{1+0{,}01V} \tag{6}$$

Where "V" is the speed of the train in Km/h, and "μ_o" is the adhesion coefficient for zero speed. The value of "μ_o" depends on the atmospheric conditions of temperature, humidity, dirt, etc, which under optimum conditions reaches a value of 0'33. Under such conditions and V=300 km/h, we obtain: μ =0.082. Therefore, we can see that the wheel-steel rail contact has limited possibilities when it comes to transmitting the tangential tractive or braking efforts "F_t",.

Below is the expression used by RENFE in Spain for the adhesion coefficient, where, as is customary, the speed is expressed in km/h:

$$\mu = \mu_o \left(0{,}2115 + \frac{33}{42+V}\right) \tag{7}$$

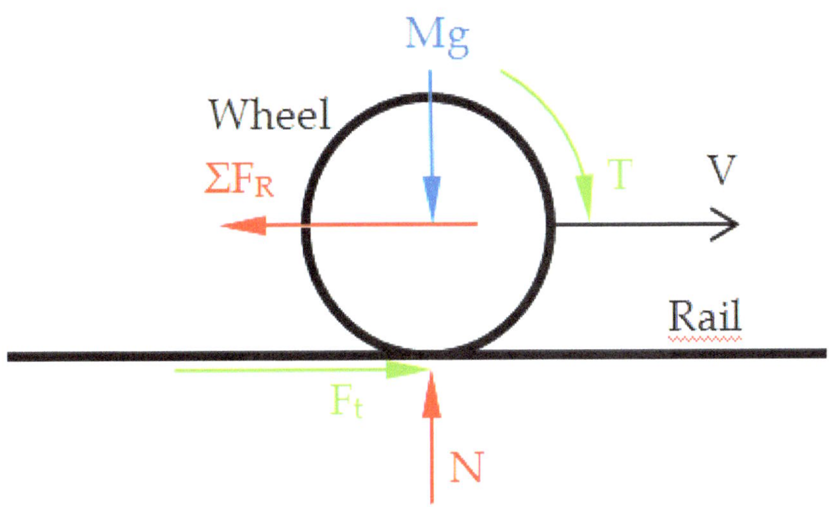

Fig. 3. Tangential adhesion forces in the wheel-rail contact.

Right from the beginnings of the railway the major challenge of railway traction has been to increase the adhesion coefficient in the wheel-rail contact. Fig. 4 shows some of the adhesion coefficient values "μ_o", used. What is surprising is the high value achieved for this coefficient in the United States of America.

Values of "μ_o"		
SNCF (France)	Electric monophase locomotives, multimotor bogies	0.33
	Electric monophase locomotives, monomotor bogies	0.35
DB (Germany)	Diesel locomotives	0.30
	Electric monophase locomotives	0.33
RENFE (Spain)	Diesel locomotives	0.22 - 0.29
	Classic electric locomotives	0.27
	Modern electric locomotives	0.31
USA	SD75MAC diesel and electric locomotives	**0.45**

Fig. 4. Some values used for "μ_o".

Some of the methods used to improve the values of the adhesion coefficient, especially during start up or acceleration, for reasons which will be made clear further on, are:

a. Introduction of sand in the wheel-rail contact. This is the traditional method using devices called "*sandboxes*", which are still frequently in use. However, this is a very aggressive system regarding the wear of wheel and rail materials.
b. Monomotor bogies that spread the tractive efforts evenly between all their shafts lead to an optimisation of the adhesion coefficient used.
c. Drawbars that connect the locomotive chassis to the bogies in a way that the locomotive's weight falls on the lowest part of the bogie at a level that is as close as possible to the wheel-rail contact. Apparently, the accelerations between the locomotive chassis and the bogie generate a force torque that aids the tractive or braking efforts, thus improving wheel-rail adhesion.
d. Electronic anti-slip, braking and traction control systems. These are similar systems to ABS, (Anti-lock Braking System), or ASR, (Anti-Slip Regulation) used in the automotive industry. The speed of the wheels is controlled and regulated by electronic devices so that there is no slip between the wheels and the rails.

In traction, the ideal situation is to have the maximum effort according to speed. As we know, the power developed is the product of force and speed:

$$\text{Power} = F_t \cdot V \tag{8}$$

Since the power supplied by the motor is approximately constant, the tractive effort available "F_t", (given by equation (8)), dependent on the train's speed "V", complies with a hyperbolic-type ratio, like that shown in Figure 5.

$$F_t = \frac{Power}{V} \tag{9}$$

Apart from the "constant power hyperbola", (in blue), Figure 5, also shows the maximum tractive effort curve constrained by the adhesion conditions, (in red), obtained from equation (5) and equations (6) or (7)). If we look carefully at the constant power hyperbole, it can be seen that at very low speed the tractive effort would tend towards infinity. However, for technical reasons, the motors are only able to supply a limited tractive effort. This is called "continuous tractive effort", and appears in green.

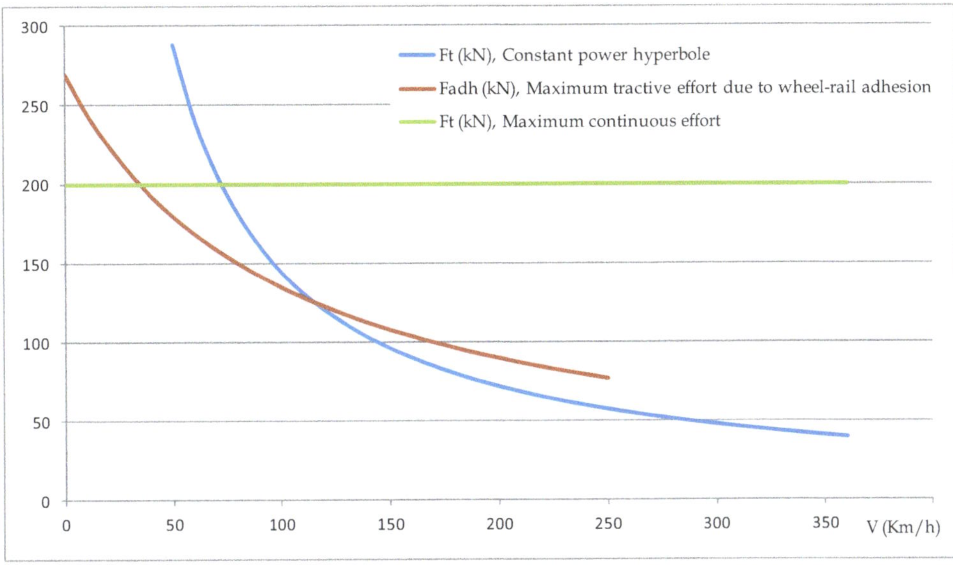

Fig. 5. Effort-speed curves

3. Railway motors

The first railways were powered by steam engines. Although the first electric railway motor came on the scene halfway through the 19th century, the high infrastructure costs meant that its use was very limited. The first Diesel engines for railway usage were not developed until halfway through the 20th century. The evolution of electric motors for railways and the development of electrification from the middle of the 20th century meant that this kind of motor was suitable for railways. Nowadays, practically all commercial locomotives are powered by electric motors (Faure, 2004; Iwnicki, 2006).

Figure 6 illustrates a flow diagram for the different types of rail engines and motors most widely used throughout their history. The first Diesel locomotives with a mechanical or hydraulic drive immediately gave way to Diesel locomotives with electrical transmission. These locomotives are really hybrids equipped with a Diesel engine that supplies mechanical energy to a generator, which, in turn, supplies the electrical energy to power the electric motors that actually move the drive shafts. Although this may appear to be a contradiction in terms, it actually leads to a better regulation of the motors and greater overall energy efficiency.

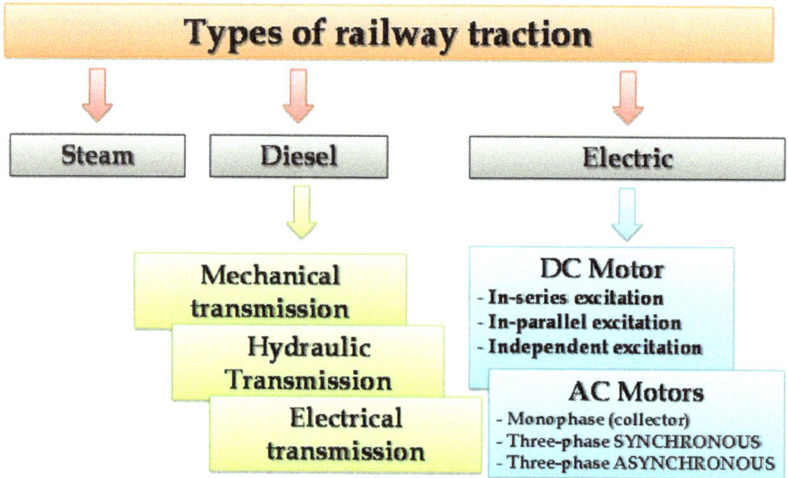

Fig. 6. Railway engine and motor types.

The major drawback of electrical traction is the high cost of the infrastructure required to carry the electrical energy to the point of usage. This requires constructing long electrical supply lines called *"catenary"*, (Figure 7). In addition, the locomotives need devices that enable the motor to be connected to the catenary: the most common being "pantographs" or the so-called "floaters". In its favour, electrical traction can be said to be clean, respectful of the environment and efficient, as an optimum regulation of the motors can be achieved. In this work, we will only focus on the functioning and regulation of the most widely used types of electric motors.

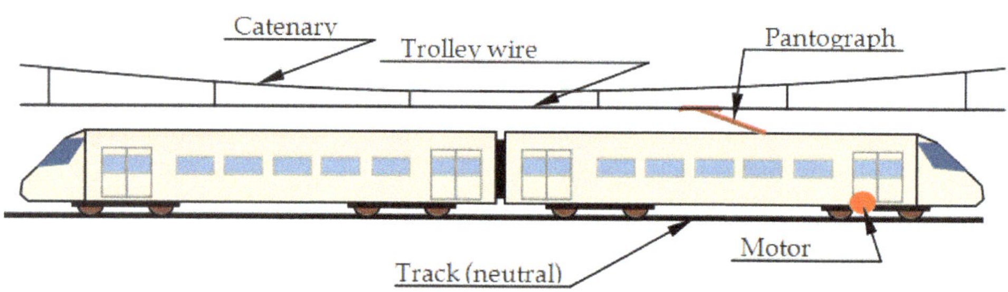

Fig. 7. Electric railway traction: General outline, catenary and pantograph.

3.1 Electric railway motors

The most widely used electric motors for railway traction are currently of three basic types, (Lozano, 2010):

1. Direct current electric motors with in-series excitation.
2. Direct current electric motors with independent excitation.
3. Alternating current electric motors.

Direct current electric motors usually work under a 3 kV supply and alternating current motors under 25 kV. Direct current motors are gradually becoming obsolescent in favour of alternating current motors. This is mainly due to maintenance problems with the direct current motor collectors and the better technological progress of alternating current motors.

This work does not aim to provide an exhaustive development of the behaviour of electric motors. It will only make a compilation of the equations and behaviour models of electric motors that have already been published and accepted, particularly those that apply the Bond-Graph Technique, (Karnopp, 2005; Esperilla, 2007, Lozano, 2010).

Direct current electric motors (hereafter DC), are mainly made up of two components: a stator or armature and a rotor or armature (see Figure 8). The stator's mission is to generate an electric field at the core of which the rotor is inserted. This magnetic field of the stator is generated by windings through which an electric current is made to flow. An electric current is also made to flow in the rotor. As this is immersed in a magnetic field generated by the stator, the electric current conductor undergoes mechanical forces that cause the rotor to rotate on its shaft. The outline shown in Figure 8 corresponds to the so-called *DC motors with independent excitation,* as the operating voltage of the stator and the rotor is independent. Also, the windings of the stator and the rotor can be interconnected giving rise to two other types of DC motors:

a. *DC motors with in-series excitation,* if the windings of the stator and rotor are connected in series.
b. *DC motors with in-parallel excitation,* when the windings of the stator and rotor are connected in parallel under the same operating voltage.

Firstly, we will deal with the modelling of DC motors with independent excitation and then go on to DC motors with in-series excitation, making some slight adaptations.

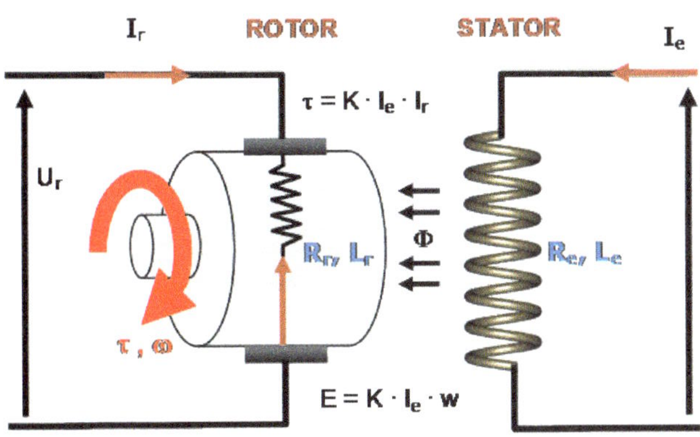

Fig. 8. Electromechanical circuit diagram of a DC electric motor.

3.1.1 DC electric motors with independent excitation

As already indicated, Figure 8 illustrates the electromechanical outline of a DC motor with independent excitation. Figure 9 represents the same model in a Bond-Graph using the Sources of Effort "SE", with ratios "U_e" for the stator and "U_r" for the rotor. The electric current voltage "U_e", is used to overcome the ohmic resistances "R_e" in the stator circuit and to generate a magnetic field "Φ" in the winding. The ohmic resistances are represented by the Resistance port "R" with parameter "R_e". The behaviour of the winding is frequently represented by an Inertial port. However, in this case, in order to be able to consider the magnetic losses produced in the air-gap and in the gap between the stator and the rotor, the electrical energy reaching the winding is firstly converted into magnetic energy. The equations governing the transformation of electrical energy into magnetic energy in the stator winding are:

$$N_b \frac{d\Phi}{dt} = U_b \tag{10}$$

$$M = N_b I_e \tag{11}$$

Where "Φ" is the magnetic flux generated in the stator winding, "U_b" is the voltage to which it is subjected and "N_b" is its number of turns. "M" is the induced magnetomotive force and "I_e" is the strength of the electric current in the winding. This transformation of electrical into magnetic energy is represented in the Bond Graph by the "GY" port with a "$1/N_b$" ratio. The magnetic field generated by the stator winding is represented in the Bond Graph by a Compliance port "C", in ratio to the reluctance of the magnetic field "R", in such a way that the relationship between the flux of the magnetic field and the magnetomotive force "M" is given by the expression:

$$\int R \frac{d\Phi}{dt}\, dt = M \tag{12}$$

The resistance port, R, with ratio "P" represents the losses of the magnetic field produced in the air-gap of the stator and in the gap between the stator and rotor.

We will now model the electrical circuit of the rotor. In this case, the electrical energy is used to overcome the ohmic resistances represented by the resistance port "R : R_r", in establishing the magnetic field of the winding represented by the Inertial port "I" with an inductance parameter of "L_r", and in overcoming the counter-electromotive force "E" induced by the stator's magnetic field and which causes the rotor to rotate. All these elements described are subjected to the same voltage as the rotor circuit "I_r", for which reason they are connected to a "1" Junction. Due to the movement of the current "I_r" in the rotor at the core of the magnetic field generated by the stator "Φ", mechanical forces appear that cause the rotor to rotate. The equations governing the transformation of electrical into mechanical energy at the core of the rotor (Karnopp, 2005), are:

$$\tau = K I_e I_r \tag{13}$$

$$E = K I_e \omega \tag{14}$$

Where "K" is a constant, "τ" is the motor torque generated in the rotor and "ω" is its angular velocity. These equations for the transformation of electrical into mechanical energy are represented in the Bond Graph by the "MGY" port with variable ratio "K I_e". The connection between the Bond Graph of the stator and the rotor is produced through the current intensity "I_e". This connection is represented by a conventional arrow with a broken line.

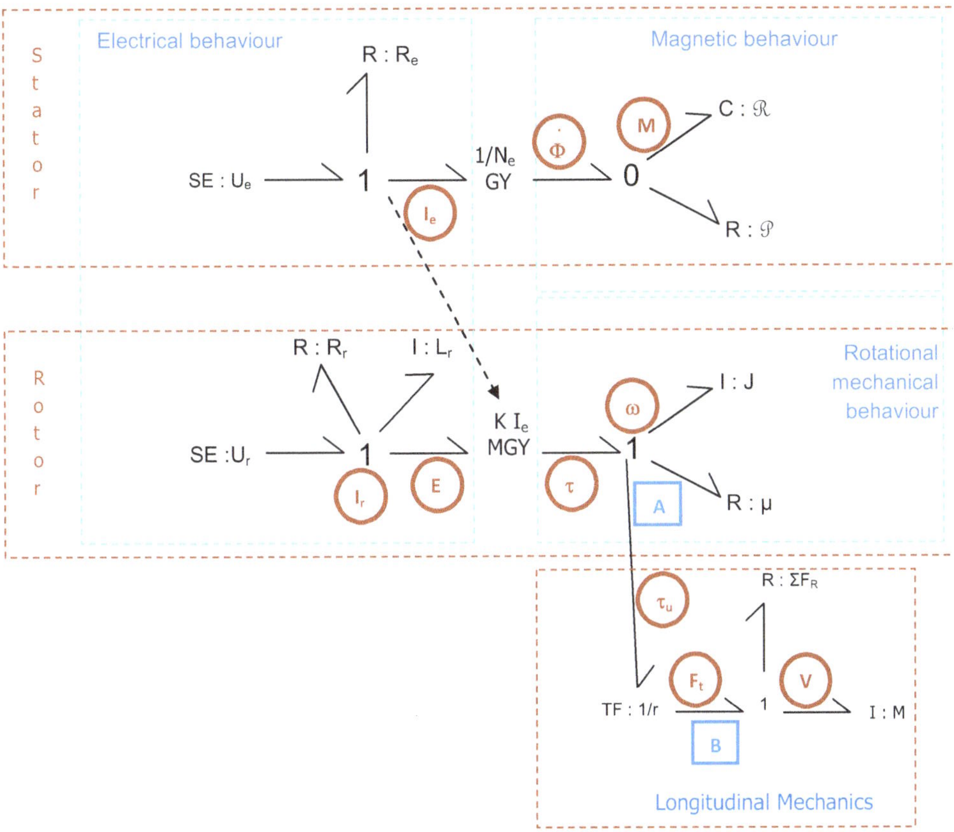

Fig. 9. Bond-Graph model of a DC motor with independent excitation.

In the mechanical field, part of the energy is inverted to overcome the rotational inertia of the rotor, represented by the Inertial port "I", the ratio of which is the moment of inertia of the rotor "J"; it is also inverted to overcome the friction losses in the rotor shaft supports by means of the Resistance port "R", whose ratio is the viscous or Coulomb coefficient "ì". In this case, these losses will be cancelled out as they are already taken into account in equation (2). As a result, what is left is the useful energy that will be inverted to power the train through the drive wheels. The useful rotational energy generated by the motor is associated with the bond marked with the letter "A" in Figure 9, which fulfils the flow "ω", the flow of the adjoining junction "1", and the effort is the useful torque "τ_u". The motor is connected to the locomotive's drive wheels which convert the rotational mechanical energy into linear energy, in accordance with the following expressions:

$$F_t V = \tau_u \, \omega \tag{15}$$

$$F_t = \frac{\tau_u}{r} \tag{16}$$

$$V = \frac{\omega}{r} \tag{17}$$

Where "V" is the longitudinal speed of the train and "F_t" is the total tractive effort supplied by the wheels in contact with the rails. This effort is the same as that already mentioned in section 2.1 and which is shown in Figure 2. The energy transformation represented by equations (15), (16) and (17), is modelled in the Bond Graph by the Transformer port "TF:1/r", which takes the useful torque "τ_u" from the electric motor output of the bond marked with an "A" in Figure 9, and converts it into the tractive effort "F_t". In this case the energy conversion is produced without losses, since the mechanical losses produced are taken into account in the Bond-Graph in Figure 2, without doing anything other than eliminate the Source of Effort "SE : F_t" which originally supplied the tractive effort "F_t", and then connect the bond marked with a "B" in Figure 9 to junction "1" of the Bond-Graph in Figure 2.

3.1.2 DC motors with in-series excitation

DC motors a with in-series excitation are very similar to those with independent excitation, except that the circuits of the stator and the rotor are connected in series. Therefore, the current flowing through both circuits is the same and is also what induces the magnetic flux that excites the rotor. As for the rest, everything stated concerning the independent excitation motor is applicable to the in-series excitation motor. Figure 10 depicts the outline of an electromechanical DC motor with in-series excitation, while Figure 11 shows its corresponding Bond Graph.

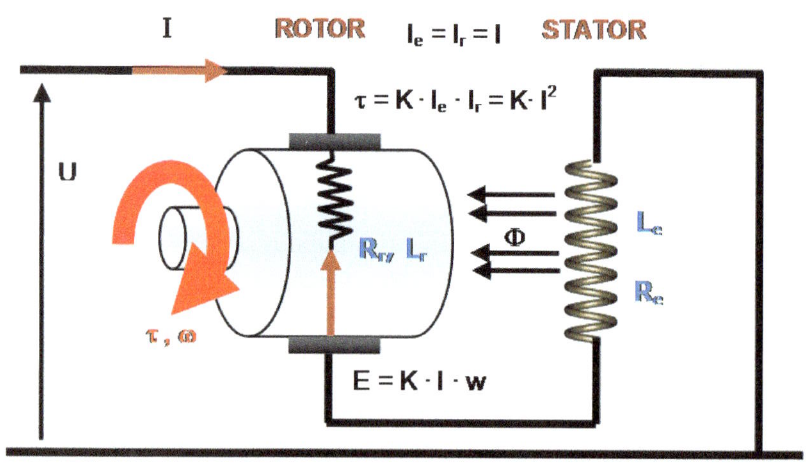

Fig. 10. Electromechanical diagram of a DC motor with in-series excitation.

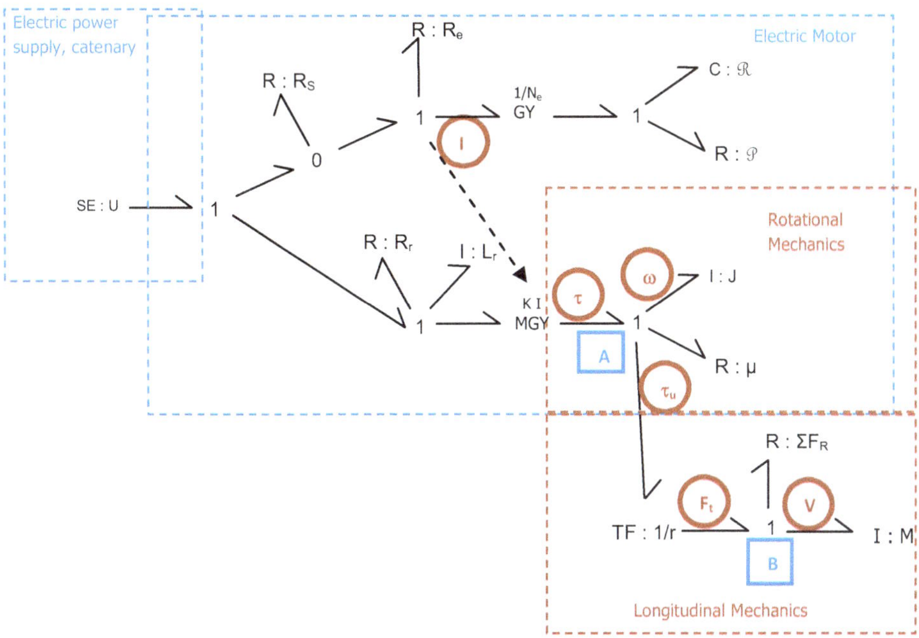

Fig. 11. Bond-Graph model of a DC motor with in-series excitation.

3.1.3 Three-phase alternating current motors

The third type of motor, which is becoming more and more used in railway traction is the asynchronous three-phase alternating current (hereafter AC) motor. This type of motor has the advantage of not having a collector and therefore maintenance is greatly reduced.

Figure 12 shows the equivalent electrical circuit for each of the three-phase motor phases (Esperilla et al, 2007). The resistance "R_e" and the winding "L_e" represent the behaviour of the stator circuit. The Resistance "R'_r" and the winding "L'_r", reduced to the stator circuit, represent the behaviour of the rotor circuit. The resistance "R_p" and the winding "L_p" represent the losses due to hysteresis produced in the air-gap and the losses due to magnetic flux produced in the stator and the rotor. Finally, the resistance "R'_C" is the equivalent load resistance that models the effect of the mechanical energy produced by each motor phase, where "s" is the slip existing between the rotational velocities of the magnetic field generated by the stator and the rotor. The electrical potential dissipated through the resistance "R'_C" is equivalent to the potential generated by the electric motor in each of its phases.

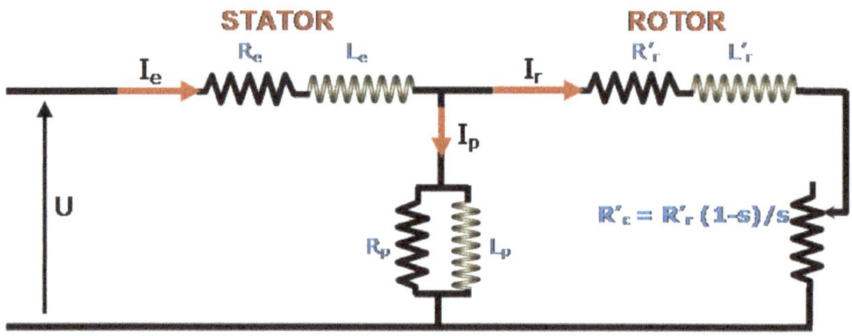

Fig. 12. Equivalent circuit of an asynchronous three-phase AC motor.

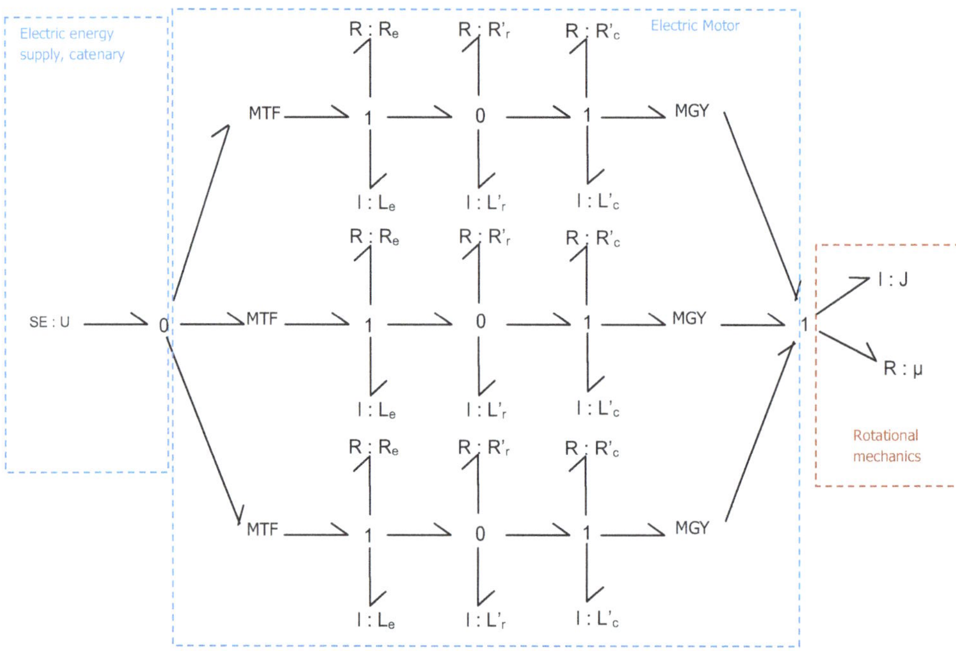

Fig. 13. Bond-Graph of the asynchronous three-phase motor.

Figure 13 illustrates the complete Bond-Graph model of the asynchronous three-phase AC motor. Each of the horizontal branches of the Bond-Graph models each of the phases of the motor, starting out from the equivalent circuit shown in Figure 4. The three phases are subjected to an alternating current "U", 120° out of phase using the Metatransformer ports shown to the left. The mechanical power generated by each phase is modelled by the "MGY" ports shown to the left. This rotational mechanical power is added to the "1" junction shown on the extreme right of the Bond Graph, so that it can be applied to the motor shaft modelled by the Inertial port "I" with parameter "J". The friction losses are also taken into account, and these are represented in the resistance port "R" with parameter "μ".

4. Motor regulation and traction control techniques

If we speak about electric motor regulation and locomotive traction control, then we also have to speak about power electronics. Even for the three motors dealt with in the preceding section, the technology used in each case varies a great deal. For the reasons already stated in the previous section, at present, research conducted in this field is focused on three-phase traction technology with asynchronous motors and a cage rotor, converters with thyristors, and, of course, control technology based on IGBT transistors and microprocessors. However, let us begin by reviewing the regulation of the two DC motors studied to then go on to the regulation of asynchronous three-phase AC motors. So as to reach a proper understanding of the logic of the regulation techniques used in each case, we will continuously refer to the behaviour equations of the motors that were dealt with in the previous section.

As already stated towards the end of Section 2.2, the ideal traction curve is a constant power hyperbole. It is said to be optimum for traction because if the motors had this operating curve their power could be put to full use, with large torques at low speeds and small torques at high speeds, as Figure 5 shows. But at very low speed, that is when starting, the motor's rotational speed is very low and the torque that it is able to produce tends towards infinity. This causes the adhesion forces to exceed their limit in the wheel-rail contact. In addition, as will be seen further on, the intensity of the current flowing in the motors takes on very high values that will lead to the destruction of the electrical circuits. For these two reasons, during the starting process the torque supplied by the motor needs to be limited so as not to exceed the adhesion limits. At the same time the intensity of the current flowing in the motors is limited so as not burn out the electrical circuits.

On the other hand, as we shall demonstrate further on, the characteristic operational curves of the motors (motor torque curves "τ", dependent on the rotational speed "ω"), do not fit exactly with the constant power hyperbole, while at high speeds, the torque supplied by the motor is less.

From the above two paragraphs two very important conclusions can be reached that affect the operational regulation of the motors:

a. During starting at very low speed, we must regulate the intensity of the current and the motor torque so that the motor will supply the maximum possible torque without exceeding the current limits or the adhesion limits in the wheel-rail contact, until the conditions of the constant power hyperbole are reached in the shortest possible time. This operation is called *"starting regulation"* or *"constant torque regulation"*.
b. Once the constant torque hyperbole has been reached, we must check how to regulate the motor function so that its characteristic "τ-ω" curve fits the constant power hyperbole. This is called *"constant power regulation"*.

The following sections will deal mainly with regulating the motors during starting and then their regulation under constant power. The conclusion to be drawn from all this is that by applying these regulation procedures, the electric traction motors are made to operate in three stages: initially under constant torque during starting: then under constant power at moderate and high speeds, and finally, following the characteristic curve of the motor at very high speeds. (see Figure 14).

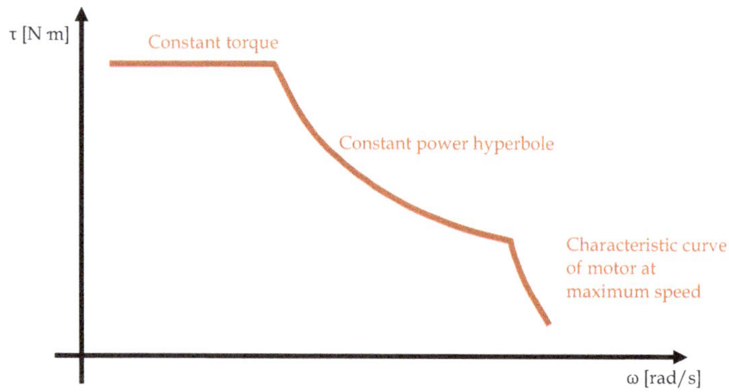

Fig. 14. Characteristic torque-speed curve "τ-ω".

4.1 DC motor regulation during starting with independent excitation

The diagram and equations (13) and (14), of the behaviour of a DC motor with independent excitation were examined in Figure 8. By simply analysing the rotor circuit of the said figure, the following equation for the rotor's operating voltage "U_r", may be deduced:

$$U_r = E + R_r I_r = K I_e \omega + R_r I_r \tag{18}$$

From which we can find the value of the current intensity in the rotor "I_r":

$$I_r = \frac{U_r - K I_e \omega}{R_r} \tag{19}$$

And by entering into equation (13), we can find the equation for the motor torque "τ", dependent on the angular velocity "ω":

$$\tau = K I_e I_r = K I_e \frac{U_r - K I_e \omega}{R_r} \tag{20}$$

By taking equation (18) the following expression can also be found for the angular velocity "ω":

$$\omega = \frac{U_r - R_r I_r}{K I_e} \tag{21}$$

By analysing equations (19), (20) and (21), it emerges that:

1. During starting, when "$\omega \approx 0$", the intensity of the current in the rotor "I_r" is very high, since "$I_r \approx U_r/R_r$", as the ohmic resistance of the rotor circuit "R_r" is very low. A very high current intensity may burn out the circuits. For this reason, current intensity must be limited during starting.
2. The motor torque "τ" during starting is also very high and gradually drops as the motor's angular velocity "ω" increases. In principal, this is sufficient as the behaviour is similar to that of a constant power hyperbole (remember Figure 5). But during starting, the torque supplied by the motor is so high that it exceeds the wheel-rail adhesion limit. To avoid slip, the starting torque needs to be limited to the allowable values of the adhesion.

3. Consequently, during starting the current intensity must be limited to avoid the circuits burning out and to limit the motor torque, thus preventing the adhesion in the wheel-rail contact from exceeding the limits. The maximum motor torque obtained during starting is called *"maximum continuous torque"*, (see Figure 16). Since the motor torque and the tractive effort "F_t" in the wheel-rail contact are directly related, the maximum available tractive effort "F_t" during starting is also called *"maximum continuous effort"*, (see Figure 5).

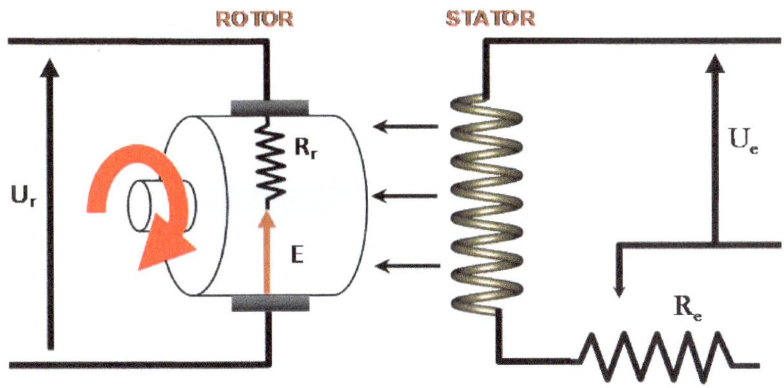

Fig. 15. Shunting of DC motors with independent excitation.

When dealing with DC motors with independent excitation the most effective way of reducing motor torque during starting is to reduce the intensity of the current in the stator "I_e", as in equation (20). To this end, a rheostat is placed in series with the stator, with a variable resistance "R_e", as can be seen in Figure 15. This is known as "Shunting". As the intensity "I_e" diminishes, the motor speed increases (equation (21), without any excessive increase in the motor torque "τ". At the beginning of starting, the value of the Shunting resistance "R_e" is maximum, producing a maximum reduction of the current intensity "I_e" and therefore, the magnetic flux induced in the rotor is minimum. This operation is controlled so that the motor torque "τ" will be the maximum appropriate one without exceeding the adhesion limits. As the locomotive gradually gathers greater speed, the intensity "I_r" and the motor torque "τ" gradually decrease (equations (19) and (20). It is then necessary to increase the motor torque to recover traction capability during starting. To achieve this, the value of the Shunting resistance "R_e" is gradually reduced in a controlled manner. Thus, the current intensity "I_e" and the motor torque "τ" again increase without exceeding the adhesion limits and recover the locomotive's traction capability (see the *"real operational curve"* in Figure 16).

Below, we have simulated the operation of a DC motor with independent excitation using the model shown in Figure 9, for different fixed values of the Shunting resistance "R_e". The motor has a power of 1200 kW and in the simulation was working under zero load. Figure 16 shows the results of the motor torque obtained for different fixed values of "R_e" as a function of the motor's rotational speed, comparing the results with the theoretical torque corresponding to the constant torque hyperbole and also to the maximum continuous torque.

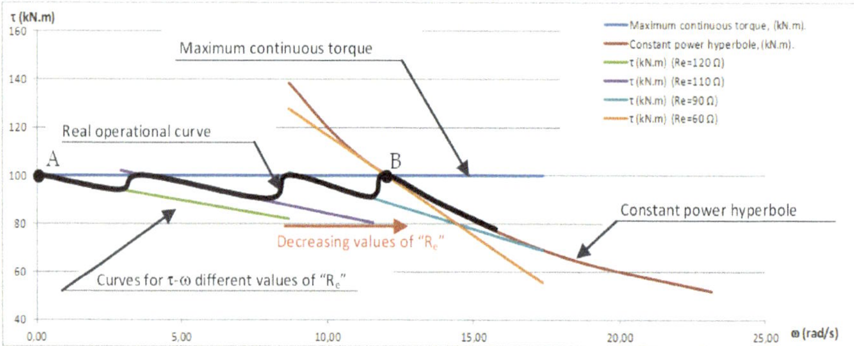

Fig. 16. Simulation results of a DC motor with independent excitation for different values of the Shunting resistance "Re".

4.2 Regulation during starting of DC motors with in-series excitation

The behaviour of a DC motor with in-series excitation is similar to that of a DC motor with independent excitation, but there is one very important difference: the current flowing in the rotor and the stator is the same (see Figure 10). Following a similar procedure to that in the preceding section, we will now find the behaviour equations for the DC motor with in-series excitation. The following equation considers a total fall in the voltage in the motor according to the intensity of the current:

$$\mathrm{U} = \mathrm{E} + R_r I = K\, I\, \omega + R_r I \tag{22}$$

From which we can find the value of the current intensity flowing through the motor "I", according to the operational voltage and the already known parameters of the motor, "K" and "R_r":

$$I = \frac{U}{K\omega + R_r} \tag{23}$$

By entering into equation (13), the equation for the motor torque "τ" can be found according to the angular velocity, "ω", since the current intensity flowing through the rotor and the stator is the same:

$$\tau = \mathrm{K}\, I_e I_r = \mathrm{K}\, I^2 = \mathrm{K}\left(\frac{U}{K\omega + R_r}\right)^2 \tag{24}$$

By analysing equations (23) and (24), similar conclusions can be drawn as from the case of the DC motor with independent excitation:

1. During starting when "$\omega \approx 0$", the current intensity "I" is very high since "$I \approx U/R_r$", as the ohmic resistance of the rotor circuit "R_r" is very small.
2. The motor torque "τ" during starting is also very high and gradually decreases as the angular velocity of the motor "ω" increases.
3. Consequently, during starting, the current intensity needs to be limited to avoid the circuits burning out and to limit the motor torque, thereby preventing the adhesion limits being exceeded in the wheel-rail contact.

In the example of the DC motor with in-series excitation, the above is achieved by inserting resistances in series with the motor, as can be seen in Figure 17. At the beginning of starting the maximum number of starting resistances "$R_i = R_1 + R_2 + R_3$" are connected to produce the maximum reduction of current intensity "I" and therefore, the maximum reduction of the motor torque "τ". As the locomotive gathers greater speed, the intensity "I" and the motor torque "τ" gradually decrease. Then it is necessary to increase the motor torque to recover traction capability during starting. By means of the connections A, B and C, the resistances are successively bridged, and, therefore, cease to function and the starting resistance "R_i" gradually decreases in value. So, in a controlled manner, the current intensity "I" and the motor torque "τ" again increase and the locomotive recovers its traction capability (see "*real operational curve*" in Figure 18).

The operation of a DC motor with in-series excitation has been simulated using the Bond-Graph model shown in Figure 11, with variable starting resistances. For this simulation, since the starting resistances are placed in series with the motor, they are taken into account by adding them to the rotor resistance "R_r, through the Resistance port "$R{:}R_r$", (see Figure 11). The motor under consideration has a power of 1200 kW and in the simulation was working under zero load. Figure 18 shows the results obtained for the motor torque according to the rotational speed of the motor for different values of the starting resistances "R_i" and comparing the results to the theoretical torque corresponding to the constant power hyperbole, as well as to the maximum continuous torque.

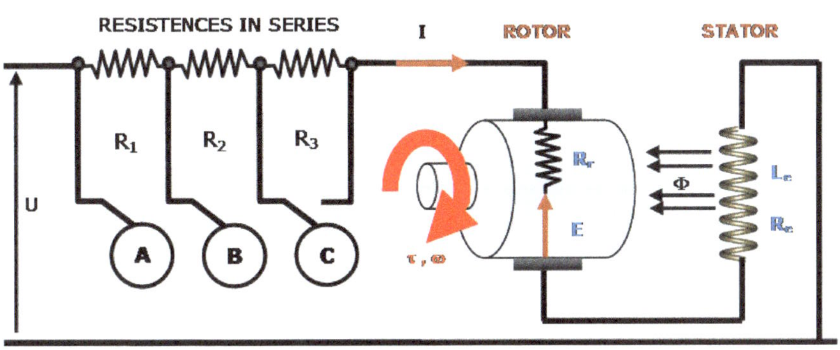

Fig. 17. Resistances in series with the motor in starting mode.

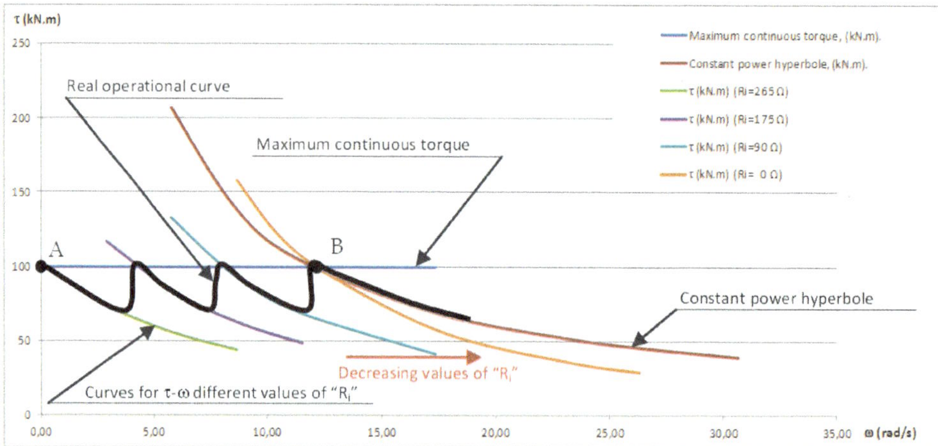

Fig. 18. Simulation results of a DC motor with in-series excitation for different values of the starting resistances "R_i".

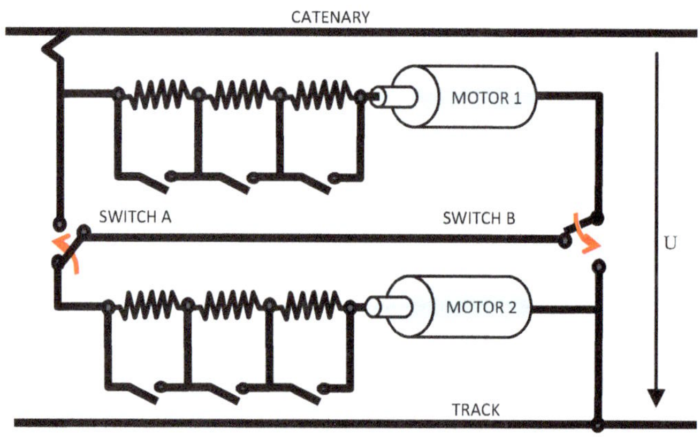

Fig. 19. Starting diagram of a locomotive with two traction motors.

Locomotives usually have several motors and starting can also be controlled by combining the incorporation of resistances in series, "R_i", of each motor with the motors being connected in series. Figure 19 shows a schematic outline of the electrical connection of a locomotive that has two traction motors. Both have resistances in series for the starting, and in addition, switches A and B are installed so as to be able to connect both motors in series or in parallel. When the motors are connected in series, the voltage in them is half the nominal voltage, and, as a result the intensity flowing through them is also half, reducing the starting torque "τ" to half. In this way, the resistances "R_i" during the starting of each motor can be less, resulting in fewer energy losses during starting. Initially, when the train is gathering speed, the starting resistances "R_i" are gradually reduced. At a given instant, the in-series connection of the motors is switched to in-parallel and the starting resistances "R_i" are slightly increased. Finally, when the train reaches a determined speed the process is concluded by cancelling the starting resistances "R_i".

4.3 Regulating the behaviour of DC motors with in-series excitation, with the constant power hyperbole

When the starting period of the motors has been exceeded (section A-B in Figures 16 and 18), we must deal with controlling the motor in order to fit its behaviour to the constant power hyperbole.

The Electromotive Force "E" induced in the rotor by the stator is proportional to the magnetic flow "Φ" and the angular velocity of the rotor's rotation "ω", (Karnopp, 2005):

$$E = K_i \, \Phi \, \omega \qquad (25)$$

Where "K_i" is a constant associated with the stator winding.

From the above equation we can calculate the angular velocity "ω":

$$\omega = \frac{E}{K_i \Phi} \qquad (26)$$

Also, from equation (22), we can obtain the following expression for the value of the Electromotive Force "E":

$$\mathrm{E} = \mathrm{U} - R_r I \qquad (27)$$

By substituting equation (27) in equation (26), we obtain:

$$\omega = \frac{\mathrm{U} - R_r I}{K_i \Phi} \qquad (28)$$

From this it may be deduced that a given voltage can increase the motor's rotational speed by reducing the flux "Φ" of the stator. As we also know, this flux depends on the current flowing through the stator, because as we are dealing with in-series excitation, it is the same as that flowing through the rotor. To change the current in the stator (and, therefore, the induced flux), without changing that of the rotor, an assembly is used like the one in Figure 20, where a rheostat is installed in parallel to the stator. As the resistance of the rheostat gradually decreases, the part of the current flowing through the stator becomes less and as a result, the flux "Φ" decreases.

The control of direct current motors, based on reducing the current passing through the stator, is called *"Shunting"* the motors. This term was used in section 4.1., (Figure 15), when dealing with the starting of DC motors with independent excitation.

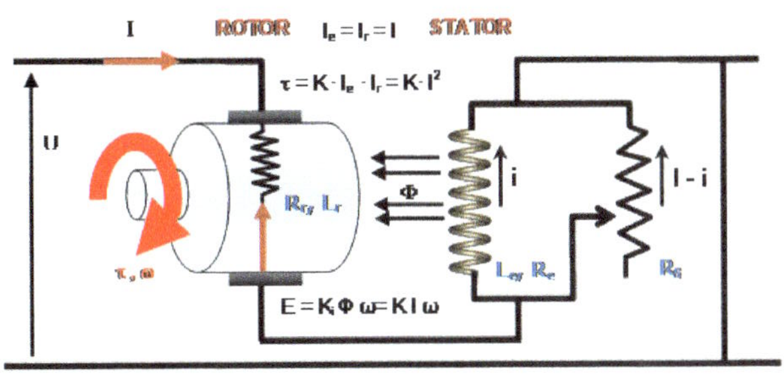

Fig. 20. Shunting DC motors with in-series excitation.

The shunting "α" of a motor is defined as:

$$\alpha = \frac{I-i}{I} \cdot 100 \tag{29}$$

By shunting the motor, its characteristic curves move towards the right, the bigger the move, the greater the shunting percentage. This can be seen in Figure 21, where the shunting curves are shown, continuing the behaviour simulation of the DC motor with in-series excitation carried out in Section 4.2, once the starting period had been completed (section A-B").

It is not possible to totally decrease the current in the stator as it depends on how the motor is constructed. It can usually only be shunted up to 55%, because with higher values switching problems arise in the motor.

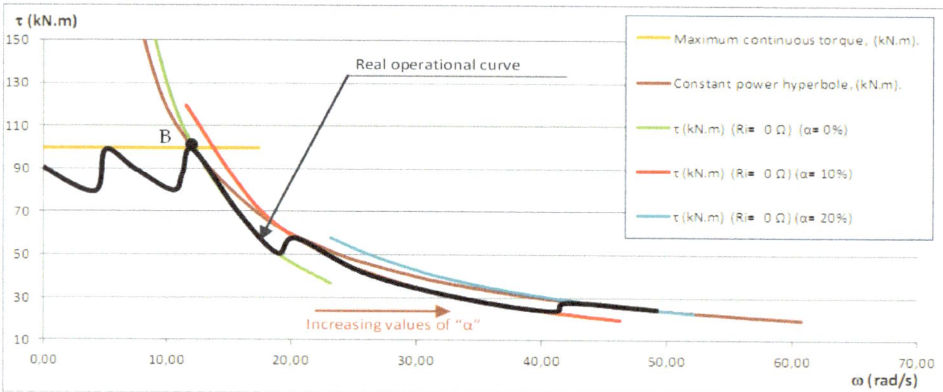

Fig. 21. Torque curves on shunting the DC motor with in-series excitation.

4.4 Electronic regulation of direct current motors

During the simulations carried out on the behaviour of the motors, the previous sections have taken account of the discrete values for the starting resistances or for the shunting resistances. This has enabled us to display the corresponding torque curves. However, the real operating curve, as Figures 16, 18 and 21 may recall, jumps up and down as it attempts to firstly fit to the continuous torque straight line and then to the constant torque hyperbole. This was how the operation of a real motor was originally regulated. It can currently be highly valid during a computer simulation for understanding the phenomena that occur. However, with the development of power electronics, these proceedings for regulating the operation of motors are not the most appropriate.

The electronic regulation of direct current motors is essentially based on regulating the feed voltage. The devices used to do this are called *Choppers*.

When dealing with a motor with independent excitation a complete regulation can be achieved by controlling the stator and rotor voltage separately. If the motor has in-series excitation, the feed voltage to the rotor and the stator Shunting must be controlled simultaneously.

The chopper's mission is to act as a switch to open and close the motor feed circuit, as Figure 22 illustrates. In this way, by starting with a fixed continuous feed voltage "U" in the catenary, a variable voltage applied to the motor is obtained which manages to regulate its speed. Since the switch is off (driving the chopper) the catenary voltage is applied to the motor to make an electric current flow through the motor. When the switch is opened the voltage applied to the motor is cancelled, cutting off the current to the switch. During the time the switch is on, the motor current flows through the diode placed in parallel. By varying the times the switch is on and off, the mean voltage in the motor can be varied and therefore, the current intensity flowing through it. This method perfectly manages to regulate the motor torque and perfectly follows the theoretical continuous torque curve (starting), and the constant power hyperbole without any steps appearing in the real torque operating curve mentioned at the beginning of this section. To achieve this phenomenon, devices called *Thyristors* are used.

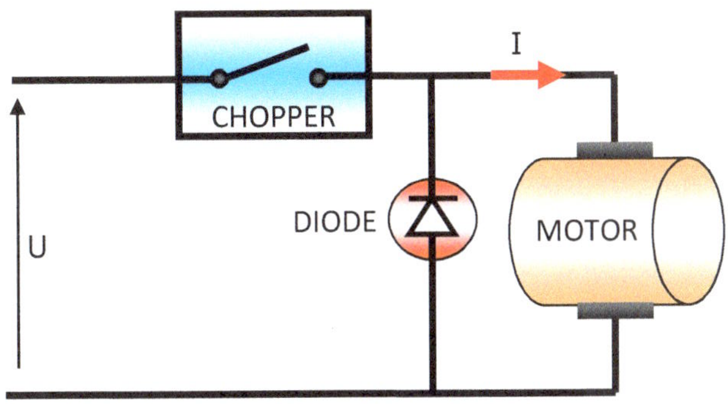

Fig. 22. Principle of how a Chopper-controlled DC motor functions.

4.5 Alternating current motors

In the last 30 years railway traction motors have ceased to be direct current motors controlled by the connection and disconnection of resistances to become asynchronous alternating current motors controlled by *IGBT transistors* for medium powers, or *GTO thyristors* through the use of pulse width techniques for high powers (Faure, 2004).

As explained in the previous sections, the first controls using rheostats for direct current motors gave way to thyristor-based control. DC motors can be better controlled by this technology by avoiding the transitory *"jerk"*, effects caused by connecting and disconnecting the starting and shunting resistances. So, thyristors led to a much better and uniform functioning of DC motors. Notwithstanding, the drawbacks derived from the *collector* still existed, which required large-size motors that needed frequent maintenance. These requirements became much less with the appearance of synchronous alternating current motors and were practically eliminated with the asynchronous motors.

The development of traction control systems has led to the use of simpler, more robust and cheaper asynchronous motors. They are more complicated to regulate than synchronous motors but this complexity became much less with the development of *IGBT transistors*. Nowadays, the major research and development projects are focussing on the technologies related to asynchronous alternating current motors (hereafter asynchronous AC motors).

The three-phase asynchronous motor is an induction motor based on the generation of rotating magnetic fields by means of the stator windings, which induce electric currents in the rotor windings. Due to the interaction between these induced currents and the magnetic fields, forces are generated in the conductors that produce a motor torque. If the rotor rotates at the same speed as the magnetic fields, there is no variation of flux passing through the turns of the rotor and the induced current is zero. For a motor torque to be produced there needs to be a difference (slip) between the angular velocities of the magnetic field and the rotor.

For a low frequency alternating electric current, the torque generated by the motor "τ" satisfies an expression that is similar to the expression for direct current motors:

$$\tau = K_i \, \Phi \, I_r = K \, I_e I_r \tag{30}$$

Where, "K_i" and "K" are behaviour constants of the motor; "Φ" is the magnetic flux generated by the stator, and "I_r" is the intensity of the current flowing through the rotor. It may be deduced from this expression that in DC motors, the torque increases along with the intensity of the current in the rotor.

However, it is also known that the current in the rotor is proportional to the magnetic flux "Φ" generated by the stator and to the slip "s" between the rotor and the magnetic field. Moreover, the magnetic flux "Φ" is proportional to the quotient between the feed voltage "U" and the current frequency "f", which means the torque can also be expressed as:

$$\tau = K_f \left(\frac{U}{f}\right)^2 s \tag{31}$$

Where "K_f" is a new constant of the motor function. The law expressed by equation (31) is valid for small slips, but for larger slips the ratio ceases to be linear. Figure 23 shows the motor torque curves "τ" compared to the rotational velocity "ω", for different values of the current frequency "f".

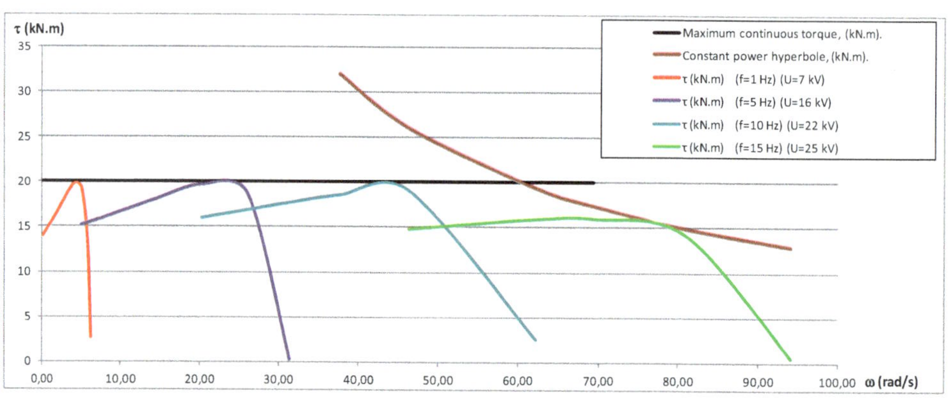

Fig. 23. Torque curves for an AC asynchronous squirrel cage motor.

By regulating AC motors and fitting their working to the theoretical torque-speed curves "τ-ω", that is to say, starting under constant torque and constant power hyperbole (remember Figure 14), the feed voltage "U" and the current frequency "f" can be varied simultaneously. To achieve this, *IGBT, GTO thyristor*-based technologies are used for high powers, which modulate the width of the electric pulses through the superposition of a triangular wave and a voltage signal that is proportionate to the signal required. Doing this will ensure that the motor describes the constant torque curve and constant power hyperbole without any steps.

Figure 23 shows the torque-speed curves for an asynchronous squirrel cage AC motor with a power of 1200 kW, simulated using the Bond-Graph model dealt with in Figure 13, for different frequency values of "f" and the operating voltage "U". Obviously, by a continuous regulation of the frequency and voltage, the real working of the motor will exactly fit the constant torque straight line during starting and the constant power hyperbole.

5. Prospects, current and future lines of research, conclusions

Obviously, for lack of space, a whole range of issues have been left undealt with, but they are of no less importance: maximum starting efforts and couplings, load capacity and in-service power, the influence of the vehicles' lateral and vertical dynamics, generation systems, energy transport and capture, pantographs and floaters, service automation and traction control, regenerative brakes and energy recovery systems, and a long etcetera.

As a final conclusion, the most current areas of progress in research, development and innovation and their future prospects are mainly directed towards achieving a railway that is better adapted to the new global needs of mobility, sustainability and respect for the environment:

a. More efficient systems for generating, transporting, capturing, transforming, utilising, regulating and recovering energy.
b. Traction control and service automation systems, regenerative braking and energy storage systems, reversible electrical supply sub-stations and rail traffic management.
c. Multidisciplinary optimisation of infrastructure and vehicle design.
d. The design implications of vehicles, infrastructures and systems in energy consumption and the environmental impact of transport.
e. Optimisation of vehicles and infrastructures for their use in multimodal systems.
f. Calculation systems, predicting and optimising energy consumption and emissions.
g. Foreseeable consequences of technological development on the innovation of vehicles and infrastructures for sustainable mobility.

6. References

Andrews, H.I. (1986). RailwayTraction. *The Principles of Mechanical and Electrical Railway Traction*. Ed: Elsevier. ISBN: 0-444-42489-X.

Coenraad, E. (2001). *Modern Railway Track*. Ed. Delft University of Technology. ISBN: 90-800324-3-3. Delft, Holanda.

Esperilla, J.J., Romero, G., Félez, J., Carretero, A. (2007). "Bond Graph simulation of a hybrid vehicle". Actas del International Congress of Bond Graph Modeling, ICBGM'07.

Faure, R. (2004). *La tracción eléctrica en la alta velocidad ferroviaria (A.V.F.),* Colegio de Ingenieros de Caminos, Canales y Puertos, ISBN 84-380- 0274-9, Madrid, España.

Iwnicki, S. (2006). *Handbook of Railway Dynamics.* Ed. Taylor & Francis Group. ISBN 978-0-8493-3321-7. London, Reino Unido.

Karnopp, D.; Margolis, D.; Rosenberg, R. (2000). *System Dynamics: Modeling and Simulation of Mechatronic systems.* Ed. John Wiley & Sons, LTD (2ª). Chapter 11. ISBN: 978-0-471-33301-2. Estados Unidos.

Karnopp, D. (2005). *Understanding Induction Motor State Equations Using Bond Graphs.* Actas del International Congress of Bond Graph Modeling, ICBGM'05.

Lozano, J.A.; Félez, J.; Mera, J.M.; Sanz, J.D. (2010). *Using Bond-Graph technique for modelling and simulating railway drive systems.* IEEE Computer Society Digital Library (CSDL), 2010 12th Congress on Computer Modelling and Simulation. ISBN 978-90-7695-4016-0. BMS Number CFP1089D-CDR. Cambridge, Reino Unido.

4

Train Braking

Cătălin Cruceanu
University POLITEHNICA of Bucharest
Romania

1. Introduction

Train braking is a very complex process, specific to rail vehicles and of great importance by the essential contribution on the safety of the traffic. This complexity results from the fact that during braking occur numerous phenomena of different kinds - mechanical, thermal, pneumatic, electrical, etc. The actions of these processes take place in various points of the vehicles and act on different parts of the train, with varying intensities. The major problem is that all must favorably interact for the intended scope, to provide efficient, correct and safe braking actions.

The purpose of braking action is to perform controlled reduction in velocity of the train, either to reach a certain lower speed or to stop to a fixed point. In general terms, this happens by converting the kinetic energy of the train and the potential one - in case of circulation on slopes - into mechanical work of braking forces which usually turns into heat, which dissipates into the environment.

At first, the rather low locomotives power and traction force allowed braking using quite simple handbrakes that equipped locomotives and eventually other vehicles of the train. As the development of rail transport and according to increasing traffic speeds, tonnages and length of trains, it was found that braking has to be centralized, operated from a single location - usually the locomotive driver's cabin and commands have to be correctly transmitted along the entire length of the train.

As a consequence, along the time, for railway vehicles have been developed various brake systems, whose construction, design and operation depend on many factors such as running speed, axle load, type, construction and technical characteristics of vehicles, traffic conditions, etc.

Among various principles and constructive solutions that were developed, following the studies and especially the results of numerous tests, the indirect compressed air brake system proved to have the most important advantages. Therefore, it was generalized and remains even nowadays the basic and compulsory system for rail vehicles.

It is still to notice that, regarding the classical systems used for railway vehicles, there are also several major challenges that may affect the braking capacity. These aspects must be very well known and understood, so as to find appropriate solutions in such a manner that the problems to be overcome by applying different constructive, functional, operational and other kinds of measures.

For example, one of these issues is the basic braking systems dependency on the adhesion between wheel and rail, which can lead to wheel blocking during braking. This determines not only the lengthen of the stopping distance, but also the development of flat places on the rolling surface of wheels, generating strong shocks transmitted both to the way and to the vehicle, with damage to traffic safety and comfort of passengers or goods transported integrity. This has generated particular concerns regarding the design and implementation of more efficient wheel slip prevention devices capable to avoid the above-mentioned phenomena with as small as possible reduction of braking capacity.

Another major problem is the friction between wheel and brake shoes, brake pads and disc respectively, which leads to severe thermal regimes and special thermal fatigue nature efforts, requiring specific constructive and operating standards.

More than that, due to the air compressibility and to the length of trains, the pneumatic commands propagates with limited speed in the brake pipe and, as a result, there always is a delay in the braking of neighboring vehicles. As a consequence, the rear vehicles are running into the front ones, producing large dynamic longitudinal reactions in buffers and couplers. The induced compression and tensile forces can reach significant levels, affecting both the rolling stock and the track, even conducting to deteriorations of safety operation of the trains.

Railway high speed operations also determined more severe requirements for braking systems, given to the necessity to develop higher braking forces and to dissipate larger amounts of energy in a short time, not to mention the problem of wear in the case friction brakes. In that case, complementary systems whose performance and reliability are safety relevant were developed to enhance the braking capacity.

These several issues, even briefly presented, reveal not only the importance and complexity of braking systems used for rail vehicles, but also the necessity of knowledge and understanding the problems in order to develop equipments increasingly more efficient and reliable.

Some of these aspects are presented in this chapter.

2. Classification of braking systems

Given the constructive, functional and operational characteristics of rolling stock, the braking systems must meet certain specific requirements, providing multiple performance exigencies. Some of the most important are then pointed out.

While achieving safe and effective brake actions to allow speed reductions, fixed-point stops and vehicle or train maintenance on slopes in complete safety, it is very important that brake operation and performance should not be influenced by environmental conditions.

A matter of great concern for the traffic safety requires for the braking systems that in cases of specific dangers to take action beyond the control or command of the driver and to perform an emergency braking action for all the train's vehicles.

Also, the centralized control of braking and release actions, as well as the transmission of braking commands along the whole train have to be simple, safe, effective and of maximum reliability. The braking systems have to allow adequate brake and release levels, giving the driver the possibility to adopt and adapt correctly the braking force targets to the instantaneous traffic conditions. Moreover, transient phenomena developed during braking should not lead to high longitudinal dynamic forces in the body of the train that may affect the traffic safety. Still, the achievable braking forces should not affect the integrity of transported goods or the passengers comfort, neither due to decelerations and shocks along the train, nor by annoying vibrations and noises, bad odor, etc.

Regarding the construction of the braking systems components, especially those mounted under the vehicle chassis or on the bogie, they must have low mass and such sizes and shapes to fit the rolling stock gauge. By design, the mobile mechanical elements of the brake installation and mainly the brake riggings have to function correctly and optimally transmit forces, regardless of the fact that the vehicle is empty or loaded at maximum capacity and as long as the wear of all constructive elements of the vehicle are within the limits allowed by regulations.

The wear of friction elements used for achieving the necessary forces to decelerate or stop the train (brake blocks, pads, etc.) have to be as reduced as possible and their action must not affect the geometry of wheels or rails profile.

It is also very important that the thermal regime developed during braking to remain within acceptable limits, without affecting the braking capacity or other elements of the vehicle or of the track.

Last, but not least, commands, achievement and maintaining the effective braking actions must neither affect the environment, nor interact in any way with other systems, elements, and circuits of the track or situated in its vicinity.

An overall image over the braking systems for railway vehicles may be achieved following some classification criteria, as follows.

Depending on how the command actions are performed and according to the mean that braking forces are basically developed, there are:

- handbrakes, which are applied by hand action to a wheel or lever on the vehicle. Nowadays are generally used for securing unattended or unpowered vehicles against unplanned movement, not for braking actions while operating. That is why are usually known as parking brakes. Recent developments conducted to a kind of automation using a spring-applied concept, which release when compressed air for the basic pneumatic brakes is available. While such braking system is mandatory for traction vehicles and passenger carriages, only a part of wagons must be provided with it;
- pneumatic brakes, which use air pressure variations both to command and to apply the brake blocks or pads, generating braking/releasing actions and forces. The vast majority of trains use compressed air, changing the level of air pressure in the brake pipe determining a change in the state of the brake on each vehicle. In the case of straight air brake system, the increase of air pressure in the pipe determines the increase of air pressure in the brake cylinders of each vehicle. In the case of indirect air brake system, braking actions are commanded by decreasing the pressure in the train's pipe, generating by special features the increase of air pressure in the brake cylinders of each vehicle. In the same pneumatic category is also the vacuum brake system where the brakes on each vehicle are actuated by the action of atmospheric pressure over a specially created vacuum in the train's pipe as long as the brakes must be released;

- electro-pneumatic brakes, that have an electrical command while compressed air is used to increase the pressure in the brake cylinders of each vehicle to apply the brake blocks or pads for generating braking forces;
- rail (track) brakes, which are usually electric commanded and the braking force is achieved due to strong magnetic forces induced by large electromagnets hung under the vehicle's bogie, over the top surface of the rails. If the braking forces are generated by the frictional forces between the electromagnets and rails it is an electromagnetic rail brake. If the electromagnetic fields generate eddy currents in the rails, creating forces acting in the opposite direction of the movement of the train, it is a linear eddy current brake system. On the same principle is the rotary eddy currents brake system, the metallic mass being either the wheels, or discs attached to the wheelsets;
- electric braking, which is based on the reversibility of the electric engine, in particular the electric traction motors are reconnected in such a way that they act as generators which provide braking effort. Practically, the kinetic and/or the potential energy (while running on slopes) are converted into electric energy. If the power generated during braking is dissipated as heat through on-board resistors it is about a rheostatic braking. On electric railways, it is also possible to convert the energy of the train back into usable power by diverting the braking current into traction supply line, this being the case of regenerative braking. Most regenerative systems include on board resistors to allow also rheostatic braking if the traction supply system is not receptive, the choice being automatically selected by the traction control system. It is to notice that regarding the sustainable development of railway transportation, the regenerative braking is suitable for hybrid vehicles (Givoni et al., 2009; Uherek et al., 2010) and developments have been done for diesel powered traction rail vehicles;
- hydraulic brakes, which act using hydraulic oil and, depending on the achievement of braking forces, there may be hydrostatic, when is due to oil pressure increasing, or hydrodynamic when the kinetic energy of the vehicle is converted in the rotor of a hydraulic pump in heat which is dissipated through the oil cooling system.

Considering the effective way to achieve the braking force:

- friction brakes, based on Coulomb type friction between specific surfaces. It is the case of brake blocks (shoe brakes), disk brakes, electromagnetic rail brakes;
- dynamic brakes, based on other processes and phenomena than friction in achieving the braking forces. It is the case of eddy current, electric and hydrodynamic brakes. It is to notice that the electric end eddy current brakes are particularly preferred for high speed railway vehicles because, in the absence of direct contact, it results a significant decrease of wear caused by friction, important aspect considering the high energy to be dissipated. For the same reason, the electric braking is also preferred in case of commuter trains, metros and tramways, due to the frequent and often quite strong braking actions, even if running speeds are usually not very high.

Considering the influence of wheel-rail adhesion, there are:

- adhesion-dependent brake systems, when the braking torque is generated directly on the wheelset. In these cases, whenever the braking force exceeds the adhesion one, either due to excessive braking, or to local poor adhesion between wheel and rail, will cause the wheel locking and skidding during braking. The main effects are an increase in braking distances and the development of flats, damage spots on wheel tread, both affecting primarily the safety of traffic;

- adhesion-independent brake systems, usually used as complementary braking systems whenever the maximum braking forces developed by the adhesion-dependent ones is insufficient to provide the necessary braking capacity.

Regarding the brake system reaction in special cases, there are:

- automate brakes, meaning that in case of important accidental drop of air pressure in the brake pipe all the vehicles of the train are submitted to an emergency brake action, independent of driver control. Also, by the operation, a direct link of general pipeline to atmosphere can be established through an alarm signal put to reach of passengers in the coach;
- no automate brakes which, in similar situations, do not determine o brake command and even they can become inactive, unable to perform braking action.

According to the air pressure evolution within the brake cylinders, the compressed indirect air brake systems may be:

- fast-acting, meaning a filling time of 3...5 s and a releasing time of 15...20 s;
- slow-acting, meaning a filling time of 18...30 s and a releasing time of 45...60 s.

Depending on the possibility to modify the braking force level during the action, there are:

- moderable brakes, which can perform various braking steps during braking or/and releasing actions;
- unmoderable brakes, which can achieve a unique braking force level that cannot be modified by the driver and release is only complete, not gradual.

3. Basic braking systems

Basic braking systems provide consistent controllable braking forces on the entire traffic speed domain, permitting speed reductions, stop at fixed point and to maintain the vehicle standstill on slopes, usually being also automate.

Compressed air straight brake system is the simplest continuous brake, both in constructive and functional terms (see fig. 1).

The installation consists of a compressor (1) as source of air under pressure, a main reservoir for compressed air storage and backup container for the entire brake system (2), the general brake pipe of the train (3), consisting of the air pipes of each vehicle (3a), linked together by flexible coupled hoses (3b), each boasting angle cocks (3c) that acts as insulation. For the centralized command and control of braking it is a driver's brake valve (4) which must be able to put into effect at least three pneumatic functions: linking the main reservoir to the general air pipe of the train for supplying it; establishing the pneumatic link between the general air pipe and atmosphere; to be able to ensure the pneumatic insulation of the train general air pipe both to the main reservoir and atmosphere.

On each vehicle it is at least a brake cylinder (5), the forces developed at the piston rod (5a) being amplified and transmitted through the brake rigging (6) to the brake shoes (7) or disc pads.

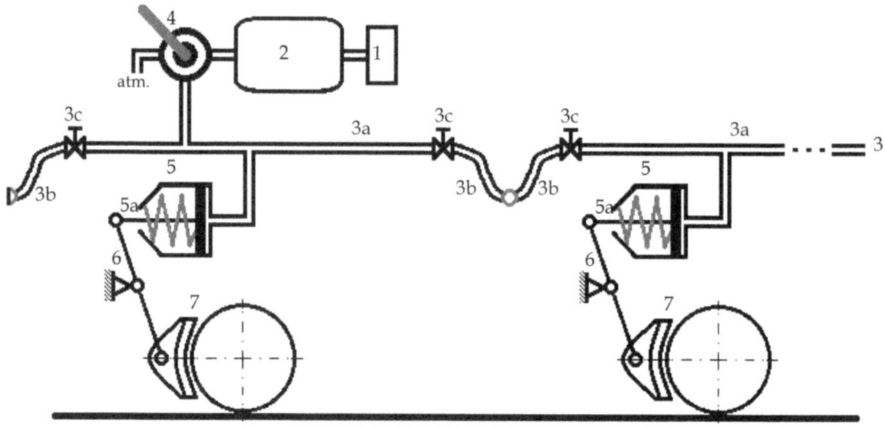

Fig. 1. Schematic of straight compressed air brake system:
1 – compressor; 2 – main reservoir; 3 – train's brake pipe; 3a – vehicle's brake pipe;
3b – flexible coupled hoses; 3c – angle cock; 4 – driver's brake valve; 5 – brake cylinder;
5a – piston rod; 6 – brake rigging; 7 – brake shoe.

The operating principle is simple: to control braking, the driver's brake valve connects the main reservoir to the general air pipe, which is supplied, implicitly increasing pressure in the brake cylinders. Brake cylinders act on the brake riggings, resulting in clamping shoes on wheels. When the train braking force is sufficient, through the driver's brake valve one stop any air supply to the general air train pipe. The system is very adaptive because, at least theoretically, by appropriate handling of the driver's brake valve it is possible to get a large range of pressure levels in the brake cylinders and, accordingly, to have a fine control of the commanded braking forces. For the control of brake release, through the driver's brake valve, pneumatic connection is established between the brake pipe and atmosphere, the air pressure coming from the brake cylinders including. By using the isolation position of the driver's brake valve, one can also get numerous release steps of the train brakes.

This type of braking control was quickly abandoned, the main reason being that a fault in the general pipeline leads to complete releasing of the brake without the driver to be warned in some way and without the possibility of restoring the action of the brakes, aspect particularly dangerous in terms of safety of the traffic.

In addition, the use of straight brake system involves a number of disadvantages, such as: a long duration of the braking propagation rate in the long of the train; high pressure differences between the brake cylinders in the transitional stages; development of large longitudinal reactions that may affect traffic safety and comfort of passengers because of slow braking wave propagation.

Also, the straight air brake system requires a large amount of compressed air when commanding braking action, which, in case of long trains, involves the use of large main reservoirs.

For these reasons, the straight air brake is now used on railway vehicles only as complementary brake for locomotives, railcars and some special vehicles. Even so, according to regulations, straight brake can be used only for individual moving vehicles and not for a multi-vehicle train, when a much safer brake system must be used.

The indirect compressed air brake system was conceived in order to eliminate the main disadvantage of straight brake system. The main operational particularity is that brakes are released as long as in the train's brake pipe pressure is maintained the regime one. Generally, almost worldwide the regime pressure in the general air brake pipe system has been established at 5 bar (relative pressure). There are also exceptions, such as former Soviet countries, that use a regime pressure of 5.5 bar, or the USA where, depending on type of train, were imposed (by AAR - Association of American Railroads) 4.8 bar, 6.2 and 7.6 bar. Braking commands are given by lowering the pressure regime within the general air pipe of the train.

Indirect air brake is a continuous brake, basically having the same subsets, with identical functions as for straight brake (see fig. 2).

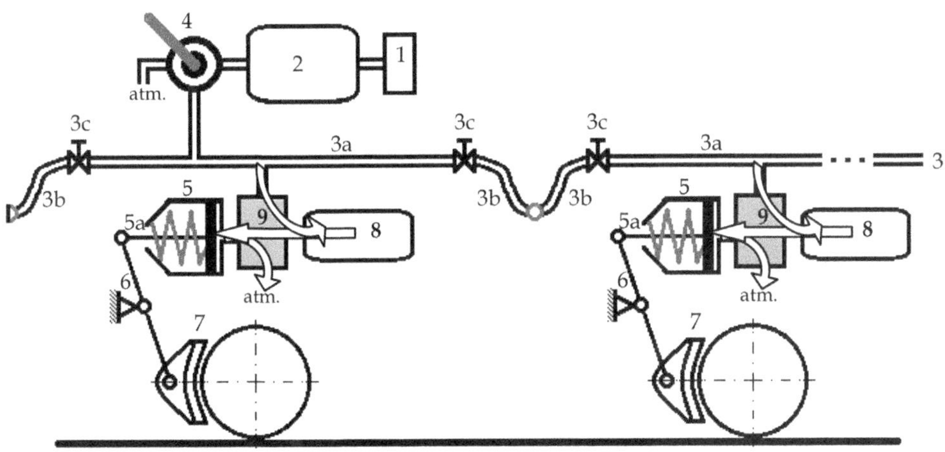

Fig. 2. Schematic of indirect (automate) compressed air brake system:
1 - compressor; 2 - main reservoir; 3 -train's brake pipe; 3a - vehicle's brake pipe;
3b - flexible coupled hoses; 3c - angle cock; 4 - driver's brake valve; 5 - brake cylinder;
5a - piston rod; 6 - brake rigging; 7 - brake shoe; 8 - auxiliary reservoir; 9 - air brake distributor.

In addition, each vehicle is equipped with an auxiliary air reservoir (8), which is the only compressed air supply reserve for the brake cylinders and an air distributor (9) that, depending on pressure variations within the brake pipe of the train, controls and commands locally the braking and release actions.

Air distributor may provide pneumatic links between brake pipe and auxiliary reservoir, between auxiliary reservoir and brake cylinder, and between brake cylinder and atmosphere.

To operate correctly, when increasing the general train brake pipe pressure, the air distributor should ensure the following pneumatic connections, in the specified order: interruption of the pneumatic link between auxiliary reservoir and brake cylinder, linking the brake cylinder to the atmosphere and establishing a pneumatic link between the brake pipe and the auxiliary reservoir.

In the case of pressure drop in the general train pipe, the air brake distributor must first interrupt the pneumatic link between auxiliary reservoir and pipeline, then to cut the pneumatic link between air brake cylinder and atmosphere and finally to establish the pneumatic link between the auxiliary reservoir and the brake cylinder.

Air pressure in the brake cylinder depends on the brake pipe pressure according to UIC leaflet no. 540 requirements that impose the characteristic shown in fig. 3.

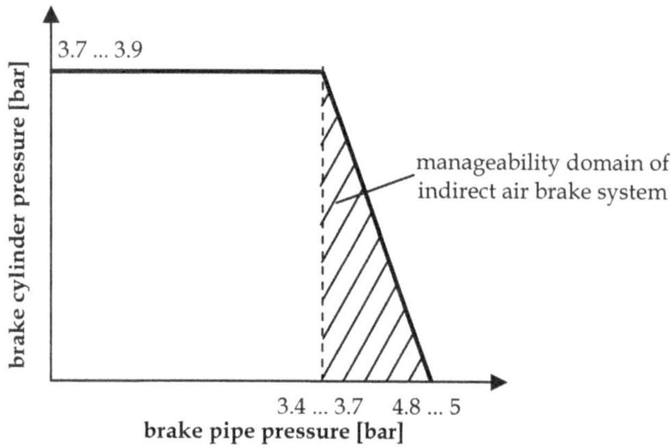

Fig. 3. Dependence of brake cylinder air pressure on brake pipe pressure.

It is noted that the manageability of this brake system is located between the values of 4.8 ... (3.4 ... 3.7) bar of relative air pressure in the general train pipe. Train driver can adjust the intensity of braking by controlling the pressure in the brake pipe between the above specified values. The maximum air pressure level in the brake cylinders is usually 3.8±0.1 bar.

The main advantage of the indirect air brake is the safety on operation, due primarily to the operating principle that makes it an automate brake, meaning that any accidental important drop of air pressure in the general air pipe determines an emergency braking command for the entire train brake system, independent of driver control. Also, by the operation, one could be put to reach a simple and safe passenger braking control of the train for emergency situations by establishing a direct link of the general pipeline to atmosphere with an alarm signal. It is to notice that modern passenger trains are equipped with an emergency brake override system, to avoid stopping the train in inadequate places and situations which tend to be more dangerous than continuing the route, at least until the reach of a much safer location. An example, for instance, is in case of an on-board fire, while the train is running through long tunnels or viaducts.

The specific pneumatic command system is recognized as very reliable and simply constructive, the command and execution actions are provided by a single air pipe. Also, compared to straight compressed air brake, the indirect one presents several advantages, including a significant increase of braking propagation rate along the train and, consequently, a decrease of dynamic in-train forces in the first stages of braking and releasing actions.

The main disadvantages of purely air brake systems rely on the transmission of the air signal which is initiated from the front of the train and has to be sent to all vehicles along the train to the rear. Due to the air compressibility and to the length of the train, there will always be a time lapse between the reaction of the leading vehicle and the reaction of one at the rear. Corresponding to the propagation rate of air pressure signal, the air distributors will come into action successively and the braking of vehicles begins at different times along the train so that, while some cars are slowing down, others are trying to push, still unbraked, from the rear. This creates conditions while transitional braking stages, immediately following the command of pressure variation in the brake pipe, to develop important longitudinal in-train reactions causing stress to the couplers and affecting passenger comfort and, sometimes, even the traffic safety. To mitigate such phenomena, it was necessary to achieve a certain delay of filling, respectively emptying the brake cylinders, accepting however an inevitable slight decrease in braking performance. That is why there are in operational use the fast-acting (or P, or type "travel") and slow-acting brakes (or G, or type "cargo").

The electro-pneumatic brake system is an improvement of the indirect compressed air brake destined primarily to overcome the longitudinal dynamic reactions generated by the gradually successive onset of brakes action due to the pneumatic command.

Basically, the electro-pneumatic system has been designed so that it can be added to the traditional air brake system to allow more rapid, practically instantaneous responses to the driver's braking commands (see fig. 4). As a consequence, vehicles were provided with braking electrovalves to evacuate the general brake pipe simultaneous along the train, according to the electric braking signal transmitted through the control wires running the length of the train. The pressure drop is present at the same time for each air distributor and the braking forces are developed simultaneous along the train. To perform the same for release commands, vehicles were also provided with electrovalves to supply the barke air pipe from a main air pipe which is directly connected to the main reservoir (8...10 bar), avoiding delays in releasing actions along the train.

Normally, the electrical control is additional to and superimposed upon the automatic air brake, although more recent systems incorporate a failsafe electrical control which eliminates the need for a separate brake pipe. Still, taking into account the safety of operating, international regulations impose that the electro-pneumatic brake must always be able to operate as a classical compressed air brake.

Usually, the braking commands are provided from the same driver's brake valve as the air brake, but using new positions to apply and release the electro-pneumatic brake. Electrical connections that are attached to the driver's brake valve send commands along the train to the electrovalves on each car. The electrical connections are added to the operating spindle so that movement of the handle can operate either brake system.

There are many types of electro-pneumatic brake systems is use today. There were developed systems that operate as a service brake while the air brake is retained for emergency use but with no compromise regarding the fail-safe or "vital" features of the air brake. Meanwhile, the main air pipe is also used for other auxiliary features, especially in the case of passenger vehicles: automatic door operation, supply of pneumatic suspension, etc.

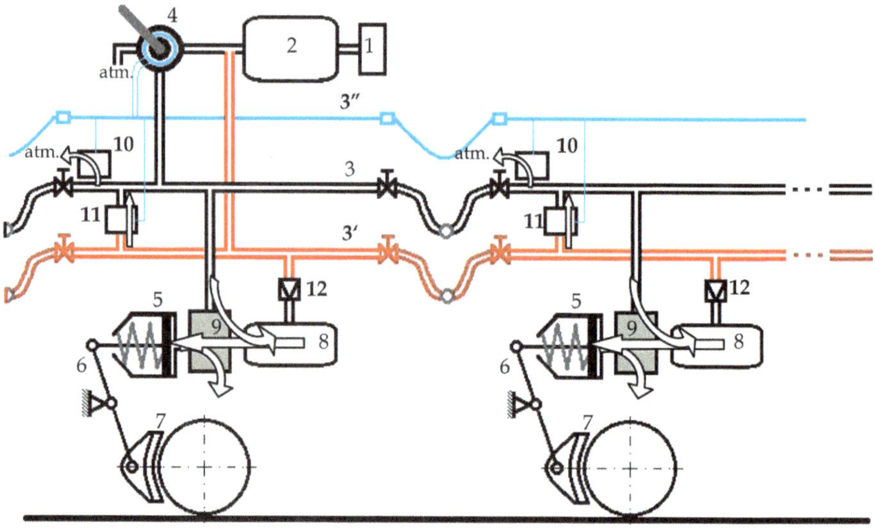

Fig. 4. Schematic of electro-pneumatic brake system: 1 – compressor; 2 – main reservoir; 3 – brake pipe; 3′ – main reservoir pipe; 3″ – electric brake command; 4 – driver's brake valve; 5 – brake cylinder; 5a – piston rod; 6 – brake rigging; 7 – brake shoe; 8 – auxiliary reservoir; 9 – air distributor; 10 – braking vehicle's electrovalves; 11 – releasing vehicle's electrovalves; 12 – one-way valve.

The main technical issues relating to the operation and design parameters to be complied with electro-pneumatic brake in order to be admitted to equip rail vehicles are regulated by UIC leaflets no. 541-5, 541-6.

Even if the main advantage of the electro-pneumatic brake is the simultaneity in brake-release operation, extremely important in the case of long trains, as the freight ones, there are not many developments because of the diversity of wagons and the cost of conversion, not to mention that getting an electric signal to transmit at a low voltage down a very long train is difficult.

It is to notice that, due to the advantages, the electro-pneumatic brakes are largely used also in combination with other complementary braking systems such as the electric and the electromagnetic brake systems, for trams and metro trains, not to mention the high speed trains.

That is why the electro-pneumatic brakes equip mainly passenger vehicles and multiple unit passenger trains, the electric command being suitable for simultaneous on-board computer braking control of multiple braking systems usually in use.

At least in historic terms, among the basic brake systems, it is to mention the vacuum one, which constructively resembles with the straight compressed air brake (see fig. 5), having a vacuum pump instead the compressor.

It operates very simple, the brakes are released while the vacuum is maintained in the air pipe of the train and, implicitly, in the braking cylinders. The braking actions are commanded by increasing the air pressure in the air pipe to the atmospheric pressure through the driver's tap.

It is to notice that, despite of the simplicity, it is an automate brake.

The main disadvantage determining its obsolescence is connected to the fact that the same braking force requires double diameter brake cylinder compared to the compressed air brake system. The system also raises operational problems related to air leakages detection.

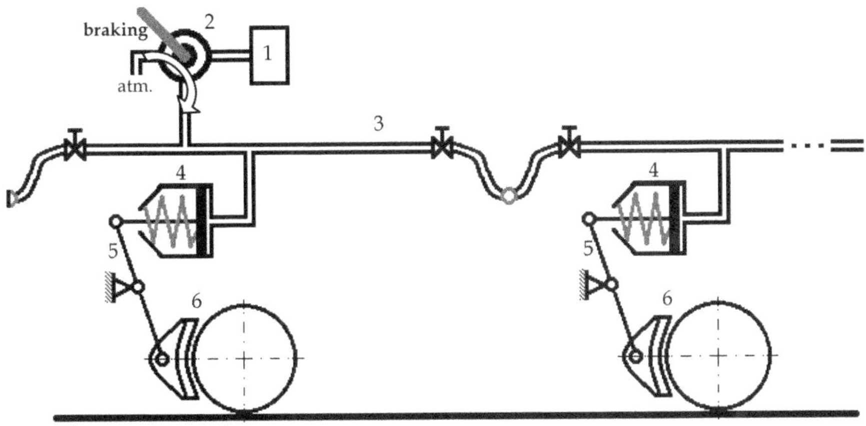

Fig. 5. Schematic of vacuum brake system: 1 - vacuum pump; 2 - driver's brake valve; 3 - general air pipe; 4 - brake cylinder; 5 - brake rigging; 6 - brake shoe.

4. Complementary braking systems

Generally the complementary braking systems provide consistent controllable braking forces permitting speed reductions but, unlike the basic ones, the braking efficiency decreases at low running speeds. As a consequence, there have to be used together with a basic braking system to ensure the capability of stopping at fixed point and of maintaining the vehicle standstill on slopes.

Complementary braking systems add braking power without having the thermal capacity limitations of the friction wheel or disc brakes that would necessitate expensive solutions or lead to excessive wear from harder use. More than that, those which are independent of wheel/rail adhesion improve safety by enabling shorter stopping distances when applied by giving a reduced dependency between stopping distance and adverse adhesion conditions caused by moisture, ice, leaves or other pollution on the top of the rails, etc.

That is why usually the decision to equip rail vehicles with complementary braking systems relies either on the incapacity of basic wheel-rail adhesion dependent brakes to ensure the braking necessary capacity for high speed trains and, in that case, there are necessary adhesion independent braking systems, or to diminish the wear of the friction based braking systems, in that case being useful the dynamic braking systems.

The magnetic rail brake operation is based on developing electromagnetic attractive forces towards the rail (see fig. 6), which causes a normal application force acting on their contact surfaces that are in relative displacement. This leads to friction forces between the magnetic brakes and rails, opposing the vehicle's direction of motion, which generate braking forces. The electromagnetic brake is used as additional wheel-rail adhesion independent braking system, generally associated to the brake disc.

There are two magnetic track brake positioned between the wheels on each side, normally mounted and attached to each bogie frame. The braking surface is of steel alloy or cast iron and is usually built up of sections with gaps between the sections and these are mounted to a sledge, which can be lowered from a parking position in the bogie. Using a rather low excitation power, about 1 kW/magnetic track brake, it is possible to obtain important application forces, about 50...70 kN and accordingly, for a normal vehicle installation (four axles coach), braking force per shoe between 4...10 kN.

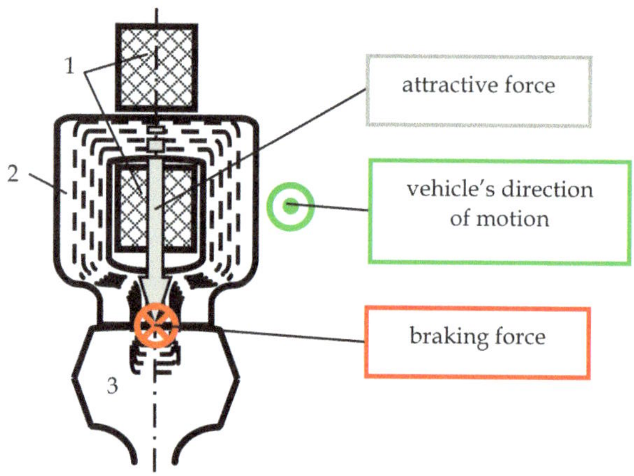

Fig. 6. Operating principle of magnetic rail brake system: 1 - coil; 2 - yoke; 3 - rail.

As stated, the main advantage of the system is the wheel/rail adhesion independence, important for the safety of operation by enhancing the braking power of classical basic braking system. Moreover, the friction between the braking surface and rail can sometimes significantly improve adhesion between wheel and rail due to vigorous cleaning of the tread rails during operation. As a result, for the classic braking systems is usually avoided the wheels slide even in adverse conditions. It is also to notice that given the mass of the magnetic rail brake assembly and its fastening system, the gravity center of the bogie is lowered, with positive dynamic effects for the vehicle, especially in high speed domain.

The main disadvantages are determined by the frictional operation of the system that lead to several drawbacks due not only to the relatively rapid wear of the braking surfaces especially for high traffic speed, but also to the increasing dependence of the shoe-track friction coefficient corresponding to the decrease of the running speed. As a consequence, the magnetic rail brake is designated only for emergency braking and is usually automatically released when the running speed is less than 50 km/h. This particular operation mode gives the complementary character of this system.

When designing the magnetic rail brake is important to consider also the interaction with the rails and generally with the track. Width and length of the magnetic brake shoe is critical in relation to the safe passage over the unguided area of switches and crossings, check/guard rails and other track design features or permanent way installations. On the other, the length is limited by the bogie wheel base, but the length of the braking surface must be kept equal to or above 1000 mm. Also, if too wide, parts of the frog can be hit outside the normal wheel-rail contact area or, in the extreme, fouling check rails. UIC leaflet no. 541-06 specifies the width of the friction plate to 65 - 72 mm, which is within the railhead width of UIC 46 to UIC 60 rails. The braking surface has to be flared at the ends in order to negotiate discontinuities in the rail head and the end elements of the brake shoes have both the characteristics of crossings with a tangent above or equal to 0.034 and the check rails. The general features of magnetic track brakes applicable to railway vehicles are stated in UIC leaflet no. 541-06.

The operating principle of the magnetic track brake by using a magnetic field may determine incompatibilities with train detection systems working on magnetic principles. Consequently it is advisable to be equipped with shields to reduce the adverse effects. Also, the friction operating may lead to abrasion of primarily shoe material, conducting to possible bridging of isolated rail joints for track circuits and to the formation of ridges on the shoe surface leading to reduced performance.

Because the magnetic track brake may develop high braking forces, one must not exceed an equivalent total deceleration of 2.5 m/s^2 over the train length, avoiding also excessive longitudinal track forces in track with low longitudinal resistance or prone to rail creep.

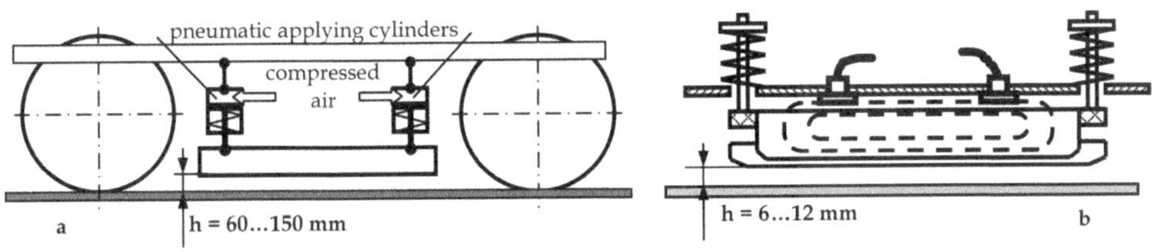

Fig. 7. Constructive solutions for magnetic track brake: a - high suspended; b - low suspended.

According to the maximum running speed, there are two constructive solutions (see fig. 7) regarding the release position: high suspended, with a distance between the braking surfaces and rails of 60...150 mm, common for running speeds exceeding 100 km/h and low suspended in the case of vehicles running up to 100 km/h, usually applied to tramways, the distance being 6...12 mm.

In the first case, pneumatic cylinders are used to descend the magnetic track brake up to the rails and only afterwards the electric circuits are activated. In the second case, due to the quite small air gap, the electromagnetic attraction forces are high enough to determine descending of the magnetic brake, too.

The eddy current brake system is a dynamic no mechanical contact one, based on the action of a magnetic field operating across an air gap between a set of electromagnets oriented successively N-S poles versus a metallic mass (see fig. 8). Consequently, it is a wear-free and silent system, requiring minimal maintenance. Eddy currents are induced by movement in a magnetic field and the kinetic and, eventually, the potential energy of the train is absorbed by the metallic mass and converted into heat that dissipates in the environment.

Retardation depends on speed - the faster the train, the greater the braking force, on the intensity of the magnetic field and on the air gap. If the air gap is maintained constant, the braking force can be accurately controlled by regulation of the magnetic field while it is created using electromagnets fed from an external power supply, offering a useful solution as a frictionless moderable braking system for high speeds. Because the braking forces have a marked decrease at low running speeds, the eddy current brake is a complementary one and cannot be used for stopping at fixed points, nor as parking brake.

Depending on the element used as metallic mass, there are two constructive solutions: rail and rotary eddy current brake.

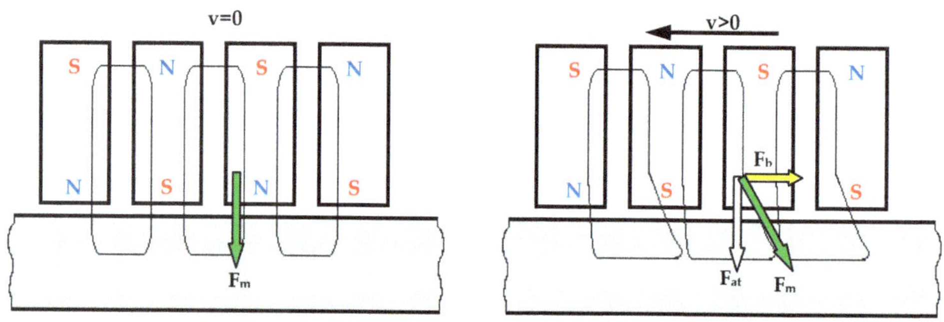

Fig. 8. Principle of eddy current brake system (F_m – magnetic force; F_{at} – attraction force; F_b – braking force).

In the case of rail brake, the braking electromagnets are disposed in a linear alignment with an alternating sequence of north and south poles, above each rail, apparently resembling to the rail electromagnetic brake. On each bogie, the magnet sets are connected through crossbeams to form a single assembly, which can be raised or lowered by a ring bellow. In order to prevent inadvertent mechanical contact between the solenoid coil housings and the track, damage to the magnet, or dampness leading to corrosion, the brake housing is fitted with an energy absorbing guard plate and sealed with a synthetic resin.

The main advantage of this system is the independence of wheel/rail adhesion, enhancing the braking power over the limits of classical basic disc brake usually associated with.

The rotary eddy current brake uses as metallic mass disks mounted on the axle or the wheels themselves, which rotate towards the electromagnets set in a housing and disposed also in alternating sequence of north and south poles (see fig. 9). The housing can be attached directly within the bogie frame, or can be supported on the weheelset, but in that case it must be secured against rotation. The first solution is simpler as design, but due to vertical and transversal relative displacements between the wheelsets and bogie frame one must achieve a large enough air gap, which consequently requires a higher excitation power. The second solution allows a smaller air gap, but the construction is more complicated and expensive and determines an increase of the unsprung weight of the vehicle, particularly undesirable in the high speed domain.

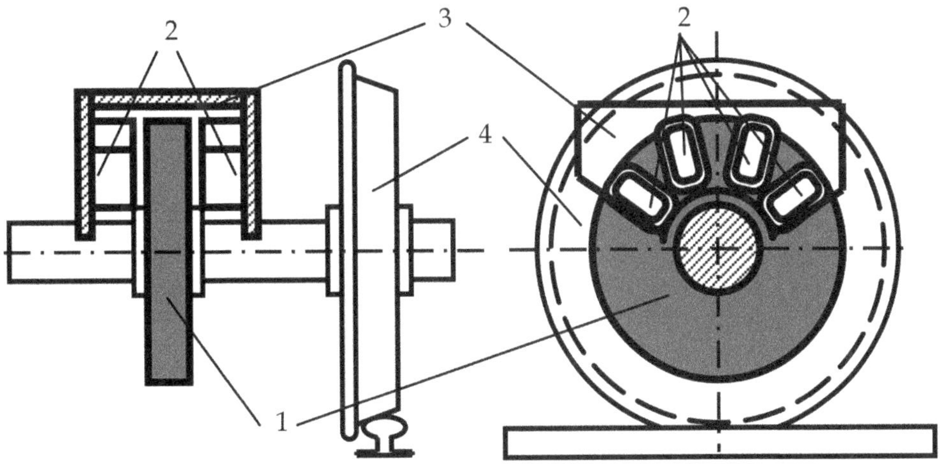

Fig. 9. Rotary eddy current brake: 1 - metallic disc; 2 - magnets; 3 - housing; 4 - wheel.

There are at least two major disadvantages of rotary eddy current brake. Due to the fact that the retarder couple is developed directly on the wheelset, it is an adhesion dependent brake and consequently unable to enhance the braking capacity of the mandatory basic system. The other is determined by thermal aspects, limitations being caused by the possibilities to dissipate the heat generated by the eddy currents in a relatively small mass, aspect enhanced due to specific aspects if the system operates on wheels.

Thermal aspects are certainly connected to the rail brake, but in case of a train, Sookawa et al. (1971), have shown that at 12 mm below the tread of the rail, in case of a 10 mm air gap, the rail temperature increase does not exceed 10^{0}C. It is obvious that the temperature of the tread of the rail attains higher values, but it was found that in approximately 10...20 minutes the temperature evens within the rail mass (Pouillet, 1974; Sookawa et al., 1970, 1971), but normally, there do not seem to raise difficulties. In that case, the issue that had to be addressed is the heating of the rails as a result of repeated brake applications made in much the same locations, which is usual due to operational and signaling reasons. So, in the case of a high train frequency there is a potential risk that the rails will not be able to cool sufficiently between brake applications that may affect the track structure, turnouts and other critical elements such as bridges. Still, only periodicities of the sequences of trains braking in the same area less than 10 minutes might appear critical (Sookawa et al., 1970, 1971).

Regarding the air gap, independent recommendations are concordant. Studies and experiments conducted by Pouillet (1974) showed that the optimum value of air gap is 7...8 mm, while according to manufacturer Knorr-Bremse and DB's practical experience on the German ICE 3 train, the air gap between the magnets and the railhead should be between 6 and 7 mm when the brake assembly is lowered into its operating position (Schykowski, 2008).

Eddy-current brakes, especially the linear ones, seem ideally suited for high speed railways, improving rail safety by enabling shorter stopping distances and reduced dependency between stopping distance and wheel/rail adhesion, particularly during adverse adhesion conditions and offering braking power, both for service and emergency actions, which is difficult to achieve with other methods.

More than that, there are also other desirable effects in their use, such as mitigating the thermal capacity problems of brake pads and discs associated with conventional friction braking systems and avoiding harder application of conventional friction brakes leading to excessive wear of the pads and brake discs.

However, due to the operating principle, eddy current brakes raises issues of compatibility with infrastructure that can potentially impair both safety and technical reliability, meaning electromagnetic and physical compatibilities with train detection installations and line side equipment for train condition monitoring. It is also to mention that longitudinal, vertical and lateral supplementary forces induced by that braking system must fit the constructive track resistance.

5. Braking capacity

The main goal in designing and operating is to provide the necessary braking capacity appropriate to the type and specific running speed of the vehicles/trains according to the traffic safety.

Braking capacity is a significant feature of any railway vehicle and train which states the overall design and functional capability to stop from a maximum running speed according to the maximum braking forces developed during an emergency braking.

The braking capacity of a railway vehicle depends on numerous factors and some of the most important are: running speed, weight, type of brakes, constructive and functional characteristics of the brake rigging, braking characteristics, thermal phenomena, etc. In assessing the braking capacity of a train, additional parameters are involved: the train type, composition and length, the braking wave propagation characteristics, etc. Therefore, a direct and consistent assessment of braking capacity is difficult to achieve.

Specific for the railway vehicles, the possible maximum braking force is critical for the basic wheel/rail adhesion dependent braking systems. The main condition imposed is that braking forces at the wheel-rail contact surface $F_{b,max}$ must not exceed the wheel-rail adhesion force F_a, for designing purposes considered in normal conditions:

$$F_{b,\max} \le F_a \tag{1}$$

Considering a vehicle having Q_v weight, relation (1) gets particular expressions for the case of being equipped with n brake shoes (see fig. 10, a):

$$F_{b,\max} = \sum_{i=1}^{n} \left(\mu_s \cdot P_{s,i} \right) \le \mu_a \cdot Q_v \tag{2}$$

respectively with n brake discs (see fig. 10, b):

$$\frac{4 \cdot \mu_d \cdot r_m \cdot \sum_{i=1}^{n} P_{d,i}}{D_o} \leq \mu_a \cdot Q_v \tag{3}$$

where: μ_a is the wheel/rail adhesion coefficient, P_s and P_d are the clamping force on a brake shoe, respectively pad, μ_s and μ_d the friction coefficient between brake shoes and wheel tread and brake pad and disc respectively, D_o the wheel diameter and r_m the medium friction radius.

The braking forces are essentially influenced by the friction coefficients involved, their dependence on different parameters having important role on braking characteristics of the vehicle. There are many factors determining the evolution of friction coefficients, among them the most important proved to be the running speed, the clamping forces, the surface contact pressure and temperature.

Orientation towards a certain friction material for braking equipments is strongly influenced by the constructive and operational characteristics of the vehicle, mainly the maximum running speed, as well as by the dependence of friction coefficient on the previous specified parameters.

It is known that the friction coefficient between cast iron braking shoes and wheel tread strongly depends on the instantaneous running speed, the applying force on each shoe and the contact pressure, while the use of plastic (composite) materials for brake shoes or pads enables an independence of the friction coefficient on the mentioned parameters (see fig. 11).

In practical calculus, for the friction coefficient between cast iron brake shoes and wheel tread there are recommended different empirical relations, determined by experiments, depending on most important influencing factors, meaning mainly the running speed V [km/h], the applied forces on a break shoe P_s [kN] or the surface contact pressure p_s [N/mm^2]. An example is UIC formula:

$$\mu_s(V,p_s) = 0.49 \cdot \frac{\frac{10}{3.6} \cdot V + 100}{\frac{35}{3.6} \cdot V + 100} \cdot \frac{\frac{875}{g} \cdot p_s + 100}{\frac{2860}{g} \cdot p_s + 100} \tag{4}$$

or Karvatzki formula:

$$\mu_s(V,P_s) = 0.6 \cdot \frac{V + 100}{5 \cdot V + 100} \cdot \frac{\frac{16}{g} \cdot P_s + 100}{\frac{86}{g} \cdot P_s + 100} \tag{5}$$

where $g = 9.81$ m/s^2.

In the case of plastic brake shoes the friction coefficient is about 0.25, while for brake pads is about 0.35.

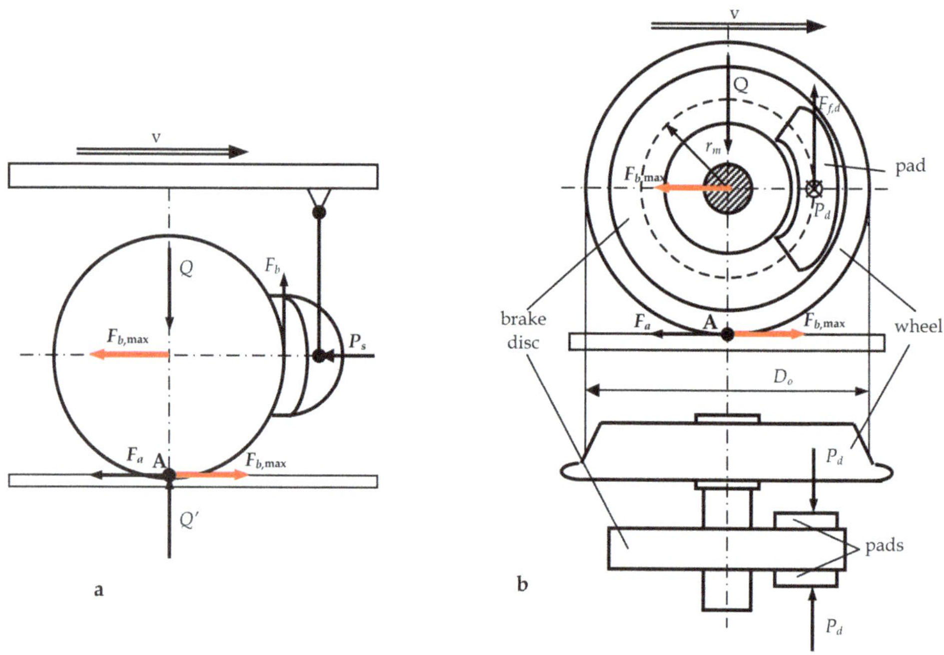

Fig. 10. Braking forces at wheel/rail limit adhesion: a – brake shoes, b – disc brake.

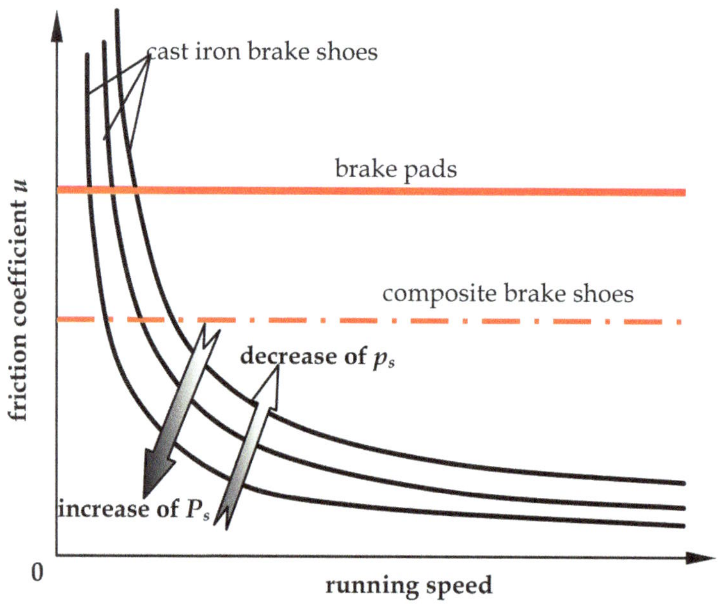

Fig. 11. Dependence of friction coefficient on running speed for different braking systems.

Due the above mentioned independency on running speed and taking into account the high value (0.35) of the friction coefficient, according to international regulations, the disc brake is mandatory as basic brake system for vehicles running with 160 km/h or more.

Sill the main problem is that friction interface between the brake pads and disc must not exceed temperatures of 350...375^{0}C, else severe and sudden wear of pads occurs. Practically, the thermal regime determines the necessity of several brake discs mounted on the wheelset, even four in the case of high speed vehicles.

The vehicle mass is one of the factors influencing the necessary braking force, being proportional with the kinetic energy to dissipate during braking actions. When the vehicle's mass may have important variations, the braking forces have to be adjusted subsequently in order to avoid wheel slip and locking of wheelsets (for empty vehicle), unacceptable lengthening of the braking distance and enhancement of dynamic longitudinal in-train reactions. This issue is specific to freight wagons and some passenger cars, such as the double-decked, the post and/or luggage and those for transporting cars in passenger trains, characterized by a maximum possible load greater or comparable to the weight of the empty vehicle.

According to the type and constructive running speed there are two technical possibilities to solve the problem: either a step, or a self-adjusting load-proportional braking system. The first is used in the case of freight wagons with constructive running speeds less than 120 km/h, while the second is compulsory for passenger cars and freight wagons running with 120 km/h or more.

A step-adjusting load-proportional brake system has a manual or automatic empty-loaded control device which usually enables two clamping force levels on the friction elements (determining two braking forces) at the same pressure command of the driver, according to the size of the actual mass of the vehicle in relation to a switching mass.

The system requires either two amplification ratios of the brake rigging, or a classical brake rigging actuated by a double brake cylinder (see fig. 12).

Fig. 12. Double brake cylinder.

The air pressure within the double brake cylinder is univocally determined by the air brake distributor, according to the driver's command in the brake pipe. Depending on the empty or loaded position of the device, the compressed air is directed to one of the segments of the brake cylinder. For the same air pressure, one may obtain two braking force levels, according to the diameter of the segment involved. The automatic empty-loaded control is based on a mechanical or pneumatic determination of the suspended weight of the vehicle based on the deformation or size of the forces acting on the suspension springs.

The self-adjusting load-proportional braking system is capable of an accurate and continuous adaptation of the braking force to the weight of the vehicle. The system requires a weighing valve situated in the suspension and determining a command pressure which is send to a pressure relay. This device operates together with the air distributor and adjusts the air pressure within the brake cylinders that is commanded through depression in the brake pipe of the train according to the command pressure proportional to the vehicle's load. That means that for the same braking command, the real pressure in the brake cylinders differs for different loads. Usually, taking into account an uneven distribution of load in the vehicle, there are at least two weighing valves placed diagonally in opposite sides and in the case of unique air distributor on the vehicle, a middle pressure valve is added to the system. Its role is to transmit to the pressure relay a mediated value of command pressures received from the weighing valves of the vehicle.

Very important is the brake cylinder pressure characteristic of the pressure relay. This shows the dependence between the maximum pressure within the brake cylinder and the weight of the vehicle and is determined based on the condition that, regardless of the load, the same braking distance to be achieved. The constrains are to respect the maximum 3.8±0.1 bar pressure within the brake cylinder for maximum load and not to decrease below 1.1 ... 1.3 bar for the empty vehicle, in order to obtain a consistent braking force even for low braking steps.

The braking distance, usually defined as the distance covered by a vehicle or a train since the command of emergency braking from the maximum running speed, to a complete stop, is the minimum distance to stop using the full braking capacity available.

The braking distance seems to be extremely appropriate for designating the braking capacity, because it is direct effect of braking forces and all other implied factors, being measurable and permitting comparisons for evaluating the braking efficiency.

For an analytical determination of the braking distance of a rail vehicle it is important to take into account numerous factors, such as the running speed when the brakes are applied, the vehicle's weight, the evolution of braking forces dependent on the type, constructive and operational characteristics of the braking systems, resistances, the geography of the track, in particular slopes, etc. For trains, there are supplementary aspects involved depending on the length, the weight and even the mass distribution in the body of the train, the brake propagation rate, etc.

For accuracy, when theoretically determine the braking distance s_b there are considered first a "braking preparation space" s_p [m] covering the phenomena immediately subsequent acting the driver's brake valve until maximum pressure is established in all brake cylinders, continued by the effective braking space s_{ef} [m] covered with full braking capacity until stop:

$$s_b = s_p + s_{ef} \text{ [m]} \tag{6}$$

Generally for a single vehicle, or for a train in case of electric command supposed to propagate almost instantaneous along the train, the braking preparation space can be calculated considering the pressure evolution during the filling time t_f [s] as a step function, at half of the duration the pressure becoming instantly maximum, determining beyond that moment the action of full brake capacity:

$$s_p = \kappa \cdot \frac{V_{max} \cdot t_f}{3,6} \text{ [m]} \tag{7}$$

where V_{max} [km/h] is the running speed at the moment of braking command and $\kappa = 0.5$. In the case of trains equipped with classical UIC air brake, the factor κ is recommended between 0.54...0.7 according to the length of the train and to the type of brakes.

Determination of effective braking space considering that the kinetic E_k and potential E_p (when running on slopes) energy of the vehicle/train are dissipated by the work of the braking and resistance forces. Because the rail vehicles have important masses in rotation, their rotation kinetic energy must not be neglected:

$$E_k = \frac{m \cdot v^2}{2} + \frac{I \cdot \omega^2}{2} = \frac{m \cdot v^2}{2} \cdot \left(1 + \frac{I}{m \cdot r^2}\right) = \frac{m \cdot v^2}{2} \cdot (1 + \rho) \tag{8}$$

considering v [m/s] the running speed, m [kg] the vehicle/train mass, I [kg·m²] the polar inertial moment of the wheelsets, r [m] the wheels radius and $\rho = I/(m \cdot r^2)$ a term accounting the rotational masses involved.

The potential energy depends on the track gradient i [mm/m] and on the travelled distance s [m]:

$$E_p = m \cdot g \cdot s \cdot \frac{i}{1000} \tag{9}$$

with $g = 9.81$ m/s². Obviously, on uphill track gradients gravity assists deceleration.

So, according to previously presented considerations, the braking effective distance s_{ef} may result from the equation:

$$(1 + \rho) \cdot \frac{m \cdot v_{max}^2}{2} + m \cdot g \cdot s_{ef} \cdot \frac{i}{1000} = \int_0^{s_{ef}} F_b \, ds + \int_0^{s_{ef}} R \, ds \tag{10}$$

where $F_b = f(v, etc)$ [N] is the maximum instantaneous braking force and $R = f(v, v^2, etc)$ [N] the resistance forces.

As previously stated, the fundamental problem in establishing a train braking capacity is thus to determine the braking distance according to the maximum running speed. As shown, this would be a basic mechanics problem unless the implication of many nonlinear parameters such as brake propagation factor, the brake cylinder pressure evolution, the characteristics and multiple dependencies of the friction laws, the different and sometimes unpredictable composition of the train, etc.

Given the presented issues, to address current problems related to the safety of rolling stock operation regarding braking, it was necessary to define a general, synthetic term able to accurately quantify the braking capabilities not only for each railway vehicle, but also to enable a rapid, correct and adequate determination of braking capacities of trains, consistent to their composition, length and all other characteristic parameters.

That specific term is called braked mass, it is expressed in tons, constitutes a measure unit for braking effect and is compulsory to be inscribed on the vehicle. It has a general and synthetic character, it may be less than, equal to or greater than the mass of the vehicle and at present has no physically correspondent.

The term's designation is traditionally preserved from the time of freight trains using generalized hand brake, when the worst case that could meet during specific operating was locking of all wheelsets due to excessive action of the braking agents. Under these conditions, the maximum braking force $F_{b,max}$ [kN] depended only on the mass M [t], considering an relatively invariant friction coefficient τ between the locked wheels and the rolling tread of rails:

$$F_{b,\max} = \tau \cdot M \cdot g \tag{11}$$

and hence the name of the term.

Currently, for determining the braked mass, there are taken into consideration not only the construction and operation of the brake equipments fitted to vehicles, but also the multitude of processes and phenomena that govern the braking action, such as: dependence of friction characteristics on instantaneous running speed, pressing forces, specific contact pressure; environmental conditions and their influence on friction coefficients as well as on the wheel/rail adhesion, being not allowed locking of wheelsets during the braking actions; specific pneumatic phenomena that determine the braking rate, the pressure characteristics of air distributors; influence of thermal phenomena and braking resultant heat dissipation; resistances of the vehicle/train, etc.

The value of braked mass consequently depends on many factors, processes and complex phenomena, which makes it extremely laborious and difficult to be established based on analytical calculation, accuracy to real braking actions proving essential given the importance on the safety of operation. Therefore, determining the braking mass of railway vehicles is mainly based on experiments and tests, but there are also relationships based on testing results fitted for some particular cases. Methodologies and procedures for determining the braked mass and are regulated by UIC leaflet no. 544-1. To be more convenient in practical use, it is also defined a specific notion, braked mass percentage ratio b, as the ratio between the braked mass B [t] and the train's or vehicle's mass M [t]:

$$b = \frac{B}{M} \cdot 100\,[\%] \tag{12}$$

It is to notice that using the braked mass B or the braked mass percentage ratio b instead of brake space s to appreciate the braking capacity has a mechanical based justification.

Considering the simplest case of a train of mass M at a running speed v_o subjected to a constant deceleration d, the braking space s and the stopping time t are related by known mechanical relationships:

$$s = \frac{1}{2} \cdot d \cdot t^2 \quad ; \qquad v_o = d \cdot t \quad \Rightarrow \quad s = \frac{v_o^2}{2 \cdot d} \tag{13}$$

The total braking force F_b of the train is proportional to the total normal applying forces acting the brake shoes on wheel tread or the brake pads against the discs $\sum P_N$:

$$F_b = \xi \cdot \sum P_N$$

The proportionality coefficient ξ might be $\xi = \mu_s$, the friction coefficient between brake shoes and wheels or, for the case of vehicles equipped with disc brake, $\xi = \frac{4 \cdot r_m}{D_o} \cdot \mu_d$ (μ_d the friction coefficient between brake pad and disc, D_o the wheel diameter and r_m the medium friction radius).

By definition, the braked mass is also proportional to the total normal applying forces:

$$B = \chi \cdot \sum P_N$$

The proportionality coefficient χ depends on the vehicle constructive and operational characteristics.

If neglecting the influence of other resistances and considering the definition of braked mass percentage ratio b given by eq. (12), then:

$$d = \frac{F_b}{M} = \frac{\xi \cdot \sum P_N}{M} = \frac{\xi}{\chi \cdot M} \cdot B = \frac{\xi}{\chi} \cdot b$$

and, taking into account (13), results:

$$s = \frac{v_o^2 \cdot \chi \cdot M}{2 \cdot \xi} \cdot \frac{1}{B} \tag{14}$$

and:

$$s = \frac{v_o^2 \cdot \chi}{2 \cdot \xi} \cdot \frac{1}{b} \tag{15}$$

Eq. (14) and (15) enhance the inverse proportionality between the braking space s and braked mass B, respectively the braked mass percentage ratio b.

6. Longitudinal dynamics of trains submitted to braking actions

As previous stated, according to international regulations, the indirect compressed air braking system is mandatory for railway vehicles because the pneumatic command is recognized as very reliable and its automat functioning principle is very important for the safety on operation.

In the case of classical UIC brake system, due to the air compressibility and to the length of the train, there will always be a time lapse between the reaction of the leading vehicle and the reaction of the rear one. Corresponding to the propagation rate of air pressure signal, the air distributors will come into action successively and the braking of vehicles begins at different times along the train so that, while some cars are slowing down, others are trying to push, still unbraked, from the rear. This creates the conditions that during transitional braking stages, immediately following the command of pressure variation in the brake air pipe, to develop important longitudinal in train reactions causing stress to the couplers and affecting passenger comfort and, sometimes, even the traffic safety.

6.1 Trains braking phases

After a braking action command, speed begins to decrease due to the kinetic and potential energy dissipation mainly through the heat developed by the action of braking systems and through the work of the resistance forces that each vehicle and, accordingly, the whole train, are submitted to. These processes develop with different intensities in various places of the train assembly. So, in the case of a train equipped with standard pneumatic brake system:

- along the train, the effective action of the brakes begins successively, according to the length of the train and depending on the braking propagation rate wave, etc.
- at each vehicle, the braking forces increase up to the commanded value is time dependent, according to the filling characteristics specific to the brake and air distributor constructive and functional types;
- the train's vehicles can be equipped with various types of braking systems;
- usually, trains are composed with different types of vehicles and consequently the resistance forces differ, while the wheelsets and masses are not uniformly disposed along the train;
- vehicles may have various masses and loads and, depending on the type of brake devices that are fitted (basic, step-adjusting or self-adjusting load-proportional braking systems), braking forces will develop in different manners, finally being more or less adapted to the total weight of the vehicle;
- if for certain reason there are vehicles with inactive brakes, then even more the braked wheelsets are unevenly placed in the train body, etc.

From the above it follows that if taken separately each vehicle, according to its particular operating, constructive and loading features, it would stop on a specific braking space, even if submitted to the same braking action and beginning at same running speed. While connected in the train body, they will have to stop on a same distance, determining longitudinal reactions, certainly amplified by the specific operating mode of the indirect air brake system. These reactions that act on the shock and traction apparatus and are transmitted through the chassis, can be important under particular conditions, determining shocks and even affecting the safety of the traffic. These aspects must be studied in order to establish specific conditions in terms of braking features and train composition, as well as constructive and operational, to diminish the in-train dynamic reactions in such a manner to avoid disturbing or dangerous levels.

To easily understand the issues, Karvatski (1950) considered that during braking actions four phases occur and, to simplify the problems, he presented them under some assumptions, such as: masses and braked wheelsets are evenly distributed in the train body, assuming that vehicles are identically constructively and loaded and equipped with the same type of brake, all active and providing the same filling characteristics. Two typical cases are presented under these assumptions: a passenger train of 20 four-axles vehicles equipped with fast-acting brakes and a freight train of 100 two-axles vehicles equipped with slow-acting brakes. The time history of certain brake cylinders pressure, representing also proportional the evolution of brake forces along the trains, are presented in fig. 13.

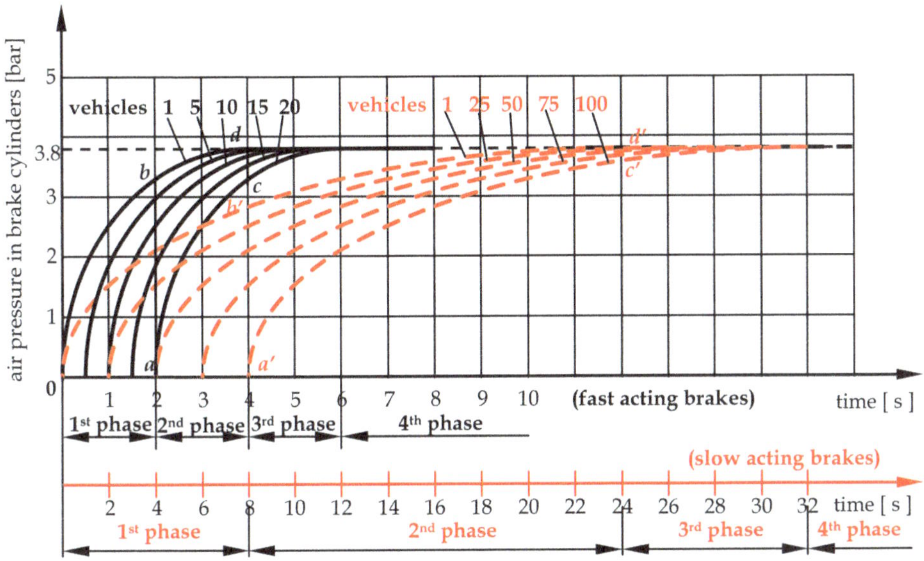

Fig. 13. Phases of train braking (continuous line for passenger train, dotted line for freight train).

The first phase is considered between the moment of commanding the brake action until the brake propagation ratio attains the last air distributor of the train. During that phase, the brakes begin to come into action successively along the train, which is submitted to a compression becoming maximum at the end of first phase, corresponding to the brake cylinder difference of pressures between the first and the last vehicle of the train (proportional to segments *ab* for passenger train equipped with fast-acting brakes, *a′b′* for freight train equipped with slow-acting brakes).

The second phase is considered between the end of the previous one until in the brake cylinders of the first vehicle in train the maximum air pressure commanded is attained. During this time, pressure continues to increase uniformly in all the brake cylinders, maintaining a decreasing pressure distribution along the train, that remains consequently compressed, the compression level being similarly proportional to segments *cd* and *c′d′* respectively. Moreover, under the assumed simplifying hypothesis, at the end of the first phase there are created all necessary condition to initiate an oscillatory motion, due to inertial in excess forces in the second half of the train, which propagates along, pushing alternatively the vehicles from the front and the end of the train. The oscillatory motion, which overlaps on the existent compression of the train, is damped according to the damping coefficient of the buffers and traction gears.

The third phase lasts from the end of the second one until the maximum pressure is established in the brake cylinders of the last vehicle of the train. During this phase the maximum pressure is achieved successively in the brake cylinders along the train. As a result of successive braking forces equalization, the potential energy accumulated in the elastic elements of the buffers during the previous phases compressions is rendered to the system. Consequently it develops a "rebound" in succession along the train, its intensity depending on the damping characteristics of the shock apparatuses.

The fourth phase is considered between the end of the previous one until the train stops or a brake release command is performed. Because during that phase the maximum pressure already existent in all brake cylinders is maintained, braking forces remain constantly to their maximum values along the train, so the deformations stop and train length remain from now on unchanged.

It is to notice that even under the simplifying assumptions, the mechanical response of the train is extremely complex, the length of the train continuously modifies during the first braking phases and the overlap of oscillatory motion propagation determines the development of important compression and traction in-train forces. Incidents such as broken couplers during braking actions, observed mainly in the case of long, heavy freight trains submitted to braking actions, constituted the evidence of practice.

6.2 Mechanical model of the train

A classical approach for theoretical studies of the dynamic longitudinal forces developed during the braking actions along trains equipped with automated compressed-air brakes is a mechanical cascade-mass-point model in which vertical and lateral dynamics are usually neglected (Pugi et al., 2007; Zhuan, 2006; Zobory et al., 2000, etc).

Assuming that the train is composed of n vehicles, these are linked to each other by couplers, traditionally based on combined use of draw-gears and buffers. Consequently, the model is an elastic-damped lumped system consisting in n individual rigid masses m_i representing each vehicle, connected through elements having well defined elastic c_i and damping ρ_i characteristics (see fig. 14).

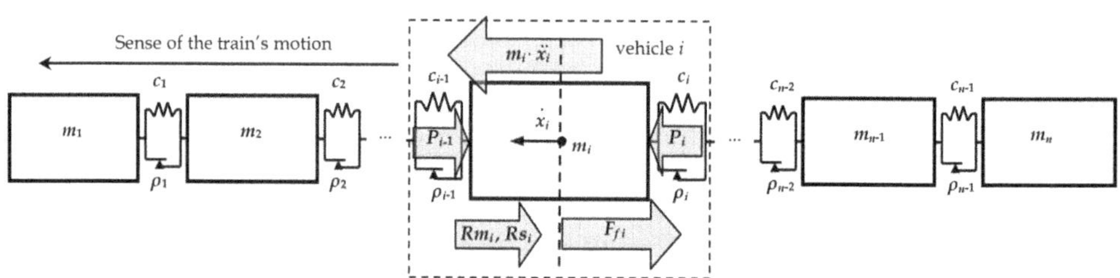

Fig. 14. Mechanical model of the train

Generally, a certain i vehicle of the train is mainly submitted to the following exterior forces: $F_{b,i}$ the instantaneous braking force of the vehicle, Rm_i the vehicle's main resistance, Rs_i the supplementary resistance due mainly to the tracking slope and curvature and P_{i-1}, P_i the in-train forces between the adjacent vehicles representing cumulated elastic and damping forces acting on the shock and traction apparatus between $i-1$ and i, i and $i+1$ respectively vehicles.

Considering x_i and $\ddot{x}_i$ the position and the instantaneous acceleration of a certain i vehicle of the train submitted to the influence of the exterior forces, the equation of motion is:

$$m_i \cdot \ddot{x}_i = -F_{b,i} - R_i - P_{i-1} + P_i \tag{16}$$

for $i = 1, 2, \dots, n$ and $P_o = P_n = 0$.

Applied to all component vehicles of the train, eq. (16) constitutes a differential nonlinear equation system of second degree.

For each m_i vehicles' mass, the covered distance, the instantaneous speed and acceleration, as well as the instantaneous braking, resistances and longitudinal developed forces in the train's body are mathematically time dependent.

According to initial applied conditions, the equation system (16) can be solved applying a numerical integration process.

Generally, the main parameters influencing the assembly of the studied problem are: train's composition, number, mass and type of the vehicles, as well as their repartition in the body of the train; the braking system functional characteristics; the elastic and damping characteristics of the shock and traction devices; the evolution of the friction coefficient between the brake shoes and wheels, brake pads and discs respectively, eventually between the electromagnetic track brakes and rail, in accordance with the equipments of the train's vehicles.

6.3 Analysis of main mechanical parameters

For the case of disc brake equipped vehicle having individual self-adjusting brake rigging, the braking force can be calculated:

$$F_{b,i} = \left[\frac{\pi \cdot d_{bc}^2}{4} \cdot p_{bc,i} - \left(F_R + R_{sa} \right) \right] \cdot i_t \cdot n_{bc} \cdot \frac{2 \cdot r_m}{D_o} \cdot \mu_d \cdot \eta_{br} \tag{17}$$

where: d_{bc} is the brake cylinder diameter [m], $p_{bc,i}$ [N/m^2] the instantaneous relative air pressure in the brake cylinder, F_R and R_{sa} [N] the resistance forces due to the brake cylinders back spring and to the self-adjusting mechanism incorporated in the piston rod respectively, D_o [m] the wheel diameter and r_m [m] the medium friction radius. The dimensionless terms are: i_t the brake rigging amplification ratio, n_{bc} the number of brake cylinders of the vehicle, μ_d the friction coefficient between brake pads and disc and η_{br} the mechanical efficiency of the brake rigging.

If the vehicle is equipped with shoe brake having symmetrical brake rigging with self-adjusting mechanism on the main brake bar, the braking force can be calculated by the relationship:

$$F_{b,i} = \left[\left(\frac{\pi \cdot d_{bc}^2}{4} \cdot p_{bc,i} - F_R \right) \cdot i_c - R_{sa} \right] \cdot i_l \cdot n_\Delta \cdot n_{bc} \cdot \mu_s (P_s, V_i) \cdot \eta_{br} \quad (18)$$

where d_{bc} [m], $p_{bc,i}$ [N/m^2], F_R [N], R_{reg} [N] and η_{br} have the same significations previously assigned. The dimensionless terms are: i_c the central brake rigging, i_l the amplification ratio of the brake rigging's vertical levers, n_Δ the number of triangular axels and μ_s the friction coefficient between brake shoes and wheel tread which depends on the clamping force on each brake shoe P_s [kN] and on instantaneous running speed V_i [km/h].

Assuming that certain terms and factors representing constructive and functional characteristics are constant for the same vehicle during braking actions, one may be put in evidence that during the filling time the brake force for the brake disc is directly depending only on the instantaneous relative air pressure in the brake cylinder $F_{b,i} = f(p_{bc,i})$, while in the case of shoe brake, the dependence is more sophisticated due to the friction coefficient between brake shoes and wheel tread $F_{b,i} = f(p_{bc,i}, \mu_s(P_s, V_i))$.

In the last case, during the filling time, while the pressure in the brake cylinder increases and the clamping force P_s increase, the friction coefficient between brake shoes and wheel tread tends both to decrease due to P_s increase and to increase due to the running speed decrease (see fig. 11).

It is to highlight that the instantaneous pressure within the brake cylinder, main variable parameter, depends on several factors and some of the most important are: the pressure's evolution during the filling time of the air brake distributor; the precise moment of reaching the maximum pressure and its value within the brake cylinder, which from that moment, all along the braking action duration, can be considered constant, except the case if antiskid equipments action occurs; the characteristics of the first time duration of braking, defined as the time period of rapid increasing of the brake cylinder pressure, up to approximately 10% of the maximum admitted value. It is considered that only at the end of the first time duration of braking begins to develop an effective brake force for the vehicle.

Railway vehicles are linked each other by different kinds of couplers that must have certain elastic and damping characteristics, because they have remarkable influence not only for the protection of the vehicle's structure and the loading's integrity, but also for the passengers comfort. Generally, the traditional couplers wide used in Europe are composed of a pair of lateral buffers, a traction gear and a coupling apparatus at each extremity of the vehicle. Their characteristics have significant influences for the longitudinal dynamics of the train, with running stability implications. There are specific types of buffers for railway vehicles, their characteristics taking into account the requirements determined by mass, potential collision shocks and passengers comfort, etc. Therefore, there are different constructive solutions, using metallic, rubber, silicon type elastomers, hydraulic, pneumatic or hydro-pneumatic elastic elements.

According to the particular constructive and operational characteristics, the behaviour of buffer and draw-gear devices is quite complex due to several non linear phenomena like variable stiffness-damping, hysteretic properties, preloads of elastic elements, draw-gear compliance, clearance between the buffers discs, etc.

Buffers and draw-gears still widely equipping railway vehicles are based on metallic elastic rings (RINGFEDER type), using friction elements to fulfil the required damping effects.

The general characteristics of these devices mainly depend on the stroke Δx representing in fact the relative displacement between neighbor vehicles and on the relative velocity $\dot{\Delta x}$ and its sign (see fig. 15). The main parameters are: the stiffness $c_{i,j}$, the precompression forces $P_{oc,t}$, the length of the stroke defining the inflexion of the elastic characteristics $\Delta x_{2c,t}$ and the precompression of the elastic elements $\Delta x_{1c,t}$ of the shock and the draw-gear devices. The elastic characteristics $c^*_{i,j}$ might be determined either by experiment, or taking into account the damping depending on the accumulated and dissipated potential deformation energy, according to international regulations.

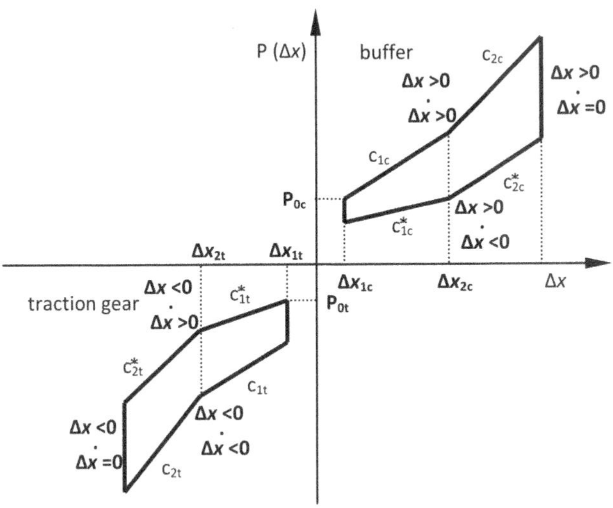

Fig. 15. General characteristics of shock (index c) and traction gear (index t) devices.

For freight wagons there are in use buffers with 75 mm elastic stroke, high capacity buffers with 105 mm stroke and high energy absorption capacity buffers with 130 and 150 mm stroke, while for coaches there are in use buffers with 110 mm stroke (prescriptions in UIC leaflets no. 526-1, 2, 3 and 528).

The resistances of trains and of each railway vehicle are determined by all forces that oppose to their movement. They depend on several factors among which the most important are the type and characteristics of rolling stock, running speed, the track characteristics (longitudinal and vertical profile), direction and intensity of the wind, etc.

Taking into account the divers causes and effects, usually there are considered two kinds of resistances. The main one summarizes all forces acting permanently whenever the train/vehicle is moving in alignment on a horizontal track. The supplementary one are intermittent opposing forces, acting only at certain times and are determined by the circulation in curves, on ramps or slopes, by the wind action, etc. and they are added up algebraically to the main resistances whenever appropriate.

To simplify calculations it is common to use the specific resistances r [N/kN] defined as the ratio between the resistances R [N] and train/vehicle weight Q [kN]:

$$r = \frac{R}{Q} \tag{19}$$

Because the dependency on many factors, for calculate the specific resistance there are usually used empirical relations, established on experimental basis.

For studies regarding the longitudinal dynamics of trains it was previously stated that in-train reactions are mainly dependent on the instantaneous variation of longitudinal forces between vehicles. It is obvious that instantaneous running speed of each vehicle in the train body during braking phases are not absolutely identical, but differences can only be very small due to the permanent interconnection. Consequently, the instantaneous main resistances Rn_i differences are expected to be almost insignificant compared with braking forces $F_{b,i}$ during the brake stages. The same observation is for the supplementary resistances Rs_i as long as the train's vehicles are running altogether on the same track and are submitted to the same atmospheric conditions.

Under these conditions, for theoretical estimations of in-train longitudinal reactions evolution during braking actions, it is expected that results would not be significant altered if not considering the running resistances, all the more for almost similar vehicles. Still, eventually the air drag affecting the first vehicle may interest more.

6.4 Analysis of main pneumatic parameters

Generally the main cause of in-train forces is the instantaneous difference of the various longitudinal forces acting between vehicles. In the case of braking actions it is obvious that braking forces are most important and their time history in the first stages following the braking command is crucial.

While the mandatory basic braking system is the indirect compressed air one, the operation is very complex in order to meet the very demanding specifications to assure the safety and interoperability between different kinds of vehicles. As previously presented in § 3, the braking command is transmitted along the train as pressure reference and the braking system of every single vehicle interacts specifically with the complete pneumatic plant of the train.

That is why an adequate study of the pneumatic processes is important for such studies.

There are two important aspects influencing the longitudinal behaviour of the train submitted to braking actions: the moment when each air distributor begins to command the filling of the brake cylinders and the subsequent evolution of the air pressure in the brake cylinders. Accordingly, there are two different aspects that are usually emphasized: the propagation of braking signal along the brake pipe and the response of the distributor.

The air movement along the brake pipe and through the calibrated orifices, valves and channels of the distributors between the auxiliary reservoir and the brake cylinders is very complex and certain simplifying assumptions are usually considered and there are taken into account factors that have essential role in the evolution of processes.

For such pneumatic system, the main basic simplifying hypotheses are: unidirectional air flows (generally small flowing sections and quite long pneumatic elements make determinant the axial component of air speed and therefore a negligible variation of fluid properties in the normal section to flow direction); air is considered to be perfect gas (permits air status parameters correlation by Mendeleev-Clapeyron equation and is accepted the adiabatic exponent invariance); the flow of air in the pneumatic brake systems is accompanied by heat transfer phenomena between air, pneumatic enclosure walls and pipes and the environment.

Air flow modeling - as compressible fluid - in air brake systems is based mainly on the application of the laws of mass and angular momentum fluid (in a volume control) conservation and on the first two principles of thermodynamics. Also, the model must include the equation of state of fluid, which allows correlation of thermodynamic properties and, moreover, to consider, when appropriate, that movement of mechanical tasks is governed by the second Newtonian mechanics postulate.

Analyzing the evolution of processes consequently a braking command performed by establishing the pneumatic connection between the brake pipe and atmosphere through the driver's brake valve, Karvatski (1950) identified the following stages: the propagation of an air wave along the brake pipe, followed by a pressure drop determining the successive actuating of each air brake distributor according to its sensitivity and the subsequent pressure increase in the brake cylinders in compliance with the filling characteristics.

He explained the air wave propagation mechanism: once opened the pneumatic connection between the brake pipe and atmosphere through the driver's brake valve located at one end of the train, one generates and starts to propagate a stream of air to exit. The opening of the valve breaks down the overall balance of the air in the outlet vicinity and so, successively, the equilibrium of each air layer is broken while losing the support of the foregoing. The beginning of air movement at each point of the pipe determines consequently a beginning of pressure drop. This process is propagating with a certain speed w_{aw} [m/s] in the brake pipe to the other end of it:

$$w_{aw} = \sqrt{\frac{\chi \cdot p}{\rho}} \tag{20}$$

depending on the absolute air pressure p [N/m^2], density ρ [kg/m^3] and adiabatic exponent χ.

After the passage of the air wave, in each point of general pipe the air pressure begins to decrease, depending mainly on the brake pipe length l [m], on the distance between the driver's brake valve and the considered point l_x [m], on the air wave propagation speed w_{aw} [m/s] and the medium pressure drop totally generated Δp [N/m^2].

Corresponding to his simplified model, Karvatzki (1950) determined the local pressure drop rate:

$$v_x = \frac{1}{2} \cdot \frac{\Delta p}{l} \cdot w_{aw} \cdot e^{-n \cdot \frac{l_x}{w_{aw}}} \tag{21}$$

where n is a logarithmic decrease factor showing the decrease of local pressure drop along the brake pipe, factor to be experimentally determined.

Analyzing the model, it appears that the main causes of the decrease of local pressure drop along the brake pipe are determined by the distributed pressure loss in the pipe and by the local pressure loss due to the air flow changes of directions. So, eq. (21) can be improved considering $n = 1$ and taking into account the pressure losses.

Distributed pressures losses occur whenever a fluid flows in a relatively long pipe having a comparatively small cross section, due to viscous friction between parallel layers of air, in this case.

The distributed pressure loss coefficient for a pipe of length l [m] and inner diameter d [m] is:

$$\xi_d = \lambda \cdot \frac{l}{d} \tag{22}$$

where λ is the coefficient of Darcy, determined experimentally according to the flow regime and the inner surface roughness or, for a laminar flow regime and smooth inner surface of the pipe, can be calculated using Poiseuille's relationship as a function of the Reynolds number Re:

$$\lambda = \frac{64}{\mathrm{Re}} \tag{23}$$

For the case of a straight circular pipe, if laminar flow regime, the pressure loss between two flow sections situated at the distance l_d [m] along pipe can be calculated based on Hagen-Poiseuille relationship:

$$\Delta p_d = 32 \cdot \eta \cdot \frac{4 \cdot l_d}{\pi \cdot d^4} \cdot \frac{\overset{o}{m}}{\rho_{med}} \tag{24}$$

where $\overset{o}{m}$ [kg/s] is the air flow rate passing through the pipe and ρ_{med} [kg/m^3], η [kg/m·s] the medium density, respectively dynamic viscosity of the air in given conditions. The temperature T [K] air viscosity dependence is:

$$\eta = \eta_o \cdot \frac{273.15 + \mathrm{C}}{T + \mathrm{C}} \cdot \left(\frac{T}{273.15}\right)^{\frac{3}{2}} \tag{25}$$

where $\eta_o = 17.09 \cdot 10^{-6}$ [kg/m·s] is the dynamic viscosity of air at 0° C, and C is Sutherland's constant C = 112 (for air).

Considering air as perfect gas and taking into account the constants values, eq. (23) becomes:

$$\Delta p_x \approx 0.017 \cdot \frac{l_x \cdot \overset{o}{m} \cdot T^{\frac{3}{2}}}{(T+112) \cdot p_{med} \cdot d^4} \ [\mathrm{N/m^2}] \tag{26}$$

permitting an approximate evaluation of the distributed pressures losses at a l_x [m] distance from the driver's brake valve along the brake pipe.

The local pressure losses mainly manifest in the zone of flexible coupled hoses, their curvature that determins the air stream to change the flow direction affecting the braking wave propagation along the train's brake pipe.

Generally, the local pressure losses depend on the local pressure loss coefficient ξ_{loc}, on the air density ρ [kg/m³] and fluid velocity v_{air} [m/s] and can be determined with the relation:

$$\Delta p_{loc} = \frac{\xi_{loc} \cdot \rho \cdot v_{air}}{2} \tag{27}$$

In the case of a pair of flexible coupled hoses, considering the interior diameter d [m] and the curvature medium radius R [m], the local pressure loss coefficient can be calculated with relation:

$$\xi_{loc} = 0.131 + 0.163 \cdot \frac{d}{R} \tag{28}$$

The medium speed of air flow in the brake pipe of S [m²] interior cross section may be determined from the continuity equation, depending on the air flow rate passing through the pipe $\overset{o}{m}$ [kg/s] and density ρ [kg/m³], resulting:

$$v_{med,air} = \frac{\overset{o}{m}}{S \cdot \rho} \tag{29}$$

Consequently, according to eq. (27), (28) and (29), considering the mean values of parameters involved, the local pressure losses for the "i" vehicle of the train are:

$$\Delta p_{loc,i} = (i-1) \cdot \frac{\left(0.131 + 0.163 \cdot \frac{d}{R}\right) \cdot \overset{o}{m}}{2 \cdot S} \tag{30}$$

On the other hand, air brake distributors are characterized by certain levels of sensitivity and insensitivity that determine the beginning of brake operation consequently to local pressure variation in the brake pipe.

Sensitivity can be defined as a minimum threshold pressure variation Δp_{sens} in the brake pipe determining the air distributor coming into action, meaning determining the fill of brake cylinders it commands. It is advisable a quite high sensitivity, so that the brakes easily begin operate and to diminish both air consumption and duration of releasing actions. Air distributors take action only if the local pressure drop gradient v_{lim} permits, in a finite time t, to get:

$$\Delta p_{\min} = \Delta p_{sens} < v_{\lim} \cdot t \tag{31}$$

Still, because technically is difficult to maintain a perfect tightness of the brake pipe, it is not advisable for air distributors to be extremely sensitive, whereas the lowest random pressure drop may cause an unsolicited brake application. Therefore, it is also characteristic a certain insensitivity level, in the sense of a maximum local rate of pressure variation in the brake pipe which does not determine brakes to come into action. Thus, the brakes do not operate if the local pressure drop gradient has a sufficient low value in the vicinity of the air distributor so that:

$$\Delta p_{\max} = \Delta p_{insens} > v_{\lim} \cdot t\,|_{t\to\infty} \tag{32}$$

For determining the moment of operation start for the distributor of "i" vehicle situated at $l_{x,i}$ [m] distance from the drivers brake valve, the time elapsed since the braking command has been performed is:

$$t_{i,x} = t_{aw,x} + t_x + \Delta t_{d,x} + \Delta t_{loc,i} \tag{33}$$

In eq. (33), $t_{aw,x}$ [s] is the duration of air wave propagation up to the considered air distributor, t_x [s] the necessary time for air pressure drop to the sensitivity level in ideal conditions, $\Delta t_{d,x}$ and $\Delta t_{loc,i}$ [s] representing the time to compensate the distributed, respectively local pressure losses.

According to previously presented processes and correspondent equations, the moment of acting for each "i" vehicle air distributor, situated at $l_{x,i}$ distance from the drivers brake valve can be determined (Cruceanu, 2009):

$$t_{i,x} \approx \frac{l_{x,i}}{\sqrt{\frac{\chi \cdot p}{\rho}}} + \frac{2 \cdot l \cdot}{\Delta p \cdot w_{aw} \cdot e^{-\frac{l_{x,i}}{w_{aw}}}} \cdot \left\{ \Delta p_{sens} + \overset{o}{m} \cdot \left[\frac{0.017 \cdot l_{x,i} \cdot T^{\frac{3}{2}}}{(T+112) \cdot p_{med} \cdot d^4} + \frac{(i-1) \cdot \left(0.131 + 0.163 \cdot \frac{d}{R}\right)}{2 \cdot S} \right] \right\} \tag{34}$$

In particular, considering the adiabatic exponent for humid air $\chi = 1.405$, the interior diameter of the brake pipe $d = 25$ mm and the radius of flexible coupled hoses assembly between two successive vehicles $R = 1$ m, eq. (34) becomes:

$$t_{i,x} \approx \frac{l_{x,i}}{1.185 \cdot \sqrt{\frac{p}{\rho}}} + \frac{2 \cdot l \cdot}{\Delta p \cdot w_{aw} \cdot e^{-\frac{l_{x,i}}{w_{aw}}}} \cdot \left\{ \Delta p_{sens} + \overset{o}{m} \cdot \left[\frac{0.017 \cdot l_{x,i} \cdot T^{\frac{3}{2}}}{(T+112) \cdot p_{med} \cdot d^4} + (i-1) \cdot 96.3 \right] \right\} \tag{35}$$

As a numerical example, relation (35) was applied for the case of a passenger train composed of locomotive (20 m length) and 10 coaches (25 m length each), coupled through flexible hoses (0.5 m length each). The air distributors were considered placed at the middle of each vehicle. The initial relative pressure in the brake pipe was considered 5 bar (brakes released) and the necessary pressure variation in the brake pipe to determine the air distributor coming into action $\Delta p_{sens} = 0.3$ bar. It was considered an emergency braking action, the medium air flow rate evacuated from the brake pipe in first phases of the process being 0.7 kg/s. The main results are shown in fig. 16. For the particular case presented, the brakes of the last vehicle in train begin to operate 1.9 s after the brake command given from the drivers brake valve situated in the front of the locomotive (first vehicle of the train).

Even if the propagation of braking wave along the train seems to be almost linear, the relative time differences show that the propagation rate slows down along the train, as expected according to previously presented arguments.

The calculated medium propagation rate of the braking wave is in that case 245.5 m/s, respectively 266 m/s if considering as reference the moment when the first air distributor begins to supply the first brake cylinder. These values are consistent with the rigors of international regulations and with general evolution of the process within the brake pipe.

It is to notice that our model does not take into account the operation of emergency braking accelerators, so it represents the minimum propagation rate for the studied case. If the main purpose is to determine the maximum values of longitudinal dynamic reactions, then the use of the model conducts to estimates that cover the real processes. This affirmation is based on the fact that shorter the relative difference in time is, lower the braking forces differences between the train's vehicles are.

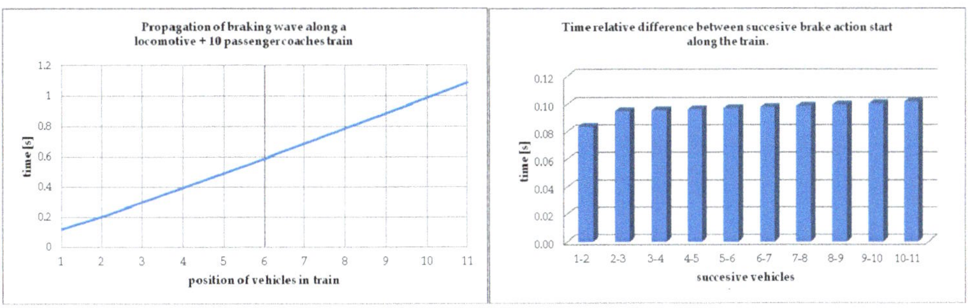

Fig. 16. Propagation of braking wave along the a short train (particular theoretical case study).

The other important target of pneumatic studies regarding the in-train forces developed during braking actions is the distributor valve, a complex pneumo-mechanic device which is devoted to control the brake response on every vehicle according to the air pressure variation in the brake pipe.

Due to the importance of the air brake distributors response to the pneumatic signals transmitted through the brake pipe, there were performed numerous theoretical and experimental studies regarding these aspects, based on more or less simplifying hypothesis, all tending to highlight mainly the brake cylinders filling characteristics as accurate as possible (Belforte et al., 2008; Cantone et al., 2009; Piechowiak, 2010; Pugi et al., 2004, 2008, etc.).

Generally, the pneumatic behaviour of the air distributor may be modelled as a lumped system of chambers and orifices, but the influences of numerous mechanic elements (pistons, valves, springs, etc.) are also important and these are more difficult to integrate in a functional model. Still, the effect of the distributor is the brake cylinders filling characteristics which are mainly influenced by the nozzles that determine the filling (and releasing) time duration.

Under these conditions, a simpler way is to emphasize the role of calibrated orifices, considering them the main pneumatic resistance of the distributor. In that case, the air mass flow $\overset{o}{m}$ [kg/s] between the auxiliary reservoir and the brake cylinder may be determined considering in a first stage the nozzles as ideal convergent nozzles and affecting with a correction coefficient α_c that takes into account that in real circular cross section nozzles the minimum flow section is smaller than the geometric orifice section (see fig. 17) and pressure losses at the entrance lead to differences between actual and theoretical air flow speed through the calibrated orifice.

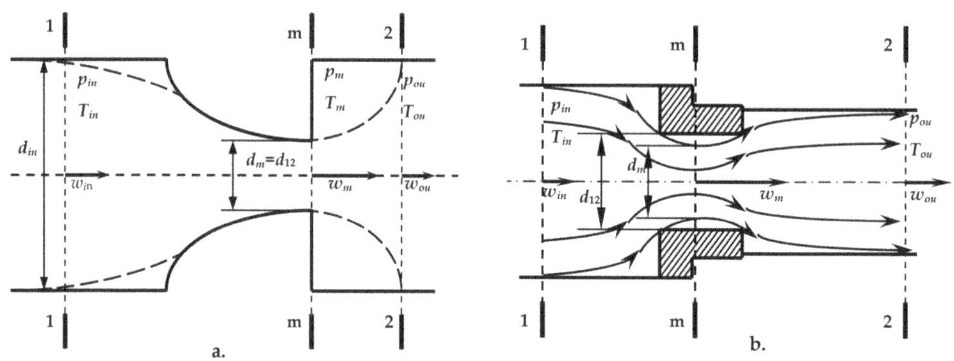

Fig. 17. Model of nozzle: a - ideal convergent nozzle; b – real circular calibrated orifice.

According to the air flow regime, the air mass flow can be determined for subsonic:

$$\overset{o}{m}_{12} = \alpha_c \cdot \frac{\pi \cdot d_{12}^2 \cdot p_{in}}{\sqrt{T_{in}}} \cdot \sqrt{\frac{\chi}{8 \cdot R_a \cdot (\chi - 1)} \cdot \left[\left(\frac{p_{ou}}{p_{in}} \right)^{\frac{2}{\chi}} - \left(\frac{p_{ou}}{p_{in}} \right)^{\frac{\chi+1}{\chi}} \right]} \quad (36)$$

and for the supersonic one:

$$\overset{o}{m}_{12} = \alpha_c \cdot \frac{\pi \cdot d_{12}^2 \cdot p_{in}}{\sqrt{T_{in}}} \cdot \sqrt{\frac{\chi}{16 \cdot R_a} \cdot \left(\frac{2}{\chi + 1} \right)^{\frac{\chi}{\chi-1}}} \quad (37)$$

where: d_{12} [m] is the minimum geometric diameter of the nozzle, p_{in} and p_{ou} [N/m²] is the air inlet, respectively outlet pressure, T_{in} [K] the inlet air absolute temperature, χ the adiabatic exponent, $R = 287.12$ [J/kg·K] the constant of the air.

The air flow regime can be appreciated comparing the outlet/inlet pressures rapport according to the critical flow point $\beta_{cr} = \left(\frac{p_m}{p_{in}}\right)_{cr} = \left(\frac{2}{\chi+1}\right)^{\frac{\chi}{\chi-1}}$ as follows: if $\beta_{cr} < p_{ou}/p_{in} \le 1$ than the flowing regime is subsonic and if $0 < p_e/p_{in} \le \beta_{cr}$ is sonic (for equality) or supersonic, in both last cases the air mass flow is maximum possible in given conditions and can be determined using eq. (37).

Generally, eq. (36) and (37) are quite accurate according to the specific filling time imposed by regulations and can be used for initial dimensioning of diverse calibrated orifices used in the construction of various braking pneumatic devices for ensuring controlled pressure variations in certain pneumatic chambers. Still, when air flow is also mechanically controlled through valves actuated by pretensioned springs and commanded through the pressure acting on the piston, the air flow rate is certainly dependent on the opening height of the valve, too.

In the case of "normally open" valves (see fig. 18, a), the instantaneous opening height h depends on: the initial $h_{\max}$ opening, the stiffness c_s of the actuating spring, the diameter of the active surface of the valve's command piston d_p, the instantaneous relative air pressure p_c in the control chamber, on the initial valve spring prestressing force F_p:

$$h = h_{\max} - \frac{\frac{\pi \cdot d_p}{4} \cdot p_c - F_p}{c_s} \tag{38}$$

For instance, in the particular situation of the maximum pressure valve that solely controls the pneumatic link between the auxiliary reservoir and the brake cylinder at emergency braking command in the case of KE air distributors, considering only the calibrated orifices, the theoretical filling characteristics determined using only eq. (36) and (37) are presented in fig. 18, b with continuous line. When considering the decrease of air flow ratio determined by the closing valve using also eq. (38), the filling characteristics become much closer to reality (dotted lines in fig. 18, b).

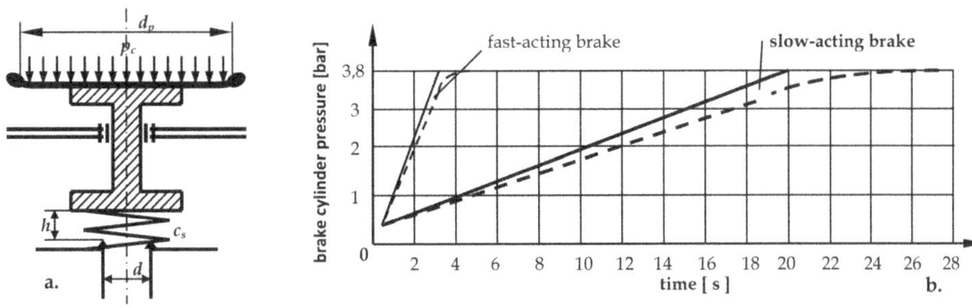

Fig. 18. Influence of valve opening on filling characteristics: a – schematic of pressure-spring controlled valve; b – theoretical filling characteristics (dotted line: valve controlled filling).

6.5 Longitudinal dynamic reactions in passenger trains submitted to braking actions

Studies regarding the longitudinal dynamics of trains during braking actions are mainly focused on long, heavy freight trains, due to the more obvious effects determined by the length of the brake pipe and numerous big masses interconnected (Belforte et al., 2008; Karvatski, 1950; Nasr & Mohammadi, 2010; ORE-Question B 36, 1972,1980; Pugi et al., 2007; Zhuan, 2006; etc).

Comparatively, issues regarding the longitudinal dynamic reactions in passenger train body seem to be less important. In fact, these are generally short, having a constant and much uniform composition than freight trains and there are sufficient arguments to support these assertions, e.g. passenger railcars are typically two axles bogies vehicles and have almost the same length, the mass difference between an empty and fully charged coach is significantly lower.

Still, there are arguments to prove not only the complex evolution of dynamic in-train reactions during braking actions, but also that there may exist circumstances in which these forces can become significant. Some relevant key features regarding braking actions in the case of passenger trains are highlighted below:

- running speed is significantly higher than in the case of long freight trains, so the energy dissipation gradient during braking is consequently greater;
- vehicles are mandatory equipped with fast acting air brake which, according to the admitted limits of filling time, determine an increasing pressure gradient in the brake cylinders of 0.7 - 1.2 bar/s, compared with 0.1 - 0.2 bar/s in case of slow action ones used in case of long freight trains. Consequently, during the first train braking phases, the instantaneous differences among braking forces become higher and are the premise of larger in-train forces;
- in the case of classical passenger trains, the locomotive weight is consistently larger compared to each passenger carriage, determining a pronounced nonlinear in-train body mass distribution: depending on train composition, the locomotive's mass may represent even 40...50 % of the whole train, constituting a large concentrated mass placed in a extremity. Therefore it is expected that longitudinal in-train dynamic reactions should be seriously influenced. More than that, it is an obvious tendency for train combinations in push-pull operations. In that case, the train can be driven either from the locomotive or the alternate cab. If the train is heading in the direction in which the locomotive of the train is facing, this is considered "pulling" and if the train is heading in the opposite direction, this is considered "pushing", the driver being located in the alternate cab. In that case, the longitudinal dynamics of the train deserves even greater attention not only in braking, but also in traction regimes, especially in pushing operations, when derailment risks are increased, mainly on switches;
- the random action of wheel slip prevention equipments, specific for passenger rail vehicles, may determine, in condition of poor wheel-rail adhesion, important and rapid braking forces variations between the train vehicles, increasing the dynamic longitudinal reactions;
- the action of the electromagnetic track brakes, as complementary system mandatory for running speeds exceeding 160 km/h, operating only in emergency braking actions, can induce supplementary longitudinal in-train forces in the case of non simultaneous releasing at the imposed running speed (see functioning principles in § 4);

- when using eddy current brakes, in particular the linear one, the air gap variation can determine random instantaneous brake forces variations between vehicles, determining an enhancement of longitudinal in-train reactions, etc.

It is also to highlight that in the case of passenger trains, the comfort is diminished during braking actions also due to the longitudinal shocks determined by the longitudinal in-train reactions.

At least these aspects conduct to the idea of studying the problem also for this type of trains. In order to determine the influence and effects of previously presented aspects, a team formed of members of the Rolling Stock Department from the Faculty of Transports of University *POLITEHNICA* of Bucharest developed certain theoretical studies regarding the longitudinal dynamic reactions for the case of passenger trains during braking actions (Cruceanu, 2009).

Dedicated software was created and numerical simulations for different case models were carried out in Matlab with the solver ode45, until numerical stability and reasonable results were reached, the relative tolerance of the solver being finally set to 10^{-9} (Cruceanu et al., 2009).

The mechanical model of the train is a classical elastic-damped lumped system according to § 6.2 and neglecting additional degrees of freedom due to bogies, suspension system, wheel-rail interaction, etc. This simplified our approach and was used to predict wagons and train compositions with the worst loading condition during the braking phases. In elaborating the general model, several initial simplifying assumptions were admitted, generally focused on considering that:

- vehicles are equipped with lateral buffers having a 110 mm stroke, according to UIC leaflet no. 528, constructively RINGFEDER type ones, meaning friction damping for these devices and the same for traction apparatus, according to UIC leaflet no. 520;
- the initial compressions of the elastic elements of the shock and traction devices were neglected;
- a tight coupling between the component vehicles, as regulated for passenger trains;
- an average steady braking wave propagation speed along the train of 250 m/s, the minimum imposed regulated value;
- exploitable braking forces develop only after reaching an approx. 0.4 bar pressure within the brake cylinder and once the pressure gets its maximum value, it remains constant during the whole braking process;
- the vehicles main resistances, which are mainly depending on the running speed, as well as the supplementary resistances, were neglected because during the braking process the relative instantaneous differences between the vehicles of the train are almost negligible in comparison with the other forces variations taken into account.

With the aim of obtaining accurate results, the conceived soft offers the possibility of using brake cylinders pressure information either directly from a complex computerized system for testing pneumatic braking equipment of rail vehicles, developed within the Scientific Research AMTRANS Program (2005/2008), or mathematical approximation functions.

For simulate diverse filling characteristics, based on the acquisitioned data (see fig. 19, a), it was first determined an interpolation polynomial function that approximates accurate enough the air pressure evolution within the brake cylinder for the interest domain of $\Delta t = 2.86$ s:

$$p_{bc,3.16}(t) \approx -0.033 \cdot t^6 + 0.31 \cdot t^5 - 1.081 \cdot t^4 + 1.53 \cdot t^3 - 0.413 \cdot t^2 + 0.88 \cdot t + 0.4 \text{ [bar]} \quad (39)$$

This function was extrapolated for the case of 4 s and 5 s filling time (see also fig. 19, b):

$$p_{bc,4}(t) \approx -0.005 \cdot t^6 + 0.063 \cdot t^5 - 0.3 \cdot t^4 + 0.59 \cdot t^3 - 0.22 \cdot t^2 + 0.643 \cdot t + 0.4 \text{ [bar]} \quad (40)$$

$$p_{bc,5}(t) \approx -0.001 \cdot t^6 + 0.016 \cdot t^5 - 0.103 \cdot t^4 + 0.26 \cdot t^3 - 0.127 \cdot t^2 + 0.49 \cdot t + 0.4 \text{ [bar]} \quad (41)$$

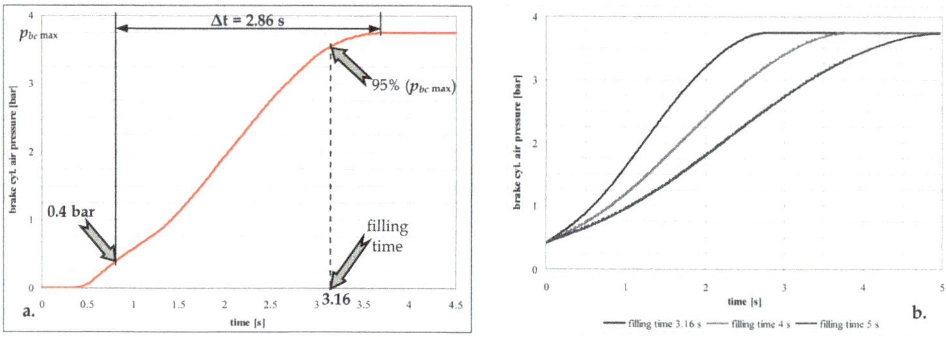

Fig. 19. Air distributor filling characteristics: a - experimentally determined; b - extrapolated functions.

The model for the shock and traction apparatus had to take into account the characteristics shown in fig. 15 which reveal the action of a friction force. Most commonly used is the Coulomb friction model which can be formulated as:

$$P = \begin{cases} F_c \cdot sign(v) \ if \ v > 0 \\ F_{app} \ if \ v = 0 \ and \ F_{app} < F_c \end{cases} \quad (42)$$

where P is the friction force, $v = \dot{x}$ the relative speed and F_{app} the applied force on the body. F_c is the Coulomb sliding force classically defined as $F_c = \mu N$. Due to the properties of the sign function, these equations couldn't be used in the simulation because the numerical method proved to be unstable. A solution was to replace tanh function with signand so the new model became valid for any value of the speed v as shown in the following relation:

$$P = F_c \cdot \tanh(k_{\tanh} \cdot v) \quad (43)$$

where k_{tanh} is a coefficient that determines how fast the tanh function changes from near -1 to near +1. Still, even if the model becomes more numerically stable, it has the disadvantage that assumes zero friction force at zero relative speed, meaning the acceptance that friction force exists only when there is a motion (Andersson et al., 2007).

More than that, as shown in fig. 15, the friction force in the shock and traction devices also depends on the applied force. This was taken into account considering that the Coulomb sliding force has the form:

$$F_c = \Delta c \cdot x \tag{44}$$

where x is the relative displacement between the cars and Δc is a constant which acts like a variation of the springs medium rigidity c_m. Finally, as the buffers and the traction apparatus have different rigidities, a ponder function p was used to modify the value of the force:

$$p = \tanh(k_{\tanh} \cdot x) \tag{45}$$

and the relation for the reactions between the cars, depending on the relative displacement x and velocity v is:

$$P = \frac{-\left[c_{mc} - \Delta c_c \cdot \tanh(k_{\tanh} \cdot v) \cdot x\right] \cdot (1-p) + \left[c_{mt} + \Delta c_t \cdot \tanh(k_{\tanh} \cdot v) \cdot x\right] \cdot (1+p)}{2} \tag{46}$$

The index c is for the buffers, while t is for the traction device. It was assumed that a negative value for x means that the apparatus is compressed (the force is given by the buffers) and a negative value of v means that the cars tend to come closer.

For example, studies were performed taking into account different passenger trains of 4...10 vehicles, for usual possibilities: train sets and classical trains with locomotive in pulling and pushing operations.

The passenger cars were considered having individual masses of 40...70 t, corresponding to usual weights for: coaches, dining, lounge, sleeping, baggage, etc. cars, double-decker included, in various in-train combinations, as for the locomotives, there were considered on four or six axles and counting on 80 and 120 t, respectively.

It was considered that trains are submitted to an emergency braking action started at a running speed of 180 km/h, being active only the classical UIC disc brake system.

It was considered that during braking action, no wheel slip prevention equipment operates.

There were considered only air distributors correctly operating and ensuring filling times between 3...5, as imposed by international regulations for fast-acting brake systems.

Some results of simulations are presented in fig. 20.

Relevant aspects regarding the longitudinal dynamic forces evolution for passenger trains submitted to braking actions emerged from the analysis of the results obtained based on simulations performed corresponding to previously mentioned cases are presented below. In order to appreciate the effects of different parameters, we referred mainly to the maximum compression, respectively traction in-forces, which have practical importance in offering an image of the maximum efforts the couplings are submitted.

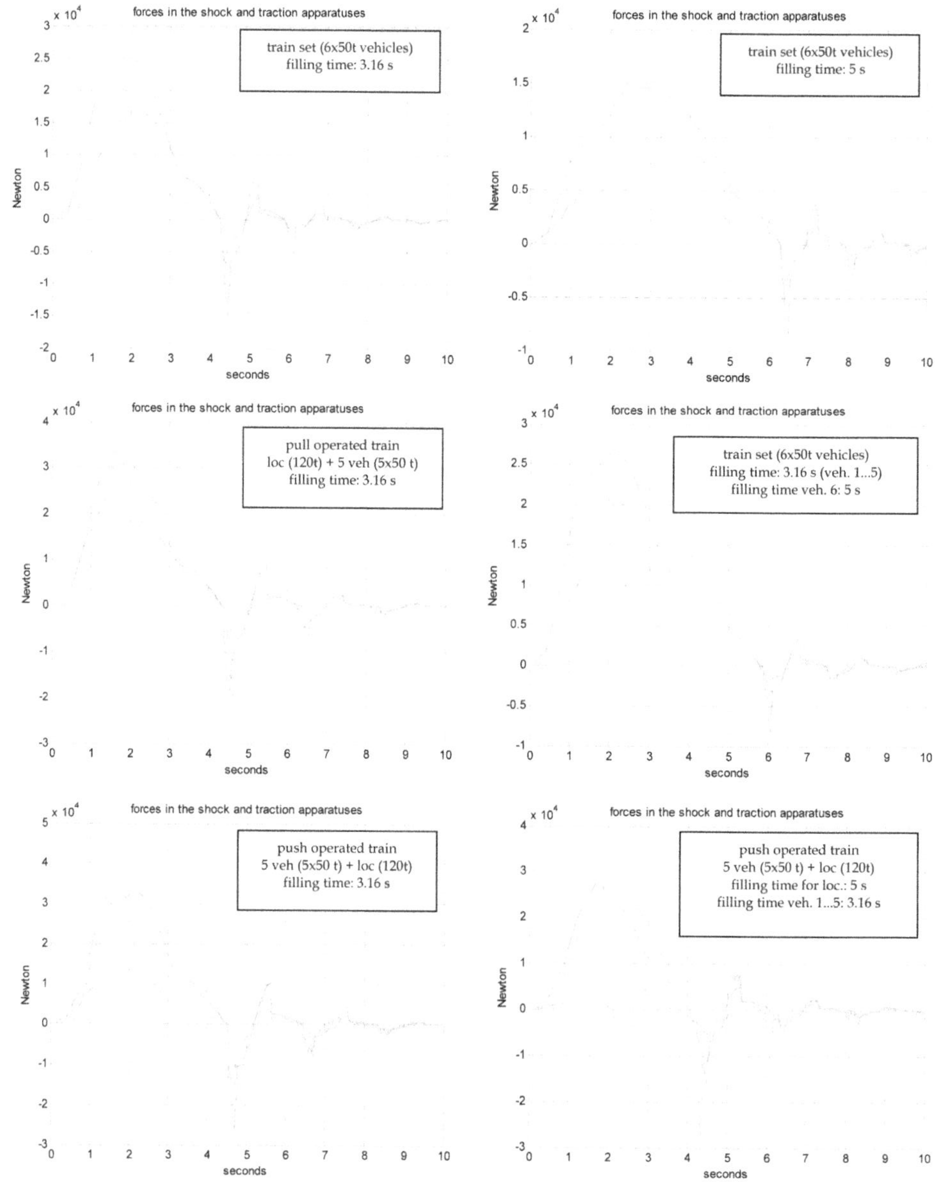

Fig. 20. Time-history of in-train longitudinal dynamic reactions between vehicles in passenger trains for different cases

So, the evolution of in-train reactions is concordant to the theoretical expectations, meaning that compression forces increase during the first braking phases, reaching a maximum value at the end of this phase, most of the times at about half of the train. In the second phase, these compression forces begin to decrease, while the braking forces are successively increasing along the train. However, in concordance with the brake cylinder filling characteristic, the differences of instantaneous braking forces along the train begin to decrease. The forces in the buffers are therefore receding, varying slowly while the relative displacements remain almost constant. The process is more accentuated in the third phase when, while maximum braking forces are achieved along the train, tensile forces increase, the maximum value exerting most of the times also at about half of the train. During the last braking phase, though the braking forces are of the same magnitude along the train, a longitudinal oscillation movement of the vehicles begins and propagates along the train, due to the potential energy accumulated in the buffers during the train's compression. So, longitudinal forces continue their evolution, acting successively on the traction and shock apparatus. The amplitude of this wave decreases due to the energy dissipation corresponding to the friction forces acting in the buffers and the traction devices.

A remarkable feature of the oscillations period is the fact that about a half of a complete cycle is characterized by considerably larger forces than the other half (see fig. 20). The reason is that when the relative displacement is negative, the stiffness of the shock and traction devices is considerably greater than the opposite situation. The friction forces increase or decrease their value, depending on the sign of the relative speed. In consequence, each cycle has a "sharp" part and a "soft" one, as a result of the combination of friction forces and variable rigidities.

It is also interesting to note that in the simulations there appear, more or, sometimes, less evident, but visible in the presented diagrams (see fig. 20), the oscillatory motion that propagates in the train, overlapping on the preexistent compression after the end of the first braking phase.

Regarding the influence of brake cylinders filling time, the first observation is that generally, higher the pressure increase gradient is, both maximum compression and traction in-train forces increase. For the cases of same filling times for each vehicle of the train, the maximum longitudinal dynamic forces diminish their magnitude in average about 24...43% while increasing the filling time from 3.16 s to 4, respectively 5 s. However, the disposition of the maximum forces remains almost the same along the train. The situation changes if the vehicles of the same train are characterized by different filling times. In that case, the magnitude of the dynamic longitudinal reactions and their layout along the train strongly depends on the position and filling time of each vehicle within the train. Longer filling times, wherever placed in the braked train, diminish the traction longitudinal dynamic forces. This assertion is also correct regarding the compression forces, but only if the vehicles having longer brake cylinders filling times are placed in the first half of the train. Otherwise, the longitudinal compression dynamic forces increase, the maximum values being attempted between the vehicles situated in the second part of the train. Anywise, at least for the studied case, in spite of the almost spectacular forces evolutions due to different filling times, their magnitude can not affect severely the shock, traction and coupling apparatuses.

Regarding the vehicles mass and length of the train, in case of uniform composition, the increase of both parameters determine higher dynamic reactions, both for compression and traction forces, maximal values exerting mainly between vehicles situated in the middle of the train. It is to notice that the evolution and distribution of in-forces along the train are similar, indifferent the masses of the identical vehicles are.

The dynamic longitudinal reactions have an almost linear dependence on the mass of the component vehicles in the considered train set. The maximum traction forces are always lower than the correspondent compression forces, in similar conditions. Still, they have a more considerable relative increase in respect to the relative growth of the vehicles mass.

For uneven composition of the train, important variations of the dynamic compression and traction reactions occur, both as magnitude and distribution among the vehicles along the train.

In the case of classical trains, the evolution and the distribution of longitudinal dynamic forces along the train are almost similar to train-sets for the middle part of the train. Still, in the extremities, a heavier locomotive (120 t) determines higher values between the front vehicles, while a lighter one (80 t) conducts to an increase of these reactions between the rear vehicles. In the case of push-operated trains, the maximum longitudinal forces are generally higher and their distribution between the vehicles is substantially modified. Forces become more important in the second part of the train.

An interesting and useful approach in analyzing the influences of various parameters on the dynamic longitudinal reactions between the vehicles of passenger trains submitted to braking actions is based on relative percentage forces variation.

For example, studying the compression and traction forces exerted in the couplers of three types of six vehicle trains (train-set, train in "pulling" and in "pushing" operation with 120 t locomotive) by reporting to the simulation results to the case of the train-set, the repartition and modification of maximum reactions in the train body become more obvious, enhancing the effects (see fig. 21).

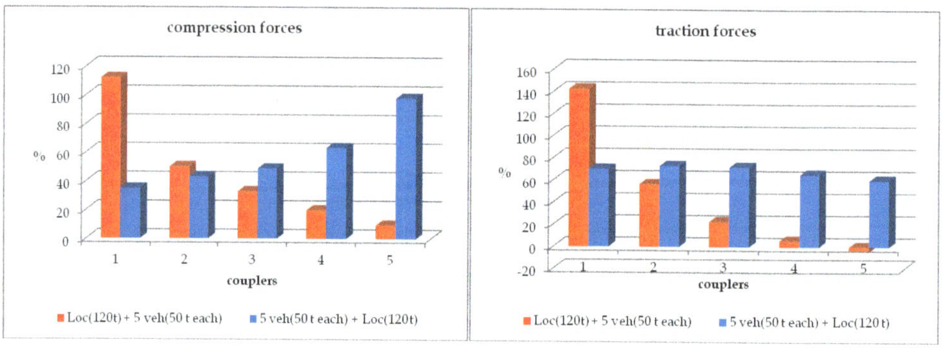

Fig. 21. Relative evolution of maximum longitudinal dynamic forces in braking actions in "pulling" and "pushing" operations reported to similar train-set

Such relative approaches indicate more clearly the important increase of in-train forces in the vicinity of the locomotive, having larger mass than the rest of the train's vehicles. Also, even if in absolute values the maximum traction forces are lower than the compression ones, their relative increase in the case of classical trains is much higher for heavier locomotive in pull operation.

It is thus obvious that the dynamic longitudinal response of passenger trains submitted to braking actions is very complex and the magnitude and distribution of compression and traction in-train forces are strongly influenced by the type, composition and mass distribution along the train. Also, the functional characteristics of the braking devices, mainly the air distributors, determine specific reactions, which are all the more influenced by their repartition in the body of the train.

Specific parameters may enhance the dynamic longitudinal reactions between the vehicles of passenger trains, either in the train assembly, or in particular sections. Results of studies regarding these problems may conduct to interesting and useful recommendations for designers and manufacturers of vehicles for passengers and not the least for the operating staff, both in terms of composition and driving passenger trains, enhancing the security of operations and comfort.

7. References

Andersson, S.; Söderberg, A. & Björklund, S. (2007), Friction models for sliding dry, boundary and mixed lubricated contacts, *Tribology International*, 40, ELSEVIER, pp. 580-587, ISSN: 0301-679X

Belforte, P.; Cheli, F.; Diana, G. & Melzi, S. (2008), Numerical and experimental approach for the evaluation of severe longitudinal dynamics of heavy freight trains, *Vehicle System Dynamics*, 46: 1, pp. 937 – 955, ISSN: 0042-3114; 1744-5159

Cantone, L.; Crescentini, E.; Verzicco, R. & Vullo, V. (2009), A numerical model for the analysis of unsteady train braking and releasing manoeuvres, *Proc. IMechE Vol. 223 Part F: J. Rail and Rapid Transit*, pp. 305-317, ISSN: 0954-4097; 2041-3017

Cruceanu, C. (2009), *Brakes for Railway Vehicles* (in romanian), 3rd edition, Ed. MATRIXROM, Bucharest, ISBN 978-973-755-200-6

Cruceanu, C.; Oprea, R.; Spiroiu, M.; Craciun, C. & Arsene, S. (2009), Computer Aided Study Regarding the Influence of Filling Characteristics on the Longitudinal Reaction within the Body of a Braked Train, *Recent Advances in Computers, Proceedings of the 13th WSEAS International Conference on Computers*, Rodos Island, Greece, July 23-25, 2009, pp 531-537, ISSN: 1790-5109

Givoni, M. & Banister, D. (2009). *Reinventing the wheel - planning the rail network to meet mobility needs of the 21st century*, Transport Studies Unit, Oxford University Centre for the Environment, http://www.tsu.ox.ac.uk/

Karvatski, B.L. (1950), *General theory of automated brakes* (in romanian), OPED-C.F.R., Bucharest, 1950

Nasr, A. & Mohammadi, S. (2010), The effects of train brake delay time on in-train forces, *Proc. IMechE Vol. 224 Part F: J. Rail and Rapid Transit*, pp. 523-534

Piechowiak, T. (2010) Verification of pneumatic railway brake models, *Vehicle System Dynamics*, 48: 3, pp. 283 – 299, ISSN: 0042-3114; 1744-5159

Pouillet, P. (1974), Expérimentation par la SNCF, d'un frein électromagnétique à courents Foucault, *R.G.C.F.*, 93-éme anné nr. 3, pp. 169-176, ISSN: 0035-3183

Pugi, L.; Fioravanti, D. & Rindi, L. (2007), Modelling the longitudinal dynamics of long freight trains during the braking phase, *12th IFToMM World Congress, Besançon (France)*, June18-21, http://iftomm.appsci.queensu.ca/proceedings/proceedings_WorldCongress/WorldCongress07/articles/sessions/papers/A822.pdf

Pugi, L.; Malvezzi, M.; Allotta, B.; Bianchi L. & Presciani, P. (2004), A parametric library for the simulation of a Union Internationale des Chemins de Fer (UIC) pneumatic braking system, *Proc. IMechE Vol. 218 Part F: J. Rail and Rapid Transit*, pp. 117-132, ISSN: 0954-4097; 2041-3017

Pugi,L.; Palazzolo, A. & Fioravanti, D., (2008), Simulation of railway brake plants: an application to SAADKMS freight wagons, *Proc. IMechE Vol. 222 Part F: J. Rail and Rapid Transit*, pp. 321-329, ISSN: 0954-4097; 2041-3017

Schykowski, J. (2008), Eddy-current braking: a long road to success, Railway Gazette International, http://www.railwaygazette.com/news/single-view/view/eddy-current-braking-a-long-road-to-success.html

Sookawa, H.; Saito, T. & Shimizu, K. (1970), Experimental Results for Temperature Rise of the Rail when Applying Eddy Current Rail Brake, *Quarterly Reports*, vol 11, nr. 1, pp. 40-41, ISSN: 0033-9008

Sookawa, H.; Sato, Y. & Tomizawa, M. (1971), Running Tests of Eddy Current Rail Brake Set to a New Test Electric Car, *Quarterly Reports*, vol 12, nr. 3, pp. 138-141, ISSN: 0033-9008

Uherek, E.; Halenka, T.; Borken-Kleefeld, J.; Balkanski, Y.; Berntsen, T.; Borrego, C.; Gauss, M.; Hoor, P.; Juda-Rezler, K.; Lelieveld, J.; Melas, D.; Rypdal, K. & Schmid, S. (2010), Transport impacts on atmosphere and climate: Land transport, *Atmospheric Environment* 44, pp. 4772-4816, ELSEVIER, ISSN: 1352-2310

Zobory, I.; Reimerdes, H.G. & Békefy, E. (2000), Longitudinal Dynamics of Train Collisions-Crash Analysis, *7th Mini Conf. on Vehicle System Dynamics, Identification and Anomalies*, Budapest, Nov. 6-8, pp. 89-110, http://www.railveh.bme.hu /Publikaciok/Zobory-Reim-Bek-Mar-Nem_Mini-7.pdf

Zhuan, X., *Optimal Handling and Fault-tolerant Speed Regulation of heavy Haul Trains*, Doctoral Thesis, Univ. of Pretoria, Oct. 2006

O.R.E., Question B 36. Dispositifs élastiques des appareils de choc et traction. *Rapport No. 9: Etudes théoriques sur l'allure de l'effort longitudinal dans les trains de marchandises avec dispositifs élasiques classiques (effort uniquement fonction de la course)*, Utrecht, avril 1972

O.R.E., Question B 36. Dispositifs élastiques des appareils de choc et traction. *Rapport No. 22: Calculs théoriques des efforts longitudinaux se produisant dans les train*, Utrecht, avril 1980

UIC leaflet 520: *Güterwagen, Reisezugwagen und Gepäckwagen - Teile der Zugeinrichtung*, 7 Ausgabe, Dezember 2003, ISBN: 2-7461-0679-5

UIC leaflet 526-1: *Wagons - Buffers with a stroke of 105 mm*, 3rd edition, July 2008, ISBN: 2-7461-1463-1

UIC leaflet 526-2: *Wagons - Buffers with a stroke of 75 mm*, 1st edition of 1.1.81

UIC leaflet 526-3: *Wagons- Buffers with a stroke of 130 and 150 mm*, 3rd edition, October 2008, ISBN: 978-2-7461-1472-0

UIC leaflet 528: *Buffer gear for coaches*, 8th edition, September 2007, ISBN: 2-7461-1323-6

UIC leaflet 540: *Brakes - Air Brakes for freight trains and passenger trains*, 5th edition, November 2006, ISBN: 2-7461-1172-1

UIC leaflet 541-3: *Brakes - Disc brakes and their application - General conditions for the approval of brake pads*, 7th edition, October 2010, ISBN: 978-2-7461-1835-5

UIC leaflet 541-5: *Brakes - Electropneumatic brake - Electropneumatic emergency brake override*, 4th edition, May 2006, ISBN: 2-7461-1027-X

UIC leaflet 541-6: *Brakes - Electropneumatic brake and Passenger alarm signal for vehicles used in hauled consists*, 1st edition, October 2010, ISBN: 978-2-7461-1843-0

UIC leaflet 544-1: *Brakes - Braking power*, 4th edition, October 2004, ISBN: 2-7461-0774-0

5

Improvement of Automatic Train Operation Using Enhanced Predictive Fuzzy Control Method

Mohammad Ali Sandidzadeh and Babak Shamszadeh
School of Railway Engineering, Iran University of Science and Technology
Iran

1. Introduction

Today traffic is one of the chief concerns in many countries and large cities. Development of public transportation is one of the key and vital solutions to this concern. Without paying enough attention to proper public transportation it becomes almost impossible to solve this issue. As a result, transportation is considered to be a development index throughout the world, and railway transportation is considered to be one of its most important and vital forms. Easy access, short waiting times, fast and comfortable trips, high safety and artistic design are all important factors in attracting more passengers to use public transportation. Recent public transportation systems move toward fully automatic operations with the least possible human interferences, so that a safer and more economic system is provided for passengers. Automatic Train Operation (ATO) plays an important role in this.

The most important role of ATO is controlling a train's movement between two stations. ATO strategy must be in such a way that it provides passenger comfort, accurate stop gap and energy saving while also adhering to the punctual schedule. Therefore the mentioned items are considered as evaluation indices of the control system. From a control point of view, ATO system has the following characteristics:

- The resolution of input data (such as velocity data) is low
- Attributes of control devices (such as brakes) are time-variant
- Route conditions such as gradient and altitude are functions of location. In other words, the power consumed in tractions and brakes alters according to the location of train.
- Evaluation indices of control system are multi-dimensional and include passenger comfort, safety, etc…

Provided that evaluation indices of control method are multi-dimensional and considering above mentioned attributes, we are facing a non-linear system with specific conditions and for controlling this system we need a suitable control method. One known method is to use PID controller to follow a target movement curve [9]. Other techniques include valuable works of Yasunobo et. al. that have created a suitable control method by means of a predictive fuzzy control which samples a skilled human operator's riding habits [1,2]. In this method, a skilled operator's strategy for train movement is used. At first, the operator's strategy is extracted in forms of sentences and then by using fuzzy and predictive control this strategy is implemented. For this purpose, passenger riding comfort, trip time, energy saving, traceability of target curve and accurate stop gap indices are defined in forms of fuzzy membership functions. Driving strategies are written as fuzzy rules. Also by modeling the real system, necessary predictions are made.

In this paper, train movement control is divided into two phases. The first phase is called constant speed control which is from the start position of the train and continues up to the point where the train enters automatic stop zone. In this phase before reaching the mentioned point, the train goes into coasting mode and coasting point is selected smartly by ATO system according to line's situation. The second phase which is called automatic train stop control has responsibility of adjusting precise train velocity until the train accurately and completely stops.

2. Modeling the train and automatic train operation system

ATO is a sub-system of Automatic Train Control (ATC) system. Figure 1 shows ATO system structure and its sub-systems.

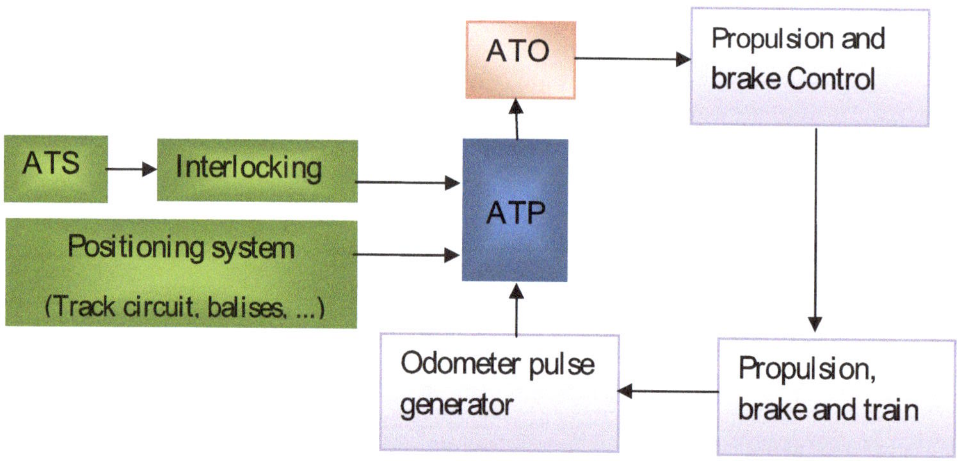

Fig. 1. ATO System Structure

According to figure 1, ATO inputs consist of:

- Distance pulses that come from speedometer or tacho-generator
- Train position marker detection signals that come from wayside equipments
- Supervisory commands that come from Automatic Train Supervision

A model of ATO system is presented in figure 2:

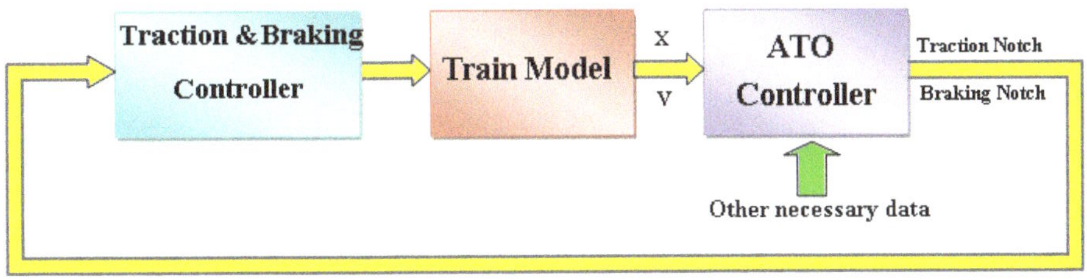

Fig. 2. Block Diagram of ATO System

Power control and brake control are respectively controlled via Power Notch and Brake Notch, which are discrete values and they create a precise force for proper train control. Today this discrete method is used in most control systems. Braking devices correct braking deceleration power in a way so that the brake notch can work without considering the train load. Even though there will be ± %30 error in real braking deceleration which is caused by the change in friction coefficient of brake pads, change in air pressure inside pipes, train weight change, etc… Train velocity is gained by the number of tacho generator pulses during a sampling time. Usually the accuracy of tacho generator pulses is about 10 cm and if sampling time is 1 *ms* then the speed detection error would be ± 0.36 km/h, which is a negligible fault [2]. The mentioned condition should be implemented in ATO and train model in order to replicate a good train dynamics [9].

Movement resistance according to Davis formula is $A+BV+CV^2$ [kg/ton] in which velocity unit is kilometer per hour. *A* is related to the axle load, *B* to the quality of track and vehicle stability, and *C* is related to area, vehicle shape, surrounding air and the tunnel air. Grade resistance is $mg \times \sin\theta$ and line gradient is in radians. Curve resistance is $mg \times 10^{-3} \times k/r$ in which *r* is radius of line curve and *k* is a coefficient that depends on line width and is 750 m for a width of 1435 mm.

In this article voltage control method is used for traction motor rev control [4]. In this method the average voltage is altered by changing the ignition angle of thyristor, although to decrease the simulation time a multi-switch with various voltages is used. For braking system a pneumatic brake system is utilized. This braking system is useful for velocities lower than 100km/h and has the ability of fully stopping the train. In this project, for the sake of simplicity, dynamic brake is omitted and pneumatic brake is equaled to resistive torque. A multi-switch with various resistive torques is used. Torques are defined and given values according to permitted brake decelerations.

Figure 3 represents train model simulation in MATLAB software environment which was used in this article. As it is shown in this figure, system inputs are voltage and torque that enter train model from power and braking controllers respectively. The resistive torque used in this simulation is gained from the total resistive torque divided by the number of traction motors. The simulation halts when train velocity becomes zero. To avoid increase in current at the moment of movement start, a resistance box is used to limit the current of traction (power) motors to the defined acceptable (1.5 times the nominal current) range.

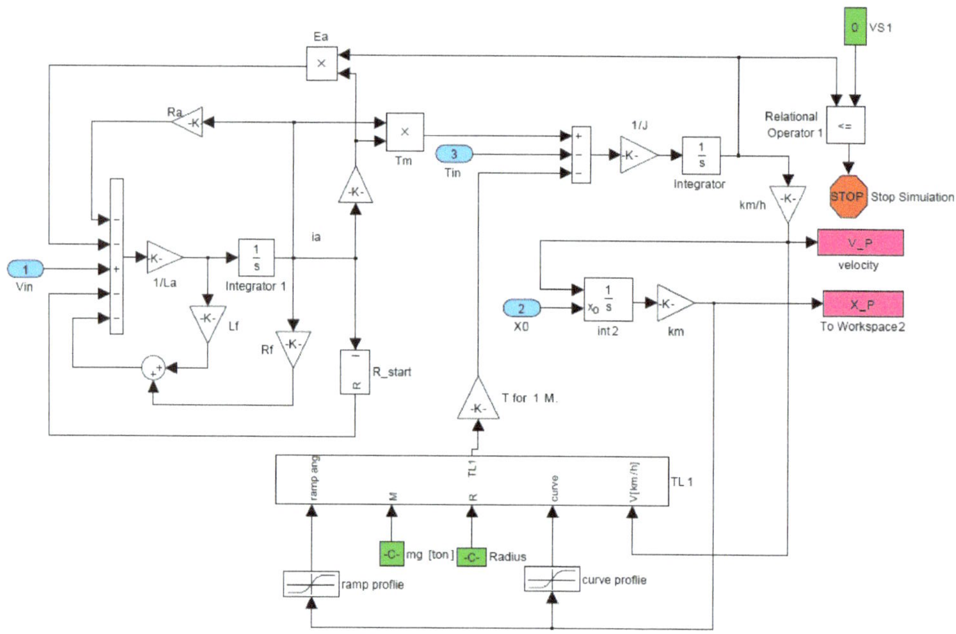

Fig. 3. Train Model Simulation

3. The proposed control method for ATO system

In this article a predictive fuzzy control method for Train Automatic Operation (ATO) is proposed. Predictive fuzzy control is a control strategy based on the system model that in each moment of sampling performs improved control action according to its current system state and simulation calculation of system model. The predictive fuzzy control has nothing to do with the mathematical model under control and this model can exist as linear, non-linear and even fuzzy model which is expressed with linguistic variable. In this article instead of having rules to improve the target function, simulation calculation of the controlled system and better estimation of the value of control rule are used [3]. In this control method, fuzzy control method and predictive control algorithm and also computer simulations are all mixed with each other to compose the predictive fuzzy control.

3.1 Fuzzy control

In 1965 Professor Zadeh proposed fuzzy logic theory in an article named "fuzzy sets". Fuzzy control is the result of applying this theory to decision making [5]. The first application in industry was performed by Mamdani on a steam engine in 1974. Fuzzy control works like this: a control instruction like u^* based on predicted control rules $R = \{R1, \ldots, Ri, \ldots, Rn\}$ is inferred from linguistic input variables { **x** is **Ai** and **y** is **Bi**} using fuzzy inference engine. One example of these rules is this:

"If x is Ai and y is Bi then u is Ui"

3.2 Predictive fuzzy control

Fuzzy control faces problems in systems that have a large delay time. To overcome this, predictive fuzzy is proposed. Just like other predictive controls, control action in predictive fuzzy control is based on predicting outputs of system.

Predictive fuzzy control is a control strategy based on system model that performs the optimum control action in each sampling time based on system's current condition and the calculations resulting from system model simulation. Predictive fuzzy control has nothing to do with the mathematical model under control, and this model can exist as linear, non-linear and even a fuzzy model expressed with linguistic variable.

In this control instead of having rules to optimize the target function, simulation calculations of the controlled system and estimation of the best amount of the control rule are utilized. In this control method, fuzzy control method and predictive control algorithm and also computer simulation are all put together according to figure 4 and create the predictive fuzzy control.

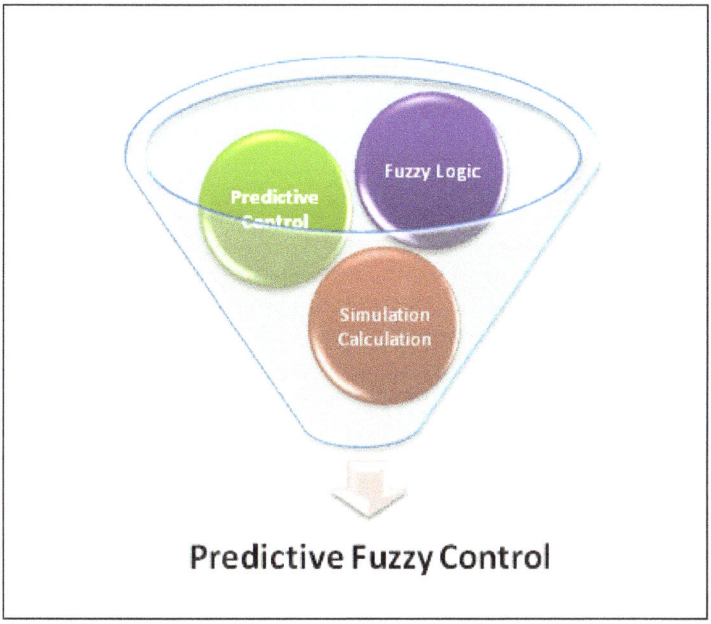

Fig. 4. Predictive Fuzzy Control System Structure

A skilled operator has extensive experience through his many experiments with system's operation and events, and he can satisfy system objectives via his high-level control method. With a small investigation on operator's control method we realized that he performs his control by predicting forthcoming system states and also by harnessing his extensive experience of the system. According to this, the predictive fuzzy control must calculate the next state of the system when selecting the control rule and propose this rule according to the next best state of the system.

This method is like this:

- Control rules $R=\{R1,\ldots,Rn\}$ are defined as :

$$Ri: \textbf{If } (U \text{ is } Ci \rightarrow x \text{ is } Ai \text{ and } y \text{ is } Bi), \textbf{ Then } U \text{ is } Ci$$

- *Ci* Control rules are selected based on the predictive results of (x, y) that show the highest probability.

Prediction of each evaluative amount (x, y) is based on the following:

"Control instruction Ci includes control rule Ri. Control rule Ri is evaluated. As a result control instruction Ci of control rule Ri with the maximum evaluation value is selected."

There is no inference or a non-fuzzy maker action in a fuzzy controller. The base of work is that after converting real variables in to fuzzy linguistic variables via Mamdani minimum, the weight of rules is resulted. The important thing in fuzzy rules is that each fuzzy rule is relegated to a real output. After giving weights to each rule, the output of the rule with the highest weight is selected as the output of the entire controller system.

4. Applying improved predictive fuzzy to ATO

Description of effects of a human operator's strategy on system functionality, definition of the meanings of linguistic evaluation indices, definition of models for predicting system operation and conversion of a skilled human operator's linguistic strategies to fuzzy control rules, are parts of designing and implementing a predictive fuzzy controller.

A skilled human operator controls the train much better than the conventional controller of ATO. This is because the human operator thinks about different control system indices and evaluates them directly. Therefore if a proper control system is designed so that it can accurately understand and mimic the control operation strategy of a human operator, it would perform the control operations even better than a skilled human operator. This is due to higher accuracy of electronic measuring devices when compared to that of a human being and also their faster response rate to commands and faster system processing speed.

4.1 Human operator strategies of train operation

Control rules based on below experience-based rules are selected in constant speed control and Automatic Train Stop control zones [1,2]:

- Constant Speed control (CSC) Zone rules:

1. For safety, if train speed goes beyond the speed limit, brake command with the maximum force is selected.
2. For energy saving, if coasting is allowed then coasting is continued.
3. For shorter running time, if speed falls too much below the limit, power notch is selected.
4. For passenger comfort, if train speed is in the predetermined permitted range, the control notch will not change.

5. For traceability of target curve, if notch is not changed and it becomes evident that train speed exceeds the permitted range, a $\pm n$ notch is selected so that the train can accurately trace the target speed.

- Train Automatic Stop Control (TASC) Zone Rules:

The operator starts the TASC operation by detecting the TASC position marker which indicates the remaining distance to the target point. In this control mode the operator selects the control notch based on tentative control rules.

1. For passenger comfort when the train is in TASC zone, the control notch is not changed if the train stops in the predetermined permitted zone.
2. For minimizing running time and maximizing passenger ride comfort, when the train approaches the TASC zone the notch is changed from acceleration to deceleration by a small margin.
3. For accurate stop gap, when train is in the TASC zone and it becomes clear that it won't stop in the predetermined allowed zone, a $\pm n$ notch is selected so that the train stops accurately at target position.

4.2 Definition of linguistic evaluation indices

In this section ATO system evaluation indices are defined which include safety evaluation index, passenger comfort evaluation index, traceability of target function evaluation index, energy saving and also accurate stop gap evaluation indices. Required membership functions for defining indices are presented here:

Tri_mf $\rightarrow$ Triangular function, which is defined by (a-b, a+b) region:

$$\text{Tri\_mf}(x,a,b) = \text{trimf}(x,[a-b,\ a,\ a+b]) \quad \Rightarrow$$

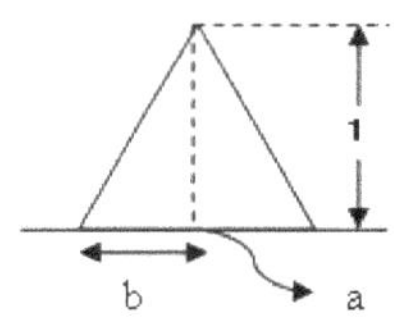

Half-triangular function, which makes values more than an equal to 1:

$$\text{Trir\_mf\_r}(x,a,b) = \text{trimf}(x,[a-b,\ a,\ \text{inf}\ ]) \quad \Rightarrow$$

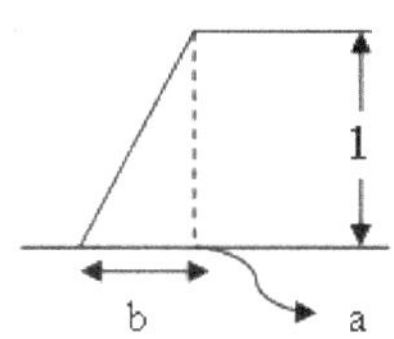

$$\text{Trir\_mf\_l}(x,a,b) = \text{trimf}(x,[-\text{inf},\ a,\ a+b\ ]) \quad \Rightarrow$$

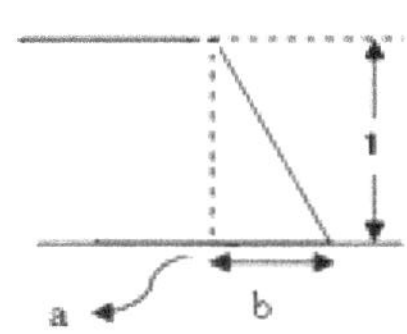

Dsig_mf $\rightarrow$ Pyramidal function, which is defined by $(-\infty,+\infty)$ region

$$\text{Dsig\_mf}(x,a,b) = 1 - 2 \times \text{dsigmf(x,[b, a, 0, 0])} \quad \Rightarrow$$

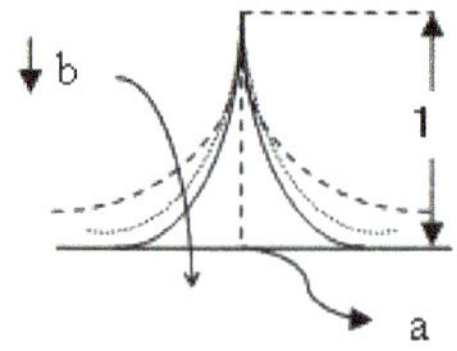

pi_mf $\rightarrow$ Trapezoid function, which is defined by $(-\infty,+\infty)$ region

$$\text{pi\_mf}(x,a,b) = \text{pimf(x,[a-2b, a-b,a + b, a + 2b])} \quad \Rightarrow$$

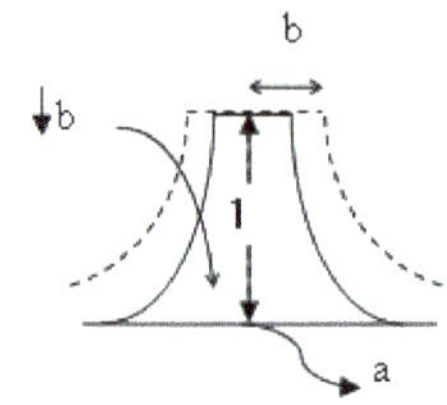

S_mf $\rightarrow$ Half-trapezoid function, which make values more than *an* equal to 1:

$$\text{S\_mf}(x,a,b) = \text{smf(x,[a ,a-b])} \quad \Rightarrow$$

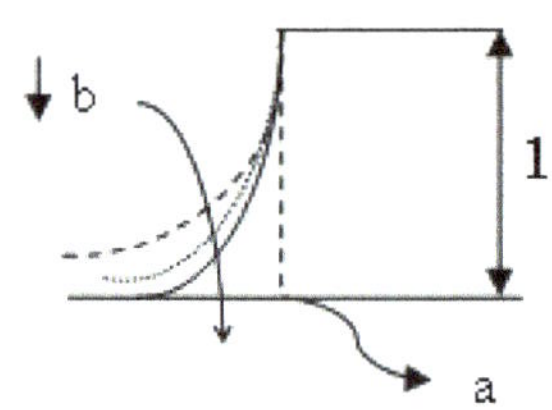

- **Safety evaluation index:** safety is calculated by the remaining time to danger or speed limit zone. Figure 5 shows the related defined membership functions :
  - Danger (SB): $\mu_{SB}(t_s) = \text{Trir\_mf\_r}(t_s, 0, T_s)$
  - Safe (SG): $\mu_{SG}(t_s) = \text{Trir\_mf\_l}(t_s, T_s, T_s)$

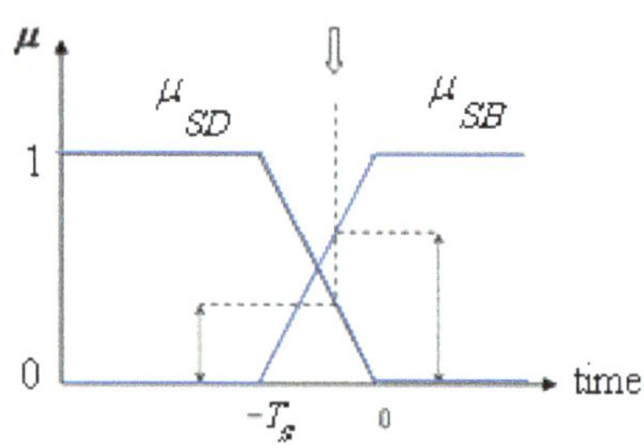

Fig. 5. Safety Index Membership Function

- **Passenger Riding Comfort performance indices:** Changing the notch frequently is associated with bad passenger riding comfort. Comfort is evaluated at N_{ch} steps and the time elapsed from the latest notch change t_{ch} :
  - Good passenger comfort (CG): $\mu_{\mathrm{CG}}(t_{ch}, N_{ch}) = \mathrm{S\_mf}(\mathrm{t_{ch}}, 1 + \frac{\mathrm{N_{ch}}}{2}, \frac{\mathrm{N_{ch}}}{2})$
  - Bad passenger comfort (CB): $\mu_{\mathrm{CB}}(t_{ch}, N_{ch}) = 1 - \mu_{\mathrm{CG}}(t_{ch}, N_{ch})$

Figure 6 shows the passenger riding comfort performance index membership function.

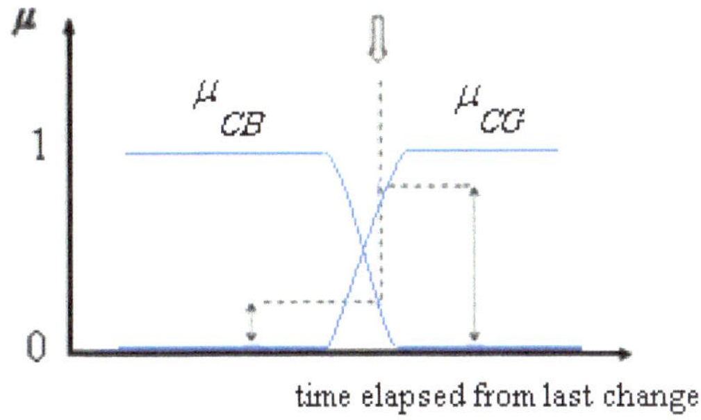

Fig. 6. Passenger Riding Comfort Performance Index Membership Function

The larger the notch change, the longer time it will take for the membership function of good passenger comfort to become 1. If the amount of notch change is 1 after ½ second passenger comfort weight becomes 1, and if notch change is 7 after 4.2 seconds passenger comfort weight becomes 1.

- **Evaluation Index of target curve traceability:** The difference between the predicted speed and the target speed is used for evaluating target curve traceability index.
  - Good Trace (TG): $\mu_{\mathrm{TRG}}(V_p(N_p)) = \mathrm{pi\_mf}(V_p(N_p), V_t, 2)$
  - Accurate Trace (TA): $\mu_{\mathrm{TRA}}(V_p(N_p)) = \mathrm{Dsig\_mf}(V_p(N_p), V_t, 2)$
  - Low speed trace (TL): $\mu_{\mathrm{TL}}(V_p(N_p)) = \mathrm{S\_mf}(V_p(N_p), V_t / 2, V_t / 5)$

Figure 7 presents the index membership functions of the above evaluation.

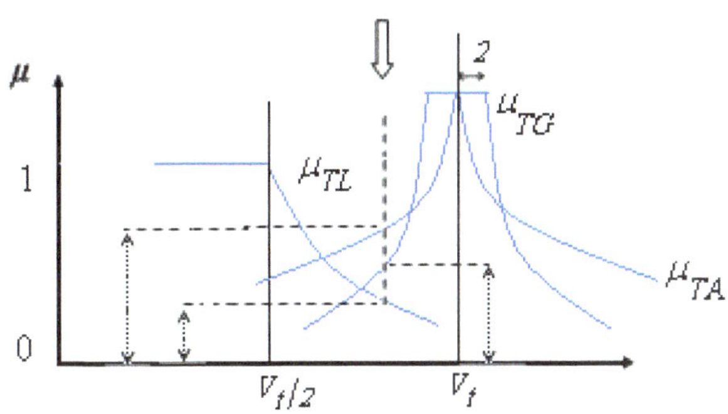

Fig. 7. Target Curve Traceability Index Membership Function

In a specific time, the position of future velocity is calculated based on the current $\pm n$ notch selected. If the input notch is supposed to be without change, then a good traceability membership function with a predicted V_t and a predetermined tolerance of V_e is specified. Based on the selected n= ±1, 2,3N and speed predictions, accurate trace membership functions with similar V_t are resulted. Finally by collision of this function with target velocity, membership function with the highest weight is resulted. It is clear in figure 8 that selection of the input notch has the highest weight.

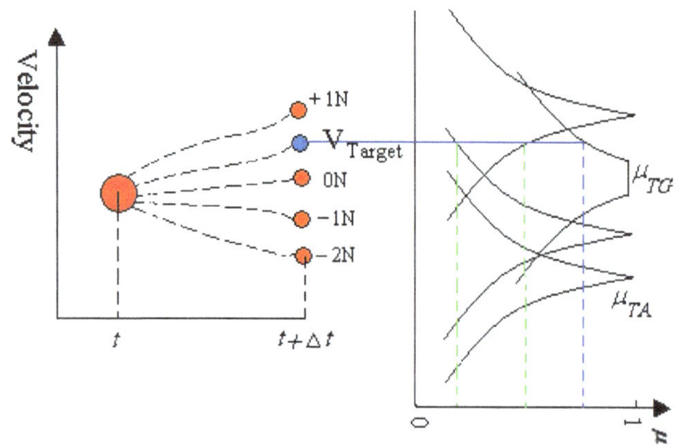

Fig. 8. Weight Of Membership Function In Target Curve Traceability Index

- **Accurate stop gap evaluation index:** Accurate stop evaluation index is defined by the difference between the predicted stop position and the stop target location.
  - Good Stop (STG) : $\mu_{STG}(X_p(N_p)) = \text{pi\_mf}(X_p(N_p), V_t, 30)$
  - Accurate Stop (STA): $\mu_{STA}(X_p(N_p)) = \text{Dsig\_mf}(X_p(N_p), V_t, 30)$

Figure 9 shows these membership functions mentioned above.

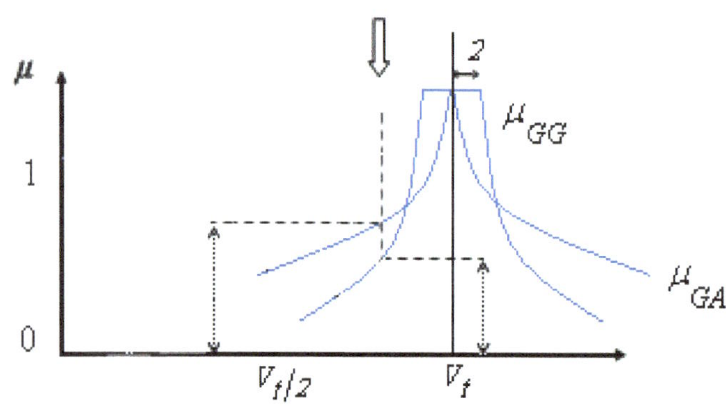

Fig. 9. Accurate Stop Gap Index Membership Function

In a specific time the control system predicts how target stop point changes by a $\pm n$ change in the current notch. Based on the input notch, good stop membership function is predicted with X_t and a determined tolerance of 30cm is specified. Then for n= ±1, 2,3N change from the current notch, accurate membership functions with similar predicted X_t are resulted. Finally weight of each membership function is resulted by colliding them with target position. It is clear from figure 10 that no change in the notch has the highest weight in membership functions and it is also visually clear that the most accurate stop gap is related to this choice.

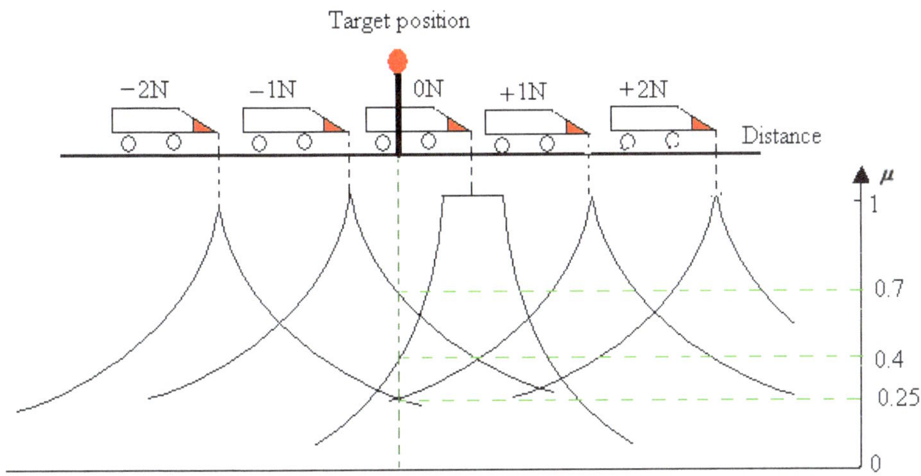

Fig. 10. Weight of Membership Function in Accurate Stop Index

4.3 Predictive model

As mentioned in previous parts, predictive fuzzy control system is based on predicting system behavior. For this purpose, a system model should be introduced that is capable of predicting system behavior with the desired accuracy. For modeling train dynamics, DC series traction motor and its resistive torque must be replicated. Each train is composed of a number of traction motors with analogous powers, and all motors should overcome the resistive torque imposed on the train. Resistive torque is gained by multiplying resistive forces by radius of wheel. The total external forces imposed on the train are calculated from movement resistance formulas (described in section 2). For modeling series DC traction motor we have to first calculate its mathematical relations. Required relations for modeling motor traction are as below:

$$E = K_a \phi_f \omega_m$$

$$V_a = E_a + \left(R_a + R_f\right) I_a + L_a \frac{dI_a}{dt} + \frac{d\lambda_f}{dt}$$

$$T_m = K_a \phi_f I_a = K_a K_f I_a^2$$

$$T_m - T_L = j\frac{dw}{dt}$$

Figure 11 represents train dynamics with DC series traction motors. From the technical characteristics of traction motors we have to fill in the undetermined items in system model. Besides, moment of inertia, radius of rotation, axis conversion coefficient, and etc should be added to the model to demonstrate system behavior properly. After modeling, the accuracy of the modeled system can be gained by experimenting on a real system.

Control system in CSC region predicts system behavior only for a few seconds, but in TASC zone system behavior up to a complete stop should be predicted. Therefore the accuracy of the predictive model should be in a way that the error tolerance of stop prediction with real error is in the acceptable range, so that the control system can perform properly based on gained results of the predictive model. In other words, if prediction results when compared to real results contain a huge error rate, then the control system loses it efficiency. In this article a train simulation model is presented in which the prediction results and the results of train dynamic simulation are in the acceptable error range.

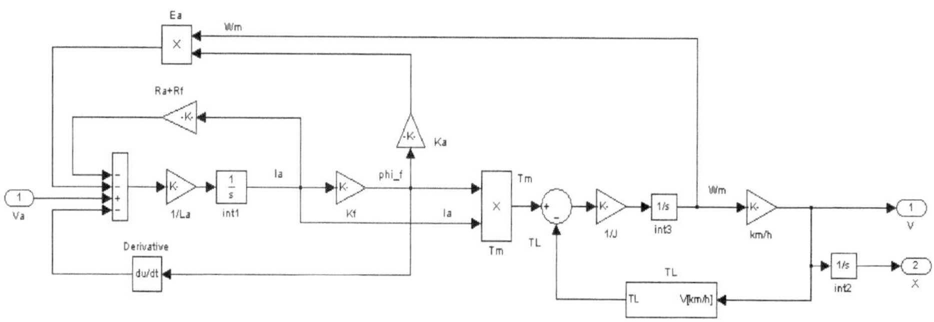

Fig. 11. A Model of Train for Prediction Traceability

4.4 Fuzzy Control Rules

DN is difference of notches, PN is power notch and BN is brake notch. The maximum power notch and brake notch is supposed to be 7 and 9 respectively. Safety evaluation index is *S*, comfort evaluation index is *C*, traceability evaluation index is *T*, and evaluation index of stop gap is represented by *ST*. Train movement strategy in constant speed control (CSC) zone is summarized in the following and figure 12 shows the fuzzy rules:

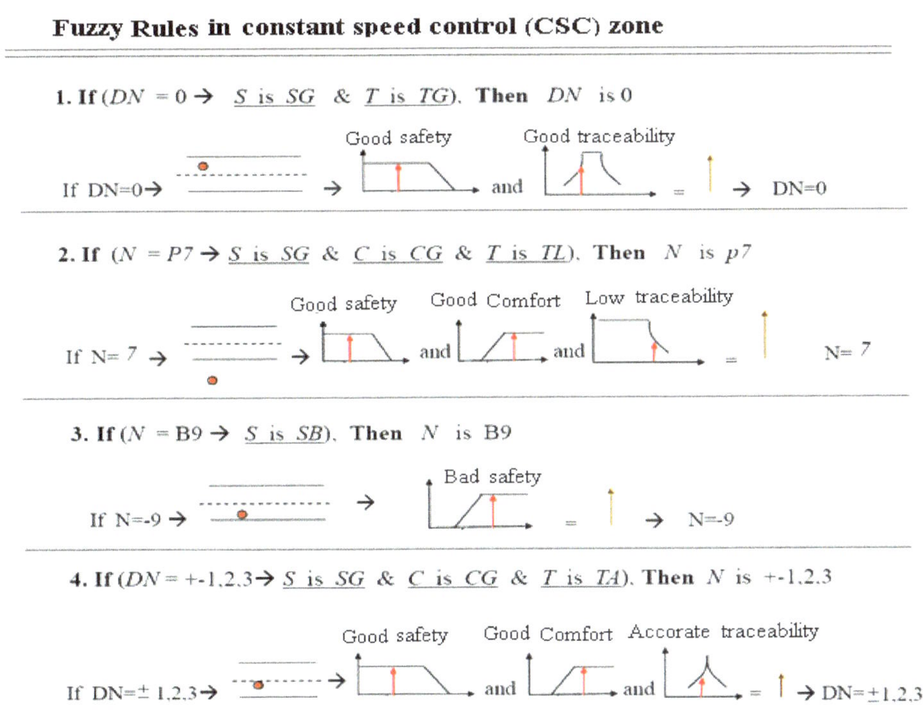

Fig. 12. Fuzzy Rules In CSC Zone

Train movement strategy in Train Automatic Stop Control (TASC) zone can be summarized by the following and figure 13 shows the rule base:

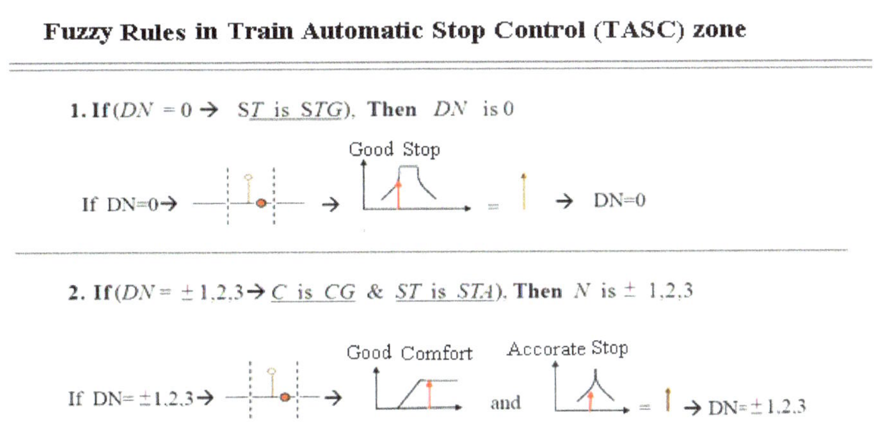

Fig. 13. Fuzzy Rules in TASC Zone

4.5 Techniques used for reducing energy consumption

One of the most important roles of ATO is reducing energy consumption. Different techniques are used for this purpose in all of which coasting is used for reducing consumed energy [6, 7, 8]. In this article two methods are used for reducing energy consumption which is presented in the following sections.

4.5.1 Using coasting in accurate stop gap zone

In Constant speed control (CSC) zone the control system always investigates a proper zone for coasting, and if all conditions are met then the control system smartly enters the accurate train stop gap (ATSC) zone. In accurate stop gap zone, control system calculates the proper situation for applying brakes up to a complete stop, and based on this the remaining time until braking zone (t_z) is gained. Based on the gained time, the following three membership functions indicated in figure 14 are calculated:

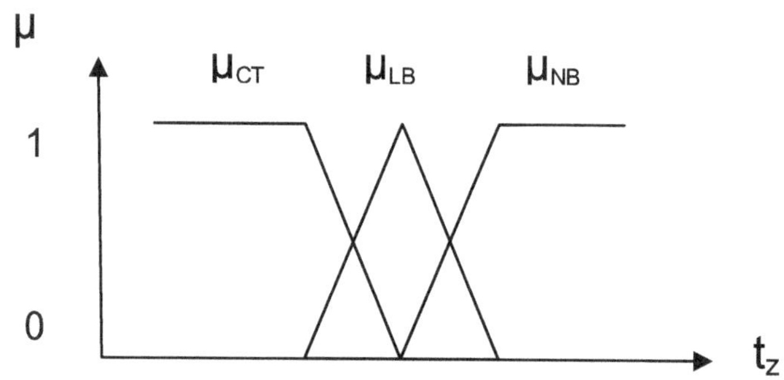

Fig. 14. Accurate Stop Gap Index Membership Function in ATSC Zone

- Coasting Time: $Trir\_m\_l = F(t_z, 0, 1)$
- Low Brake Time: $\mu_{LB} = Trir\_mf(t_z, -1, 1)$
- Normal Brake Time : $\mu_{NB} = Trir\_m\_r(t_z, -2, 1)$

When the control system moves from constant speed control zone to accurate stop gap zone, coasting time membership function gains weight of 1 and the train moves in coasting mode until low braking is applied. Coasting time depends on two factors: One is the braking time and the other is the time control system moves from constant speed control zone to accurate stop gap zone. Based on route conditions and estimation of brake capability, control system calculates the braking point so that the train applies brakes with average deceleration. In this article the average deceleration is gained in the fifth notch. Therefore the time achieved for braking depends on route conditions and braking capabilities of train, and it is not changeable, though the second element can be changed by system designer.

In constant speed control zone, system conditions for moving into the stop zone are always supervises and investigated. In constant speed control zone, the control system predicts train speed conditions in the braking position with notch zero and smartly calculates the best point for entering the constant speed zone. Required conditions for entering constant speed zone are as follows:

1. 1- Train velocity in braking position is in a distinctive area of maximum velocity.
2. 2- Safety condition is acceptable while coasting
3. 3- Passenger riding comfort is convenient

As a result of using this method the control system always smartly and based on route conditions-before entering constant speed zone- moves in coasting mode. Due to the fact that there is less than %2 time increase in this technique, ATO will always use this method.

4.5.2 Using coasting in constant speed control zone

If in constant speed control zone and by using coasting the train is kept in the permitted region, then for reducing energy consumption coasting is used. For this purpose energy saving membership function is represented like figure 15.

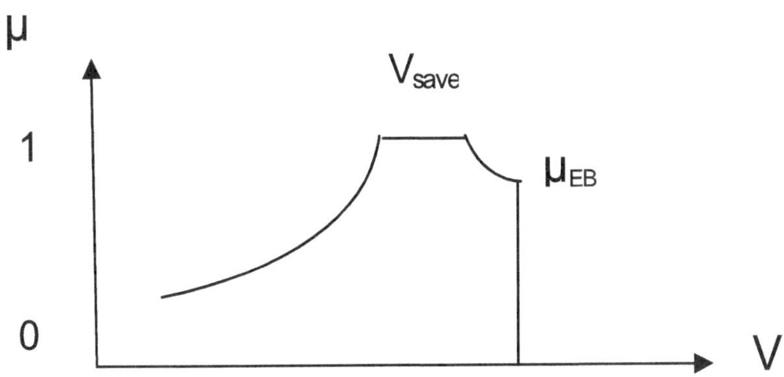

Fig. 15. Accurate Stop Gap Index Membership Function in CSC Zone

The amount of V_{save} represented above determines the amount of coasting. The more it increases, the higher coasting value becomes and energy consumption is reduced more. It should be noted that coasting increases the running time. Therefore there should be a balance between running time and coasting. It is better to define running time as a penalty function for energy saving, so that by taking the permitted running time into account, coasting is continued (previous methods). The problem with this method is that coasting is continued only until the time schedule allows it and it saves less energy when compared to the time coasting is used alongside the moving route. Therefore coasting level as a variable gives out different times and coasting level or V_{save} is selected according to the assumed time.

5. ATO system controller working method

In this method, the control system estimates the braking capability. By estimating the braking capability, accurate stop gap sensitivity to braking capability changes is reduced. Then control system indices that were introduced in the previous section are defined. Finally according to indices and energy saving conditions the best control rule in each part is gained, so that according to where the train is, rules relating to that location are extracted and the relating notch is applied to power or braking control system as the output notch. The mentioned Procedure is executed each 100ms which is the sampling time.

Figure 16 shows the general diagram of predictive fuzzy control for ATO presented in this article. As it is exhibited in this figure, control system's required inputs are fed into an s-function in which the control algorithm is written. Some inputs are used for predicting system dynamic and some others for defining indices. A coefficient for converting control region from CSC to ATSC is generated via a latch memory, so that the control system uses this coefficient to know in which region it has to operate.

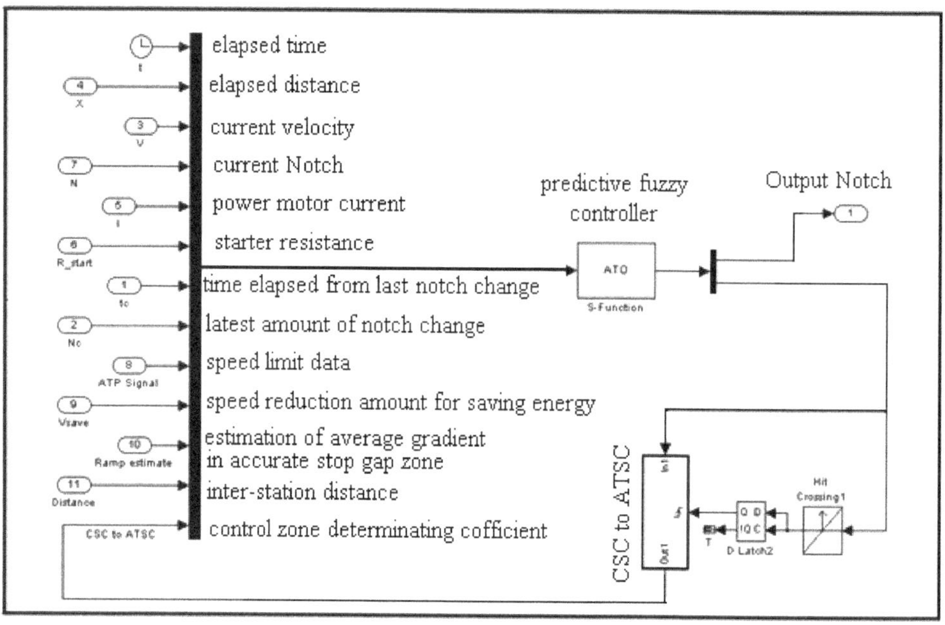

Fig. 16. General Diagram Of Predictive Fuzzy Control For ATO System

6. Simulations and results

In this part of this article ATO is simulated and is presented in a graphical simulator. Capabilities of improved predictive fuzzy control method are revealed in the next part.

6.1 Simulation assumptions

Train travel route is assumed to be a path between two stations with a distance of 1000 meters. Speed limit is assumed to be 40 km/h between distances of 550 and 650 meters. Curve radius is assumed to be 200 meters along the path. Gradient and altitude are assumed to be %2 and %5 respectively along the path. The assumptions considered in simulation are presented in Table 1.

Train weight	190 Tons
Movement Resistance	**$1.97+0.016v+0.00084v^2$**
Number of power notches	7 steps
Number of brake notches	9 steps
B_m **Maximum Brake Deceleration**	min =3.6 km/h/s norm=5.14 km/h/s max =6.68 km/h/s

Table 1. Simulation Assumptions

6.2 Implementing the simulation

Figure 17 displays the simulation in a path with the length of 1000 meters. As it is shown, with the same velocity profile and route attributes, time and energy consumptions are both reduced compared to previous methods [1, 2]. In the previous method running time was 97 seconds and energy consumption was 0.65 KWh, however in this method time and energy consumptions are reduced by 1 second and %16 respectively, and this is because of the techniques introduced and explained in section 4-5 of this article.

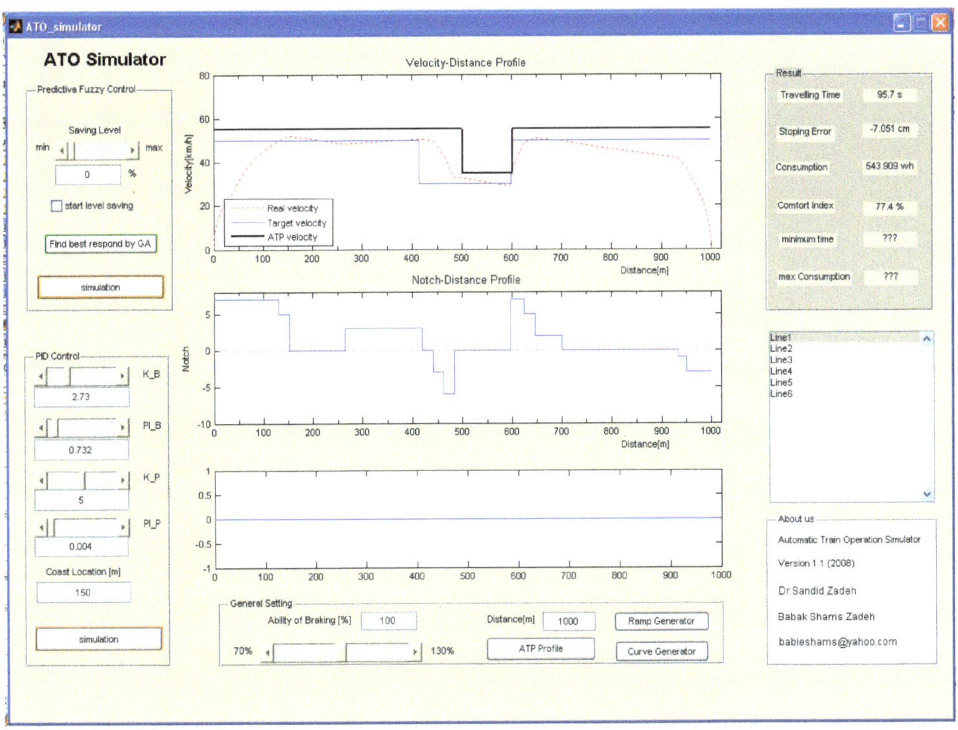

Fig. 17. Result of Simulation without Coasting Level in ATO Simulator

Table 2 also displays energy consumption based on coasting level (which is relevant to the increase in time), for a sampled line profile and velocity. As it is shown, in a slide with a small increase in time the amount of energy consumption increases significantly.

	Coasting Level %20		Coasting Level %40	
	Time Increase	**Energy Consumption Reduction**	**Time Increase**	**Energy Consumption Reduction**
- %5 inclination for movement path	0.2 seconds	%6.5	3.6 seconds	%18.6
+ %5 inclination for movement path	0.7 seconds	%1.8	2.9 seconds	%4.9

Table 2. Energy Consumption Based on Coasting Level

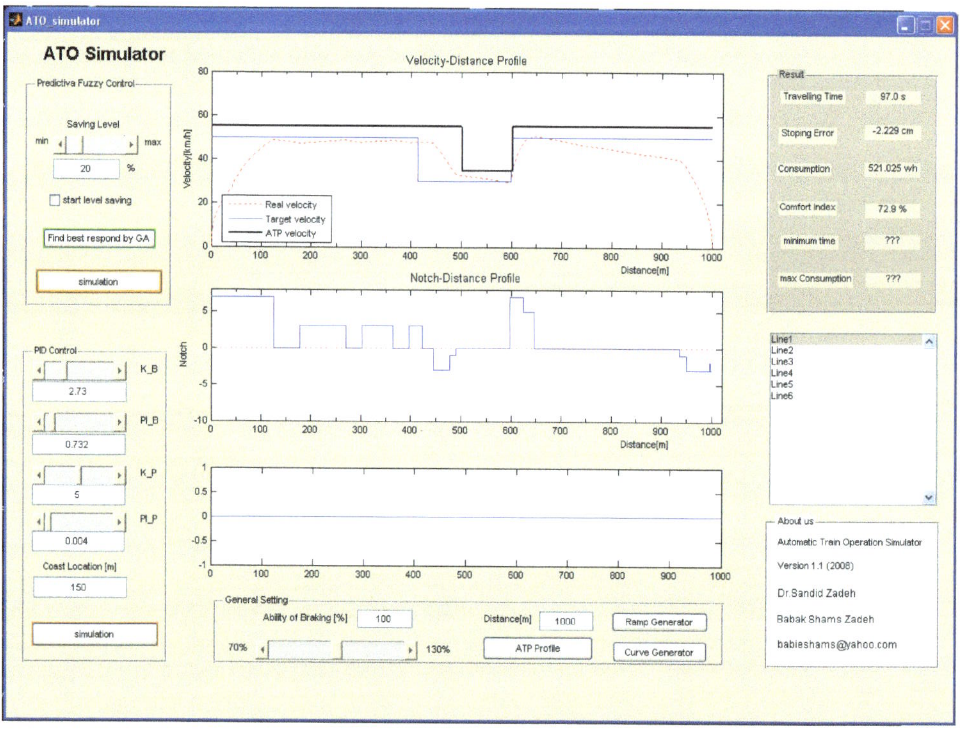

Fig. 18. Result of Simulation Based on Coasting Level in ATO Simulator

Therefore we can reach a convenient energy consumption rate by tuning the coasting level appropriately. Figure 18 clearly exhibits this fact. As it is displayed, running time is increased by 97 seconds, and energy consumption in comparison to similar methods is reduced by %20.

7. Conclusion

In this article, the predictive fuzzy control method has been implemented to automatic train control with new techniques for energy saving. As explained in this paper, the applied strategy of energy saving has two parts. The first part is in the zone of constant speed control (CSC), which regulates the coasting level in whole running of train in this region. This technique offers more energy saving in comparison to previous methods which use the coasting particularly in defined and permitted section of train running. In addition, considering the tradeoff between running time and coasting level, an optimized point with respect to running time and energy saving could be obtained which will be possible to be used in normal operation. The second part of applied strategy in this paper for energy saving is for train automatic stop control (TASC) which will improve the accurate stop, passenger comfort and energy saving by intelligent definition of TASC zone. By applying the above techniques and simulation, it is observed that the energy consumption is lower than previous suggested methods by other researchers e.g. 16% whereas more improvement in energy consumption could be gained by optimized regulation of coasting level and minor variation on running time. The results of simulation in this paper present the conclusion.

8. Acknowledgement

The authors express their thanks and appreciations to the School of Railway Engineering of Iran University of Science and Technology for their support to carry out this research. The same feelings will be applied to Tehran metro company for their data to do the simulation. The efforts of Mr. Hamed Zafari from Khatam Equipment & Development Company in Tehran are needed to be mentioned and appreciated for his cooperation to present this paper in English language.

9. References

[1] S. Yasunobu et al, Fuzzy Control for Automatic Train Operation System, 4th IFAC/IFIP/IFRS int. Conf. on Transportation Systems, (1983), 39-45.

[2] S.Yasunobu & S. Miamoto. & H. Ihara, Application of Predictive Fuzzy Control to Automatic Train Operation Controller, Proc. Of IECON , (1984), 657-682.

[3] S.Yasunobu, A fuzzy control for automatic stop control, Trans. of the society of instrument and control engineers Vol E-2, No. 1 , 1/9,2002.

[4] F. Xiaoyun , J. Junbo, L. Zhi, The Research of Fuzzy Predicting and Its Application in Train's Automatic Control, IEEE Trans. SMC-3-1, 2000, 28-44

[5] L.A. Zadeh, Outline of a new approach to the Analysis of Complex System and Description Processes, IEEE Trans. SMC-3-1, (1973) 28-44

[6] G. Horwood and D. Kearney and Z. Nedic , A fuzzy controller using floating membership function for the braking of a long haul train , first International Conference on Knowledge-Based Intelligent Electronic System, Australia, 1997

[7] K.K. Wong. T.K. HO, Coast control for mass rapid transit railways with searching methods, IEE Proceedings-Electrical Power Applications, Vol. 151(3), 2004, 365-376.

[8] B. Mellit , G. Goodman, N.B. Ramabukwella, Optimization of Chopper Equipment for Minimizing Energy Consumption in Rapid Transit System, IEE conference Railway in the electronic age, London,1981, 34-41

[9] Fujikura, Nohomi, Yasunobu et.al, Automatic Train Operation Method, The Proceedings of The Inst.of Electrical Engineers of Japan, 1977, 2205-2206

Part 2

Safety and Reliability in Railway

6

A Systems Approach to Assurance of Safety, Security and Sustainability in Railways

A.G. Hessami
Innovation Director, Vega Systems
UK

1. Introduction

The transportation network constitutes the artery of economic activity and growth in modern economies. Whilst challenged by telecommunications and internet technologies, the movement of goods and people is still an indispensable aspect of social and economic life contributing around one tenth of the GDP in the developed world economies[1]. It is not surprising therefore to find transportation on the social and political agenda and any faults, failures and consequent accidents, being given a high degree of publicity and exposure. Traditionally, the key mantra in transportation especially railways has been safety followed by reliability, punctuality, cost, journey time and quality of travel. This has held true so far for most modes of transport until recently when malicious intent with the aim of disrupting the network, victimising its customers and inflicting large economic losses has added a new ingredient to the traditional concerns of the industry. The malicious intent broadly falls into a number of categories comprising;

- Vandalism & Unlawful Adventure
- Robberies, Assaults
- Illegal Access
- Unauthorised Use of Property/Facilities
- Theft, Fraud
- Intimidation and Extortion
- Disruption, Sabotage
- Terrorism

Whilst vandalism is of limited consequence and often related to adventure seeking youth, the other categories of concern specifically terrorism pose a largely new sinister development often beyond the powers of transportation authorities to predict, prevent or contain. This is where the power of scientific structured approaches and methodologies principally applied in safety engineering can be exploited to render assurance in integrated transportation safety and security in Road, Rail, Shipping and Aviation including the Transport Hubs.

---

[1] U.S. Department of Commerce, Bureau of Economic Analysis

The rapid development of technology generates new products, systems, services and process knowledge often with significant potential to improve technical, commercial and environmental performance and enhance the overall quality of life. However, the new innovations especially those with embedded intelligence and adaptability are plagued by uncertainty about their overall characteristics including the concern about the risks arising from their adoption. To this end, a systemic assurance process and associated methodologies are required to underpin verification, validation and enhanced confidence in the desired performance of industrial & technological systems and innovations.

With the advent of high-speed rail transportation technology, the industry is now poised to compete with short haul aviation and deliver enhanced economic and social benefits across the world. The technology that was first introduced in Japan in the 1960s is now advanced and pervasively employed in much of Western Europe, South East Asia and the People's Republic of China. As a notable case, China that currently boasts a standard and high speed rail network of 53000 miles plans to spend an estimated $300 billion to meet a 2020 target of 75000 miles with the world's largest high speed network (Green Leap Forward, 2010). In spite of high economic cost, there are a number of compelling reasons in favour of high-speed rail namely incessant demand for mobility, accelerated economic development and more sustainable transport using electric trains. Whilst high speed rail networks are deemed to transform economic geography by bringing cities closer together, enabling higher business productivity, supporting employment growth and regeneration, the inherent complex technologies adopted require extensive scrutiny, verification and assurance to ensure desirable levels of safety, security and potentially higher attained sustainability.

1.1 Background

Aided by the transportation, communications and computer technologies, the global village presents many opportunities and benefits to mankind from rise in international trade and growth to redistribution of wealth. However, alongside these emerging opportunities, innovations, scientific discoveries and burgeoning complexities in products, processes and indeed human relationships pose a challenge to mankind threatening health, safety, security and the natural habitat. In a recent international survey, DEMOS the think tank organisation (DEMOS, 2007) identified six major global drivers for change namely:

- Pervasive Complexity;
- Matrix versus Functional Restructuring;
- Riskier Markets;
- Transition to Knowledge Economy;
- New Business Practices, Offshoring & Outsourcing;
- New Accountabilities, Corporate Governance and Corporate Social Responsibility.

Prudent exploitation of the opportunities and credible assessment/management of potential adversities in the face of these influential drivers demands a more systematic and potent approach to tackling the emerging global challenges.

1.1.1 Safety

Safety is synonymous with freedom from unacceptable levels of harm to people and is a highly desirable property of products, systems, processes and services. However, in view of ever increasing complexity, faster pace of development and change, safety is often difficult if not sometimes impossible to entirely predict, manage and guarantee. At the same time, rising social awareness and the more stringent legal requirements almost globally demand higher levels of safety performance from products, processes, systems, services and the duty holders. Safety problems are characterised by unintended yet harmful incidents and accidents that apart from acts of nature are mainly traceable to our shortcomings in concept, design, development, deployment or maintenance of products, systems and services. Safety is heavily regulated and health, safety and welfare of people are under legal protection in most developed and increasingly in developing countries.

1.1.2 Security

Security is synonymous with freedom from unacceptable levels of harm to people, damage to business operation/property or the natural habitat. Unlike safety, security problems are characterised by often malicious intent (threat) which aligned with inherent or intrinsic vulnerabilities cause incidents and accidents with significant potential to cause harm and consequent loss. However, complexity, rapid change, global geopolitics and novelty pose increasing threat to the security of products, processes and systems requiring an enhanced degree of proactive assurance. Security, as yet, is not generally regulated and freedom from intentional harm to people and property is still largely a commercial decision by duty holders.

1.1.3 The Environment & sustainability

Since the dawn of industrial revolution, the scale of mankind's influence on the natural habitat has increased significantly. Apart from the depletion of non-renewable resources and generation of waste and heat, industrial and man-made disasters often involve the natural environment causing damage, contamination or major change in the ecology to the detriment of plants, wild life and potential for human habitation. Given man's destiny and quality of life on the planet are strongly related to the health and balance in the natural habitat, the environment is now protected through regulation enforced through laws and government agencies.

1.1.4 Synergies

Safety and security possess a significant synergy in that safety is characterised by unintended and security by intended errors, faults, failures and acts leading to accidents. Apart from differences arising from the nature of intent, the prediction, prevention and successful risk control in both contexts can be carried out in one integrated regime due to similarity of escalation processes and much of remedial actions. The concurrent identification, assessment and mitigation of safety, security and environmental issues render enhanced integrity whilst posing significant savings in costs and time scale for assurance on these fronts.

The regulatory regime is the key instrument in the overall certification and deployment of new innovations in products and services. Many developments including the safety case regime mandated within nuclear, offshore (Offshore Safety Case Regulations, 2005) and rail transportation (Railway Safety Case Regulations, 2000) in the UK are intended to pave the way to enhanced confidence as well as rapid deployment of modern innovations. In this context a systematic and principled approach to identification, control and management of risks is fundamental to the achievement, maintenance and improvement of the overall confidence and performance of products, processes, services, systems and undertakings.

Products, processes and systems exhibit a number of facets in their performance that are either inherent or perceived by the relevant stakeholders. These generally comprise:

a. Technical/operational;
b. Commercial;
c. Safety & security;
d. Environmental & sustainability;
e. Reliability, availability & maintainability;
f. Quality;
g. Perceived value.

Amongst these often inter-related aspects of performance only safety and environmental dimensions of products, processes, services and systems are currently subject to regulation (Hessami, 2004). Understanding the key factors influencing the overall safety and security performance of various services, industrial and infrastructure systems will lead to the development of policy initiatives to promote safer, more secure and cost-effective solutions at the enterprise and industry level. It will also simplify regulation while providing transfer of knowledge and expertise from more successful domains and states to those that have evolved at a slower pace. With the advent of high-speed rail transport, the industry requires higher degrees of confidence and assurance in the advanced products, systems and services deployed to avoid costly accidents. To this end, a systematic framework for identification, evaluation, assessment and management of risks founded in systems theory is called for.

2. Risk and assurance

2.1 Derivation of principles

A principle is regarded as a fundamental truth or proposition on which many other propositions depend. It is also regarded as a fundamental assumption forming the basis of a chain of reasoning. It is argued that a management regime founded on a suite of principles will be superior in terms of its stability, integrity, effectiveness and its capacity to be adaptable and scalable for multiplicity of circumstances and stakeholders since it is constructed using a set of fundamental & universal truths. A framework for management of risks should inherently address all life-cycle phases and issues comprising:

- Definition and characterisation of the system of concern and its environment of application;
- Identification/recognition of fundamental threats, faults and failures (causes of hazards);
- Prediction of realisation/occurrence of hazardous states arising from threats, faults and failures;
- Assessment of potential escalation of hazardous states into accidents/loss scenarios;
- Coverage of post-accident scenarios, actions and recovery processes;

- Human organisation, capabilities, resourcing, procedures and competencies;
- An inherent monitoring, measurement and enhancement regime.

On the other hand, assurance is synonymous with gaining increasing confidence about the performance of an often complex product, service, process or system so that;

- It delivers an optimal level of essential and desirable properties/performance
- It is free form an unacceptable level of undesirable properties/performance

A systems framework based on a complete and inter-related set of principles for performance assurance would enhance the degree of confidence that apart from the delivery of required functionality, the product, service, process or system is free from potentially harmful properties and behaviours hence assurance.

A key aspect of the current approaches to understanding and managing desirable and undesirable properties is the disjointed and unsystematic treatment of the issues in these domains (Hessami, 1999). Apart from lack of joined up approach in even one of these domains, most experts operating in one domain operate independently often unaware of the issues, processes and solutions in the other.

It is argued that a comprehensive scrutiny, objective evaluation, assessment, understanding and management paradigm encompassing a systems world view would result in enhanced assurance and surety, in the face of complexity, uncertainty and change.

3. The systems approach

We propose two complementary and advanced sets of systemic principles and processes as the underpinning backbone to tackling the challenges of safety, security and sustainability in all products, processes, services, systems and undertakings. This is particularly pertinent in the modern railway environment in view of the pervasive deployment of advanced technologies to deliver higher speed and improved efficiencies. Taking a life-cycle perspective, these comprise items I and III below;

i. Assessment: This comprises proactively recognising the need, defining the system, specifying and identifying/understanding of key properties, behaviours, hazards and vulnerabilities, evaluating and assessing expected impact;
ii. Realisation: This is ultimately aimed at developing the product, process, system, mission or undertaking whilst incorporating the desirable properties and avoiding the undesirable behaviours thus achieving the optimal performance;
iii. Management: this comprises taking the outcome of assessment and realisation into consideration and ensuring deployment, delivery of requisite performance, continued monitoring and control through a responsive and holistic suite of strategies and actions.

Whilst Realisation is specific to a given domain, context and technology, the Assessment and Management aspects as a suite of principles constitute a meta-knowledge framework that can be abstracted and developed for almost universal application across many domains and disciplines. The systemic framework of assessment and management is equally applicable and effective within the context of desirable as well as undesirable properties of products, services and systems. This is contrary to the current conventional wisdom where specification, delivery and continual monitoring of desirable aspects of performance is regarded as an essentially domain expertise whereas the undesirable and unintended emergent properties (hazards and vulnerabilities) are the forte of so called risk management. The +Safe3 extension (Australian Defence Materials Org, 2007) to the renowned CMMi model (Chrissis et al, 2007) also distinguishes between Safety Engineering & Safety Management, which are mainly synonymous with Risk Assessment and Risk Management advocated here.

Whilst presented as a dual and complementary suite of principles and processes, assessment and management are iterative and systemic in the sense that processes inherent in the management framework employ assessment activities at requisite points to support judicious decision-making and ensuring optimal performance. These are collectively referred to as Systems Assurance and labelled as Surety Framework.

3.1 Risk assessment

This key facet of Surety framework depicted in figure 1 is proposed as a backbone to the identification, specification, evaluation and assessment of the undesirable events or properties adversely affecting technical functionality, cost, reliability, safety, quality etc. The risk assessment process (Railtrack plc, 2000) comprises seven systemic aspects such as:

a. Hazard Identification;
b. Causal Analysis;
c. Consequence Analysis;
d. Loss Analysis;
e. Options Analysis;
f. Impact Analysis;
g. Demonstration of Compliance.

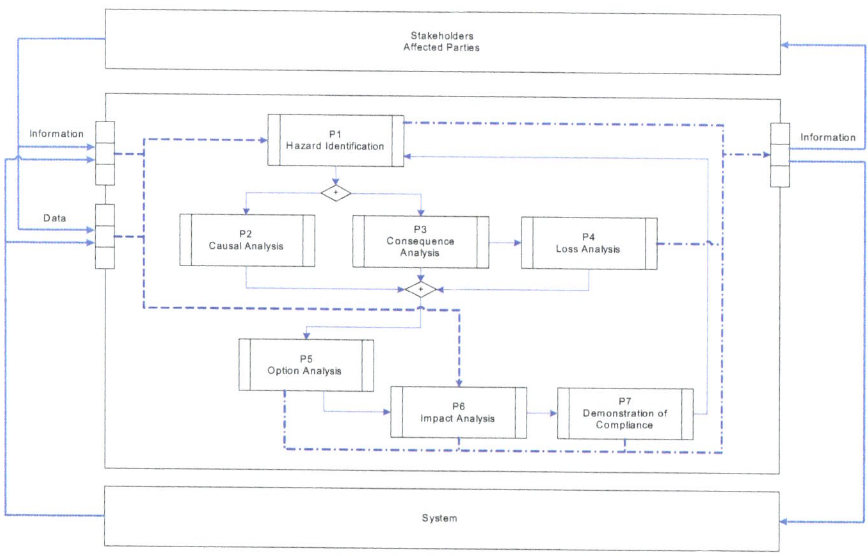

Fig. 1. The Systematic Framework for Risk Assessment, its interfaces and interactions

The principles of risk assessment are general and equally applicable to the qualitative as well as the quantitative approaches to this discipline. They constitute a systematic framework within which, a broad spectrum of situations hazardous to health, safety and security of people and detriments to the environment or an enterprise may be identified, analysed and assessed.

The qualitative risk assessment process broadly relies on expert judgement and empirical experience sometimes within a subjective and coarse quantitative process. It is worth noting that mere use of quantification and numbers does not necessarily qualify an assessment as quantitative. These are however mainly a reflection of judgement and lack the objectivity and accuracy to generate a detailed and reliable measure of risks.

The Risk Assessment process highlighted above satisfies the following requirements;

- Potential for use of modelling;
- Predominate application of objective and validated data;
- Treatment of uncertainty associated with input data and results;
- Treatment of dependency between significant factors;
- Use of statistical simulation where appropriate.

Modelling predominately represents a simplification and generalisation of reality but, enhances our understanding of causal relationships, highlights important factors and provides a useful tool for anticipation and potentially prediction of future.

Advantages

The quantitative framework for assessment of risks arising from hazards of undertakings, services, products and processes, yields a number of major advantages over its qualitative counterpart;

- generates a quantified measure of risks in complex situations;
- capable of addressing uncertainty and statistical variations in input data;
- capable of addressing dependencies in the input parameters/data;
- capable of generating confidence intervals for the quantified risks;
- capable of demonstrating compliance with ALARP and other Industry Benchmarks;
- auditable objective process with scope for review and improvement;
- does not employ arbitrary tolerability criteria popularised by risk matrices;
- does not require customisation or a specific form of a ranking matrix;
- provides an auditable and traceable approach to decision support;
- employs the same framework and principles as in the qualitative approach.

Disadvantages

The constraints and dis-benefits of the quantitative approach must be borne in mind however, namely;

- complex hence unsuitable for low risk systems and undertakings;
- requires expert resource in knowledge elicitation and risk modelling;
- need for extensive range of objective data and the requisite pre-processing;
- need for formidable computing resource and know-how;

- resource intensive, costly hence inappropriate for applications where a qualitative approach may suffice;
- lack of readily available, robust and comprehensive computer based tools.

The quantitative approach to assessment, recording and management of risks strives to generate a systematic framework for decision-making and demonstration of legal and professional duty of care. In contrast with the qualitative approach and in compliance with the spirit of the Safety Case regime (CENELEC, 2003) and Regulations, the approach and methodologies of the quantitative process are more stringent and thus germane to the nature of significant risks.

The systems framework comprising seven key principles highlighted above is equally applicable to qualitative and quantitative approaches to the assessment of risks arising from products, processes, services, undertakings and systems. The guidance for the required processing at each one of the seven stages of the systematic risk assessment framework is given below, commensurate with the requirements of the quantified process.

3.1.1 Hazard identification

Circumstances with a potential to lead to loss, i.e. harm to people, financial detriment or environmental damage are associated with most activities and undertakings. Whilst it is relatively straightforward to identify these within the context of familiar day-to-day tasks and experiences, more complex products, processes, services and undertakings generally pose a more arduous if not insurmountable challenge in this respect. Rapid development and widespread exploitation of cost saving or performance enhancing technologies generally exacerbate the situation and increase the scope for larger potential losses in the event of unforeseen or unprotected errors and failures.

The structured comprehensive identification of hazardous circumstances arising from threats or unintended failure of products, services, systems, processes or human error/action is fundamental to any safety and security process. It is however even more pertinent to large scale or complex undertakings with a potential to lead to significant losses in the event of hazardous occurrences. In the absence of a systematic and robust hazard identification phase, all the subsequent safety analysis processes amount to no more than an exercise in vain, creating an illusion of safety/security and a false sense of confidence and comfort. This is particularly pertinent to circumstances where, due to a poor process, a number of significant hazards remain un-identified hence dormant within the system posing intrinsic vulnerabilities.

The determination of the domain of influence of a product, process, service or undertaking is another by-product of the systematic hazard identification process. This is essential in establishing the scope of the subsequent assessment and should be employed in preference to the traditional approach based on the physical boundaries of the subject under consideration. The radiation of electro-magnetic interference typifies instances where the domain of influence extends well beyond the physical boundaries of a poorly designed and constructed system.

The systematic identification of hazardous circumstances entails two key stages at the outset;

- Empirical Phase;
- Creative Phase.

In view of the extensive resource requirements, the approach described here is more appropriate to products, services, processes and undertakings that are likely to lead to significant losses due to their scope, scale or novelty.

3.1.1.1 Empirical phase

Traditionally, the knowledge and experience of the past, in the form of Check-lists have been applied to the determination of the potential hazardous circumstances in new products, processes and undertakings. This approach is seldom adequate in isolation especially, when there are novelties or significant changes in the functionality, technology, composition, environment (time / space) or the mode of exploitation of the matter under consideration. It is essential therefore to compile and maintain a Check-list of hazardous circumstances pertinent to specific products, processes or undertakings, in order to facilitate a simple first cut identification of the likely problem spots and where possible, avoid the errors, failures and losses of the past.

Where the product, process or undertaking lend themselves to a more detailed scrutiny, Failure Mode and Effects Analysis (FMEA) for equipment / systems and its human related counterparts, Action Error Analysis and Task Analysis may be applied in order to identify the particular component failures or errors conducive to hazardous circumstances. These however require a detailed knowledge of the failure modes of the components and sub-systems, including human actions and the likely errors.

The application of Check-lists, FMEA, Action Error Analysis and Task Analysis are generally not resource intensive and may be carried out by suitably competent individuals and appropriately recorded for further analysis. The hazards identified through the application of these techniques generally constitute a sub-set of the total Potential Hazard Space that should be further explored with the aid of the complementary Creative techniques.

3.1.1.2 Creative phase

The systematic and creative techniques have an established pedigree in the analysis and resolution of complex problems. These generally capitalise on cognitive diversity through a team based approach, comprising members with diverse and complementary knowledge and backgrounds. Furthermore, in view of their reliance on lateral perception, divergent thinking and imaginative creative faculties, the structured and systematic variants of these techniques generally share a numbers of key characteristics namely:

- planning and process management;
- study panel (team) selection and briefing;
- hierarchical decomposition and graphical representation of the problem domain;
- high level probing of the key elements of the system and coarse determination of the critical sub-systems and interfaces;
- comprehensive, step-by-step probing of the sub-systems and interfaces with a more meticulous scrutiny of the critical areas;

- identification and recording of the hazardous circumstances including causes, consequences and potential mitigation and control measures;
- expert driven ranking of the identified hazards employing an appropriate frequency/consequence matrix;
- maintenance, update and management of the records throughout the life of the product, process or undertaking.

The hazards identified through the empirical processes must be reviewed at appropriate stage(s) during the creative phase and recorded together with the other attributes alongside the newly identified items in a log. The empirical phase is sometimes employed as a completeness test or means of detailed probing of specific hazards and failures, subsequent to the creative identification phase. Whichever the temporal order, the empirical and creative phases must be applied in a consistent and complementary manner to re-enforce and increase confidence in the hazard portfolio.

The two-phase process enhances the integrity and coverage of the potential hazard space, increasing the effectiveness and confidence in the safety and security process. It has to borne in mind that the hazard identification exhibits an essentially non-linear gain and a creative identification of a single significant hazard may outweigh the contribution of a large number of less severe items. In this spirit, it is the quality and not the quantity of the identified hazards that is of the essence. The methodologies that generate an unrealistically large number of mostly trivial hazards are wasteful of resource, misleading and unproductive and should be avoided wherever possible. Furthermore, the subsequent analytical treatment of hazards as detailed in this chapter should be applied on a prioritised basis, beginning with the highest-ranking hazards.

3.1.2 Causal analysis

Upon systematic identification and ranking of hazards arising from a product, process, service, system or an undertaking, it is often constructive and sometimes necessary to further explore the logical relationship between the basic errors and failures that could potentially realise the hazards. The aim is to address each hazard at the root cause level with a view to preferably eliminate and where not feasible, reduce the frequency or likelihood of its realisation (occurrence).

3.1.2.1 Process

The causal analysis is a mainly empirical process requiring domain knowledge of the product, process, service or undertaking. The techniques of Causal Analysis are generally applied recursively in a top-down mode to a given general state (hazard or threat) until all low level specific causes, errors and failures are arrived at. This deductive approach generally produces a number of intermediate states, each potentially caused by lower level causative factors. The general heuristic is to continue with the decomposition of each intermediate state until all fundamental causal factors such as basic component failures, unintended human errors or malicious acts are arrived at or it proves impracticable to acquire reliable data pertaining to lower level factors. The causal analysis techniques are predominately applied within reliability engineering and are generally supported by mathematical foundations and a suite of computer based tools.

3.1.2.2 Modelling

The causal analysis techniques generally employ graphical modelling which constitute a potent form for the capture and communication of the inter-relationship of the primary errors and failures leading to a hazard or vulnerability. Whilst predominately employed qualitatively, the causal models often lend themselves to quantification that ultimately generates a probability or frequency for the hazard or threat under analysis. The key issues to bear in mind during the causal modelling process are:

- correct logical relationships;
- decomposition commensurate with data availability;
- common cause failures;
- redundancy;
- inter-dependency of some errors and failures.

It is also important to ensure that different variables expressed in probabilities or frequencies are combined appropriately to generate consistent results for example, ensure that two frequencies are not multiplied to yield units in terms of per time squared! An illustrative causal model for a railway hazard is depicted in Figure 2.

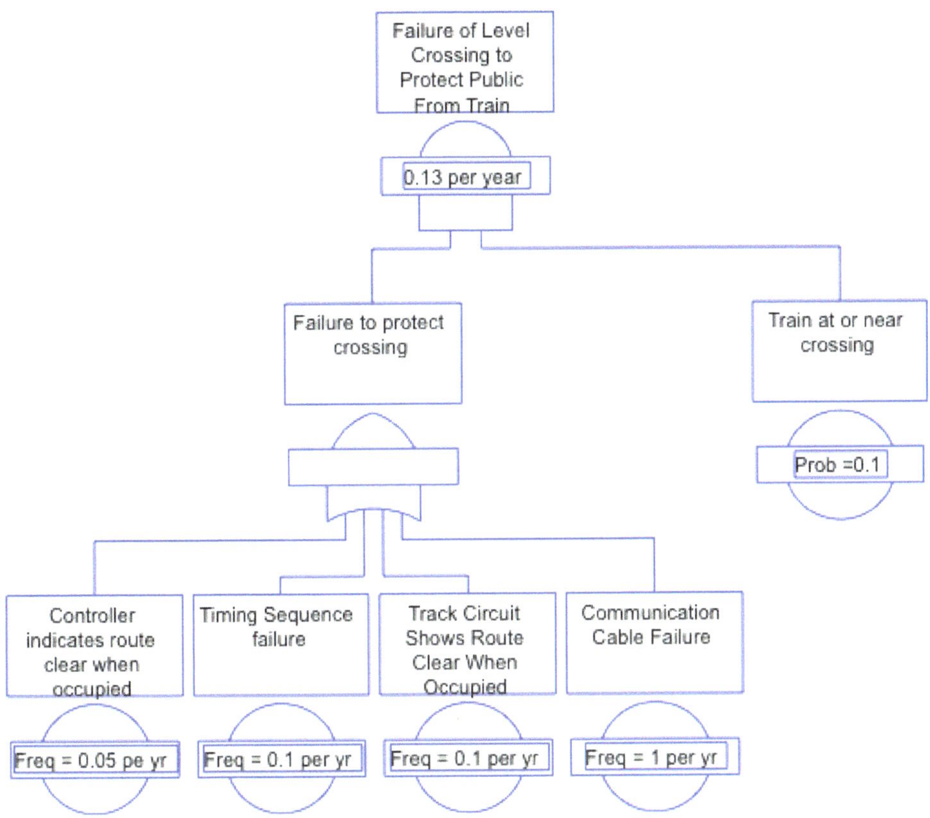

Fig. 2. Illustrative Causal Model for a railway hazard

3.1.2.3 Quantification

The quantification of causal models entails an objective assessment of the potential frequency or likelihood for the causal factors. These are combined according to the rules of probability calculus and Boolean logic to generate a normalised or absolute measure for the realisation of the hazard or threat often referred to as the top-event in view of the top-down nature of causal modelling. The key issues to bear in mind during the provision and statistical processing of data for the quantification of causal models are:

- reliable and objective sources for the basic errors and failures;
- consistent application of compatible data types;
- appropriate pre-processing of the data e.g. mean over a number of years;
- uncertainty and non-linearity in the data;
- sensitivity and importance criteria for the errors and failures.

Where input data is specified with confidence intervals or a significant sample size is available, the use of statistical simulation techniques is essential in generating a probability or frequency forecast for the hazard or threat.

3.1.2.4 Constraints

The causal modelling techniques are generally incapable of addressing temporal variations in data and only apply if frequencies and probabilistic errors and failures remain constant over time. Furthermore, causal models are often generated by individual domain experts and it is essential to subject these to peer review in order to enhance confidence in their integrity and correctness.

3.1.3 Consequence analysis

Whilst the causal analysis is aimed at establishing the factors leading to the realisation of a hazard or threat, consequence analysis is concerned about what may potentially follow the occurrence of a hazardous situation. This is the least understood and exercised mode of analysis to the extent that most established criteria for safety and security in vogue in industry are only concerned with the occurrence of a hazard and implicitly assume each occurrence necessarily equates directly with an undesirable catastrophic accident or loss. The notions of Wrong Side and Right Side Failure and their application as criteria for safety performance are indicative of this misunderstood discipline. In truth, the occurrence of a hazard may potentially lead to a broad range of consequences, some of which may probabilistically be undesirable events. The correspondence between the hazard and a catastrophic consequence/accident is seldom at parity i.e. it rarely follows that the existence of a hazard or threat can be assumed to correlate 100 per cent with the worst likely accident.

3.1.3.1 The process

The consequence analysis is a largely probabilistic and potentially creative process requiring domain experience pertaining to the application of the product, process, system or undertaking. The techniques of Consequence Analysis are generally applied recursively in a bottom-up or forward inference mode to a given specific state (hazard or threat) until all potential general consequences (incidents and accidents) are arrived at. This inductive approach generally produces a number of intermediate states, each probabilistically leading to a number of other likely intermediate states or consequences. The heuristic in this mode is to continue with induction at each intermediate state until all known barriers to the escalation of the hazard or threat are exhausted and all potential incidents, accidents or safe states are identified. The consequence analysis techniques are predominately applied within decision theory.

3.1.3.2 Modelling

The consequence analysis techniques generally employ graphical modelling which constitute a potent form for the capture and communication of the incidents, accidents and other benign states potentially arising from the realisation of a hazard or threat. Consequence models often in the form of trees lend themselves to quantification that ultimately generates a probability or frequency for each predicted incident and accident. The key issues to bear in mind during this modelling process are:

- clear understanding and definition of the hazardous or threat state to be analysed;
- existence of physical barriers (protection systems) to the escalation scenario ;
- existence of procedural barriers to the escalation scenario;
- existence of circumstantial barriers to the escalation scenario;
- the strength of each barrier's capability in preventing further escalation;
- the escalation path upon success or failure of the identified barriers;
- uncertainty and non-linearities in barrier strength;
- the inter-dependencies between various barriers to escalation scenario .

Where barriers to escalation are non-existent, it is possible to identify the need for protective measures in the course of consequence analysis i.e. the need for a non-existent detection system or procedure. Furthermore, if a hazardous situation is experienced and knowledge of its rate or likelihood of occurrence is at hand, consequence analysis would prove more beneficial than its causal counterpart in establishing the likely consequences and extent of potential losses. This might obviate the need for causal analysis in some circumstances.

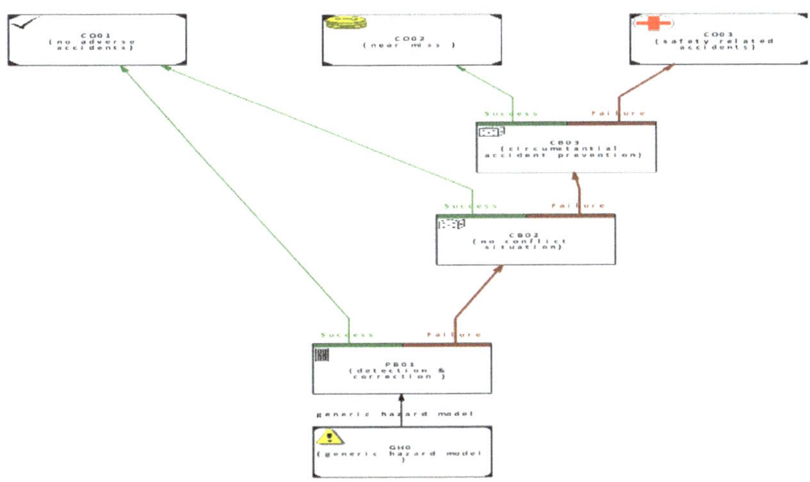

Fig. 3. Illustrative Consequence model for a railway hazard

It is prudent to explore the existence and effects of physical, procedural and circumstantial barriers to escalation scenarios associated with a hazard or threat in a systematic and ordered manner to ensure all potential safeguards are identified and incorporated in the consequence model. An illustrative consequence model for a railway hazard is depicted in Figure 3.

3.1.3.3 Quantification

The quantification of consequence models entails an objective assessment of the potential strength (likelihood for success) for all identified physical, procedural and circumstantial barriers. These are either based on historical data, result of specific causal analysis or expert judgement where no objective data can be traced. The key issues to bear in mind during the provision and processing of data for the quantification of consequence models are:

- reliable and objective sources for the barrier strength (success probability);
- appropriate pre-processing of barrier data;
- uncertainty and non-linearity in the data ;
- dependency of barriers;
- sensitivity criteria for the barriers within a model.

Where input data is specified with confidence intervals, the use of statistical simulation techniques is essential in generating a probability or frequency forecast for the consequences.

3.1.3.4 Constraints

The consequence modelling techniques are generally incapable of addressing interdependency, spatial and temporal variations in data and only apply if identified barriers to the escalation scenario retain a constant strength over time. Furthermore, in view of the probabilistic and creative nature, it is prudent to develop consequence models with the aid of a team comprising diverse domain experts as opposed to resorting to a single analyst.

3.1.4 Loss analysis

Loss comprises various degrees of harm to people, commercial/operational detriment to an enterprise or contamination/damage to the ecology of the environment or a combination thereof. It is associated with most undesirable consequences arising from the hazards of products, processes, systems and undertakings. Loss analysis constitutes the final stage of intrinsic hazard evaluation prior to adoption of reduction and containment strategies.

The statutory legal framework is mainly focused on prevention and regulation of harm to people and more recently, the environment. Commercial/operational losses on the other hand remain the prerogative of the business to avoid, transfer, mitigate, reduce or tolerate. In view of the diversity of needs and requirements, it is prudent to evaluate the losses associated with undesirable consequences in three distinct categories and aggregate these at a later stage. Loss analysis comprises the systematic investigation of the adverse outcome associated with all incidents and accidents identified through consequence analysis. The key processes in the evaluation of loss comprise:

- Safety Loss Estimation;
- Commercial/operational Loss Estimation;
- Environmental Loss Estimation.

Since the totality of loss is of the essence in the decision making process, upon evaluation, these need to be converted into a common currency and aggregated. The scale, scope and treatment of loss is context sensitive and in view of the inherent complexity, these are often treated subjectively.

3.1.4.1 Safety loss evaluation or estimation

The evaluation or estimation of measures of harm to people arising from undesirable consequences such as collisions, derailments, fires and a whole host of man-made and natural disasters is dependent on a large number of context sensitive factors. The significance, causal relationships and dependencies between these factors are not adequately understood and most industries currently resort to published historical data for the estimation of safety losses. This is often in the form of statistical Means over a number of years and is fraught with a number principal difficulties namely:

- irrelevance of a historical mean to specific circumstances under study;
- distortion of means caused by rare catastrophic incidents and accidents;
- insufficient data regarding causal and contributory factors;
- variability due to introduction of different generations of technologies and infrastructures;
- secondary effects e.g. fires, derailments subsequent to a collision or exposure to harmful substances;
- poorly understood relationship between circumstances and loss severity;
- multiplicity of the types and classes of harm.

It is prudent therefore to establish an objective process for safety loss estimation that is capable of generating forecasts for the specific circumstances under consideration. The current practice severely undermines the effort spent in causal and consequence analyses of significant hazards in the industry and reduces the accuracy of the overall assessment process. It is also incompatible with the systematic framework depicted in this chapter. In the interim however, historical Means have to be appropriately scaled and processed to take account of specific circumstances predicted by consequence analysis, in order to give a semblance of reality and systematicity.

Safety loss should be measured in Minor Injuries, Major Injuries, Fatalities and Equivalent Fatalities. This is a process through which, various degrees of estimated harm to a given group of people exposed to the consequences of a hazard, is aggregated into an equivalent fatality figure for decision making purposes. The current convention is to aggregate fatality, serious injury and minor injuries in 1 : 0.1 : 0.005 ratio respectively in order to generate an estimate for safety loss in Equivalent Fatalities.

3.1.4.2 Commercial/operational loss estimation

In addition to the potential safety implications, most incidents and accidents entail a measure of loss to the enterprises involved in terms of:

- disruptions to services causing delays;
- damage to movable assets;
- damage to infrastructure and equipment;
- loss of goods and material;

- loss of goodwill;
- loss of stake-holder/consumer trust and decline in custom;
- claims and potential legal fines;
- premium increases and other consequential losses.

An objective measure of these pertaining to the specific circumstances predicted by consequence analysis should be estimated, converted to a common currency (money) and aggregated to generate an overall figure for commercial/operational loss. Also note that whilst it is difficult to delegate safety and environmental duty of care, in the short term, their consequent losses including those due to commercial and operational loss can be largely transferred through contractual agreements and insurance.

3.1.4.3 Environmental loss estimation

Apart from the commercial implications, release and dispersion of harmful substances in the environment as a result of incidents and accidents poses threats to health and safety as well as the eco-system. These may typically involve any combination of:

- fuels, oils, flammable substances;
- liquefied gases, explosives;
- caustic, corrosive and reactive chemicals;
- minerals and reactive material;
- radio-active materials;
- bio-toxins.

In addition to the immediate effects, further damage may be caused through dispersion into the atmosphere and contamination of land, water tables and rivers. The specific circumstances should be identified through consequence analysis. The environmental loss estimation may potentially involve an evaluation of the costs associated with:

- clean-up operations;
- containment strategies;
- emergency services;
- fines by Environment Agency, Rivers Authority etc.;
- Claims by other affected parties.

Systematic causal and consequence analysis may also reveal the need for further barriers to scenario escalation including protection systems, damage containment policies and emergency preparedness measures. It is prudent therefore to develop an objective process for the estimation of likely effects of incidents and accidents on the environment and convert and aggregate these in a common currency (money) to generate an overall figure for the environmental loss.

3.1.4.4 Loss integration

The three broad categories of Safety, Commercial and Environmental loss may be realised as a result of incidents and accidents pertaining to hazards associated with the products, processes, systems and undertakings.

The evaluation of Health and Safety losses are required under the UK and most statutory frameworks in order to establish the tolerability and reduction of these to within reasonably practicable levels. Further to legal compliance, the knowledge of the extent and scope of the safety, commercial and environmental losses provides the objective data for prudent business decision-making. However, it is useful for all three components to be converted and expressed in a common currency such as money for potential comparison and aggregation in order to provide a coherent view of the totality of potential loss associated with a hazardous situation. This ensures that safety and environmental issues become integral to often largely commercially driven decision making, enabling a realistic and balanced perspective on risk management within the enterprise.

The commercial and environmental losses or risks associated with each hazard are generally expressed in monetary terms. The safety loss or risk on the other hand is measured in terms of harm to people generally in the form of estimates or statistics pertaining to injuries and fatalities. A convention exists for normalising injuries and converting these to an Equivalent Fatalities (Lives). It is then possible to add injury forecast or statistics to fatality forecasts/statistics and produce a single estimate for safety risks in terms of Equivalent Lives. For aggregation with other mainly monetary losses, safety loss forecasts can further be converted into their equivalent monetary value employing the concept of Value of Preventing a Fatality (VoPF or VPF). This mainly statistical concept is sometimes referred to as Value of Preventing a Statistical Fatality (RSSB, 2006) and is purely employed to support safety related decision making and should not be misinterpreted as putting monetary value of lives of individuals. It is customary to employ the product of equivalent fatalities estimated for a product, process, system or undertaking by the industry benchmark for VoPF to develop an objective measure of total safety losses as a basis for further safety investment and prevention of such losses. This is intended to transform safety based investment and decision making from a fundamentally moral imperative to a rational process that can be contrasted and enforced globally with the key variant being the VoPF for the given circumstances under consideration. The adoption of a systematic risk framework and setting of a global value for VoPF to underpin enforcement of safety considerations in major undertakings is an imperative for transparency, fairness and demonstration of duty of care as witnessed in the controversy surrounding the large scale North American oil exploration disasters (BP, 2010).

3.1.5 Options analysis

The hazard identification process reveals a portfolio of circumstances that are subsequently prioritised and analysed through causal, consequence and loss analyses. Depending on the consequent losses, the hazards may subsequently require risk elimination, mitigation, transfer, control or an appropriate combination thereof. The identification, ranking, evaluation and management of viable pro-active hazard rate reduction (causal level) and largely re-active containment (consequence level) strategies constitute option analysis.

The identification and ranking of options is carried out within a process analogous to that defined for the hazards although, causal and consequence analyses of a hazard also serve to characterise appropriate rate reduction and containment strategies.

A number of options should generally be identified and recorded for each hazard or groups of synergistic hazards, taking into account established and emerging technologies. The options portfolio comprises those that precede the occurrence of a hazard (RO type) and those that are effective post hazardous event (CO type). The options that precede the hazard horizon are primarily aimed at elimination or rate reduction hence labelled as Reduction Options (RO). The RO type measures are generally aimed at the prevention or retardation of the causal factors and are usually evaluated with the aid of causal analysis tools.

The options that are effective post occurrence of a hazard are mainly aimed at loss Containment (CO) and constitute further barriers to the escalation scenario. The CO type measures are usually identified or assessed with the aid of consequence analysis. Irrespective of the type mix, the options portfolio must be reviewed at reasonable intervals in order to ensure compliance with the ALARP principle in the context of management of safety risks in the UK or other statutory criteria.

A sensitivity parameter may be derived for the RO and CO type options through the causal and consequence models in order to ascertain the most effective measures for risk reduction and containment.

For each option, the annualised or the Net Present Value of the associated costs over the effective life must be evaluated and assessed as appropriate and recorded for comparison against potential benefits derived during through impact analysis.

3.1.6 Impact analysis

Upon identification and recording, it is essential to estimate the likely effects and potential benefits of each option on the consequent safety, commercial and environmental losses in order to establish the objective and systematic criteria for selection and implementation. This is a requirement of the statutory legal framework in the UK to ensure and demonstrate that the safety risks arising from a product, process, system or an undertaking are reduced to As Low As Reasonably Practicable (ALARP) levels.

Impact analysis comprises a systematic analysis of the beneficial and any detrimental effects of implementation of an option with a view to eliminate, reduce, mitigate, transfer or control the risks rising from a given hazard.

3.1.6.1 RO type impact

The pro-active elimination or Reduction (RO) options are generally the preferred type and require treatment within the context of causal analysis. This generally involves incorporation of the option within the causal model or an assessment of its likely effect on the causal factors and appropriate adjustment of the rates or probabilities for each affected error or failure. The consequent safety, commercial and environmental Losses are subsequently re-evaluated through consequence and loss analysis. The Equivalent Lives differential thus evaluated pre and post implementation of an option should be recorded together with corresponding commercial and environmental loss differentials. These collectively constitute the Impact Parameters associated with the option and are employed in conjunction with the cost estimate in order to derive the safety and business criteria for the implementation of the option.

3.1.6.2 CO type impact

The mainly re-active Containment (CO) type options comprise detection and protection systems and procedural barriers to further escalation of a hazard. This class of options are generally effective in the post hazard horizon in that they will not affect the realisation of a hazardous state but assist with reducing the likelihood of a hazard transforming into accidents or the consequent accidents causing as much loss. The CO type measures should be evaluated through the consequence model of a hazard with their probability of success judiciously set to reflect their potential effectiveness on demand. In view of the time and resource implications, the CO type options should preferably be incorporated into a consequence model during the knowledge elicitation and capture of consequence scenarios. In this case, their effectiveness should be defaulted to zero until impact analysis provides the necessary criteria for implementation or dismissal.

In a similar process to that for RO type options, the evaluation of CO type measures entails the derivation of resultant safety, commercial and environmental Losses/risks subsequent to the adjustment of the effectiveness parameter. The Equivalent Lives differential evaluated pre and post implementation of the CO option should be recorded together with corresponding commercial and environmental loss differentials ideally computed in net present value terms if the effects are considered over a period of time. These Impact Parameters are employed in conjunction with the cost estimate arrived at during options analysis, ideally computed in net present value form, in order to derive the safety and business criteria for the implementation of the option.

3.1.7 Demonstration of ALARP and compliance

The demonstration of compliance with the regulatory requirements and the ALARP principle in the UK (HMSO, 2001) necessitates an assessment of individual risks arising from the undertaking, product, process or system for the members of the affected groups (Employees, Customers and the General Public). Individual risk represents an average across a group and its assessment is contingent upon the knowledge of the totality of risk and the size of the exposed group within the population. It is also customary to consider the most at risk amongst the groups affected since the exposure patterns will differ even for the members of the same group. However such detailed differentiation is only justified when patterns of risk exposure in a given population or group are vastly different to the average and supporting data justifies such elaborate considerations.

3.1.7.1 Demonstration of ALARP

Within the context of UK regulatory legal framework, a duty is imposed on those who create a specific risk to health, safety and welfare of their employees, customers and the general public to ensure and demonstrate that these are reduced to As Low As Reasonably Practicable (ALARP) levels. The criteria for tolerability of risks has been published by UK's Health and Safety Executive (HSE), in terms of numerical targets for the individual risk of fatality for a specific group of people, exposed to the risks arising from a product, process, system or an undertaking. The HSE criteria effectively define an upper quantitative limit for individual risk of fatality beyond which, risks should not be tolerated, save in extraordinary circumstances. Risks falling below the upper limit of tolerability are expected to be subject to mitigation on a cost benefit basis, unless these are around two orders of magnitude smaller than the upper limit. It is important to note that the tolerability concepts apply at a holistic level i.e. to the totality of risks and not generally at the individual hazard level.

The demonstration of compliance with the legal duty of care and ALARP principle entails the following stages:

- identification of the hazards and the exposed groups potentially associated with the application of a product, process, system or an undertaking and treatment of the hazard portfolio under the qualitative and/or quantitative framework as appropriate with a view to assess the likely safety losses/risks associated with each hazard;
- development of a total risk profile for the product, process or undertaking in safety terms (the evaluation of the commercial/operational and environmental risk categories should also prove valuable);
- identification of elimination, rate Reduction (RO) or Containment (CO) option(s) for each hazard;
- determination of the net present value cost and impact of option(s) on the safety loss associated with the corresponding hazards;
- determination of cost effectiveness for the identified options and derivation of Cost Safety Benefit (equivalent cost of saving a fatality through application) for each option;
- Implementation of all options for which Cost Safety Benefit is smaller or equal to the Value of Preventing a Fatality (VoPF) or other industry criteria (Hessami, 1999). The concept of gross disproportion should be applied to disparity between the Cost Safety Benefit of an option and the VoPF convention, depending on the magnitude of the total risk;
- recording of the data, assumptions, calculations and consequent decisions.

Whilst this process accords with the guidance given for the qualitative assessment of less significant risks, it is insufficient within the context of major risks that may violate the tolerability criteria. However, it ensures that all risk elimination, mitigation and control options are assessed and implemented, thus reducing the totality of risks to As Low As Reasonably Practicable level, but cannot determine tolerability against the published benchmarks. Furthermore, for new products, systems and processes, by focusing on individual hazards, the approach only ensures compliance with ALARP for the adopted design or approach. It would not guarantee the optimal low risk solution that might involve a different hazard portfolio. Optimisation of risks arising from products, processes, systems and undertakings is beyond the scope of the current discussion.

The determination of the tolerability of risks and the significance of gross disproportion as a criterion for the implementation or dismissal of the options requires a comparison of the totality of risks against apportioned industry benchmarks. This is achieved through the complementary demonstration of compliance stage.

3.1.7.2 Demonstration of compliance

Whilst the achievement and demonstration of lowest practicable levels of risk is broadly sufficient for the demonstration of legal duty of care, it is not suitable for determination of the tolerability, against industry performance benchmarks. Furthermore, in dealing with major risks within a quantitative framework, implementation of risk reduction and containment options cannot be carried out in isolation from the knowledge of the position of overall risk within the tolerability scale.

The industry safety performance benchmarks in the UK are generally derived from the Health and Safety Executive's guidance on industrial risks and criterion for tolerability. However, the benchmarks represent an annual average for the individual risk, influenced by a vast and diverse range of products, processes and undertakings. These benchmarks generally lie within the middle of the tolerability scale for each affected group, which is bounded by upper and lower numerical limits.

The comparison of the aggregated risks of a product, process or an undertaking with the published benchmarks requires an assessment of the contribution of the particular item under consideration to the industry's annual safety performance. This is known as apportionment, which in the absence of a systematic dynamic model for the whole of an industry is an un-productive and unsystematic exercise. In the absence of such a model, a simple rational argument and calculation for apportionment is preferable to the often wasteful and expensive efforts in manipulating historical data.

The demonstration of compliance with the industry safety principles and performance benchmarks entails the following stages:

- A review and justification of all identified hazards and mitigation options against industry or regulatory safety principles;
- aggregation of the safety loss of the hazards in the portfolio generating a total risk estimate for the product, process or the undertaking for each affected group;
- estimation of the size of population exposed to the risks in each group;
- calculation of the average risk per person in each group;
- apportionment of the industry benchmarks to the specific contribution of the product, process or undertaking;
- comparison and determination of tolerability against apportioned benchmarks;
- if the risk is intolerable, i.e. it exceeds the upper level of tolerability, it shall be reduced to within tolerable levels or the product, process or the undertaking abandoned, save in extraordinary circumstances;
- if the risks are tolerable, follow a process as for demonstration of ALARP bearing the following in mind:
  - if the computed individual risk is close to the upper limit of tolerability, a gross disproportion between the Cost Safety Benefit and VoPF should be the criterion for implementation of RO and CO options.
  - if the computed individual risk is close to the lower limit of tolerability, the parity between the Cost Safety Benefit and VoPF should be the criterion for implementation of RO and CO options.

In view of the current uncertainties and inaccuracies inherent in the apportionment process, the demonstration of compliance with the industry benchmarks should be treated as a coarse and relative indicator of safety performance of products, processes and undertakings. It is imprudent therefore to treat the individual risk calculations and the apportioned benchmarks as the sole dependable absolutes for decision-making.

3.2 Risk management system – Principles

Compliance with the requirements cited above requires a systematic scrutiny of defect/error-failure-accident scenarios to ensure a comprehensive risk perspective. In reality, adopting a hazard and threat based approach to risk assessment and management generates a more systematic framework for coping with varieties of risks. A defect-error-failure sequence is proposed to address the processes leading to the realisation of a hazardous state or event in a product, process or system. Consideration of the post hazard horizon in this approach involves identifying the potential escalation scenarios, the defences against accidents, the range of accidents that arise due to the failure of defences and optimal response and recovery regimes for each major accident scenario. In a similar manner to the assessment regime, the systematic framework for risk management comprises the following seven principles:

i. Prediction and Proactivity;
ii. Prevention;
iii. Containment & Protection;
iv. Preparedness & Response;
v. Recovery & Restoration;
vi. Organisation & Learning;
vii. Continual Enhancement.

These principles collectively address the total risk landscape and are inter-related in a systemic fashion. They also relate to the framework for risk assessment in a consistent and demonstrable way. The principles are detailed below.

3.2.1 I – Prediction and proactivity principle

The primary principle in systematic assurance is that of "prediction" which involves analysis and identification of credible system modes and potential loss/hazardous states, anticipation of escalation scenarios, evaluation and assessment of the baseline risks and taking hard and soft risk control measures in advance of foreseeable accidents. This by necessity involves developing and implementing methods and procedures to assess the risks and establish the baseline performance in order to support the case for further risk reduction, control or mitigation as appropriate.

The principle is the focal point for the identification (prediction) of foreseeable activities, modes and states within a system that adversely affect performance (safety, operational, environmental, RAM etc.) comprising Normal, Degraded, Failure and Emergencies and the triggers and transitions for these.

The administrative, strategic and implementation facets of performance are addressed through "proactivity" comprising policy, planning, resourcing and determination of strategy and plan for compliance with existing, emerging and modified directives, regulations, rules and mandatory standards. Proactivity also implies setting the ground rules and the scene for Prevention, Protection, Response and Recovery policies (see other principles).

Establishing communications channels between internal and external stakeholders including the production of a Safety Case for the organisation/undertaking, a Safety Management Manual, a Document Management System and a Configuration Management and Change Control System also fall within the scope of Proactivity.

3.2.2 II – Prevention principle

Once the baseline performance is established through "prediction" and the need for risk reduction is identified, the "prevention" strategy provides the most logical and prudent approach to the realisation of this objective. Prevention principle addresses the analysis of the known and new hazards/threats, understanding of their causation chain and identification of the measures capable of eliminating or reducing the likelihood of occurrence of the threats/hazardous states.

Prevention strategies are best attempts at reducing defects, errors and failures and comprise a broad range of technical, procedural and human competence related measures. This is the cornerstone of most industries' traditional approach to ensuring safe/secure states through design and implementation of fail safe systems, inspections, preventative maintenance, selection, training and briefing of staff. However, whilst prudent, these measures fail to completely eliminate or control the threats/hazardous states thus assurance of desirable performance of the overall system cannot be relied upon the success of preventative strategies alone.

The "prevention" focus ensures all causations and escalation routes to the threats/hazardous states are identified, analysed and all credible and reasonably practicable elimination and control measures are evaluated and implemented. This includes scheduled and preventive maintenance activities aimed at maintaining the functionality and integrity of the system.

3.2.3 III – Containment & protection principle

The thrust of the classical approach to performance assurance of systems, services and operations is embodied in the designs, architectures, rules, processes, systems and behaviours that are mainly based on the Prevention philosophy as cited before.

Whilst allocating resources and focusing attention on *prevention* is rational and prudent, it should not be at the expense of the mitigating risks, once undesirable hazardous events occur or threats are realised. The aim here is to determine the escalation mechanisms/scenarios for hazardous conditions and establish strategies, responsibilities and timely responsive action aimed at containing the energy or potential of hazardous states/threats in such a manner that they would not escalate into accidents potentially causing commercial, environmental and human harm/loss.

The preference here is to set up effective barriers to escalation and where possible, turn loss prone or hazardous occurrences into incidents or lower severity accidents. The second aspect to this is to attempt to "Protect" the people/property at risk against potential injuries, fatalities and collateral damage should accidents occur or attempt to reduce the severity of such harm/damage.

The Containment and Protection Principle is developed and proposed in recognition of the fact that in spite of major efforts by duty holders, hazardous states do occur and threats do materialise in any system or environment often driven by complexity and change or adoption of unproven yet promising technologies. It is prudent therefore to have strategies, plans and measures in place to reduce the harm which would otherwise be caused by the escalation of these states if not controlled in a timely and effective manner.

The Protection focus ensures that the escalation paths for credible loss prone/hazardous states are recognised and reasonably practicable measures (barriers) are identified, assessed and adopted/strengthened to detect and rectify the hazard/threat escalation and where not possible mitigate the consequences.

3.2.4 IV – Preparedness and response principle

The essence of risk management lies in the success of the Proactivity, Prevention and the Protection strategies and prudent risk control initiatives. However, in view of the complexities inherent in the many industrial, infrastructure and service sector operations, accidents do occur from time to time. In the same spirit, a high degree of anticipation and preparedness for responding to accidents, emergencies and degraded modes of operation is an integral facet of ensuring the impact is kept to a minimum.

The *preparedness* is an aspect of organisational and resource planning and provision which entails anticipating, planning, resourcing, training and clarifying roles, responsibilities, communications, command structure and resources to address critical classes of degraded, failure and emergency states occurring within the operational environment. This by necessity requires a degree of learning from past experience as well as anticipating new scenarios when changes are enforced to the organisation, composition, structure or the operation of the systems being managed.

The *response* dimension of the principle is mainly concerned with the implementation of the Preparedness plans comprising:

- mobilising resources for presence on the scene and in support roles;
- protecting the site;
- evacuating the affected parties and the public;
- determining a command structure to manage each event;
- informing relevant civil authorities and emergency services with a view to protect and rescue those exposed or involved in the circumstances and minimise the degree of harm which would otherwise be sustained;
- minimising overall harm and loss arising from an accident.

The *preparedness* and *response* principle also addresses contingency scenarios i.e. new/unexpected degraded, failure and emergency aspects and circumstances for which, a general class of reaction is required as a safety net against all unforeseen cases. The *preparedness* and *response* focus ensures optimal reaction to accidents, catastrophes and security related losses is recognised and attained with a view to minimise safety and property losses in such circumstances.

3.2.5 V – Recovery & restoration principle

The timely and appropriate response to incidents and accidents ensure that people and collateral exposed to threats, hazardous states or accidents receive optimal help and support with a view to minimise any harm/damage which would otherwise be incurred in the circumstances. However, depending on the severity and nature of the degraded, failure or emergency state, a degree of anticipation, advance planning and resourcing is required to initiate timely and efficient *recovery* activities on the affected system or infrastructure.

Recovery after incidents and accidents essentially begins after *response* process has resulted in securing the safety of the affected or exposed people and is mainly concerned with the processes and resources to repair the damage incurred in a safe, timely and efficient manner working towards the *restoration* of the system to normal state/service. It may also arise from disturbances to the system including preventive or reactive maintenance when the system is being brought back to normal operational state. Depending on the nature of the degraded, failure or emergency, the *recovery* activities may additionally impose various risk control restrictions on the functionality, infrastructure or the operation of the system.

The *restoration* addresses the rules, processes, roles, tests, competencies and authorities required to ensure the state of the infrastructure or operations after the *recovery* activities are technically sound, efficient, affordable and acceptably safe for return to restricted or normal service. In this spirit, *recovery* and *restoration* are assurance related activities. Restoration may be achieved in a number of phases culminating in the full resumption of the normal operational state.

The *recovery* and *restoration* focus ensures the repairs to the infrastructure and the service/production system post disturbances (including maintenance) and accidents is carried out in a safe and efficient manner and the subsequent deployment is subject to a systematic test, verification and validation process.

3.2.6 VI – Organisation & learning principle

The achievement, maintenance and improvement of the overall performance of any system or operation is contingent on timely appropriate actions assured through a learned and competent human organisation.

The Organisation Principle addresses the entire spectrum of human resource issues pertinent to the maintenance and improvement of performance. These include recruitment, induction, deployment, training, development briefing and communication of critical issues, qualifications, physical fitness, certification and regular verification and validation of the capabilities and competence.

Traditionally, assurance is treated as a specialist discipline and relegated to a particular group of staff solely concerned with this objective. However, whilst performance assurance like other disciplines has its specialist niches, its recognition, understanding of the underlying concepts, care for other people's health, safety and welfare constitute a broad suite of beliefs, values and practices referred to as organisational culture. The recognition, promotion and nurturing of this culture is a crucial factor in the success of policies and initiatives within an organisation. Assurance culture promotes the notion that apart from specialist activities, knowledge, practices, beliefs and values in accident prevention should be common to all who have a role in the provision of service or systems with a potential to cause harm to the customers, employees and the general public or damage to the environment and property.

The organisation does not necessarily imply a dedicated arrangement for risk management, fundamentally separate from other functions of the business or service organisation, infrastructure management or other stakeholders. Apart from specialist activities, a supportive and pervasive assurance culture must be developed and promoted throughout the enterprise including education, briefing and establishment of a confidential channel for communication of observations, suggestions and feedback on all performance related matters. In this spirit, the principle underpins all other aspects of the framework since it provides the human motive force for realisation of all other principles inherent in assurance management.

The other facet of the organisation principle is the ability to learn and capitalise on the new and emerging knowledge for improving performance. A key instrument supporting the learning process is development, implementation and maintenance of a corporate memory to underpin the recording, retrieval and processing of relevant knowledge and resultant learning. The corporate repository of performance information must include an up-to-date directory of infrastructure, systems and operational threats/hazards that needs to be initiated at system level whilst being updated for local conditions. This repository must be made accessible to all stakeholders to inform them about all pertinent issues which may relate to their roles, tasks and undertakings within the system.

The repository of performance information should additionally include records of reported failures, threats, incidents and accidents and any analysis establishing causation, escalation mechanisms and the degree of harm or damage caused. It is crucial that these are captured, shared openly and employed actively to enhance systems and processes with a view to prevent future occurrences (prevention principle). This is a costly but essential aspect of learning from what in principle amounts to the failures of the Management System.

Finally, the organisation principle must cater for the relationships, reporting structure, licensing and responsibilities of various organisations involved in the design, installation, operation, maintenance and disposal of the infrastructure, service delivery, production system and its constituents.

The focus on organisation and learning ensures that competent people are recruited, trained and tasked with assurance related activities and lessons are learnt from faults, failures, incidents and accidents with a view to eliminate or minimise future occurrences.

3.2.7 VII – Continual enhancement principle

The principles and their inherent activities cited earlier can underpin achieving and sustaining a desirable performance in the context of a product, process, service, system or undertaking. However, improving quality of life, advancing social values and consequent emerging legislation, rules and standards tend to demand more stringent targets, more responsive behaviours and improving overall performance. The other key driver is the rising consciousness in the society about duty of care and negligence by people and organisations delivering services and products and the consequent criminal and civil claims in the event of accidents causing harm to victims or financial loss to the stakeholders.

The inherent complexities of the infrastructure, production systems and operations in industrial and service sectors as well as the increasing demand for incorporation of novel technologies pose a challenge to the maintenance of performance levels during the transition. A rational, systematic and scientific approach to the traditionally empirical treatment of assurance matters in the industry is called for. Identification of key performance indicators, measurement and proactive control of risks are key instruments in the new approach.

The *continual enhancement* of various facets of performance necessitates an objective appreciation of the existing drivers, actors, faults, failures, hazards, threats, targets and existing performance levels before reasonably practicable options are identified, assessed and adopted for improvement. To this end, a comprehensive approach to identification, monitoring and measurement of precursors to accidents, agreement on relevant performance criteria and normalising factors, audit of safety and security processes and culture, review of targets and making a case for performance improvements constitute the essence of this principle.

The enhancement of performance may arise from the introduction of novel feature/functionalities, identification and strengthening of the barriers to causation or escalation of the hazardous states or complete elimination of hazards through adoption of new materials and technologies. The extent and scope of the performance improvements may be driven by revised targets, new standards or emerging lower cost technologies making risk reduction reasonable when contrasted against the likely gains.

The corporate repository of performance information cited under the *organisation* principle should also be actively reviewed for detecting trends in the underlying causes and breaches, precursors to accidents and near hits (strangely referred to as near misses). This information should be communicated with all stakeholders and employed as a potent tool to systematically eliminate the unacceptable levels of faults, failures and errors arising from human or automation sources, thus preventing accidents.

The focus on *continual enhancement* ensures attainment of tolerable levels of overall performance is treated as a dynamic and evolving objective subject to a systematic and on-going measurement and assessment regime to support credible understanding of the performance thus underpin the need and quest for sustaining good performance and enhancement.

3.2.8 Risk management framework

The seven principles inherent in the performance assurance of products, processes, services, systems and organisations fall into three broad categories;

The first principle, *proactivity*, is mainly concerned with establishing an environment and a baseline for the product, process, service, system or organisation in terms of its desirable properties and performance. It represents an antithesis to reactivity in facing the potential of accidents. In this spirit, *proactivity* is fundamental to the achievement and improvement of performance since it emphasises that plans and resources must be devised, secured and applied in advance of incidents, threats and accidents to enable the duty holders to eliminate, control or mitigate the risks.

The second group comprising *prevention*, *protection*, *response* and *recovery* are mainly associated with causation and escalation of accidents and the optimal preparedness in responding to these and emergencies with a view to minimise losses.

The third and final group of two principles relate to the significant role that the human organisation, communications, responsibilities, competencies, certification, regulation and corporate memory/learning play in the attainment and improvement of overall performance. This includes a drive for continual enhancement based on an audit, measurement and feedback loop to ensure a set of common indicators are continually monitored to empower the duty holders to take effective remedial and improvement actions as appropriate.

The seven fundamental principles collectively constitute a systematic and systemic framework for assurance of overall performance in the face of threats and risks. These are outlined in Table 1.

Principle	*Scope & Intent*
I. Prediction & Proactivity	Setting Policy and Strategy, identifying all stakeholders and interfaces, Hazard/Threat Identification, planning, resourcing and data collection. Modelling, assessing baseline risks, identifying key performance indicators and implementing policy. Developing Safety, Security & Sustainability Cases and relevant Management Manuals
II. Prevention	All measures, processes, activities and actions including maintenance aimed at eliminating or reducing the likelihood/frequency of threats/hazardous states with a potential to cause harm and loss
III. Containment & Protection	All measures, processes, activities and actions aimed at reducing the likelihood/frequency or severity of potential accidents arising from the hazardous states or security breaches
IV. Preparedness & Response	All plans, measures, processes, activities and resources relevant to managing degraded and failure modes and emergencies, investigation of the causes, collection, maintenance & sharing of records
V. Recovery & Restoration	All plans, measures, processes, activities and resources relevant to recovery from planned and unplanned disturbances, degraded and failure modes and emergencies towards full resumption of production/service including the criteria and organisation for authorising the system back into service post disruptions and emergencies
VI. Organisation & Learning	Structuring, communications, training, certification, competencies, roles & responsibilities and validation for human organisation as well as ensuring lessons are learnt from incidents and accidents and key points recorded, shared and implemented
VII. Continual Enhancement	All processes associated with setting and reviewing targets, measuring/assessing, processing, auditing, reviewing, monitoring, regulating and sustaining/improving performance including decision aids and criteria

Table 1 The Systemic Assurance Framework of seven Principles

The framework depicted in Figure 4 represents a constellation of complementary and inter-related principles which when applied collectively, can systematically underpin the attainment, maintenance (principles I-VI) and improvement (principle VII) of overall performance. A framework founded on systemic principles is more fundamentally credible, stable and universally applicable than specific context related suite of actions, processes or methodologies.

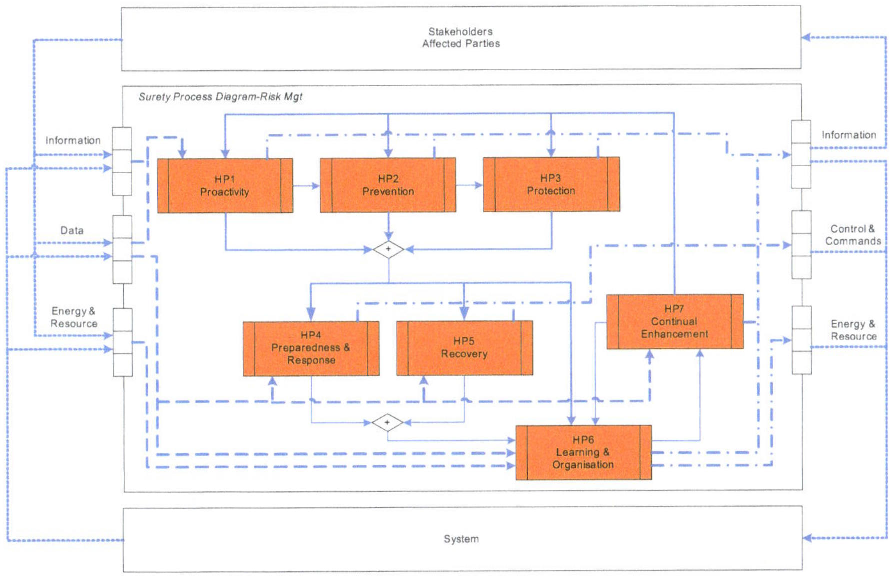

Fig. 4. The Systematic Framework for Risk Management, its Interfaces and Interactions

3.2.9 Systemic characteristics of the management framework

Whilst holistic and complete, the proposed framework for risk management possesses essential properties such as simplicity, rationality and a level of abstraction that lends it adaptable to any context, scale and organisation. These are crucial to the stakeholders understanding, adapting and applying it to optimal effect.

The framework transforms the traditional focus on accidents and loss to understanding, control and management of threats and hazards. This fundamental shift of emphasis yields a more profound knowledge on the root causes of faults, errors and failures thus resulting in a more effective management of business and operational risks.

The framework sets out all the building blocks for systematic risk management starting with establishing the environment and baseline performance (principle I) leading to four focal points (principles II-V) for actualising plans and policy. A major emphasis is also placed on the organisational facets from performance focused structure, roles, responsibilities, accountabilities, competencies and communications to the more subtle cultural aspects (principle VI). Finally an active learning ethos and actualisation of learning in improvement of overall performance is emphasised in principles VI & VII. The intangible human dimension related to buy-in, motivation, participation, conflict resolution and taking people and property into account in everything we do is often ignored or not given sufficient prominence in existing management frameworks and standards.

The principles are not things to do per se. They constitute a complete strategic perspective and roadmap providing the essential focal points for the requisite activities and processes inherent in the systematic assurance of performance in products, processes, services, systems and undertakings. In this spirit, each principle also constitutes a focal point for measurement, benchmarking and determination of the status, success or shortcomings of the specific aspects of the Risk Management System.

The principles within the framework are goal-oriented and apart from guidance on the purpose and nature of essential activities, are designed to allow specific stakeholders to adapt these to their roles and circumstances and innovate to improve performance. This is particularly relevant to the historically diverse nature of the international trade with different cultural and structural underpinnings to the participants and stakeholders.

The four key focal points (principles II-V) on the actualisation of the plans and policies empower duty holders to collaboratively contribute to the overall performance of their operations. These principles would naturally involve a different set of activities for an each stakeholder organisation but none-the-less remain equally applicable at the framework level hence the need for scalability and adaptability.

The proposed principles are valid at any stage of the life-cycle (ISO/IEC, 2002) therefore, they are equally applicable to any group or organisation involved in the provision of service, products or management of infrastructure, production and operations. These can provide proactive indicators to assist the duty holders with their tasks as well as those responsible for the supervision and regulation of the relevant industry.

It would therefore be feasible to audit, assess and score an organisation's processes, capabilities and maturity in Proactivity, Prevention, Protection, Response, Recovery, Organisation and Continual Enhancement as appropriate to the nature of the undertaking. These scores and proactive criteria when benchmarked, will signify the status, strengths and shortcomings of an organisation in their systemic approach to the management of risks (Hunter & Hessami, 2002).

Apart from audit, assessment and scoring of the individual principles, it is also possible to generate an overall index of merit for the performance of the whole framework, thus giving a holistic indication for the capabilities and maturity for an organisation in its risk management endeavours. This provides an objective and constructive framework for intra-industry benchmarking, comparisons and enhancements.

The proposed framework is founded on seven systemic principles that can underpin performance assurance when applied in aggregate. In this spirit, the architecture of the proposed framework is entirely scalable and can be adopted to manage risks at the level of a product, process, team, project, department, organisation, an alliance of organisations and an industry as a whole or any larger aggregate of these constituent entities. At every level of the application, the essential invariant aspects of the framework i.e. the seven inter-related principles, require mapping and adaptation to the nature, scale, context, tasks and the application.

4. The way ahead

Our systems approach to the holistic treatment of risks recognises the need for examination, understanding, characterisation and assessment of principal threats and hazards followed by a requisite suite of principles as a focal point for monitoring, supervision and management of resources to sustain performance. This has resulted in two systemic frameworks, one focused on identification, evaluation and assessment of risks and the other comprising seven principles on the performance assurance and management of risks.

The seven principles underpinning the assessment of risks cited above constitute a comprehensive and disciplined framework capable of rendering a thorough understanding of the key threats, hazards and the magnitude of potential risks associated with these in a given context. However, these are not adequate to maintain effective control and assurance.

The approach to the holistic management of risks is best served through a systemic framework comprising principles that hold true in different sectors, levels of hierarchy, contexts and circumstances. The principled approach generates consistency, integrity and a familiar harmonised process to underpin assurance activities. However, the principles in a framework only constitute focal points for allocation of resource and energy and require mapping to the specific characteristics and demands of an environment, sector, system or undertaking.

A framework of seven principles developed and proposed for risk management addresses the risk management requirements comprehensively and holistically. The framework is equally applicable to security issues pertinent to the malicious intents and can provide one consistent and systemic environment for successful management of safety, security and potentially sustainability risks pertinent to products, processes, services, systems and undertakings in railways.

Because of its principled constitution, the two frameworks are scalable and can be applied at any level and within any industrial, infrastructure and service sector context. The risk management framework has been adopted by the EU - Project "Safety and Reliability of Industrial Products, Systems and Structures" (SAFERELNET), funded by the European Commission under the "Competitive Sustainable Growth" programme (SafeRelNet, 2006). This has recently been published in an accompanying book to the work of the SafeRelNet network (Guedes Soares, 2010).

5. References

A Look at China's High-Speed Rail Investments, http://www.solarfeeds.com/, Tuesday, 04 May 2010 The Green Leap Forward

Chaudhury, Abijit. Jean-Pierre Kuilboer. (2002). *e-Business and e-Commerce Infrastructure.* McGraw-Hill, ISBN 0-07-247875-6.

Chrissis M.B., Konrad, M., Shrun S. (2007) *CMMI Second Edition, Guideline for Process Integration and Product Improvement*, ISBN 0321279670.

Engineering Safety Management Issue 3 (Yellow Book III), Volumes 1 & 2, Fundamentals and Guidance, Railtrack PLC UK, January 2000, ISBN 0 9537595 0 4.

European Standard EN50129 *Railway Applications – Communications, Signalling and Processing Systems – Safety Related Electronic Systems for Signalling*, CENELEC - February 2003.

Hessami, A. (1999) *Risk Management a Systems Paradigm*, Systems Engineering-The Journal of the International Council on Systems Engineering, Volume 2 Number 3, pp156-167.

Hessami, A. (1999). *Safety Assurance, A Systems Paradigm*, Hazard Prevention- Journal of System Safety Society, Volume 35 No. 3, pp8:13.

Hessami, A. (1999). *Risk, A Missed Opportunity*, Risk and Continuity Journal, pp2:17-26.

Hunter, A. and Hessami, A.G. (2002). *Formalization of Weighted Factors Analysis*, Knowledge-Based Systems.

Hessami, A.G. (May 2004) *A Systems Framework for Safety & Security - The Holistic Paradigm*, Systems Engineering Journal USA Volume 7, Issue2. http://www.demos.co.uk/

ISO/IEC15288 (October 2002), *System Life Cycle Processes* - ISO/IEC.

Miller, R. *The Legal and E-Commerce Environment Today* (Hardcover ed.). Thomson Learning. pp. 741 pages. ISBN 0-324-06188-9.

Palmer, C. (December 1998). *Using IT for competitive advantage at Thomson Holidays*, Long Range Planning Vol21 No6 p26-29 Institute of Strategic Studies Journal. London. Pergamon Press.

Reducing Risks, Protecting People, HSE's Decision Making Process, HMSO 2001, ISBN 0 7176 2151 0.

Report on the Causes of the Gulf of Mexico Tragedy, BP, 08 September 2010. http://www.bp.com/genericarticle.do?categoryId=2012968&contentId=7064893

+Safe Version 1.2, *A Safety Extension to CMMi-DEV Version 1.2*, Defence Materials Organisation, Australian Department of Defence, March 2007.

SafeRelNet European Network of Excellence, www.mar.ist.utl.pt/SAFERELNET Guedes Soares, C. (editor 2010), *Safety and Reliability of Industrial Products, Systems and Structures*, CRC Press, PP:21-31, ISBN 978-0-415-66392-2.

The Offshore Installations (Safety Case) Regulations 2005, Reprinted 2006, The Stationery Office Limited.

The Railways (Safety Case) Regulations 2000, The Stationery Office Limited.

T430: *The Definition of VPF and the Impact of Societal Concerns*, Railway Safety & Standards Board, RSSB 2006, UK. *http://www.rssb.co.uk/*

7

Icing and Anti-Icing of Railway Contact Wires

Liu Heyun[1], Gu Xiaosong[1] and Tang Wenbin[2]
[1]*School of Communication and Control Engineering, Hunan University of Humanity, Science and Technology, Hunan Province,*
[2]*School of Energy and Power Engineering, Changsha University of Science and Technology, Hunan Province*
China

1. Introduction

The icing of overhead electrical transmission lines, or railway contact wires, may cause a series of problems such as overloading, non-uniform icing, and wire galloping. In countries with cold climates such as the U.S.A., Canada, China, glaze and rime deposits on power transmission lines and electrical railway contact wires are crucial problems for engineers and scientists working in these fields. Power cables and towers have been damaged or destroyed on numerous occasions due to the added burden of the ice or an increase in aerodynamic interaction leading to unacceptable movement such as galloping. The research on the mechanisms of icing and anti-icing is both a basis of transmission line design and a new objective of the environmental thermophysics study.

Atmospheric icing may have a great impact on the overall design and safe running of power transmission lines or railway contact wires. For engineers attempting to determine ice loads in a specific area, knowledge of historical ice events and ice type for that area would be extremely important. However, there is no systematic ice accretion database that can be used to help engineers. The prolonged period of ice storm of Central and Southern China in 2008 caused extensive damage to the electrical installations and electrification railways. As a result, the transmission provider of the state has deployed substantial efforts to mitigate the effects of future ice storms.

This chapter mainly introduces the icing of railway contact wires and corresponding anti-icing technologies and deals with the following four aspects: (1) the hazards of the icing of railway contact wires and significances of this research; (2) the characteristics of the icing of railway contact wires and affecting factors of the icing; (3) experimental research into the icing and anti-icing of railway contact wires; (4) expert system of the icing and ice melting of railway contact wires. Through the above introduction, the paper aims to make researchers, engineers and managers who are interested in this field have a comprehensive understanding of the causes, hazards and contributing factors of the icing of railway contact wires, to let them have a basic knowledge of icing and ice melting and the current technologies of anti-icing and ice removal and finally to provide readers with an expert system that can anti-ice and melt ice.

1.1 Background of research into icing of railway contact wires

Large, heavy ice accretions may form on objects exposed to freezing rain. These can damage or destroy structures and cause great economic and social hardship.

The United States, Canada, France, Russia, Japan, Korea, etc. are frequently attacked by icing in the world. In February 1994, large areas of southeastern United States had suffered heavy ice accidents, resulting in the direct economic losses of 30 billion. In 1998 and 2003, the United States and Canada had a large-area wire icing accident once again, causing greater economic losses.

In China, Hunan, Hubei, Jiangxi, Yunnan, Guizhou, Sichuan, etc. are areas where heavy icing is very common. Since Hunan and Hubei had an icing accident in 1954, Central China Power Grid experienced the worst ice disaster in the history in February 2005. Thousands of miles of transmission lines were covered with ice, and Hunan Power Grid became a "Severely Afflicted Area ". The maximum thickness of ice covered on wires is up to 70 ~ 80mm, which is extremely rare. This heavy icing resulted in the collapse of 21 towers with 500KV transmission line, causing the electric power of Three Gorges not to be sent to Guangdong, which incurred heavy economic losses. In January 2008, the ice storm which lasted for half a month caused unprecedented damage to transmission lines, communication lines and electric railways in 10 provinces of Southern China. The power supply in Hunan, Hubei, Guizhou, Guangxi and other provinces were seriously affected. There have been no power and water supply for two weeks in Chenzhou. The Beijing-Guangzhou electrification railway was cut off by icing in Chenzhou, Hunan province and traffic was severely interrupted.

Therefore, the researches of icing mechanism and solutions to the problem have good engineering application prospects, and significant economic and social benefits.

1.2 Research progress about the railway contact wire icing

Few papers about the research of railway contact wire icing can be found. In the limited literature, there are more ones about the study of the roof insulator flash by icing, and fewer ones about the analysis of the cases and countermeasures of the contact wire icing. Generally speaking, they simply make a description of the phenomena or a simple analysis of the causes of icing. Very little literature is concerned with the theory of contact wire icing forecast models and mechanism analysis.

The research of contact wire icing has not been paid enough attention to, perhaps due to the following reasons:

First, the development of electrification railways is relatively late and large-scale construction has just been started. The contact wire has not been extended to the areas where ice accretion are apt to happen (for China, such areas mainly refers to mountains, lakes and other similar places in Hunan, Hubei, Jiangxi, Guizhou, Sichuan, Yunnan, Anhui, etc.);

Second, because of global warming, disastrous icing events have not happened in the freezing rain areas for the last 30 years. This concealed the problem of the contact wire icing, or we can say the problem of the contact wire icing has not been highlighted.

Third, the built electrification railways are busy lines conveying passengers and cargoes. The leisure time is short. In the limited leisure period, the growth of icing is difficult to reach the thickness which can affect the operation, so the problem is concealed again.

However, with the continuous extension of the electrification railways, the topographical features of the regions where railways pass will become increasingly complicated. In addition, the leisure time of railways will be extended, so the icing problem will be highlighted.

Now de-icing and anti-icing methods of practical application at home and abroad are mainly manual removal, contact wire thermal running, chemical removal methods, resistive wire de-icing technology, etc.

1.2.1 Manual de-icing

Removing snow and ice through human or machines is the most simple and primitive method. It is a time-consuming, inefficient and unsafe operation. The environmental conditions are always harsh during the period of icing, which brings a lot of difficulty for manual de-icing. Currently, many large domestic railways use this method to remove ice covered on the overhead contact wires.

Fig. 1-1. Manual de-icing on overhead contact wires

1.2.2 Contact wire thermal running

The ice situation is observed by staff at the railway before the operation of overhead contact wire. The staff determines the seriousness of icing. If the ice thickness reaches the warning level, then the control center starts de-icing operation, letting the electric current flow through the overhead contact wire. The electric vehicles get power from the collect strip, letting the collect strip run to keep touching the contact wires in idle mode, and ice is removed.

1.2.3 Chemical de-icing

This method uses chemical de-icier to decrease the freezing point of water in order to prevent icing. The most common and most effective de-icier is CaCl. The Railroad Car Company in Bremen, Germany, has successfully used the de-icier coating device produced by the Stemmann Technik Company to apply antifreeze to the contact wires. Currently, domestic airports, highways have a practice which is throwing de-icing agents directly. But applying the de-icing agents to contact wires has not been in use. Moreover, the environmental pollution of scattering chemicals is a problem to be considered.

1.2.4 Resistive wire heating de-icing

Alston, France and Hitachi, Japan developed a de-icing system of overhead contact wire using built-in wire with insulation resistance characteristics, and applied it to railway and tram systems in Japan, France, South Korea, and the UK. The de-icing system is similar to composite conductors used in de-icing of high-voltage transmission lines.

In China, Harbin subway station uses a resistive wire heating to melt ice. Supply power to the power plate, and connect power plate to the resistive wire to heat the contact wire. Although the project which uses the contact wire de-icing program can achieve the purpose of de-icing theoretically, but the de-icing system is lack of experience in successful application and is also lack of the corresponding specifications which are dependent on weather conditions, ice thickness, ice melting current and time, etc.

1.3 Major hazards of contact wire icing

1.3.1 Icing reduces the performance of the contact wire

The added weight to the contact wires because of icing can increase the tension of the wires, cause the dropper clamp for messenger wire to break, deform the mid-anchor clamp and damage the swivel clip holder. Especially in glaze ice conditions, icicles may grow at the bottom of the clamps for suspension, which may result in the increase of the variation of the elasticity, and lead to the increase of the vertical contact force between collect strip and contact wire. This may reduce the performance of the contact wire. This performance includes two meanings: one is the higher smooth, and the other is the higher kilter. The "higher smooth" means the actual distance between the contact wire and the orbits plane is smaller than the distance of the ideal parallel state; Second. The "higher kilter" means that the contact wire itself is flat and straight all the way, so that the collect strip can run smoothly with a high-speed without any blocks. If there are icicles at the bottom of the contact wires, a hard wear may occur when the collect strip runs. The contact wires can not meet the ideal smoothness requirement, which is the main reason for being off-line of the collect strip, volatility of collect current, and flash. The rising rate of electric locomotive collect strip off-line not only damages the high-speed running locomotive, but also shortens its lifespan.

1.3.2 Icing causes divergence of the contact wire

The changed shape of cross-section after icing makes aerodynamic characteristics of the contact wire and lines change. Then the added loads under wind may make the contact wire diverge. In order to make the contact wire wear in a uniform manner in the design, the installation of the overhead contact wire should be in the shape of a "Z". The "Z" shape pull-value is generally (480 ± 10) mm. If the contact wire diverges, it may break the smooth contact between the collect strip and the contact wire. This not only affects the current collection quality, but also speeds up the wear of collect strip and contact wire.

1.3.3 Icing causes an electric arc

Ice covered on the contact wire surface will reduce the conductivity when current flows from the contact wire to the collect strip, and produce electric arc, which can burn out the contact wire and the collect strip. If a heavy icing happened, and it is glaze or hard rim, the thickness of icing is so big that the contact wire is apart from the collect strip, resulting in no current flow, then the locomotive has to stop. There is a huge difference between icing of contact wire and icing of high voltage line. Because as long as the tower won't collapse and the line isn't broken, it still can transmit power with ice on lines.

1.3.4 Icing causes insulator flash-over

When it snows and is foggy in winter, ice accretion will coat on the high-speed locomotive's roof surface insulators, and will deposit mainly on the upwind side. The ice is generally 5-10 mms thick, and the maximum thickness can reach over 30 mm. The higher the speed is, the more serious the icing will be. When the locomotive pulls in, the ice will melt due to corona heating, and the contaminated medium in the ice will form a conductor, which can cause short circuit which in turn will cause flashover, cause damage to electrical equipment and affects the normal transport. For the northern electrification railway, the major problem is flash-over of insulator on the roof of power locomotive.

1.3.5 Icing leads to galloping of the contact wire

Despite the fact that galloping of the overhead contact wire caused by icing is rare, its hazards are massive. The damage caused to the parts of contact wire in the area influenced by the galloping mainly includes the breakage of positive feeder cable, steadyarm, and dropper clamp, deformation of mid-anchor clamp, and the breakage of cantilever insulators, etc. From 20:43 to 12:22 on February 9th, 2003, on the Xuchang—Mengmiao section of Beijing-Guangzhou Line and Mengmiao-Eastern of Pingdingshan section of MengBao Line, a galloping occurred in the overhead contact wire equipment which was a rare phenomenon in the electrification railway history of China. The galloping amplitude is 1.5m; horizontal amplitude is about 0.6m. It caused large amounts of damage to equipment.

Fig. 1-2. Icing of overhead contact wire

2. Characteristics and affecting factors of the icing

2.1 Types of icing

Atmospheric icing of structures, which is also called icing or ice accretion, often happens in cold-damp regions. It can be classified into three categories, i.e. Precipitation icing, In-cloud icing and Sublimation icing.

Icing is influenced by factors, such as micro-climate, micro-terrain, wind speed, temperature and content of supercooled water in air, etc. Influenced by the above factors, icing on aerial wires can be divided into several categories according to different taxonomy.

2.1.1 Classifications of ice based on the appearance

Glaze: transparent vitreum, unbreakable and strong in texture with density between 0.6 and $0.9g/cm^3$, can also be called ice slush or clear ice which covers wires with strong adhesive power and never easy to shed.

Granular Rime: ivory opaque with density between 0.1 and $0.3g/cm^3$, loose and crisp in texture with air bubble voids inside, sinuous surface and irregular shape.

Crystalline Rime: white crystal with many air bubbles inside, loose and soft in texture with density between 0.01 and 0.08g/cm3, weak adhesive power on wires and easy to shed.

Wet Snow: ivory or offwhite, usually soft in texture with density between 0.1 and $0.7g/cm^3$. Wet snow on wires will turn into hard frozen body when the temperature continues to decrease.

Mixed Rime: ivory, large with many voids, is formed by the alternate freezing of glaze and rime on wire surface, and the density ranges from 0.2 to $0.6g/cm^3$.

2.1.2 Classifications of ice based on formation mechanism

Precipitation Icing: Icing or snow cover formed by freezing rain (supercooled water) or snowflake falling on the wire surface whose temperature is close to 0°C or even below. The supercooled degree of water drops has something to do with the size of water drops. Usually the bigger drops are, the lower degree is. The supercooled degree of little water drops can be several centigrade, while the degree for fog droplets can even be more than ten degrees. Once touching the wires, the supercooled water droplets would freeze. Because the speed of releasing latent heat is slow during the process of freezing, a film of water would appear on the wire surface, and thus glaze is formed. Glaze formed by precipitation icing is the most dangerous to aerial wires because of its high density and strong adhesive power. Freezing rain often happens in America, Canada, Russia China, etc., while snow covering is more common in Japan and the Alps.

In-cloud icing: ice frozen by the supercooled cloud or fog in the air on contact with the wires. Without rain or snow, icing mainly depends on humidity, air velocity etc. This type of icing happens with high frequency in many places and is easy to do simulation study with artificial weather, which is one feature of in-cloud icing. Another is that in-cloud icing can be formed as long as supercooled water drop exists. Small in size, the fog droplets can release the latent heat quickly when freezing. Thus, it won't form a water layer on wire surface. Therefore, in-cloud icing usually produces rime.

Sublimation Icing: a frost formed when water vapor in the air freezes on the surface of things directly. It is also called crystalline rime. Formed through sublimation, it is called sublimation icing. It won't get large because of its weak adhesion and being easy to shed. Therefore, it won't pose a big danger to the aerial wire.

When it comes to transportation system, glaze, mixed rime and wet snow, etc. can be of great danger to it.

2.2 Main affecting factors of icing

Generally speaking, the icing of aerial wire is mainly influenced by micro-climate and micro-terrain. The size, rigidity and shape of the aerial wires may also affect the icing, which, however, are not the premier influence.

Micro-climate mainly includes the temperature, wind speed, wind direction, and liquid water content in the air (LWC). These parameters are used in the prediction model of icing. They mainly control the flow and heat transfer in icing process.

Micro-terrain mainly includes the altitude of the wire, the local topography, water distribution, etc. These parameters may not be manifested in the prediction model of icing. Their major functions are to affect the heterogeneity of flow field and therefore causing different icing.

Temperature is the key influencing factor of icing. Generally speaking, the temperature of icing is always below 0°C. However, if it is too cold (below -10°C), icing won't occur. If the supercooled degree of water drops is high, icing would happen even over 0°C (about 1°C), and this will form dense glaze. The general tendency is that the colder, the heavier of icing.

Wind speed is also an important influencing factor of icing. It will ice only when the supercooled water droplets touch the wire surface. The wind sends the supercooled water drops to the wire surface, and the wind speed also affects the heat transfer in the process of icing. Therefore, the higher the speed of wind is, the more supercooled water drops can touch the wire surface; the more easily the latent heat released in the process of icing can be flew away, and the more icing on wire surface will be.

Direction of wind also has certain influence on icing. The wind direction mainly affects the effectiveness of supercooled water drop delivery. If the wind direction is perpendicular to the wires, the delivery is the most effective; if the wind direction is parallel to the wires, the delivery is the least effective, hence less icing.

Liquid water content (LWC) exerts an important influence on icing. It will not only influence the speed of icing but also the types of icing. Generally speaking, if LWC is low and diameter of the droplets is small, rime is formed and icing develops slowly, whereas if LWC is relatively high and diameter of the droplets is comparatively big, glaze is formed and icing develops quickly. If the LWC is very high and diameter of the droplets is very big, rain is formed.

2.3 Comparison between icing on contact wires and on aerial high-voltage transmission lines

2.3.1 Differences in structure

High-voltage transmission wires are steel-cored aluminum strands while most contact wires of electrification railway are copper-alloy hard wires with glossy finish. Therefore, with difference in wire surface, the icing might also be different. Nowadays, the research on the influence of wire surface on icing is seldom touched.

The hard contact wire has many bearing points and short in span (less than 65m) in which the suspension is well-distributed, often with an interval from 8m to 12m, and therefore it can be regarded as a straight line which won't twist or wave when icing; high-voltage transmission line is catenary with huge span (300-500m), which will twist when icing and wave drastically in wind.

Fig. 2-1. Icing of high-voltage line

Fig. 2-2. Icing of contact wire

2.3.2 Differences in height

There is a huge difference in height between electrification railway contact wire (usually about 6m) and high-voltage line (ranges from tens of meters to hundreds of meters). Therefore, the influence of wind speed varies. Influenced by air-flow boundary layer, the speed of surface wind is weaker than that of high-altitude wind. Since wind dominates the delivery of supercooled water drops to the wire surface, generally speaking, the icing of contact wire will be weaker than that of high-voltage line under the same meteorological parameter and geographical and topographical conditions. The influence of other meteorological parameters makes no difference between the two.

3. Experimental study of icing and ice melting on contact wires

3.1 Experiment setup

This experiment is made in an artificial environment chamber in a key laboratory of high efficiency heat exchange technology and equipment of Hunan province, China. Low-temperature Freon solution produced by the chillers flows through a wall-hung heat convection tube which makes room temperature as low as -18 degrees Celsius.

A circular wind tunnel is set in the environmental room with a test section which is made of Plexiglas and whose cross section size is 350×350mm (fig. 3-1, fig. 3-2). Three specimens are arranged in the test section: a real CTS120 contact wire with a calculation area of 121mm2, the unit of mass for reference 1082kg/km, and a density of 8.94g/cm3 is in the middle. Its electric resistance in 20°C is 0.01777Ω/m. The other two are cylinder specimens. The cylinder specimens have the same diameter and shape as CTS120 (with a built-in electric heater, fig3-3). This experiment studies the influence of wind speed, temperature and liquid water content on icing and ice melting process.

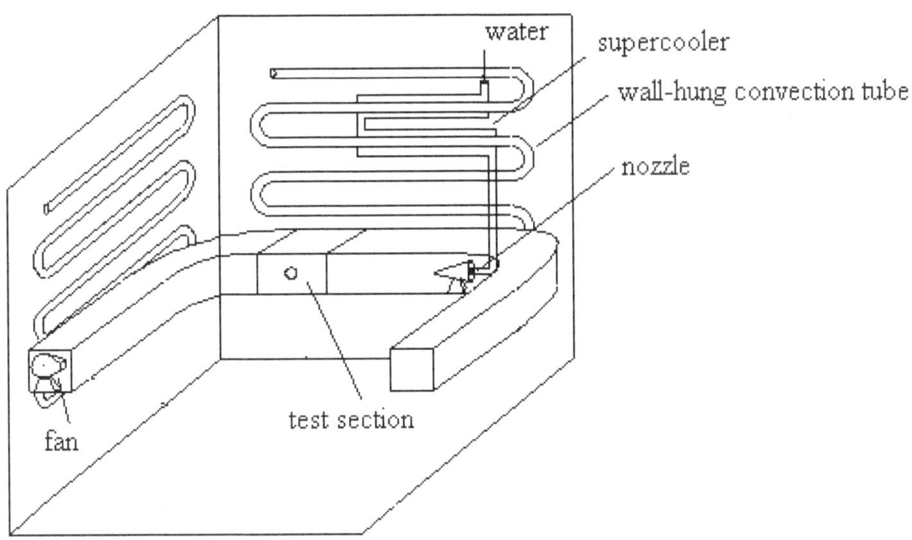

Fig. 3-1. Experiment system schemes

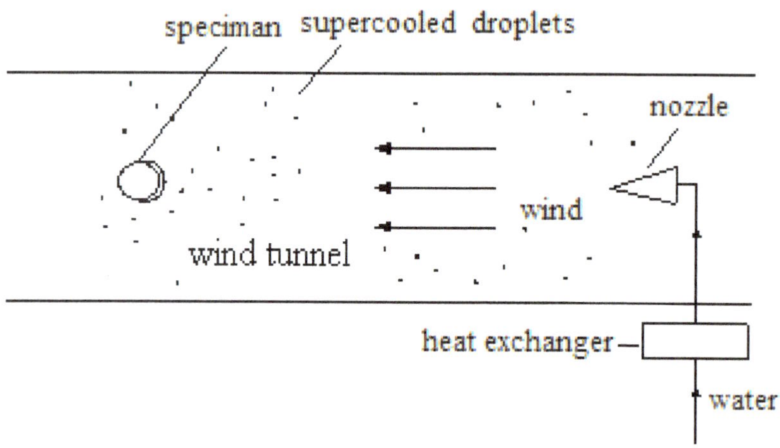

Fig. 3-2. Isometric view of wind tunnel

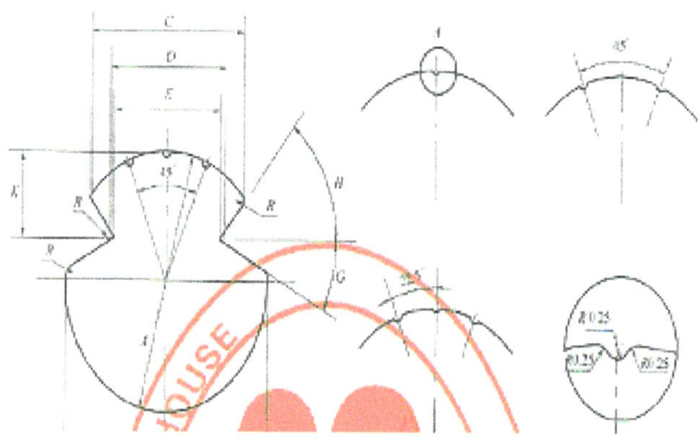

Fig. 3-3. CTS120 contact wires shapes

3.2 Fitting formula of ice accretion

According to the summary of Henry, there are over 20 models used to predict such icing problems as glaze and rime. These include the Imai model (wet growth process), Lenhard model (a simple formula to calculate the weight of the ice), Goodwin model (dry growth process), Chaine model (dry growth process, but shape of the icing is the uneven oval), and Makkonen model (considering the existence of icicles).

The formula is based on the Imai model determining ice load by heat transfer process on wire surface and the icing intensity is proportional to negative air temperature section (-T), and is irrelevant to precipitation.

$$\frac{dM}{d\tau} = C_1\sqrt{VR}(-T) \tag{1}$$

Where

V : wind speed, m/s;
T : air temperature, °C;
T : time, hours (h);
M : ice mass, kg.
C_1 : constant

Equation (2) can be drawn form (1)

$$M = C(V(-T)\tau)^{4/3} \tag{2}$$

The fitting formula for ice mass calculation based on experimental data (T=-2°C, V= 2m/s) is as follows

$$y = 0.00621\tau^{4/3} \tag{3}$$

Comparing equation (1) with (2), we can get

$$M = 1.7927 \times 10^{-3} \sqrt{V(-T)}\tau^{4/3} \tag{4}$$

Fig.3-4 shows the curves of experimental data and fitting data under different conditions (T=-2°C, V=2m/s; T=-2°C, V=6m/s). Experimental data are very much in line with that of fitting formula. Ice load on contact wires with random length can be calculated by:

$$M = 8.9635 \times 10^{-3} \times l \times \sqrt{-VT}\tau^{4/3} \tag{5}$$

Where

l: wire length, m。

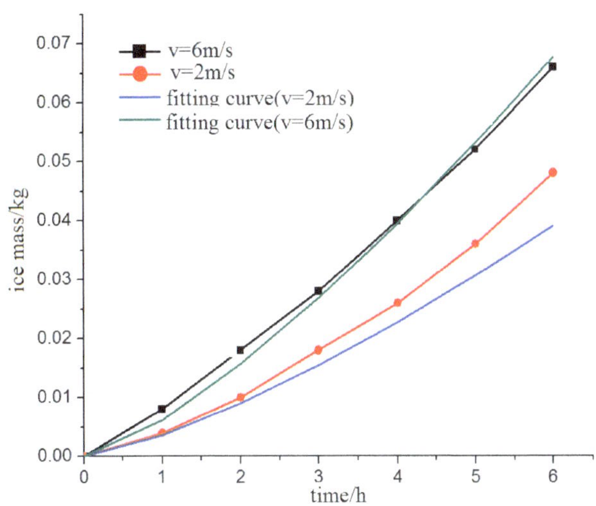

Fig. 3-4. Ice growth with time

3.3 Experimental results and analysis of ice accretion

This experiment mainly concerns the influence of air temperature, air speed and liquid water content, etc, on ice accretion.

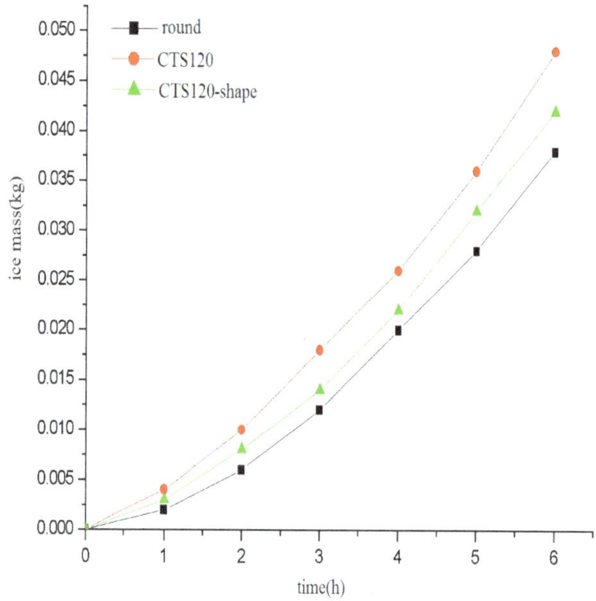

Fig. 3-5. Ice growth of different specimens

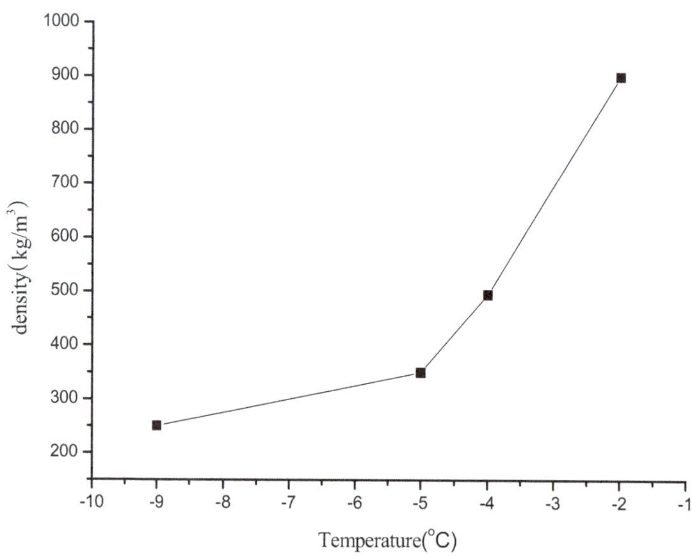

Fig. 3-6. Ice density variation with time

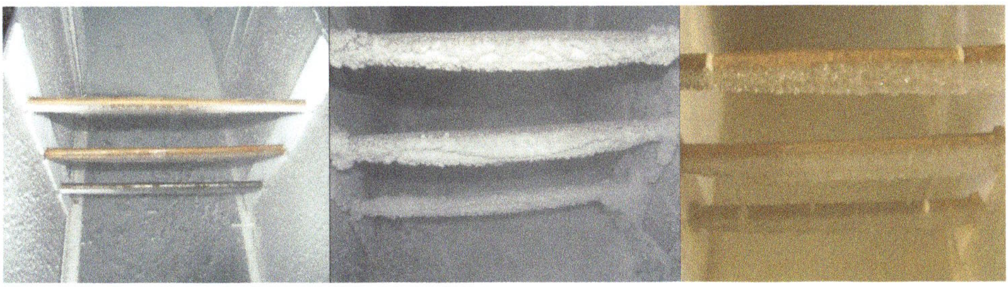

Fig. 3-7. Photos of ice accretion

From fig.3-5, 3-6 and 3-7, we can draw the following conclusions:

1. The wind speed has significant influence on icing and the higher the wind speed is, the bigger the ice density will be.
2. Increasing air temperature leads to increasing ice density but decreasing ice thickness.
3. Ice type begins to change in -4°C. When the temperature is higher than -4°C, large mixed glaze ice or glaze ice will form; when the temperature is below -4°C, rime ice will form.
4. Within experimental wind speed limits, the influence of wind speed on ice type is not obvious. Ice type is mainly affected by air temperature, and, the increase of wind speed leads to the increase of adhesion between the ice and contact wires.

3.4 Icing melting process

Fig.3-8 shows uniform cylindrical glaze ice on the round specimen built in electrical heater. Table 3-1 presents the ice melting time under different ice thickness with the electrical power 100W, and the air temperature -5°C. The ice melting process is divided into two stages. The first stage begins to melt ice sleeve to the top and the second stage begin from the end of the first stage to complete falling off the specimen.

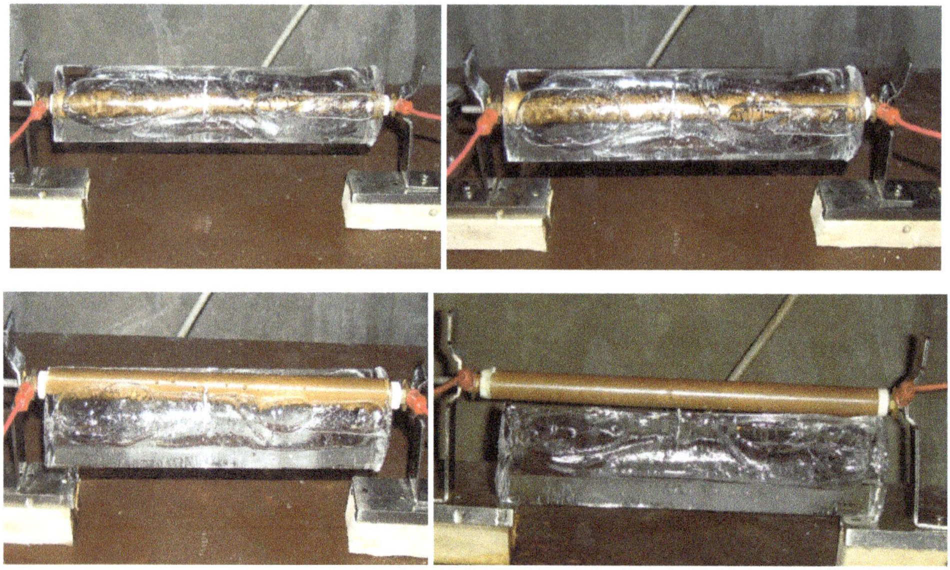

Fig. 3-8. Photos of ice melting process

Ice thickness (mm)	Wire type	Ice melting time (s)		Ice density (kg/m³)
		The first stage	The second stage	
4	Round	116	18	884.36
	CTS120-shape	330	45	901.25
7.5	Round	382	34	912.56
	CTS120-shape	628	96	932.18
12	Round	554	53	905.68
	CTS120-shape	804	124	914.24
16.5	Round	728	33	879.89
	CTS120-shape	856	58	886.47
29	Round	1425	152	901.56
	CTS120-shape	1621	231	907.30

Table 3-1. Ice thickness and ice melting time

The experiment results show that in the same thickness, ice melting time of CTS120-shape specimen is longer than that of round wire and the possible reasons are:

1. The CTS120-shape specimen is made from steel tube and its heat or contact resistence is higher than the round specimen.
2. The two grooves of CTS120-shape are filled with water, causing the ice melting process relatively slow, and affecting the upstairs of ice sleeve.
3. Ice on the two specimens has different density.

The fitted curves by the least square method are shown in fig.3-9.

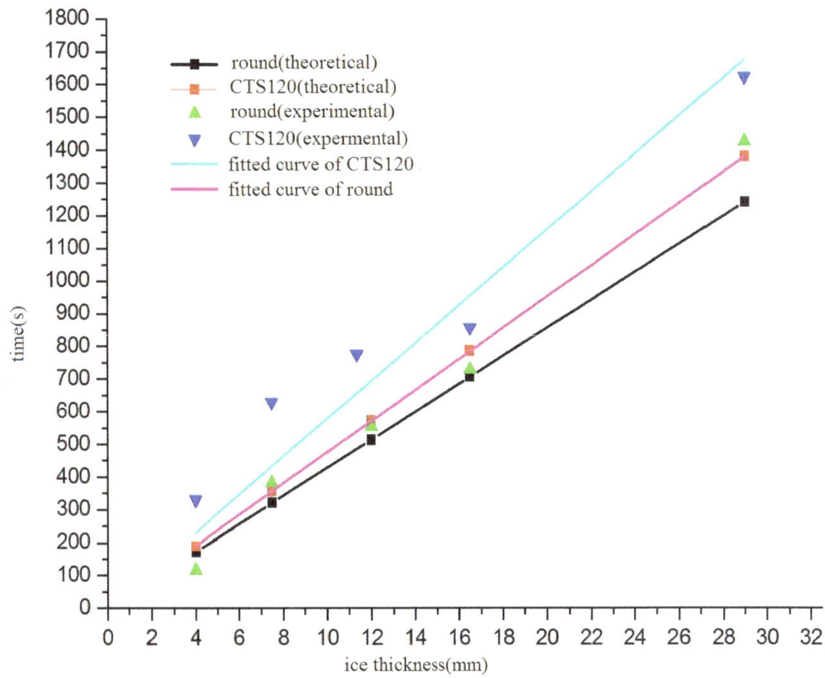

Fig. 3-9. Fitted curves of ice melting time &ice thickness

The linearized expression of ice melting time of round specimen:

$$\tau = 47.62\,x \tag{6}$$

The linearized expression of ice melting time of CTS120-shape:

$$\tau = 57.77\,x \tag{7}$$

τ: ice melting time of the first stage(s) x: ice thickness(mm)

The total ice melting time is equal to the sum of the first and second stages ice melting time. It is found that the ice melting time of the first stage of round specimen account for 8.7% of the total time and that of CTS120-shape specimen accounts for 13% of the total time.

By correction, the formula for ice melting time of round specimen is:

$$\tau_{total} = \varepsilon_1 \times \tau = \frac{2\varepsilon_1 R_i \rho_s \cdot h_{sf}^{'}}{I^2 R_e} \cdot (R - R_i) \tag{8}$$

The formula for CTS120-shape specimen is:

$$\tau_{total} = \varepsilon_2 \times \tau = \frac{2\varepsilon_2 R_i \rho_s \cdot h_{sf}^{'}}{I^2 R_e} \cdot (R - R_i) \tag{9}$$

where $\varepsilon_1 = 1.22$, $\varepsilon_2 = 1.39$.

For the existence of grooves in CTS120-shape specimen, in the same ice thickness, its ice cover is more than that of the round specimen and its ice melting time is 1.18 times longer than that of the round specimen.

4. Expert system of icing and icing melting

4.1 Design principles of ice melting

Due to railway's structural characteristics, the automatic ice melting device cannot be used on the charged contact line for the time being.

1. According to the train schedule, the start instruction of ice melting is given by an expert system before the operation of the maintenance line.
2. Control the ice melting time within 20 minutes.
3. In order to guarantee no interference with railway communications and no damage to communication equipment, railway tracks and the corresponding ground equipment can not be used in ice melting grounding contact.
4. Start ice melting devices when the ice thickness is over 5mm.

4.2 Ice melting schemes

4.2.1 Melting by direct short circuit method on AT

Because of the small leakage resistance of AT traction nets and self coupling transformer, the short-circuit current is great. The short-circuit current on AT set by JEC of Japan is 25 times than the normal current. Therefore, the use of a short-term (2 min) short circuit on AT for ice melting is available.

4.2.2 DC ice melting

DC ice melting devices are not affected by stray inductance and capacitance of contact wires is highly efficient. The main equipment includes a step-down transformer and single phase linear DC stabilized voltage source connected to contact wires during ice melting.

4.3 Expert system of icing and icing melting

This expert system is based on electrical current de-icing strategy and Labview language. The main objectives of this study are:

1. To estimate ice type according to the meteorological parameters such as air temperature, relative humidity, wind speed, which are obtained by corresponding sensors.
2. To calculate the current required to melt ice in a given thickness and a given time interval or the time required to melt the ice buildup for a given current according to the air temperature, wind speed, air humidity, and ice density.
3. To predict ice accretion types, and control ice melting process based on both theoretical and experimental studies.

With the help of video cameras, it can make real-time analysis, monitor, control on-line work station and stop the melting process in advance in case of cable overheat. Labview serving as a good bridge between hardware and software makes weather parameters collection easy. In some circumstances, it is necessary to use the technology of Labview& Datasocket to realize data share between the data collection and control room.

4.3.1 Software structure

This expert system has two key functions, one is the ice type estimation (according to meteorological parameters), and the other is the control of ice melting strategy (provide values of the current required to melt a given buildup of ice in a given time interval or the time required to melt the ice buildup for a given current). The man-machine conversation (file management), data acquisition, data display, data analysis, preservation, and other functions, are also achieved in this software. The structure of the software is shown in Fig.4-1.

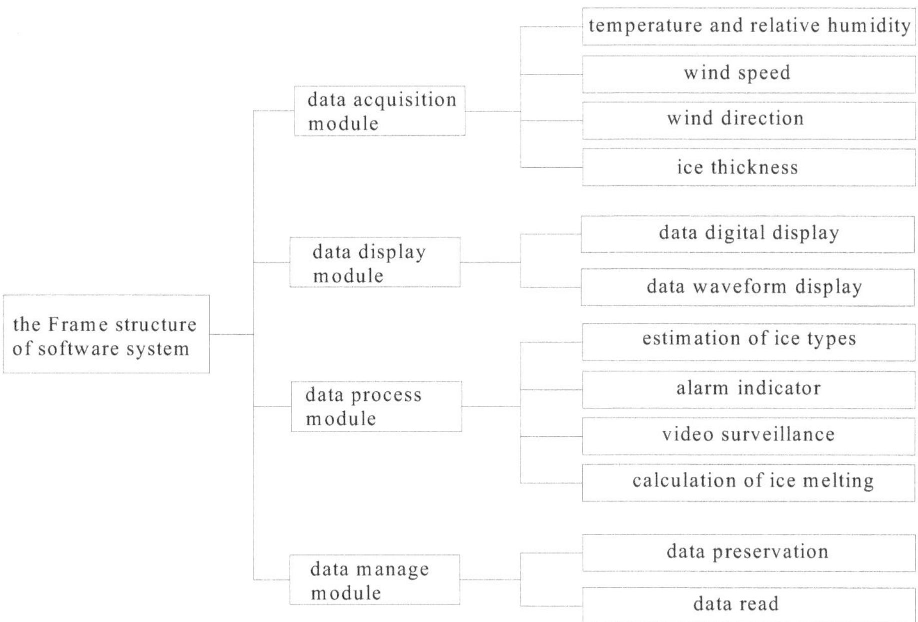

Fig. 4-1. The software's structure

Data acquisition module can collect air temperature, relative humidity, wind speed, wind direction, ice thickness and the real-time video of the scene. Ice type estimation module can estimate ice type according to wind speed, air temperature and air humidity via experimental investigations. Electrical de-icing calculation module can provide values of the current required to melt ice in given thickness and given time interval or the time required to melt the ice buildup for a given current according to meteorological parameters. All data will be recorded in computers for subsequent search, analysis, display, print, and research in data management module.

4.3.2 Logic structure

The logic chart of the software is shown in Fig.4-2. If the weather is likely to cause icing, the system will sound the alarm to warn the operator. When ice thickness exceeds the specified value, the light or voice alarm indicator will remind the operator that melting is needed. If we choose the off-line analysis, we need to enter the needed data manually, which helps the operator to estimate when the cables are iced, and whether the sensors in the scene are broken.

With the help of the video camera, we can control the melting process in real-time. For example, when the given melting time is up, but the video shows that the ice has not fully shed yet, the system will give an instruction to continue melting the ice. However, if, within the given time, the video shows that the ice has been shed, the system will give a stop instruction to stop the ice melting process so as to prevent the wire from being burnt.

The interfaces of the software are shown in fig.4-3.

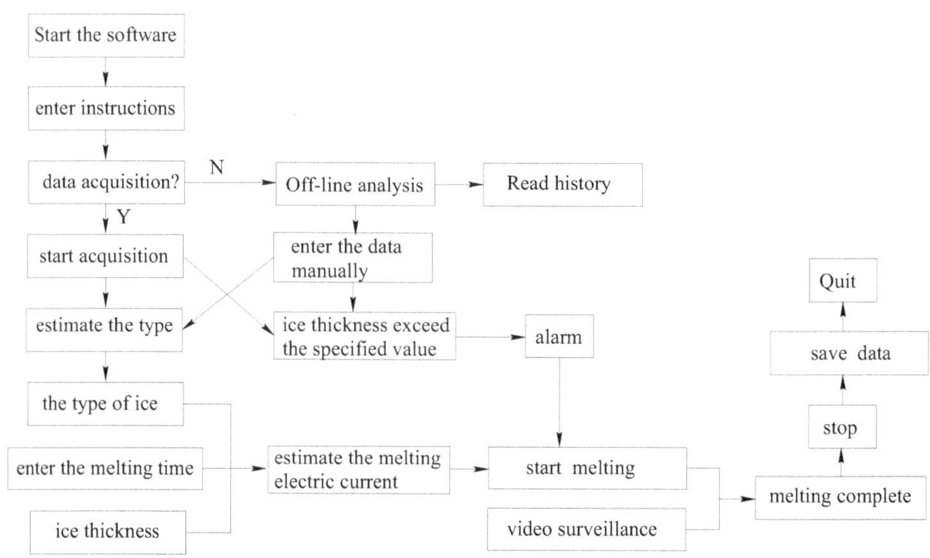

Fig. 4-2. Logic chart of the software

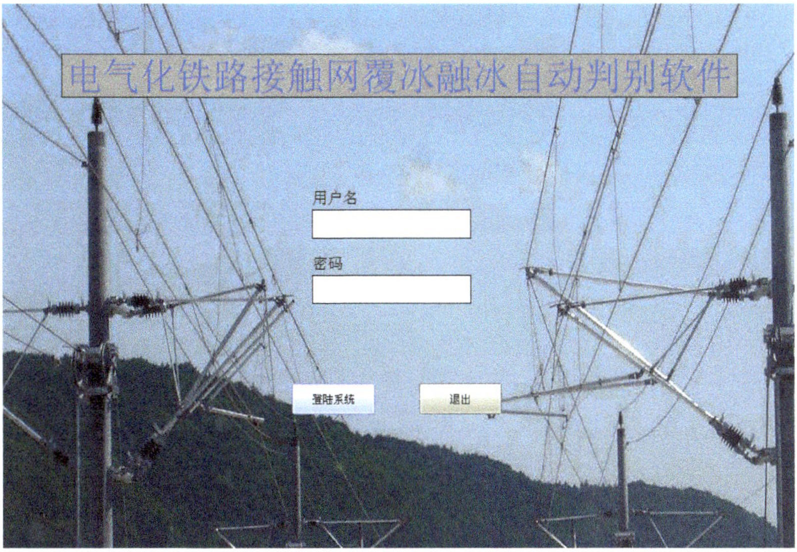

Fig. 4-3. Interface of the software

5. References

[1] Liu heyun, Theories and Applications of Ice Accretion and Anti-icing on Overhead Transmission Lines. China Railway Press, 2001(In Chinese).

[2] Zsolt P. Modeling and simulation of the icing process on a current-carrying conductor[D]. Universite Du Quebec, 2006:125-131

[3] T.W. Brakel, J.P.F. Charpin, T.G. Myers, One-dimensional ice growth due to incoming supercooled droplets impacting on a thin conducting substrate[J]. International Journal of Heat and Mass Transfer 50, 2007,1694 - 1705

[4] Gu Xiao-song, Wang Han-qing, Liu heyun. Experts system of ice prevention on overhead transmission lines, 2010 International conference on intelligent computation technology and automation, 2010.273-276.

[5] Gu Xiao-song, Wang Han-qing, Liu heyun. Calculation model of ice melting joule heat for overhead lines, 2010 International conference on electrical and control engineering, 2010:5079-5082.

[6] Ma xiaohui, Liu heyun. Experimental Researches on Micro-droplet Freezing on a Solid Surface under Atmospheric Conditions Part I, 2008 Proceedings of the ASME Micro/Nanoscale Heat Transfer International Conference, January6-9, 2008, Tainan, Taiwan

[7] Tang wenbing, Liu heyun. Study on the Expert System of Overhead Lines Icing and Icing Melting, Proceedings of the Intelligent System and Knowledge Engineering (ISKE 2008, Xiamen)

[8] Liu heyun, Huang shuyi, Zhou di. Heat balance analysis on icing of wires and the current of preventing icing, Proceedings of the International Conference on Energy Conversion and Application, 2001.6, Wuhan, China

[9] Guo hua, Liu heyun. Numerical Simulation of the Local Collection Efficiency on Icing Conductor, The International Conference on Electrical and Control Engineering (ICECE 2010), 2010.6.27,Wuhan,China

[10] Liu Heyun, Gu Xiaosong and Wang Hanqing. Expert System of Icing and Anti-icing on Wires in Freezing Rain , The 14th International Workshop on Atmospheric Icing of Structures, 2011.5.18, Chongqing,China

8

System for Investigation of Railway Interfaces (SIRI)

Sanjeev Kumar Appicharla
[1]*Institution of Engineering and Technology*
[2]*The International Council on Systems Engineering, UK*

1. Introduction

This chapter presents an abstract system framework called "System for Investigation of Railway Interfaces" (SIRI), to study potential or past railway accident(s). The aim of the study is to learn about the multiple causal factors (elements or conditions) which represented together can be called a cause leading to the undesired state called potential or accident situation. Safety studies like SIRI can be used in conjunction with the quantitative risk estimation method (PRA) to help highlight or uncover decisions leading to assumption of unreasonable risk or human error in engineering and management factors needs to be studied. Author accepts the viewpoint of George E.Apostolakis on the utility of quantitative risk analysis (QRA) or probability risk analysis (PRA) techniques in general (E.Apostolakis 2004). The questions of human error and organisational learning are clarified later in the chapter in the context of acceptance of QRA method.

The SIRI Framework uses *synthetic* mode of thinking as opposed to *analytical* mode of thinking. Analytical mode of thinking is like decomposing water which does no longer contains anything liquid and has taste. The SIRI Framework synthesises multiple study methods into a cohesive process represented as a system. The study methods used in stages to arrive at the decision of a potential accident situation in an unambiguous manner are Hazard identification method (HAZOP), Event Causal Factor Analysis (ECFA), Energy Barrier Trace Analysis (EBTA), accident investigation technique (MORT), and cognitive human factors framework (SRK) and systems thinking integrated into a cohesive system framework. This is to facilitate the conceptual work of defining an operational system and inquiry into causal factors (individual, technical and organisational factors) to get an unambiguous feedback on the potential or actual accident situation. In this way, it is hoped that decisions, which might take operational situation outside the safe envelope due to groupthink bias or individual decision maker bias, may be detected at the planning stage or pre-design stage itself. Readers can gain access to the MORT user manual and related information from the NRI Foundation (Noordwijk Risk Foundation 1998). Need for analytical framework or multiple study methods are noted in the safety literature, but author wishes to cite two articles in support (Hovdon, Storseth and Tinmmanvisk 2011), (Hale, P.H.Lin and Roelen 2011).

This paper verified two theses accepted within the SIRI Framework. First, fallible decision making is the starting point of the accident sequence and is connected with the failure of foresight and/or not heeding warning signals, or where hindsight bias or groupthink bias dominates or lessons learnt are not applied or no lessons are learnt or not performing system safety analysis (B. A. Turner 1976), (J. Reason 1990), (Johnson.W.G 1974), (Wei 2008), (Kletz 2002). Second, it is possible to gain insight into the hazardous conditions or events that pre-disposes a normal act into an unsafe act in conjunction with local less than adequate defences in a complex system (Johnson.W.G 1974), (Briscose .G 1990), (IEC 2001), (Kingston, et al. 2004). Knowledge of past outcomes is not necessarily a good guide to future outcomes was established by Fischoff (1975) and this phenomenon was named as *hindsight bias*. The effect of *hindsight bias* and its two forms is cited by James Reason in his study of human error (J. Reason 1990). The concept of *impossible accident,* promoted by Wagenaar and Groeneweg (1988), is used to convey the idea that accidents appear to be the result of highly complex coincidences which could rarely be foreseen by the people involved (J. Reason 1990). The SIRI analyses of the Herefordshire Accident show that notion of *impossible accident* is not true in the case of level crossing accidents. Why? Because people involved in the decision making situation are prone to group think bias and prone to blame others in the projects and/or different organisations and rarely look at own actions which lead to bad policy or decision making (Goodwin 2006), (Whittingham 2004), (Weyman 2006), (S. Appicharla 2010), (S. Appicharla 2011). Re-evaluation of data and hypothesis is necessary to avoid confirmation bias. This can be seen in the case of scientists who assumed that the thesis nothing can travel faster than light. This thesis was falsified in the OPERA experiment. The summary of this OPERA experiment can be found in the science and technology section of the Economist (The Economist 2011). The two ideas learnt from Albert Einstein are: a) that time sequence of events experienced cannot be equated with the order of experience in time in the context of acoustical and visual experience; and b) physicists endeavour to eliminate psychical element from the causal nexus of existence (Einstein 1920). Contrary to the physicist(s) approach in the Einstein tradition, it is necessary to include psychical element and conduct the evaluation of technical as well as organisational aspects individually and re-evaluate them together in the safety studies (both accident investigation and project safety studies). However, no such re-evaluation was seen on the part of UK railway signalling industry in the case of the Herefordshire accident cited in this chapter. Author presented the Herefordshire railway accident case study based upon the principles of line side signalling perspective to verify the thesis accidents are due to "satisficing behaviour" displayed by the railway organisations. No case study is presented from a cab signalling perspective as an incident on a level crossing installed on the ERTMS signalled railway is under RAIB investigation. Details are given in the chapter later on.

The main thrust of this chapter is on the topic of taking a 'system approach to railway Safety' and is designed to:

a. help railway signalling engineers and managers utilise the framework as an independent system safety analysis methodology to help identify potential accident scenarios (system hazard) , detect and analyse the hazard causal factors and enable take preventive actions;
b. help post-accident/incident investigators utilise the framework to facilitate learning of lessons and help draw correct conclusions from a single event (incident or accident)

which occurred in the recent or remote past to identify and verify the thesis that conjunction of management and engineering oversights and/or omissions or the undue acceptance of risk were the causal factors behind the occurrence of the incident or accident.

The conceptual basis of the chapter is based on author's three published papers in the IET International System Safety Conferences and unpublished consultation commentary provided to the UK statutory body, the UK Law Commission in October 2010 (S. Appicharla 2006), (S. Appicharla 2010) (S. K. Appicharla 2010), (S. Appicharla 2011). Author's work experience, and learning from the past RSSB Research projects and study of related literature from domain of systems engineering, decision making, risk management, accident analysis and investigation, psychology, mathematics and philosophy have also provided necessary inputs. The concepts associated with the 'system approach to safety' which author wished to promote in conjunction with interested members of public are publicly available on the Wikipedia website (Wikipedia 2011). Demand for system approach is described in the safety literature as well (Elliot 1999).

This chapter highlights the application of the third step of the SIRI Framework. The aim is to support efforts to identify system hazard(s), and select amongst alternative solution(s) to deal with the identified hazard(s). Earlier application stages of the SIRI Framework were elaborated in the IET International System Safety Conference publications in 2006 and 2010 (S. Appicharla 2006), (S. Appicharla 2010)

The process of dealing with the hazards that arise with the implementation of the selected option can be dealt with by applying the same procedure or the procedure developed by Stephen Derby and Ralph Keeney (L.Derby and Keeny 1981). Stephen Derby and Ralph Keeney argued that the question of 'how safe is safe enough' cannot be answered using the subjective utility criteria (experts judgement) or using risk quantified in 10^{-7}/person/year risk or by performing value trade-off analysis as they do not satisfy the needs of collective decision making. They reckon that collective decision making is a problem riddled with ethical constraints. Any analysis of decisions on acceptable risk must ponder on social, technical, political and ethical dimensions. A similar observation has been made by the Royal Academy of Engineering on the matter of engineering ethics in practice. They have issued a long and short version of documents discussing the complications involved and the short version was accessed by author (The Royal Academy of Engineering 2011). Author has noted that there is a growing interest in the subject matter of philosophy in engineering domain.

The rest of the chapter is organised in this way. Section 2 defines the concepts used in the SIRI Framework. Section 3 presents a case study using the SIRI Framework to help understand its application. Section 4 states the problem statement which the solution has addressed. Section 5 summarises and draws conclusions on subject matter of the chapter. Section 6 acknowledges the help received from others. Section 7 provides the references.

2. Definitions of concepts in the SIRI framework

2.1 Cognitive systems engineering, affordance of harm and reality

The notions of ''system'' and 'systems engineering' are used in the way as they are defined by Benjamin Blanchard (Benjamin.S.Blanchard 2004). Systems engineering is taken to mean the orderly process of bringing a system into being. A "system" comprises a complex of combinations of resources (in the form of human beings, materials, equipment, software, facilities, data, information, services, etc.) integrated in such a manner as to fulfill a designated need.

A system is developed to accomplish a specific function, or series of functions, and may be classified as a natural system, human- made system, physical system, conceptual system, closed-loop system, open-loop system, static system, dynamic system, and so on. Readers can refer to the work of Benjamin S. Blanchard and Wolter J. Fabrcky to learn the distinctions between analytical and synthetic mode of thought (J.Fabrcky and S.Blanchard 2005). For want of physical space, author cannot reflect upon the two modes of thinking in this chapter.

Author got familiar with the application of system approach in the social science field from reading the works of an economist, F.A. Hayek apart from its application in the thermodynamic field[1] and has argued an application in an unpublished paper in 1998 (S. Appicharla 1998). Subsequent to this, author has learnt that Herbert A.Simon's concepts of "bounded rationality" and "satisficing behavior" have influenced the works of Irving L. Janis, Barry Turner, Charles Perrow and James Reason as well (L.Janis and Mann 1977), (B. A. Turner 1976), (Perrow 1984), (J. Reason 1990). Herbert A. Simon's work was influenced by F.A. Hayek's concepts is gathered from the quote by Herbert Simon that no one has characterized market economy better than F.A.Hayek. Gary Baker, an empirical economist, acknowledges the influence of F.A Hayek in his work on Human Capital but notes that centrally planned and other such economies that do not make effective use of markets and prices raise co-ordination costs thereby reduce incentives for investments in specialized knowledge. Baker states that F.A. Hayek stated that… the problem of a rational economic order is….the utilization of knowledge which is not given to anyone in totality (Baker 1964). The division of labour is greater, in economies that make effective use of prices and markets to co-ordinate tasks and skills across firms. However, the case study presented in this chapter found that market economy destroys divine capital (it is assumed in this chapter that life is the work of divine capital) and no necessary investment into human capital is needed to prevent this destruction from occurring. In other words, as Charles Perrow argued the cost of transactions are borne by the wider society (Perrow 1984).

The concept of *cognitive systems engineering* is introduced by members of human factors engineering community, as an approach to describe and analyse man-machine systems. Daniel Woods, ERIK Hollnagel (1983) described the concept in a paper titled, "cognitive systems engineering; new wine in new bottles" (Hollnagel and David.D 1983). In that paper, they quoted Criak (1943) who remarked: "If the organism carries a "small-scale model" of external reality and of its possible actions within its head, it is able to try out various alternatives, conclude which is the best of them, react to future situations before they arise, utilize the knowledge of past events in dealing with the present and the future, and in every way to react in a much fuller, safer, and more competent manner to the emergencies which face it". An extension of this idea that a machine must possess a logical model of its

[1]Examples of system approaches such as heat flow balance modelled as differential equations can be gathered from standard text books on control systems engineering or from text books on Bayesian mathematical functions.

environment in multiple levels was discussed in the paper by Erik Hollnagel and David.D. Woods.

Prior to this, Barry A. Turner relying upon a similar concept of collective adoption of simplified assumptions into a framework of 'bounded rationality' helped deduced the fact that large scale intelligence failures are seen to occur in the organisational and inter-organisational practices prior to the occurrence of disaster. Drawing upon three case studies of public inquiries in United Kingdom, Turner hypothesized that a set of cultural beliefs about the world and its hazards in the social context and associated pre-cautionary norms set out in the laws, codes of practice, mores and folkways are the starting point of events in a process made up of 6 stages ending up with cultural re-adjustment. One of the public inquiry studied was the Hixon accident which is relevant in the railway context (B. A. Turner 1976). Earlier to this period, the concepts of organism, adaptive behavior and regulation in an environment were studied by Ashby (W. Ashby 1960), (W. Ashby 1960), (Ashby and Conant 1970). The concepts of representation of external reality as an object and the four roots of the principle of sufficient reason were discussed by Arthur Schopenhauer (Schopenhaeur 1820/2006). These set of ideas were followed by Leo Tolstoy (Tolstoy 1887/1930), and Alfred North Whitehead (Whitehead 1927/1978). Alfred North Whitehead traced the origin of these concepts back to the Buddhists whereas Arthur Schopenhauer traced the origin of these concepts back to the Upanishads. The principle Upanishads were translated from Sanskrit to English by Valerie Roebuck (Valerie 2003). Valerie Roebuck in the contemporary period traces the origin to pre-buddhistic Upanishad era in the fifth or sixth centuries BCE (Valerie 2003). Author has learnt from Jens Rasmussen and others that the concept of system coming into being can be traced back to ancient Greek times represented by Aristotelian notion of causation ((Rasmussen, Pejtersen and Goodstein 1994). The concepts of self or soul and external reality are mentioned by Aristotle in the ancient Greek times in the book VII on Politics (Aristotle 323 BC/1951).

The concepts of the Platonic world of mathematical forms such as squares, cubes, circles, spheres etc were recognized to be distinct from the corresponding approximate entities (substantial forms) in contemporary Greek physical world,and giving considerations to them in abstract form may give rise to a doubt whether the Platonic world of mathematical forms is 'real' (Penrose 2004) (Plato 375 BC/1995). However, these doubts can be dispelled when the fact that RSA algorithm relies on mathematics to support electronic communication between sender and a recipient in the public domain in modern communications is recognized (Singh 2001). The idea of relativity of time can be gathered from the BBC news headline that engineers can learn from the way slime mould searches for food resembled the 100 year old Tokyo rail network and the way it solved the problem of finding an efficient way through the maze (The BBC 2010). Mathematics cannot determine absolute motion since everything determined by it ends in relations by stating perfect equivalence between theories as in astronomy is a fact noted by G.W. Leibniz in his New System of the Nature (G.W 1695/1998).

According to Benjamin S. Blanchard and Wolter J. Fabrcky(2006), social groups organizing themselves possessing inherent abilities and the knowledge to maintain its stock of technology can be said to have civilization. They assert that modern civilizations possess pervasive and potent technical systems that provide products, systems, structures and services. In this sense, author asserts that both ancient and modern civilizations possessed potent technical systems that can afford harm and benefits in the way they followed the norms. As an example, in the ancient Vedic civilizations, people carried out the sacrificial ceremonies which belong to the ritualistic portion of the Vedas to the letter without comprehending the spirit of sacrifice. This is learnt from reading the verses 3.10 to 3.13 of the Bhagavad Gita as translated and commented by Swami Nikhilananda (Nikhilananda 1944). It is accepted in this paper the truth which is stated in the Upanishad that there are three kinds of adversity: fever, headaches etc arising from disorder of the body(internal) , arising (from) external objects, such as tigers, snakes; arising from the action of great cosmic forces, such as those cause rain, storms, or earthquakes. Similarly, prosperity is of three kinds (Nikhilananda 1944). For explanatory purposes, it is assumed that the actions cause results (the desired or undesired) observed which arise from the combination of three causes acting together: material cause, efficient cause and formal cause to give rise to the final cause (the desired or undesired) effect. This is Aristotle theory of causation (Aristotle, Ethics 350BC/1955). The Aristotle theory of causation bears a very close resemblance to the deductive science of Gunas (three modes of material modifications) articulated in the fourteenth chapter of the Bhagavad Gita and reading of the verse 3.27 which states that all work is performed by Gunas of Prakriti and the idea that Self is an agent is false knowledge (Nikhilananda 1944). Scientists and philosophers citing or drawing inspiration from the Sacred Scriptures is not an uncommon phenomenon. Sir Issac Newton held the view that both Nature and Scripture were presentations of God's message to man, which was to be learned scientifically or through the study of God's revelation as presented in the Bible and/or Koran. Newton in his Principia Mathematica, offered a version of the argument of design to show that we (critical reader may disagree) could know of God scientifically (Popkin H. R 1969).

Some examples relevant to the misery faced by the railway customers can be comprehended from the real world examples. Reading the news items about signalling cable thefts or the rats eating away the signalling cables, lightning causing signalling circuits to fail highlight nature of problems faced (Wainwright, 21 September 2011), (The BBC News, 27 September 2011), (The BBC News, 12 August 2011). To prevent erroneous conclusions that can be drawn based on the foregoing that wireless or radio communication based signalling systems provide better safety and security, a process with a pattern similar to the SIMILAR process is needed. The customer or beneficiary can be assumed to be members of local or wider society. Author had conceived the SIRI Framework in response to a request from the signalling engineer's request to help identify the duty-holder interfaces. The problem statement author started working from is given in the section 4.

2.2 Systems thinking process

A.Terry Bahill and Bruce Gissing asserted that humans (as individuals, on teams, and in organisations) employ simple processes to increase their probability of success. They argued that Peter Senge's Fifth Discipline, Shewart's Plan-Do-Check-Act cycle, Covey's 7 habits of highly effective people, Katzeban and Smith's The Wisdom of Teams, INCOSE Fellow Consensus on Systems Engineering Process and IEEE 1220 Systems Engineering Process shares the common roots of the SIMILAR process. These processes were mapped to the SIMILAR process to show them sharing common roots in systems thinking in the IEEE article (Gissing and Bahill 1998). Thus, the SIMILAR process would serve as a comparative benchmark for the SIRI Framework.

The SIMILAR process is stated in a very brief manner. A picture of the SIMILAR process is shown in Figure 1. It is concerned with logically consistent and effective means of planning and problem solving. The process starts with the stage of developing a system in an engineering environment to address the current deficiency and description of what function must be done or satisfied by the system. At this stage *what is* determined. This is State the Problem stage. With completion of the problem definition stage, the process moves to the Investigate Alternatives where the functional alternatives are searched to satisfy the problem statement. After the match has been made between the need and the solution to satisfy the need, a model is developed and is analysed to determine what is *to be*. At the Integrate Stage, the developed model or simulations etc are checked for the compatibility with the sub -systems to assure that interfaces exist between sub-systems for transferring the outputs/data/information as the case may be.

Inherent feedback loops between inter-connected sub-systems must be checked to minimize the exchange. Architects of the SIMILAR process assert that well-designed systems integrate sub-systems such that they contribute to whole direction of the system. Launch the System means running the system and producing the outputs. Assess Performance is the stage where the metrics are used to measure the performance. Terry Bahill and Bruce Gissing did not mention the qualitative aspects at this stage, but author reckons that both quantitative and qualitative aspects must be taken into account. The Re-evaluate is the feedback stage at each of the process stage to assess and evaluate the performance. Terry Bahill and Bruce Gissing reckon that repeated application of the process to systems, sub-systems, components in an iterative manner produces outputs similar to a fractal process. Further, the Re-evaluate Stage runs parallel to the main work streams of the SIMILAR process. Author wishes to clarify that the representation of the SIMILAR process does not represent causality. It should be noted the notion of causality applies to physical systems rather than social systems.

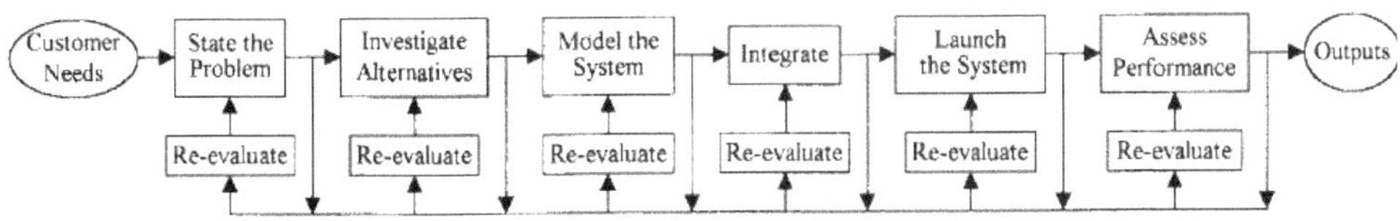

Fig. 1. A graphical representation of the SIMILAR process.

The perception and definition of a particular system, its architecture and its constituent elements depend on an observer's interests and responsibilities. One person's system - of - interest can be viewed as a system element or product in other person's system - of- interest. Conversely, it can be viewed as being part of the environment of operation for another person's system of interest. The basic definition of a system at the modelling stage can give rise to disputes. Mathematically inclined people would not agree to a definition of a system given in the new formed (during 1970s) soft systems engineering tradition (I.Mitrani 1982). The notion of a system and its definition can give rise to multiple interpretations can be evidenced from a recent publication in the UK railway domain as well (G.J.Bearfield; R.Short September 2011). However, author notes and agrees with I.Mitrani observation that motivation for modelling objective (from any kind of modelling activity) is that process of observing and learning from the real system in operation is too difficult, too hazardous or too expensive. This is the case when a system is not yet built. Alternatively, the objective may be to assess the performance of the effect of a proposed major changes in an existing system (for example, addition of European train control systems, moving block systems, renewal of fixed block signalling etc).

The SIMILAR process does provide a basis on which such problems can be sorted out by describing a new process bench marked to a tested and agreed systems thinking process model. This is line with the thoughts of the architects of the SIMILAR process. Given that human behavior is fragile and fallible in nature, it is necessary that a systematic method that can transcend or overcome the errors in faculties of perception, cognition and judgments based upon limited, narrow professional expertise is necessary.

The SIMILAR process appears to cater to that need. However, in the case of safety problems, no stakeholder is in a position to outline the problem in a manner as it is expected by the SIMILAR process. Further, it is a matter of every day experience that man -made and natural system (s) do produce unwanted outcomes in the form of incidents or accidents. However, the SIMILAR process calls for multiple methods to be used is evidenced from the graphical representation of the process in the Fig. 1.

Many accident investigators or researchers have used models or methods to explain 'what' happened afterwards. However, system developers, owners and operators and maintainers are more interested in learning lessons and taking preventive actions in a cost effective manner. This paper presents a proactive approach by revealing the gaps in the knowledge of parameters that sit on the boundaries of the systems, sub-systems and components. These give rise to notion of a 'system hazard' which given the rights conditions can escalate to an accident.

From an energy perspective, harm arises from inherent danger in the sources of energy which have not been diverted into safe channels in the performance of work and its unsafe flow of energy is triggered by a change in the circumstances due to lack of awareness and/or risk taking behavior on the part of social elements of the system. Energy is a necessary ingredient for attainment of success of any system of work. This perspective focusses attention on the organisational factors, workplace factors and the individual factors and the state of barriers in the analysis of the sequential progress of accident sequence. The decisions taken at the work group level (or organisational level) provide for latent failure pathway if the hazard is not recognized at the planning of the work or in the standards that regulate the work processes. The Swiss Cheese Model SCM is used to trigger the analyst's awareness to check for organisational factors in the M Branch of the MORT analysis.

James Reason (2007) in his 18th Westminster Lecture on 'recurrent patterns in transport accidents' stated that there are three levels of accident contributors: universals (the ever present tension between protection and production), conditions and causes (the local factors that combine with certain conditions to breach defences in unforeseen and unforeseeable ways) (Reason J. , 2007). Reason's lecture was intended to pose more questions than provide answers. He noted that answers to the 'why' question are often found well 'upstream' in both time and space from the event: indicting the organisation, the regulator and sometimes the entire transport system. Author could not determine why this method of explanation should not be pursued from Reasons' perspective. Author's speculation in this area is that James Reason does not think re-engineering of a social technical system is a legitimate and feasible activity. Author notes that social technical system is a small part of wider geographical society which is greater than a local social technical system.

Author thinks that James Reason's research is countered by 17th Westminster Lecture given by Phil Goodwin on 'determination' and 'denial: the paradox of safety research and traffic policy (Goodwin, 2006). He gave three arguments for not taking action aimed at improving safety in the case of road transport. First, the trends will take care of the problems. This is the idea that the technical advance will solve all problems. Second, road driver behavior is unsafe and cannot be influenced. This is an example of 'law of unintended consequences'. And third, reducing collisions in one place is not due to anything due to human intervention, but a random effect counteracted by increases somewhere else. In the extreme form, this is an example of the view that universe operates randomly, and human agency is ineffective. He concluded his lecture by saying that his research led to the recurrent conclusion that the effects of policy on behaviour are bigger than has been conventionally assumed -behaviour does change, and substantially. This has been obscured from general awareness by biases in the form of data and models which have been influential, which has led to a continual underestimate of the potential both for making things better by good policies and making them worse by misguided policies (Goodwin 2006). The role of various biases in the failure of Incident Reporting Systems in UK has been studied by Chris Johnson (C. Johnson 2002). Author wishes to argue that the 'SIRI Framework' and 'STAMP' accident models fall under this category of models for safety improvement from cosmological perspective(analysis of cause-effect relations) (S. Appicharla 2011) (N. Leveson 2011).

Author notes that accident modelling from ontological perceptive (analysis of what is - ought to be relations) is studied by Why Because Analysis method which is advanced by Peter Ladkin based upon David Hume's philosophy (Ladkin 1995).

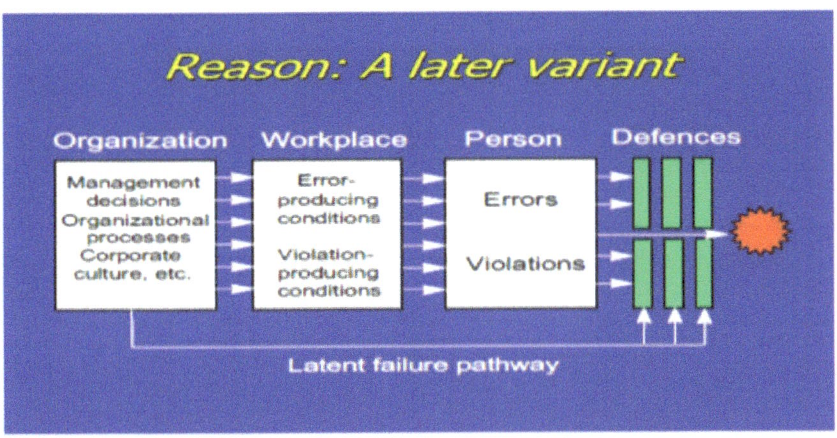

Fig. 2. A reference Model for Accident Analysis. Sourced from (Wkipedia 2011).

With basic notion(s) outlined above, the mapping of the SIMILAR process onto the SIRI Framework is provided in the Table 1. This is to argue the case that without sacrificing the core notions of the SIMILAR Process, it is possible to establish identity between two systems thinking processes. Author has learnt that the notion of systems thinking was part of the deliberations when Management Oversight and Risk Tree (MORT) was being developed by William Johnson and his team. Both concepts of system as composite entity made up of entities and as a systematic method for investigation as well were noted in the original MORT documentation (Johnson.W.G, 1974). Prior to this notion of system made up of barriers, threats, regulator and essential variables was articulated by Ashby (Ashby W., 1956/1999), (Ashby W., 1960) (Ashby and Conant 1970). The SIRI Framework does re-use several of these concepts developed by Ashby and MORT team. Ashby used the example of adaptive behavior on part of a train driver in his treatise on Design for a Brain (W. Ashby 1960).

Claiming completeness of knowledge is a herculean task; however, using the principles of abstraction, refinement and information presentation in a careful manner, UK railway industry can make a claim that railways are safe to operate even under changing conditions. This is achieved by detecting all safety critical deviations possible in the system at the operational time and ensuring that safeguards are available to prevent those variations escalating into accidents. Author rejects the idea of safety case approach based upon formal approaches such Bayesian belief networks or compliance with railway CENELEC norms is sufficient unless the safety and electro-magnetic compatibility (EMC) is designed into the technical or operational systems based upon understanding[2] (B.Bateman, S.W. Hatton 2006), (Hughes,D , Saeed A, 2009). The idea that TBTC command for service braking distance function can be given lower safety integrity level SIL 2 than the TBTC command for emergency braking distance SIL 4 can be seen in the paper by S.D.Turner and a safety case has been prepared and accepted on that basis (S. Turner 2011). The way SIL targets are assigned to the command functions appear to be correct at first sight. However, a bit of thinking would reveal that if the output command for service brake function has failed then it is clear from the logic of IEC 61508 to note that distance computation or information available on train location is in error at the input stage of processing or algorithms for speed and location determination are in error. Over time, the distance to a danger point from a reference point on the physical track space at which the following train must necessarily stop does not always grow in size. Common cause failure analysis (CCFA) is a must as single point failure (SPF) destroys independent redundant designs is noted fact in the safety literature (Clifton 2005), (The UK Health and Safety Executive 2003) . In line with author's argument, a paper by Tim Kelly calls for a cautionary approach when preparing safety cases (Kelly 2008). EMC must be designed into system is argued by Armstrong (Armstrong 2006). Interpretation is necessary when safety cases are prepared strictly in accordance with the CENELEC norms is noted by an independent safety assessor (Skogstad 1999). Author asserts that analysis of safety property in the form of safety cases cannot be like banker's note which is issued by a firm which has nothing other than paper obligations to back it with. The notion of comparing safety case analysis with a banker's note is gained reading from of text on the page 123 of Arthur Schopenhauer's book (Schopenhaeur 1820/2006).

2.3 Emergent property, perceptions, causation, system safety viewpoint

The SIRI Framework adopts the cognitive science tradition of Rasmussen's skill-rule-knowledge framework and argues that measures to eliminate affordances for errors and harm are feasible in the railway context. The term affordance refers to the basic properties of

[2] Understanding is a technical term denoting a faculty which means causation from David Hume's perspective and this definition is accepted in this chapter. The Bayesian Belief Network, if used, should model engineering and managerial factors as well to meet the requirements of the risk management model advanced by Jens Rasmussen (Rassmussen 1997).

objects that shape the way people react to them. An artefact that is well designed should, through appropriate use of invariant features, make obvious what is for and how it should be used. Author is of the view point that the signalling systems and railway infrastructure in the context of rail-road interfaces or man-machine interfaces should be designed for errors assuming that active human errors do occur and eliminate error inducing situations from operations (J. Reason 1990), (S. Appicharla 2010). From a human factor perspective, a 24 element model made up of 5 elements of perceptive organ system, 5 elements of sensory motor organ system, 4 elements of cognitive system with 5 elements of vital systems of respiration etc is used to represent the human element in the context of an operational environment. This is a universal model which is supported in the domain of systems engineering, philosophy, behavioral science as well found in ancient Sanskrit texts like Kausitaki Upanishad III (Valerie 2003).

The SIRI Framework through the application of HAZOP/EBTA/ECFA/MORT studies is able to focus attention of the analyst on the where the cognitive mis-match between the task and the person is occurring or could occur by taking into account the organisational and inter-organisations perspectives and their impact on it. The concept of affordance of harm directs attention to the areas where the lack of awareness is creating safety problems. Affordance in Gibson's ecological approach is the direct perception -action mode of cognitive control which is depending upon a human being actively engaging in a goal directed activity in contrast to a human passively making judgments about a given environmental situation as in the case of knowledge based sematic interpretations(e.g. of a work of art). Jens Rasmussen cited two comments from Gibson (1988) as being relevant-"affordance links perception to action" and "learning about affordances entails explanatory activities" (Rasmussen, Pejtersen, & Goodstein, 1994). This is demonstrated through a simple example.

Engineers experienced in the area of design of power distribution systems can grasp the concept of affordance for harm by studying the case of electrocution of several farmers when working on agricultural watering systems. This case study by Casey (1993) was cited by Jens Rasmussen (Rasmussen, Pejtersen, & Goodstein, 1994). When moving the system from one location to another, apparently some workers occasionally raised the 38-ft. long thin-walled pipes to a vertical position and touched the high voltage lines. No one expected farmers to raise the long pipes to a vertical position for transporting them to the next field. Only after analysts actually visiting the location and interviewing people did it become clear that farmers usually raised the pipes to a vertical position to release rabbits hiding in the pipes and lethal consequences resulted when done below the high voltage lines. Thus, it is argued in the case study that it demonstrates the limit of empirical safety control and the expresses the need for shorter pipe lines as defence against electrocution under high voltage lines. The idea of teaching about human error categories was refuted as a barrier (Rasmussen, Pejtersen, & Goodstein, 1994).

From the perspective of Energy Trace and Barrier Analysis(EBTA), the transfer of energy across the air-gap was, in the case of agriculture water system, triggered by reduction in the distance between the charged conductor and pipe which raised the potential difference above the natural dielectric strength of air acting as an insulator (barrier). The design failure in the situation was not to provide earth-wire grid beneath the overhead conductors to safe guard against the charged conductors falling to the ground or objects lifted up to them. The accident situation would manifest in the SIRI Framework when the designer's emergent property of Sustain_ the Dielectric Strength would be tested against possible variations during the HAZOP study. The application of guide word-No or Less would reveal the potential accident to reveal the farmer's lack of perception of electrocution hazard (system hazard) and the failing barrier of safe distance (height) above ground.

The dynamic interfaces between the system hazard, barrier and potential accident form the essence of the SIRI Framework. The loss of protection to the farmer together with dis-functional interaction between S branch (technical system) and M branch (risk management factors) of the MORT would be revealed at the ECF/EBTA/MORT stage of analysis. The design engineer's knowledge of clearance and creepage requirements for protecting against earth potential rise for touch, step and transferred voltages for various working conditions and states of environment would be tested during the HAZOP and EBTA studies. Inspection of author's log book from initial days of work experience reveals that awareness of such conditions is essential competence on the part of the electrical power engineer and is still a valid idea as evidenced by entry on the public data base (Wikipedia 2011). This pattern of reasoning leads to "how" question implicating conjunction of technical failures in the planning stage and local influences at the sharp end of operations. The 'why' question implicating the engineering and management factors would be revealed from the use of question set in the M branch of MORT studies.

The IEC 15288 noted in the Annex D (informative section) the essential concept which the International Standard is based upon (ISO/IEC 2002). It noted that humans contribute to performance and characteristics of many systems for numerous reasons, e.g. their special skills, the need for flexibility, for legal reasons. Whether they are users or operators, humans are highly complex, with behavior that is difficult to predict, and they need protection from harm. Author notes that the rational actor model assumed by behavioral school of economics is refuted implicitly in the Annex D of the IEC 15228 Standard. Why? Because the empirical theory of human capital propounded by Gary S.Baker, assumes that investments into human capital usually are rational responses to a calculus of expected costs and benefits (Baker, 1964). Case study in the paper would show that this assumption is not true in the case of the UK railway signalling industry and the concept of 'bounded rationality' developed by Herbert A. Simon and acknowledged by James Reason persists (J. Reason 1990).

The foregoing notions require system life cycle process to address human element factors in the area of human factors engineering, system safety, health hazard assessment, man power, personnel and training. These issues are addressed by particular activities and iteration in the life cycle, and are described in more detail in ISO 13407 and ISO/TR 18529 (ISO/IEC 2002).

The concept of emergent property is important to be grasped in the context of systems thinking. An emergent property is a property which a collection or complex system has, but which the individual members do not have. Three examples are given to illustrate the concept. First, ammonia is a gas and so, is hydrogen chloride. When both gases are mixed, the result is a solid. The property is not possessed by either of the reactant. Second, carbon, hydrogen and oxygen are tasteless but the particular compound, sugar', has characteristic taste possessed by none of them. Third, The twenty (or so) amino-acids in a bacterium have none of them possess the property of being 'self-producing', yet the whole with some other substances has this property (Ashby W. , 1956/1999).

The SIMILAR Process	The SIRI Framework
State the problem	Derive the Emergent property using signalling layout, system diagrams and stakeholder's consensus on the norms
Investigate Alternatives	Assess the variations in the selected emergent property of the user and identify the potential accident situation using the HAZOP study
Model the System	Construct the narrative of expected operations using the ECF notation. If the potential danger situation is identified in the HAZOP study, then identify the trigger from the ECF diagram using MORT Decision Model in the operational situation.
Integrate	Generate EBTA diagram and perform MORT analysis , identify the interfaces between the potential hazard, barriers (missing) and targets, and decide upon the root of the problem using MORT accident process model
Launch the System	
Assess Performance	Compare with SRK model with Jens Rasmussen SRK Decision ladder and Reason's SCM and prepare the SIRI Hazard Causal Analysis Report
Re-evaluate	Release to stakeholders for consultation and peer review

Table 1. The SIMILAR Process mapped onto the SIRI Framework.

It is noted by author that concept of 'emergent properties' is not easily understood in the UK railway signalling domain. Author found a paper by E.Goddard which considered the subject of emergent properties in a very brief manner in the context of mass transit systems. But E. Goddard did not go far enough to include the concept of affordance of harm or system safety as an essential property of the railway system (Goddard, E 1998). Based on the concept of emergent properties, author rejects the J.S.Mill explanation perspective that it is possible to reason the property of the whole from the properties of parts.

This view of J.S.Mill is countered by Chris Johnson in the discussions on the theme of complexity following the workshop held on Complexity in Design and Engineering during March 2005 (C. Johnson 2006). Author has noted that the contents of the document by Chris Johnson were not peer reviewed. However, latest research from the behavioral science domain indicates that 90% population of the human subjects tested by researchers did not confirm to the expectations of the ethical theory of utilitarianism promoted by J.S.Mill and Bentham. The goal of this ethical theory is encapsulated in Bentham's aphorism that "the greatest happiness of the greatest number is the foundation of morals and legislation. The results published in Cognition were cited in the science and technology section of the Economist (The Economist 24 September 2011)". Author's assertion is that the idea of utilitarianism and associated ALARP judgments without taking into account the concept of unreasonable risk is verified by this research which generated the evidence that the antisocial personality traits predict utilitarian responses to moral dilemmas. The principle of correspondence has been used to draw similarities between control group and non-control group of human subjects in the laboratory conditions and outside of it. On comparison of ideas between Mill's theory of causation and the Aristotelian theory of causation, it appears that Aristotelian theory is more logical (M.Copi and Cohen 1998).

A key insight from the systems theory is that different individuals and organisations within a problem domain will have significantly different perspectives based on different histories, cultures, and goals. These different perspectives need to be integrated and accommodated if effective action is to be taken by all the relevant agents (Chapman 2004). But hope of acquiring true information from all the agencies to learn about the local as well as global interactions, from a top-down perspective is remote, and therefore, a bottom-up approach with no accident is acceptable policy is adopted to reflect upon the disturbing causes leading to system hazard and efficient barriers to eliminate them from operations. The idea of bottom-up approach in safety literature has been thought of by John Adams as well (S. Appicharla 2006), (Adams 2009). However, in contrast to John Adams approach, author does not recommend a safety action to be legislated as an effective procedure in the absence of efforts to detect policy and regulatory oversights and omissions that are occurring and impacting the standards in an adverse way (S. Appicharla 2010).

Literature in the decision making domain often calls upon readers to imagine typical scenarios to draw their attention to the subject matter. These may involve an imaginary scenario of tough decision making on the part of an old and poor clerk to replace an old worn out coat and trigger an action to start saving for a new coat. Under the critical judgment of a young tailor, old and poor clerk needs to depart from a previous approach of incrementally repairing the patches of the worn-out coat. This school of thought may lay emphasis on the concept of psychological stress upon decision making in conflict situations

(L.Janis and Mann 1977). Sociologist(s) may demand the reader to consider the plight of a person who misses a crucial interview due to a coincidence of foreseeable but imaginary events or circumstances (Perrow 1984). Other researchers in business decision analysis would like to focus attention away from black swan events which are impossible to predict. Instead, they offer advice on risk management and recommend try to reduce the impacts of the threats we don't understand (Taleb, Goldstein and W.Spitznel 2009).Some authors, would like the readers to consider the fact that human information processing might be subject to various kinds of biases due to use of thumb rules (heuristics) and due to things like representativeness, anchoring, and availability effects (Tversky and Daniel 1974).The process of multiple valued decisions requiring trade off approach developed by Benjamin Franklin of assessing pros and cons of any given situation as he called it as moral or prudential algebra is another perspective on the decision problem. This method is cited by (L.Janis and Mann 1977).All of these articles may direct the attention towards a process called humble decision making. Successful decision making is all about avoiding decisions with no sense of overarching purposes reckons another researcher in decision making (Etzioni 1989). In the medical health sector, Gerd Gigerenzer and J.A.Muir Gray argue that inability to make informed decisions by the doctors and patients using the laws of conditional probability wastes lot of public money. They urge that better use of condition probabilities and statistics would help the matters in the case of delivering patient care in a more economical way (Gray 2011). Charles Handy noted in the chapter on the 'working of groups 'that it has been shown in experimental studies that groups who attack a problem in a systematic manner perform better than groups who 'muddle through' or 'evolve'. He suggests that the decision making procedure is also of great importance (Handy C., 1999). The same suggestion is made by Derby and Keeney that there is no single solution to the how safe is safe enough problem and social, political and ethical aspects of the problem must be addressed by the analysis explicitly (L.Derby and Keeny 1981). In the wake of Japanese earthquake disaster, few observers like Ekekwe call for better risk communication (Ekekwe 2011).

As detecting subtle influences on decision making which cause behavior (rational or irrational) is an elusive phenomenon, the decision principle embedded in the SIRI Framework is the inherent safe design principle for the whole system. Thus, the system is biased towards no accident policy. Author notes that biases can be seen even in the case of founding father of modern economics, Adam Smith and other economists.

Adam Smith writing in Wealth of Nations in the chapter IV on the origin and use of money stated that once division of labour being established, every man lives by exchanging, or becomes in some measure a merchant, and the society itself grows to be what is called a commercial society.

Further, after elaborating that barter trade must have been in place before the origin and use of money, he notes that in a village in Scotland during his time, it was not uncommon for a workman to carry nails instead of money to the baker's shop or the ale house. Adam Smith, it is reasoned by author, thought it is odd on the part of nailer to do that. This is reasoning is validated by commentary of a later editor of the Wealth of Nations in 1805. The commentator gave explanation of this fact in this manner: Factors furnish the nailers with materials, and during the time they are working give them credit for bread, cheese, and chandlery goods which they pay for in nails when the iron is worked up. The fact that nails are metals is forgotten in the text the following paragraph. At another place, Adam Smith remarked in the chapter on real and nominal price that at the same time and place, money is the exact measure of the real exchangeable value of all commodities (Smith 1776/1937). Author finds price system to be invalid and untrue in the case of ordinary commodity like jam. On a comparative basis, a brand can be cheaper on the supermarket shelves even after being better on account of energy and fat content. The difference in price cannot be attributed to time it takes to make them as David Ricardo (1772-1823), asserted in labour theory of value. Another instance of an assumption made by economist(s), which author finds is invalid and not true is on the need for a government.

Max Lernet, editor of 1937 edition wrote that Adam Smith assumed that there is 'divine hand' which guides each man in pursuing his own gain to contribute to social welfare and therefore, government is superfluous except to preserve order and perform routine functions. Author contends that this assumption is negated by the David Hume's picture of a man not as a religious creature, nor as a machine, but as a creature dominated by sentiment, passion and appetite (Hume 1739/1984). E.L.Woodward in his History of England wrote about new types of business-men arose who were' without scruple and without pity'-free, but lacking in any sense of obligation to their fellow men (Brown 1958). This mode of thought of sense of self within a community agrees with the thought of devotionalism or Bhakti Yoga expressed in texts like the Bhagavad Gita(BG) –with its emphasis on 'service', 'grace', 'humility', and 'love'. This devotion is often compared to the anvil in a black smith's shop in the Vedanta literature. In spite of repeated blows the anvil remains unshaken is learnt from the verse 10.7 of the BG (Nikhilananda 1944) . This thought appears to contradict the Vedic saying of prey-predator logic of eater is eaten away is noted by another Sanskrit Scholar, Wendy Dongier (Manu unknown/1951). The point of these discussions is that for each and every thesis, it is not difficult to find a contradiction in the form of anti-thesis which were called Antimonies by Immanuel Kant. Both antimonies which can be validly proven and since, each makes a claim that is beyond the grasp of spatiotemporal, neither can be confirmed or denied by experience (McCormick 2005). The way to resolve Antimonies is to grasp the Principle of Sufficient Reason. The Principle of Sufficient Reason is expressed generally by the idea that our knowing consciousness, which manifests itself as outer and inner sensibility (or receptivity) and as understanding and reason, subdivides itself into Subject and Object and contains nothing else is accepted in this chapter. To be the Object for the Subject and to be our representation are the same thing. All our representations which may be determined apriori and on account of which nothing existing separately and independently, nothing simple or detached can become an Object for us. This concept is due to Arthur Schopenhauer (Schopenhaeur 1820/2006).

Unless, we represent an idea to ourselves we cannot compute. A cognitive animal, which carries in it a model or image of the environment, is a sort of animal which can form ideas about environment and process information. In other words, cognition is a computation on a representation. Let us consider the issue of external world which contains many diverse features, and objects which cannot occupy the tiny size of human brain. Therefore, it is reasonable to assume that we store image copies of these objects or subjects which we derive out of experience (S. K. Appicharla 2010). From the forgoing, it is argued that reasoning from an economic point of view is fallible in some sense and does not provide a coherent, valid, logical and consistent explanation. On the contrary, the spiritual perspective as expressed in the Vedanta texts is more rationale and valid in the modern context as well. This viewpoint finds supported by Arthur Schopenhauer (Schopenhaeur 1820/2006). Author notes that Adam Smith wanted to attack the feudal and mercantilist institutions of his age. Author learns from the BG Verse 5.18 that wisdom lies in seeing the same in all – wether it be a brahmin endowed with learning and humility, or a cow or elephant or a dog or an outcaste. This verse suggests to the author that subjective biases can enter into safety assessments due to economic or social evaluations and therefore, a sense of equality is necessary when dealing with the question of energy that is harmful.

It is noted by the author that Hume's perspective and Kantian perspective on causation is challenged by Schopenhauer (Schopenhaeur 1820/2006), (Wicks 2007). David Hume accepted that reasoning about cause-effect relations results in a sequence of events whereas Immanuel Kant reasoned that a sequence of events presupposes cause-effect relation. However, Arthur Schopenhauer argued that empirical reality is a complex of space and time in which objects (things which are represented) co-exist with the subject (the representer) in space. For example, when a substance catches fire, for instance, this state of ignition must have preceded by a state which is made up of conjunction of affinity to oxygen, of contact with oxygen, and of a given temperature. Ignition must necessarily follow upon this state and as it has just taken place that cannot always been there, but must on contrary, have only supervened. This supervening is called a change. Therefore, law of causality applies exclusively to changes.

Arthur Schopenhauer argued an explanatory account of anything does involve reasoning by a human subject using the connection of four independent kind(or parallel) of objects of material things(law of causality) , abstract concepts(law of logic) , mathematical and geometrical(law of mathematics and space), and psychological motives(law of intention).

From author's perspective, these connections form the basis of motivation in seeking explanation of multiple factors for hazard occurrence and present true explanation of accident phenomenon.

The early railway companies seemed satisfied with a philosophy that a big steam engine at the front of a train could easily push a horse-drawn vehicle on a crossing out of its way (Hall.S & Mark, 2008). No decision was enforced on the railway companies to avoid or eliminate the danger. It is a regrettable fact of phenomenal reality that Hume's perception of a man dominates the social psychological perceptive. If any examples of divinity are to be seen in the railway world then one can observe it in the personalities of Col.W. Yolland, Captain Laffan, Captain Sir H.W.Tyler and several others who were actively involved in promoting railway safety in the period between 1851-1871. These individuals promoted the concepts of interlocking the signals and points, blocking the route and braking as essential principles of railway safety. However, the concept of braking distance was taken for granted by them.

An expert in the UK railway domain thinks differently from a lay person on the matters of risk is demonstrated by the RSSB Research report T517 (Risk Solutions 2006). Author has found this observation to be valid even in the case of safety experts. Author inquired from his lecture audience on 21 September 2011 as to whether they saw one or two women in the Figure 3. Author was surprised to learn that at least two persons out of about 12-15 persons in the room did not re-cognize both woman figures in the picture. Author did not perform any further examination to make both groups to communicate with each other to establish reasons for not seeing the double figures. The audience was made up of railway and road safety experts at the sixth IET International System Safety Conference held at Birmingham (S. Appicharla 2011).

Fig. 3. The picture shows an old crone or a 19th century young girl, depending upon the perspective of person looking at the picture.

As Charles Handy noted that the order of presentation of information is important in the case of perception. He noted that people who were first conditioned to see young girl first saw young girl but did not see the old woman and vice versa (Handy 1999). The idea which author wants advance from his foregoing observation is that it is a case of selective perception of objective evidence rather than a framing effect and this may bias the safety assessor or safety authorities as well. Why? Because all cases of human perception involve selective perception and therefore, it is necessary that perceptual illusions and cognitive errors are eliminated to grasp the reality of the hazard situation. Mirage is one example of perceptual illusion (S. K. Appicharla 2010). The point of the above figure perception is to make clear that consciousness is spatially multiple but in temporal terms it has unity which has been stated by Baroness Susan Greenfield (Graham Walker 2007). The argument advanced by Baroness Susan Greenfield is that there are two requirements for the consideration of morality from a scientific perspective. These are a sense of self and a sense of consequences of one's action. In line with Roger Penrose's thinking, Baroness Susan Greenfield argues that synthetic brains will never be conscious because they do not have, amongst other things, intuition or common sense. It is clear from the foregoing brief discussion that the R.L. Maguire's assertion that everyone has same mental model of safety or accident investigation is not true (R.L.Maguire and Brain 2006). Author accepts Ashby's assertion that every creature has same physical brain is both acceptable and true (W. Ashby 1960).

Unless, the idea of treating 'a person as a machine' is abandoned, the concept of taking the systems view of the self cannot take root in our consciousness. This idea is borrowed from Lynn M. Rasmussen (2004). Lynn M. Rasmussen states that the idea of the systems view of the self with its simple description of surrounding systems, purposes, functions, and processes, shows us how we are all the same, that everyone is "like us." It transcends belief systems that divide us, links our inner functions with universally held values and ideals, and gives us a clear means for increasing our own consciousness and the consciousness of people of the systems in which we live (M. Rassmussen 2004). The UK railway industry body, RSSB, did conduct research into the topic of ethics in relation to safety is noted by author (Elliott 2003) (Wolff 2002). However, the research findings were not reflected upon by the industry.

This pattern of thought aligns with concepts of systems thinking as they are represented in the systems engineering standard, IEC15288, Annex D (ISO/IEC 2002). However, author would like to raise a concern here that this idea of relative self does not by itself extend to Karl Marx's worldwide view of dialectical materialism (Rupert Woodfin 2004).

To overcome the limitations imposed due to cognitive economy, system thinking is deployed to consider the emergent property of system safety by becoming aware that two kinds of awareness are present in the process of empirical perception. According to the system of Vedanta philosophy which this chapter accepts is that any object which is conditioned by the law of cause and effect is not absolutely real; for every effect is a change brought about cause, and every effect is temporary. According to the system of Vedanta philosophy the unreal never is. The real never ceases to be. The only Reality is the Atman, Consciousness, which is unchanging Witness of changes in the relative world(Samsara). The Absolute Reality is not conditioned by causality as stated in the Bhagavad Gita verse 2.16 (Nikhilananda 1944).

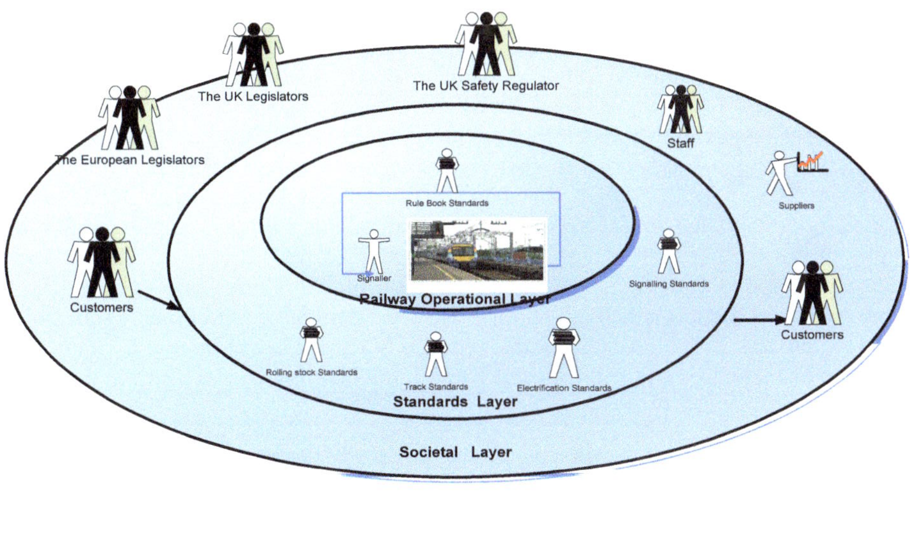

Fig. 4. A layered view of the UK non- ERTMS Railway Transportation Process

A schematic example of the multiplicity and complexity of perceivable systems -of -interest in the traditional UK railway operational situation and its context is given Figure 4. This schematic shows:

a. importance of defined boundaries that encapsulate meaningful needs and practical solutions;
b. layered perception of the system physical structure ;
c. an entity or element at any level of the layers can be viewed as a system;
d. a system comprises of a fully integrated, defined set of sub-ordinate systems,
e. characteristics properties of a system boundary arise from the interactions between system elements;
f. humans can be viewed as users external to a system (e.g railway user) or as elements within that system (e.g. train drivers or signallers) or as regulators or controllers of the system (e.g ORR/DfT) or as suppliers (e.g. signalling or rolling stock suppliers)
g. a system can be viewed in isolation as entity, i.e. as a product or as an ordered collection of functions capable of interacting with its surrounding environment, i.e a set of services
h. rules and regulations in the form of standards constitute a system with explicit purpose of documented and agreed procedures governing the interactions (lateral and vertical) within and across railway organisations, which can be used to support the self-regulation of system safety by the duty holders directly.

Terry Bahill and Steven Henderson (2005) discussed 23 famous failures and identified putative cause of those failures of system designs in terms of important system engineering categories of requirements development, requirements verification, requirements validation, system verification and validation where these activities were done correctly and incorrectly. They discussed that the Tacoma Narrows Bridge disaster can be taken as relevant to the railway domain was a case of system validation error. Validating a system means building the right system: making sure that the system does what it is supposed to do in its intended environment. They did not advance any scheme or framework by which the safety failures could have been foreseen. They hoped that the model System Requirements Classification Model (SCRM) and its divisions would help to improve understanding and compliance in the five systems engineering tasks (Bahill and J.Henderson 2005). Author learnt that 64% of the failures studies were B1category of unverified or invalidated systems with valid requirements but poor design realisation. Examples of this type of system design are the Tacoma Narrows Bridge disaster or war in Vietnam and Super Conducting Super Collider. 12% of the failures belonging to the B2 category of system designs which fail to adhere to their designs or fail to satisfy stakeholder needs in the process. System designs of this type are Mars Climate Orbiter and Titanic. This research can be taken to support the idea of Groupthink bias as a crucial factor in the 76% of the cases studied.

From a systems engineering perspective, in accordance with the IEEE 1471, it is assumed that each stakeholder would hold a perspective relative to the system behavior, its elements and/or attributes (W.Maier, Emery and Hillard 2004). In addition to traditional Three Viewpoints of Requirements, Structure and Allocation of the Hatley Pirbhai Method, the system and related concepts are defined in the Viewpoint Method as indicated in Table 2.

Viewpoint Name : Safety Analysis and Requirements.
Stakeholders: HAZOP Chair, HAZOP study members, MORT/ECF/SRK Analyst(s), Concept/Functional System Design Team, Operational and MaintenaceTeam, Risk Management Team, Human Factors Engineering team, Rolling Stock engineers, Signalling engineers, Track Engineers, Operational and Maintenace Staff, Accident Investigators, Data team, Configuration team, Electrification Team, Hazard Analysis Team, Asset Engineer, Software Team, Safety Policy Team, Reliability Team
Concerns: Potential or actual accidents, Root Causes, Hazards, Barriers, Targets, Controls Factors, Management System Factors, Energy Flows, Vulnerability, System elements, Interfaces, Behavior, Biases, Safety Risk, Data Analysis
Modelling Language: Entity Relationship diagrams, SIRI Diagrams

Study Methods: HAZOP/Managmenet and Oversight Risk Tree /Events Causal Factors Analysis/Engery Barrier Trace Analysis.

Consistency and Completeness Analysis Methods: Documented series of consistency rules and compliance of the safety studies with the basic concepts defined in the standards and guidelines such as IEC 61508/IEC 61882/IEC 15288/UK HSE Guideline 238/ BS EN 50126/IEEE-STD-1233. Some of the rules and process may seem redundant but it is necessary to assure that different persons check of the same rules inside and outside railway domain provide diverse means of checking the reports produced by the SIRI Framework. The underlying concept is to perceive the harm afforded by the system. (IEC 2001).

Table 2. System Safety Analysis and Requirements Viewpoint

3. SIRI case study: Herefordshire level crossing accident

The analysis started with the recording of facts connected with the accident which is treated as the top event or loss event in the MORT diagram.

3.1 RAIB report on herefordshire level crossing accident

The UK rail accident investigation agency, Rail Accident Investigation Branch (RAIB), published results of its findings into the Herefordshire level crossing accident in February 2011. This accident occurred on 16 January 2010 when a passenger train collided against two cars at the Infrastructure Manager(IM) staff managed manually controlled barrier type of level crossing. A woman passenger in one of the cars died as a result of this collision at the hospital while the car driver was seriously injured (RAIB 2011). The occupants in the other car suffered no injuries. Fig. 5 shows a BBC photo of the accident scene at the time when the barriers were up and the red lamp was lit.

The RAIB accident investigation process identifies, as a general procedure, a causal structure of an accident made up of immediate cause, causal and contributory factors, and underlying factors. This structure explains how a particular accident came into being. In this particular accident, the report identified several causal factors.

The immediate cause of the accident was the signaller located in the adjacent signal box who raised the barriers when train IV75 was approaching the MCB type of crossing allowing the cars to move in the path of approaching train 1V75. The causal and contributory factors that led to the accident identified were a) unrecovered human error in the operating situation caused by signaller being distracted by a call from a user of a User Worked Crossing (UWC) and engaged in monitoring the progress of another train ; b) working out the time available to allow sheep movement across UWC; c) mentally distracted to work out the rules by which UWC situation to be managed; and d) the lack of engineering safeguard such as approach locking to protect against signaller's error.

The possible underlying factor stated in the report was the absence of requirements to consider safety benefits of such a measure in the Group or company standards (industry requirements) or in the UK Government regulations.

Finally, there were no cues available in the operating situation which could draw signaller's attention to the fact that danger was imminent in the operating situation. A further reading of the report reveals information that RAIB found the lack of regular communication between the operational risk team and the signalling team within the Infrastructure Manager (IM) organisation, Network Rail. This communication failure compounded the problem of determining the true level of risk. Human error that could occur in the level crossing situation was not included in the risk calculations due to the lack of regular liaison.

Fig. 5. A BBC photo of accident scene. Accessed on http://news.bbc.co.uk/1/hi/england/hereford/worcs/8465412.stm

3.2 SIRI event causal factors, energy barrier trace analysis and MORT analysis

The actual pre-cursor events which were triggered the crash are shown in the hand sketch using the Event Causal Factor notation in Fig 7. Author has used hand sketches for the analysis of the accident situation in the tradition of soft systems engineering started by Peter

Checkland (Checkland 2000). This diagram can be created using Microsoft Visio as well. The oval shapes denote conditions (perceptions or abstract rules) connected to the events and one to many and many to one relationship are shown on the diagram. To facilitate ease of comprehension, conditions which extend over time are shown in dotted lines. The analysis begins with the situation which is normative pattern/ situation of functional sequences of events. User arriving first at the level crossing, the train later (bearing in mind relativity of time at play) and train departing first and user leaving the crossing, later. This forms the core operational layer of the Railway System which is regulated by the Standards Layer which is surrounded by Societal Layer (refer to Fig. 4).

The emergent properties involved in the actions taken by various people involved in the accident situation which were directly connected in the perception -action mode in the sense of affordance are as follows:

a. Car driver's action of perceiving (event UD/E/2) the lifting barrier afforded the information that it is safe to go across the crossing space after having waited for the barrier to lift(event UD/E/1). This is in line with the connection between perception and action mode advanced by Gibson.
b. Train driver's perception of the Stop signal ML 42 changing aspects (RU/E/2) afforded the information that it is not safe to go as train driver became aware that some obstruction is expected. The action of brake application is directly connected to the perception as argued by Gibson.
c. Signaller's perception of the lifting barrier and the train (IM/E/3) afforded the information that there was an error on his part which could not be recovered despite his best intentions. This is in line with the connection between perception and action mode advanced by Gibson.
d. Despite the application of the service and emergency brake, the event in the RU domain (RU/E/3) occurred affording the information to analysts or observers that signalling distance beyond the Stop Signal ML 42 was less than adequate for the train to stop relative to the train speed. This is an indirect inference which is not directly perceived by us (author + reader). This is based upon the expected engineering and cultural norms that a signal shall be provided with a sufficient braking distance in case of emergency conditions obtaining in the operations.

Based upon the above four premises, it is reasoned that affordance of harm was due to failure to provide adequate braking distance at the Stop Signal ML42. By the method of counter factual reasoning, it can be deduced that three elements of information went missing in the System failure scenario case. User was not afforded information as to whether it is safe to cross or not? The signaller had no direct access to the information as whether the train passed the crossing space or not directly from the environment. The train driver had no direct access to information at a point in space from where the train could have been braked to safety. The RAIB report provides complete information with respect to the signaller's error.

The analysis of the operational scenario depicted in the ECF diagram requires from a behavioral science perspective application of the generic human factors framework of Jen Rasmussen Skill-Rules-Knowledge (SRK) or the MORT decision model of the accident process. Fig. 6 shows the MORT decision model which author used for making decisions.

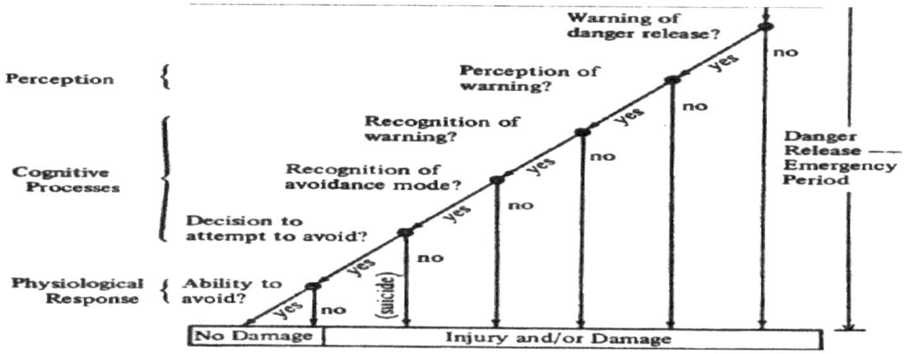

Fig. 6. A decision model of the accident process inherent in the MORT method.

The same model of the accident process can be used to represent danger at the design decision stage where decisions are taken to assume risk by calculation or by ignoring harmful safety outcomes where the end product of such decision making is that potential danger is embedded into the operational situation.

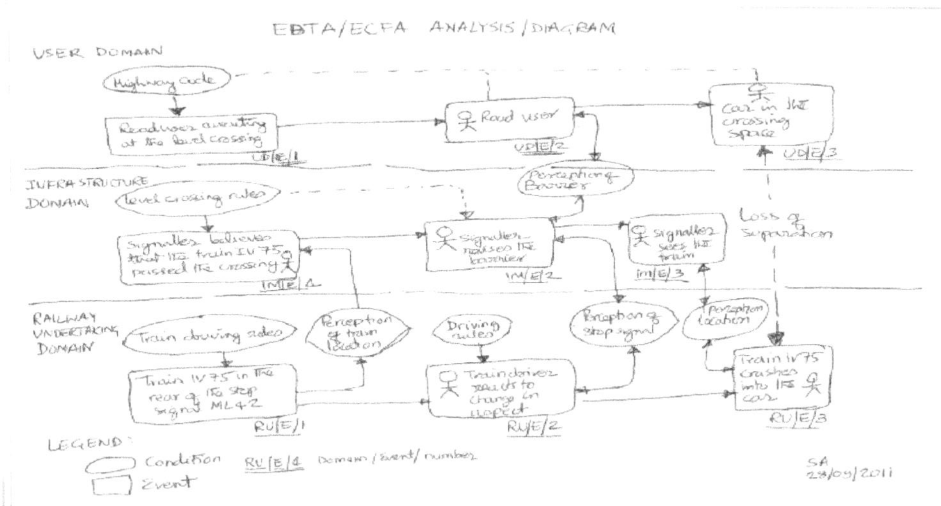

Fig. 7. A hand drawn sketch showing the pre-cursor events and conditions leading to the Herefordshire Crossing Accident on 16th January 2010. This is based upon the Schopenhauer's method of explanation discussed in section 2.1.

Based upon the perspective of energy barrier trace analysis (EBTA), author has listed barriers and controls for the purpose of evaluation of the alternatives which can protect the road user. These have been collected as part of the desk -top search of the ORR and other railway websites for the initiatives underway. These are shown in

Table 3. An evaluation of these barriers is conducted using the MORT questionnaire and as per the flow chart for conducting the investigation in the MORT User Manual. This is freely available for downloading from the NRI website (Johnson.W.G 1974). The results of the MORT application are stated in the Table 4 as per the instructions in the literature available (Gunderson 2005).

Harmful Energy Flow or harmful Agent, adverse environmental condition SB1	Target Vulnerable person or thing SB2	Barrier & Controls to separate Energy and Target SB3
Kinetic hazard (train movement into the crossing space) when it is occupied	Car drivers and passengers	Full service braking distance
		Restriction on train speed
		Obstacle detection
		Lifting barriers
		Road traffic light signals
		Active audio-visual alarms
		Passive visual signs
		Approaching locking
		Interlocking system
		Railway Protective signal
		Bridges, underpass etc
		radio communication systems to private user worked crossing users

Table 3. Energy Barrier Trade Analysis

MORT Branch Description	Problem statement/comments	Evidence
S/M Oversights and Omissions		
Specific Control Factors LTA	SA1: A passenger killed in Herefordshire Level Crossing Accident when a train IV75 struck two cars at Morten -on-Lugg near Hereford. The movement of the train 1V75 into the crossing space when the car is in the crossing space is regarded as not being functional part of the level crossing when considered as an Operational System. The judgement that a blame culture is prevailing the UK railway industry is deduced from the fact that full service braking distance is not provided at the Stop Signal ML42 to facilitate train braking to halt before entering the crossing space in hazardous situation. Such a requirement is not stated by the Accident investigator, the Regulator, the Infrastructure Manager, the Railway Undertaking and the Standards body which form the Social Layer. The Office of Rail Regulation (ORR) is the independent safety and economic regulator for Britain's railways. Following is the extract from the Office of Rail Regulator Website (The Office of Rail Regulator 2008): The on-going safety of level crossings ultimately depends on you, the users recognising the hazard and obeying instructions. The UK's level crossing safety record is among the best in the world. Over a third of all accidents involving a train are at a level crossing. 95% of the train accident risk arises from incorrect use of crossings by road vehicle drivers, such as attempting to 'beat the barriers' or run red lights. Less than 5% of train accidents at level	1. The RAIB Report Summary. (The RAIB 2011). 2. The train driver applied full service braking is evidenced from the paragraph 80, 156 of the RAIB report. 3. More than 8% of accident risk is within the industry control at level crossings is stated by the RAIB (paragraph 167). This data contradicts the ORR information given in the adjacent column. 3. The risk of an accident involving a train and vehicle does not fall into Assumed Risk Category under the MORT Questionnaire as risk to be properly assumed it has to meet the decision criteria of adequacy of cost-benefit analysis, uncertainty about risk themselves, tolerability of risks, adequacy of information and interpretation provided to the person making decision, and finally whether decision to assume risk was made by an appropriate person. This question set can be seen from page 46 of the NRI MORT User Manual. 4. The signaller error is a skill-based performance error. The road user error is a rule-based performance error.

MORT Branch Description	Problem statement/comments	Evidence
	crossing are as a result of a level crossing failure. Pedestrian fatalities and major injuries are most associated with footpath crossings and automatic half barriers (a type of level crossing).	
SA2: Stabilisation and Restoration LTA	Not considered due to the nature of the MORT desk top study. It is assumed that these branch events were adequate.	The RAIB Report Summary.
SB3 Branch Events LTA: : This branch is judged as being less than adequate (LTA) due to the following reasons		
	SD1 Technical Information Systems LTA b1.Knowlegde LTA d1. Application of knowledge from codes and manuals LTA	The RAIB report detailing lack of approach locking in paragraphs 95,130 and 136.
	d2. Was the list of experts(to contact for knowledge) adquate	Page 15 of Level crossings (Hall.S and Mark 2008).
	d3. Was any existing but unwritten knowledge about the work flow/ process known to the ''action' 'person? d4. Was there research directed to the solution of known work flow/process problems and was this adequate?	The RAIB report paragraphs 35, 80, and 89 citing the events of signal aspect change and application of full service braking.
	Action person is the individuals (or individuals) undertaking the work task/process. The signalling engineering renewal works did not use approach locking to prevent the raising of barriers. Rule 119 of the Rule Book did not indicate how the Gate keeper of level crossing would satisfy himself that no train is near before opening the gates to the road traffic. The signalling engineering renewal works in 2009 did not provide sufficient braking	The RAIB report paragraph 15, 35,48,79,95 and 133. Paragraph 38 provides clear indication that ML42 and ML5 are protective signals in opposite directions. ML5 and ML42 fitted with the TPWS indicate that TPWS were fitted (at least at this location) without paying attention to the fact the fitting TPWS toML43 does not provide any safety benefit. The list of research projects conducted by RSSB can be accessed from their website freely at the following URL:

MORT Branch Description	Problem statement/comments	Evidence
	distance at the stop signal ML42 as the RAIB report stated that the braking system on the train was functional, and the driver of the train IV75 applied full service braking when Stop Signal ML42 changed status. No SPAD risk is considered when the installation of the TPWS equipment at ML 43 and ML5 signal in 2003 was considered. This clearly indicates that there was awareness among the project signalling engineers that ML42 did not have sufficient braking distance. Either this information was either not shared with higher management or management has accepted that fact that engineering or management error cannot be compensated. The industry body, RSSB, and the European body, UIC conduct huge amount of research. None of the research had identified non-provision of sufficient braking distance as a risk factor at the MCB type of crossing. The SD1 branch is set to LTA based upon the foregoing problem set.	http://www.rssb.co.uk/SiteCollectionDocuments/pdf/reports/research/T907_guide_final.pdf The European research efforts can be accessed from the UIC Website freely at the following URLs: http://www.uic.org/com/article/european-commission-workshop-on?page=thickbox_enews http://www.iva.ing.tu-bs.de/levelcrossing/selcat/ All of the above represent knowledge based performance errors.
	d5. Previous Investigation and Analysis LTA b4. Independent organisation and person review the work/process to identify high potential hazards LTA. The signalling engineering renewal works did not consider the operational scenario during the project planning stage in which the train might encounter the stop signal being replaced to danger after passing the distant signal for the crossing in the clear position. An identical accident took place on 22 September 1965 at Roundstone level	1. Stanley Hall and Peter Van Der Mark give detail of the similar occurrence and note that this typical of several accidents of this type in pages 33-4, (Hall.S and Mark 2008). 2. The RAIB paragraphs 157 to 167 detailing risk due to manual operations. The RAIB observation that AHB are safer in comparison to manually operated barrier crossings is

MORT Branch Description	Problem statement/comments	Evidence
	crossing near Angmering, on the Brighton to Portsmouth line Col Reed, who led the public inquiry into the Roundstone accident did not consider the psychological pressure felt by crossing keeper and falsely believed that automatic half barriers would provide safer alternative. He did not inquire whether the sufficient braking distance was available. In this occurrence, the signaller lifted the barrier under the distraction from once in 20 year kind of an event, user (farmer) of the adjacent level crossing distracted the attention of the signaller. The RAIB notes 12 near miss incidents and 37 incidents of user abuse. Author had reviewed the work of ABCL/AHB level crossings as a HAZOP Chair and made the results available in the public domain.	contradicted by data given by Stanley Hall and Peter Van Der Mark on page 78 that between 2000 and 2006 16 fatalities have occurred on the AHB level crossings and none have occurred on MCB type of crossings (Hall.S and Mark 2008). 3. The risk data provided by RSSB in the Annual Performance Report in the form of histogram shows that user worked crossings and Active (automatic controlled crossings) pose more risk than manually controlled barrier or gated crossings (RSSB 2010/11). 4.Past MORT study showed that group think bias has an adverse impact on the safety outcomes (S. Appicharla 2010). All of the above represent knowledge based performance errors.
	The signalling engineering renewal works interpreted the term 'absolute' in the Absolute Block System to mean that only one train in a section is permitted between signal boxes on the same line at the same time. This interpretation did not include crossing space as a part of the line or route where trains and vehicles and/or passengers can be on the line at the same time on the crossing space. The signalling engineering works did not provide overlap or safety margin according to the Absolute Block Regulation 3.4 of BR	Chapter 13, pages 48-53 discusses the Absolute Block System Rules and Regulation of the Modern Signalling Handbook (Hall 2010). All of the above represent knowledge based performance errors.

MORT Branch Description	Problem statement/comments	Evidence
	300062/2, 1992 The civil engineering works did not provide bridges or underpass according to the section 13 of the Railway Regulation Act 1842. The 1842 Act also gave the Board of Trade powers in Section 13 to authorise companies to construct bridges in place of level crossings at their own expense, although it should be noted that it did not give the Board of Trade powers to compel them to do so.	Page 7 of Level Crossings (Hall.S and Mark 2008). All of the above represent knowledge based performance errors.
	The civil engineering, signalling engineering works and operations departments of the erstwhile BR organisation did not find satisfactory solution to problems of reduction of manning costs at the gated crossings, reduction of delays to road traffic and to improve safety as they adopted unsafe automated half barrier crossings According to study published by Andrew Evans (2010), on automated crossings accidents rates are higher because the primary responsibility for the safe operation of automatic crossings rests with road users in observing the warnings indicating approaching train. But in this accident, loss of protection took place as the train did not come to stop before striking the cars.	Chapter 7 of Level Crossings details the major reappraisal of level crossing policy (Hall.S and Mark 2008). The accident process theory developed by Stott and used by many accident investigators in the level crossing risk studies is rejected by author in his previous publication (S. Appicharla 2011). All of the above represent knowledge based performance errors.
	The civil engineering works, signalling engineering works and operating rules and regulations did not specify restrictions on train speed when the route contained the hazard of stop signal being placed to danger in the Absolute Block Signalling system. Active human error in the ABS was perceived for the first time in January 1885 when a signalman gave 'Train out of Section' bell signal to the previous signalman before the train cleared out of section.	Page 52, The North Staffordshire Railway Accident near Stroke on Kent, January 1885 and The London, Chatham& Dover Railway Signalling Arrangements (Hall, Railway Detectives 1990). All of the above represent knowledge based performance

MORT Branch Description	Problem statement/comments	Evidence
		errors.
	The signalling engineering works did not consider any hazard review of the UWCs type level crossings which have potential to distract the signaller from monitoring the progress of trains. This type of error which has occurred is similar in nature to signaller error that emerged in the context of track worker safety HAZOP workshops author has chaired in 2006. This HAZOP workshop demonstrated that signalling engineers possess false beliefs about the equipment performance and outcome of the degraded scenarios. The results of HAZOP study were published in 2010 by (S. Appicharla 2010).	Rule 119 of the Rule Book did not indicate how the Gate Keeper of level crossing would satisfy himself that no train is near before opening the gates to the road traffic. The common belief that as long as the rules were followed, safety would be maintained is also seen in the case of the Astra Train crash in 2000 (Halvorsrud 2002). Author from the Norwegian Railway Inspectorate reported that the signalling engineers and technicians reacted with disbelief to the suggestion that re-engineering of the signalling system is necessary. Why? Because railway signalling systems are assumed to be fail safe and it just not fail unsafely.
	In the case of Herefordshire accident signaller's error is located at the skill based level where the necessary condition for the occurrence of a slip of action is the presence of the 'attentional capture' associated with either distraction or preoccupation. The actions of road user, train driver and signaller in the SB2 branch are judged adequate as per the cognitive science tradition. All active human errors were triggered by external causes.	The defective historical rule 119 and defect in the signalling layout and planning are likely to provoke similar reactions of dis-belief to the idea re-engineering of the signalling system and operational rules is necessary as it can be seen from this study.

M Branch Events: This branch is judged as being less than adequate (LTA) due to the following reasons.

Problem statement/comments	Evidence
From the perspective of organisational behaviour, management control is the process through which plans are	1. As per the SIMILAR process discussed in the section 2.2. 2. Barry A. Turner (1976)

MORT Branch Description	Problem statement/comments	Evidence
	implemented and objectives are achieved by setting standards, measuring performance, comparing with actual performance and then deciding necessary corrective action and feedback. However, it is important that standards should prioritise safety. Case studies of King's Cross Underground fire published by Reason (1990) and discussions on automatic train protection system by Whittingham (2004) do not lay emphasis on why policies failures in safety matters continue to occur. No lessons have been learnt from Barry A. Turner study of the Hixon Accident.	published his findings on the problem of failure of foresight. Refer to section 2.2.1.
Policy LTA	The ORR guide on level crossings did not call upon the duty holder organisations (RU/IM) to provide full service braking distance when the stop signal is replaced to danger in the MCB type protected crossings or barrier crossings with obstacle detection As the ORR guide did not ask for full service braking distance at the protective signal, this is considered as a causal factor. When signalling schemes with full service braking distance at level crossings or changes to line speed or constructing bridges or underpass etc when proposed (as a risk reduction measures) these proposals must meet the process requirements defined by ORR as below. As these measures are technically and practically feasible as per MORT terminology and therefore, they are not logged as Assumed Risks in the MORT study. Further, the concept of duty of care from engineering perspective demands safety must be designed into operations. The definition of Risk and Hazard which is accepted in the UK Case Law is noted by	1.COMMISSION REGULATION (EC) No 352/2009 of 24 April 2009 on the adoption of a common safety method on risk evaluation and assessment as referred to in Article 6(3)(a) of Directive 2004/49/EC of the European Parliament and of the Council contradicts the UK HSE Case Law 5. The conflict of philosophy between European Union legislation for inter-operability certification and United Kingdom case law, rules and regulations is noted by Andrew Rae and Mark Nicholson (Nicholson.M and Rae.A Septembe 2010). 2. ORR, Railway Safety Publication 7, Guide for level crossing managers, designers and operators does not call for sufficient braking distance to

MORT Branch Description	Problem statement/comments	Evidence
	author and there is a unresolved problem between the UK Case Law definitions and EU legislations on inter-operability, Safety Directive (Appicharla S. , 2010).	be provided in case stop signals are replace to danger after showing clear aspect. (The Office of Rail Regulator 2011)
SFAIRP 'so far as is reasonably practicable' Policy	The ORR internal policy guidance on safety related investment decisions did not expect the duty holder to perform cost benefit analysis when the risk reduction action is to be taken based upon the relevant good practice as a baseline. When the relevant good practice is not good enough it recommends rough CBA to be undertaken and along with a correction for 'optimism bias'. This is to make adjustments for overconfidence in the project estimates to account for cost overruns in capital projects. Where risks are difficulty to quantify, the guidance documents suggests using qualitative techniques such as structured workshop assessments supported by expert judgement. The RSSB guidance on taking safe decisions uses reasonably practicable policy. The argument advanced is that predicting accident risk in inherently uncertain. Similar accidents may give rise to different fatalities: 31 fatalities (Ladbroke Grove) or 7 fatalities (Southall) and therefore, low frequency high fatality accidents cannot be predicted. Quantitative risk assessment is considered to be useful as high frequency and low fatality incidents can be easily predicted as there is plenty of historical data. This argument is not in accordance with the best practice of safety management. When the information is uncertain, the precautionary principle should be invoked. The principle of inherent safe design of signalling or any other engineering works is perceived but not cognised.	1. This document can be accessed here. http://www.rail-reg.gov.uk/upload/pdf/risk-CBA_sdm_rev_guid.pdf. 2. The scrutiny of account the engineering safety management process followed by the UK railway industry which is biased towards operational reliability by taking into the number of years of reliable operation. The Yellow book does not contain any process for performing system hazard causal factor analysis as identified in the informative clause, 4.4.2.12 of BS EN 50126 (CENELEC 1999). 3. E.L.Woodward in his History of England wrote about new types of business-men arose who were' without scruple and without pity'-free, but lacking in any sense of obligation to their fellow men (Brown 1958). 4. R.B.Whittingham (2004) argued in the page 188 that there is a lack of will to make necessary investment into automatic train protection by the railway industry using arguments of high cost of ATP

MORT Branch Description	Problem statement/comments	Evidence
	It may be argued that the recent commitment by the railway industry to install ERTMS removes this concern. Author wishes to draw readers' attention to the fact ERTMS technology cannot be considered as a barrier in the same sense as TPWS. This judgement is arrived at by reading the technical review of the ETP project by the UK HSE Research Report 0067 (2003) where the reviewers have expressed concern that all potential accident scenarios have not been examined. Further, the recent incident which occurred at the level crossing at Llanbadarn, near Aberystwyth, Dyfed, on the ERTMS signalled railway between Aberystwyth and Machynlleth, on Sunday 19 June 2011 has drawn author's attention. This incident raises a concern that integration and commissioning of national signalling elements into the Inter-operable sub systems to form a coherent and consistent operational system may not have been preceded by any hazard and safety analysis. Selective attention to optimism bias without considering other biases which can operate in the decision making process is a policy error. Author has already published the results of past HAZOP and MORT studies which show that expert judgement is compromised by group think bias in 2010 (Appicharla S. , 2010). The question of blind spot does not arise as information and cognition of that fact that signal ML42 did not give sufficient braking distance has been there since 2003. Thus, question set is marked LTA	per fatality averted but the situation is exacerbated by a lack of consistent policy by successive governments (Whittingham 2004). Author accepts the definition of internal and external causes of human error defined by Whittingham (S. Appicharla, Analysis and modelling of the Herefordshire Accident using MORT Method 2011) 5. Blame culture prevails in the parts of the UK railway industry is noted in the following paper on Organisational Dynamics and Safety Culture in UK Train Operating Companies (Weyman 2006). 6. The railway projects do not consider all accident scenarios and include safety concerns is seen from the following papers on the Train Protection - Technical review of the ERTMS Programme Team report, The UK HSE Reserach Report 067 and performance of ERTMS system. (The NEL Consortium 2003), (D. Hicks 2004). 7. Risk in management systems is a cause for concern is concluded in the RSSB Research Project T169 (Ansper Consulting 2004). 8. Error in policy is a knowledge based performance error.

MORT Branch Description	Problem statement/comments	Evidence
MA2. Implementation of Policy LTA	ORR notes in its annual assessment, "Safety - weaknesses in Network Rail's safety culture have been recognised including the exposure of flawed injury reporting. ORR is often frustrated by the slow pace of necessary safety improvements, and a number of enforcement notices followed failure to make timely progress". This admission by ORR (ORR/14/11) suggests failure in general to the lack of thinking about alternative counter measures for minimising the problems.	ORR Annual Assessment Report. All ORR documents can be freely accessed from their website directly. This represents knowledge based performance error.
	The question of budgets LTA does not arise as Network Rail is a profit making enterprise with annual profit after tax of £313 million in the year 2010-11. This profit can fund replacement of public level crossings and implement communication systems for the private level crossings in the signal box area. A rough estimate of £1 million per unit bridge cost is assumed. The case of private level crossings can be solved by a communication system which can activate and communicate train arrival message to this set of users. Argument from social cost benefit analysis does not arise as it is evident that there is no shortage of funds for investment and the risk falls into intolerable zone. Failure to set an example by ORR is reflected from the above admission. Thus, this question set is marked LTA.	
MA3. Risk Assessment and Control System LTA	This branch is judged as being less than adequate due to the following reasons.	
	• MB1 Hazard Analysis Process LTA	RAIB Report paragraph 167.
	The like for like replacement project in 2009 did not recognise the need for hazard analysis to identify potential accident scenarios. The RSSB and Network rail risk	This represents knowledge based performance error.

MORT Branch Description	Problem statement/comments	Evidence
	assessment process did not include the task of hazard analysis. The RSSB Topic Report says that, ''level crossings are safe when used correctly. Over 90% of risk in the previous ten years has resulted from user misuse in the form of error or violation (the remainder being due to other causes, such as equipment failure, reduced visibility or railway operator error). The analysis of S branch suggests hazard causal factors such as less than adequate control of work process are not modelled in the accident risk equation and therefore, answers to this set of questions are set to LTA.	
	• MA3.Standards LTA There are no requirements to consider alternatives to current work process controls (approach locking) suggested in the Railway Group Standards or NR Company Standards or in the ORR Government Regulations. This conclusion is based upon examination of the ORR Risk Profile Topic Strategy for Level Crossings, HMRI Safety Principles 4 and 23 and Railway Group Standard GI/RT7012. The RGS GI/RT 7012 did not contain any requirement for the full service braking distance to be provided at the protective signal. It is noted that it sets a limit between 50m to 600 m for the stop signal's location from the crossing space but it does not specify whether it applies when the stop signal replaced to danger for the crossings operated by IM staff. The ORR Guidance on level crossings did not call for full service braking distance to be provided in the case of barrier crossings operated by Infrastructure Manager Staff. The guidance does not consider the fact	1.The Railway Group Standard GI/RT7012 (RSSB 2010) 2. ORR Risk Profile Topic Strategy for Level Crossings (Office of Rail Regulator 2008-09 to 2009-10) and ORR Guidance on Level Crossings (Office of Rail Regulator Aug 2011) A paper by X. Quayzi (2011) argued that bounded rationality biases our decision making and leads us to a false sense of safety when using best practices. We are naturally inclined to use best practices without taking a critical view of culture, regulatory systems not defined in the best practices. This approach could lead to an increase of the level of risk (Quayzi 2011).

MORT Branch Description	Problem statement/comments	Evidence
	when a train passes the protective signal at Stop it might lead to an accident scenario. However, provision of full braking distance in other type of crossings is defeated due combination of the driver error and road user error. To be useful, where protecting signal is provided, it is necessary to provide full service braking distance with TPWS protection.	
	• <u>Data Analysis LTA</u> From the inspection of the data in Figure 8 and Figure 9, modelled by Andrew Evans of Imperial College, London show that the trend of high frequency and low fatality events continues and reveals that risk is not ALARP. The statistical overall frequency of accidents is estimated to be 2.63 per year in 2009 causing 3.71 fatalities per year. The product of two figures gives 9.73 accident fatalities per year. The corresponding figure for 1967-2007 was 10.321. This is unreasonable risk as per ALARP classification of risk and does not meet policy requirements for risk management from ALARP perspective as per the guidance clause 3.7 and 3.8 listed on the Guidance Note on the website (The UK Health and Safety 2003). It is vital to consider decision errors in statistical process control domain where it may lead to a situation where vital clue about the phenomena under observation may be missed and out of control process is continued in the operations. The RAIB report provides the evidence that risk analysis was LTA. The risk analysis did not cover information about human error in the operating situation. Further, that the risk analysis procedure used by IM internal procedure for risk assessment i.e. All Level	1. RAIB Report paragraphs 55, and 147 to 155. 2. Reading of the following paper reveals that possible wrong side failure of barriers was not foreseen. The idea of condition monitoring using the fault tree and FMEA techniques detected only one third of the failures and none of the detected failures were causal factors in this accident (Roberts, Márquez and Tobias 2010). 3. The Infrastructure Risk Modelling (IRM) undertaken by Railtrack, predecessor of Network Rail in 1997 did consider the dangerous failures in the consequence analysis and this modelling showed that some branches of the event tree did not contain any safety barriers and directly led to accident scenario due to automatically failure of barriers in the case of CCTV/AHB level crossings. But the modellers and analysts of the event trees did not consider engineering and managerial errors as types in

MORT Branch Description	Problem statement/comments	Evidence
	Crossings Risk Management Model (ALCRM) did not generate any trigger for hazard analysis is evident from the RAIB Report. In 2009, a mean of about 5 per cent of fatal accidents were at railway controlled crossings, 52 per cent were at automatic crossings, and 43 per cent were at passive crossings. 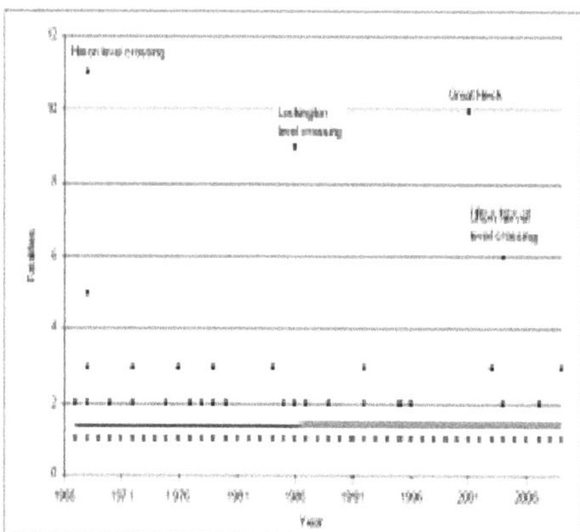 Fig. 8. Fatalities in collisions between train and road vehicles collisions 1967-2009. Source: Andrew Evans (Evans 2010)	the fault or event tree analysis and restricted themselves to operator or user error which appear as external causes to the engineers and managers involved in the decision making on the standards and facility designs. The IRM later developed into the Safety Risk Model (SRM) which was reviewed by the Health and Safety Laboratory (HSL) in 2002 and noted that root causes of human error are not modelled in the SRM. The report noted a concern that SRM fault tree modelling might not support a detailed assessment of root causes of some failures. The report did not express any concern that engineering and management errors are not considered in the pre-cursor model. (Human Factors Group, Health and Safety Laboratory 2002). The above comments show that probabilistic risk assessment was less than adequate to show contribution made by human failures in engineering, management and organisation levels. The event tree analysis in the context of nuclear power is discussed by James Reason (J. Reason 1990).

MORT Branch Description	Problem statement/comments	Evidence

Accident Location	Estimated rate of change in accidents per train-km (with standard error)	Accidents per year in 2009	Fatalities per accident	Fatalities per year in 2009
At level crossings		2.39		3.37
Not at level crossings		0.25		0.35
All	−3.2% (se 0.6%) p.a.	2.63	1.41	3.71

Fig. 9. Fatalities in collisions between train and road vehicle in 2009. Source: (Evans 2010).

- b1Technical Information LTA -b2Defintion of ES& H goals LTA -Trigger to Hazard Analysis LTA -Sensitivity LTA	The flow chart used for the decision making in the IM organisation on the proposed changes did not draw any attention to the hazardous nature of the activity. The existing method of ALCRM is insensitive to changes in the real circumstances concerning pre and post-accident risk modelling are facts read from the RAIB report (paragraphs 55,149,150). The need to perform hazard causal analysis along with risk analysis is stated in the IEC 61508 at phase 3 before the allocation of requirements. This basic safety standard can be used for non-programmable technologies as well. A paper published by another railway administration regulated by the ORR gives an instance of this kind of conceptual error of not considering system hazard factor casual analysis. Conceptually, BS EN 50126 describes the idea of hazard causal factors analysis in clause 4.4.2.12, Figure 7 of the standard, but this analysis is not mandatory for the regulatory or system development process. The UK Railway Safety Risk Model does not integrate the fault tree and event trees correctly as it is required for the proper estimation of the risk. The top event of the	1. The concepts of system, system hazard, and probabilistic risk analysis are not understood in the UK railway industry. This acknowledgement is made in a paper presented by G.J.Bearfield and R, Short of RSSB and Atkins Rail in September 2011 (G.J.Bearfield; R.Short September 2011). 2. The RAIB Report Paragraphs 55,149,133 and 150. 3. Hazard Management with DOORS: Rail Infrastructure Projects (Hughes,D , Saeed A, 2009). Hazard management is taken to mean management of hazard log rather than concrete action to eliminate the unsafe situations. Metro railways undertake multi-method of analysis is learnt from the published literature on São Paulo Metro (Joao Batista Camargo Junior 1999).

MORT Branch Description	Problem statement/comments	Evidence
	fault tree is used as an input to the event tree in the case of the Railway Safety Risk model. This conceptual error does not arise with other PRAs where there is correct integration. The Installation of the TPWS equipment at ML 43 and ML5 signal in 2003 clearly indicates that there was awareness among the project signalling engineers that ML42 did not have sufficient braking distance. This information was either not shared with higher management and management failed to act or was suppressed locally. Lack of communication between the risk team and the signalling team is noted in the RAIB report. Thus, answers to risk data analysis, setting of EH& S goal setting, trigger to hazard analysis etc in this sub-section are set to LTA.	4. The Infrastructure Risk Modelling carried out by Railtrack using the Cause Consequence Analysis method the scenario crossing open before train has passed is recognised as an accident scenario (automatic function) with no barriers in place to safe guard road user life in the case of manually controlled barrier crossing CCTV type (Ref Railtrack/S&S/IRM_CCA/18 dated March 1998). This document can be searched on the web using the above reference.
•Safety Program Review LTA	ORR guide did not call up for any safety program review in the guide. There is no evidence provided in the RAIB report which gives the assurance that a safety program review exists in the IM/RU organisations. British railway did not have any cohesive plan for safety management is gathered from David Maidment's account (Maidment 2002). The details can be seen at this URL: http://www.davidmaidment.com/railways.htm. In hindsight the claim made by David Maidment is wrong as BR did not implement MORT method even after being aware of its existence. This fact Is cited in the literature (S. Appicharla 2011). Thus, MA3 branch events are set to LTA.	There is no single, clearly defined assurance process and formal safety assessment in the railway industry is concluded in the RSSB research reports T219 (DNV Consulting 2004) and T220 (DNV Consulting 2004).

MORT Branch Description	Problem statement/comments	Evidence
Conclusion	This paper verified the MORT thesis that the Herefordshire accident occurred because affordance of harm posed by MCB level crossing was not eliminated by the signalling layout, less than adequate signalling rules, less than adequate operational rules due to oversights and omissions.	1.The MORT and Swiss Cheese Model
		2. Latest experiment published in the in the journal of Experimental Social Psychology reveals that society looks upon the role of producer with respect and admiration and looks down the role of worker (The Economist 2011). Society rewards risk taking but punishes safety risk taking is the inference author draws from the demands for public inquiries after every major accident. If human agency causes accidents, then human agency can prevent them as well.
	The MORT study concluded and reconfirmed that safety critical decision making suffers from individual as well as group think bias. Internal decisions taken by the industry attribute human error to external causes and this attribution provides a latent pathway to erode the barriers as per the SCM.	
	The safety interventions can be initiated if the two proposals of bridges replacing public level crossings, radio communication with the private crossing users (UWC and other private crossings) can be established to provide information on the arrival of trains at the crossing space. Obstacle detection without provision of stopping distance for the train is not design which complies with inherent safe design principle.	The idea that latent errors that precede a major disaster in defended systems is analogue to resident pathogens in the human body is refuted by this case study as there were no multiple defences in the design of the operational system. The notion that fallible decision making cannot be detected is not true is learnt from this case study.

Table 4. MORT Table for the explanation of the Herefordshire Level Crossing Accident

4. Problem Statement

Problem Statement electronically created 13 December 2005

How do we identify the Interfaces?

4.1 Background

The standards strategy is centred around the filter process which takes existing measures and determines whether they are defining a duty holder interface. The assumption is that within our existing standards all the interfaces are sufficiently defined.

Within CCS & ENE the feeling is that this assumption is not valid. Many duty holder interfaces are either not covered or at best implied. If the existing measures where the only source of material for the new standards the fear is that many gaps would be left.

What is required is a means of identifying all the relevant interfaces with a high level of confidence that omissions do not exist. Modelling of the railway systems is proposed as the means of achieving this. Furthermore, modelling may offer additional benefits. Note that modelling may not be the only means of achieving the objectives.

4.2 Solution objectives

The primary objective is to identify all interfaces (at the product level) between duty holders for each part of the CCS & ENE railway disciplines. These must be sufficiently detailed to ensure that all known technology implementations are described. In some cases the various technologies will create different interfaces and therefore need separate means of identification, in some cases the known technology solutions will not require to be identified separately.

Where the choice of technology creates different interfaces, it will be helpful (but not immediately essential) to represent a non-technology dependent system since this would appear to provide assistance for the future development of new solutions.

As a secondary objective, it may be helpful if the solution could provide assistance with the safety justification needed. Achievement of this objective is not essential.

4.3 Exclusions

It is not anticipated that the wording or the measurement values of the final control measure will be generated by the solution.

5. Conclusion

This chapter advanced presented a framework for the conduct of independent accident analysis and system safety analysis and assessment by taking cognitive systems engineering perspective in response to a problem statement expressed in 2005. This takes into account organisational, technical, individual and regulatory factors. The application of the SIRI Framework to the case study of Herefordshire level crossing accident showed groupthink bias is active in UK railway industry. Why? Because material cause (less than adequate signalling rule regarding stop signal location), efficient cause (less than adequate operational rule 119) and formal cause (managerial policy of relying on numerical risks and their pre-cursors and subject expert information and decision based upon a false erroneous grasp of ALARP principle) have been demonstrated in the case study presented in the chapter. Thus, it is necessary that re-engineering action on the current signalling system, rule-making and procedures of decision making is taken to manage the rail-road interface safety. This is to tackle the problem of group think and individual bias which has negative impact upon safety outcomes.

6. Acknowledgment

Author wishes to acknowledge the efforts of following organisations and/or individuals for their indirect help rendered in the preparation of this paper.

- The MORT team and supporting team at the Noordwijk Risk Initiative Foundation, Netherlands, for hosting MORT related documentation.
- The Trustees of the estate of Dr. W. Ross Ashby for making available Ashby's documentation.
- The team(s) at the Health and Safety Executive, United Kingdom for making available HSE related research reports.
- The team(s) at the Delft University, Netherlands, and their collaborators for the invitation to submit a paper.
- Reviewers for their helpful suggestions and comments.

7. References

Adams, J. "Risk Management: the Economics and Morality of Safety Revisited." London: Proceedings of Safety-Critical Systems: Problems, Process and Practice, 2009. 23-37.

Ansper Consulting. *T169, Risk in Management Systems.* London: RSSB, 2004.

Appicharla, S. "Analysis and modelling of the Herefordshire Accident using MORT Method." *The 6th IET International System Safety Conference.* Birmingham: Institution of Engineering and Technology, 2011. 10.

—. "System for Investigation of Railway Interfaces." *The 5th IET International System Safety Conference.* Manchester: Institution of Engineering and Technology, 2010. 6.

—. "System for Investigation of Railway Interfaces." *The 1st IET International Conference on System Safety.* London: Institution of Engineering and Technology, 2006. pp.7-16.

Appicharla, Sanjeev. "Economy, a thermodynamic perspective." *Unpublished Manuscript.* New Delhi, 12 December 1998.

Appicharla, Sanjeev Kumar. *Response to the Consultation on Level Crossings , UK Law Commission* . London: Unpublished draft, 2010.

Aristotle. *Ethics.* Translated by J.A.K.Thompson. Aylesbury : Penguin Classics, 350BC/1955.

—. *The Politics.* London: Penguin Classics, 323 BC/1951.

Armstrong, K. "Why EMC Testing is Inadequate for Functional Safety -and what should be done Instead." *The 1st IET International Conference on System Safety.* London: Institution of Engineering and Technology, 2006. pp.179-83.

Ashby, W.R. *Introduction to Cybernetics.* London: Chapman and Hall Ltd, 1956/1999.

—. *The Design of a Brain.* London: John Wiley & Sons, 1960.

Ashby, W.R., and C.R Conant. "Every Good Regulator of a System must be model of the System." *International Journal of System Science* Vol 1, No.2 (1970): 89-97.

B.Bateman, S.W. Hatton. "The Increasing Role of Structured Methods in Arguing Safety." *1st IET International Conference on System Safety.* London: Institution of Engineering and Technology, 2006. pp.158-63.

Bahill, A.Terry, and Steven J.Henderson. "Requirements Development, Verification, and Validation Exhibited in Famous Failures." *Systems Engineering* (INCOSE and Wiley Subscription Services) 8, no. 1 (2005): 1-14.

Baker, Baker S. *Human Capital* . London: The University of Chicago Press, 1964.

Benjamin.S.Blanchard. *Systems Engineering Managment.* Vol. 3rd Edition. New Jersey: John Wiley& Sons,Inc, 2004.

Briscose .G, J. *MORT Based Risk Managment.* Idaho Falls: System Safety Development Centre, 1990.

Brown, J.A.C. *The Social Psychology of Industry.* Middlesex: The Penguin Books, 1958.

CENELEC. *EN 50126 Railway applications- The Specification and demonstration of Reliability, Availability, Maintainability and Safety Speicification* . Brussels: CENELEC , 1999.

Chapman, J. *Systems Failure.* London : Demos, 2004.

Checkland, Peter. "Soft Systems Methodology." *Systems Research and Behavorial Science,* 2000: pp.11-58.

Clifton, Ericson II . A. *Hazard Analysis Techniques for System Safety.* New Jersey: Wiley& Sons, 2005.

D. Hicks. "Performance modelling for the National ERTMS Programme (NEP)." *The IEE Seminar Railway System Modelling.* London: The Institution of Electrical Engineers, 2004. 61-74.

DNV Consulting. *Review of the efficacy of the safety assurance processes in preventing catastrophic accidents.* London: The RSSB, 2004.

DNV Consulting. *T220, Applicability of Formal Safety Assessment Process Approach to Rules and Standards Developmetn within the Railway Industry.* London: RSSB, 2004, 28.

E.Apostolakis, George. "How Useful is Quantative Risk Assessment." 24, no. 3 (2004): Risk Analysis .

Einstein, Albert. *Relativity.* London: Routledge, 1920.

Ekekwe, Ndubuisis. *Better Risk Communication.* 11 May 2011. http://blogs.hbr.org/cs/2011/05/better_risk_communication.html (accessed September 29, 2011).

Elliot, John. "Systems Approach to Safety-related Systems." London: Springer-Verlag London Limited, 1999. 75-98.

Elliott, Chris, Taig,T. *T230 a Ethical Basis of Rail Safety Decisions.* London: RSSB, 2003.

Etzioni, Amitai. "Humble Decision Making." *The Harvard Business Review,* 1989: 122-26.

Evans, Andrew. "Fatal accidents at railway level crossings in Great Britain: 1946-2009." *Accident Analysis and Prevention,* 2011: 1837-1845.

—. "Fatal Train Accidents on Britain`s Main Line Railways: End of 2009 Analysis ." *Center for Transport Studies, Imperial College London* . March 2010. http://www.cts.cv.ic.ac.uk/html/ResearchActivities/publicationDetails.asp?PublicationID=1330 (accessed October 04, 2011).

Fischhoff, Baruch. "Setting Standards : a Systematic Approach to Managing Public Health and Safety Risks." *Management Science* 30, no. 7 (1984): 823-43.

G.J.Bearfield; R.Short. "Standardising Safety Engineering Approaches in the UK Railway ." *The Sixth International System Safety Conference.* Birmingham: The Institution of Engineering and Technology , September 2011. 5.

G.March, James, and Zur Shapira. "Managerial Perspectives on Risk and Risk Taking." *Management Science* (The Institute of Management Science) 33, no. 11 (1987): 1404-418.

G.W, Leibniz. *Philosophical Texts.* Translated by R.S.Woolhouse and Richard Franks. London: Oxford University Press, 1695/1998.

Gissing, Bruce, and A.Terry Bahill. "Re-evaluating Systems Engineering Process Using Systems Thinking." *Systems Engineering* (The Institution of Electrical and Electronic Engineers) 28, no. 4 (November 1998): 516-27.

Goddard, E. "Supervision and operation of mass transit system." *Seventh Vacation School on Railway Signalling and Control Systes.* Glasgow: The Institution of Electrical Engineers, 1998.

Goodwin, Phil. *Determination and Denial: The Paradox of Safety Research and Traffic Policy* . London: The 17 thParliamentary Lecture on Transport Safety, PACTS , 2006.

Graham Walker. "The Science of Morality ." Edited by Graham Walker. London : RoyaL College of Physicians, 2007. 130.

Gray, Gigerenzer and J.A.Muir. *Better Doctors, Better Patients, Better Decisions; Envisioning Health care.* Cambridge, MA: MIT Press, 2011.

Gunderson, Scott. "A Review of Organizational Factors and Maturity Measures for System Safety Analysis." *Systems Engineering* 8, no. 3 (2005): 234-44.

H.Popkin, Richard, and Avrum Stroll. *Philosophy* . Oxford: Butterworth Heinemann, 1969.

Hale, Andrew., P.H.Lin, and A.L.C Roelen. "Accident Models and Organisational Factors in air tranport: the need for multi-method models." *Safety Science* 49 (2011): 5-10.

Hall, Stanley. *Modern Signalling Handbook.* Skipton: Ian Allan, 2010.

—. *Railway Detectives.* London : Ian Hall, 1990.

Hall.S, and Peter Van Der Mark. *Level Crossings.* Hersham: Ian Allan Publishing, 2008.

Halvorsrud, Gunhild. *The Asta Train Crash, its Precursors and Consequences, and its Investigation.* London : The Safety-Critical Systems Club, Springer , 2002.

Handy, C. *Understanding Organisations.* 4th. London: Penguin Books, 1999.

Hollnagel, EriK, and Woods David.D. "Cognitive Systems Engineering: a New Wine in New Bottel." *International Journal of Man-Machine Studies* 18 (1983): 583-600.

Hovdon, J, Storseth, and R,N Tinmmanvisk. "Multilevel Learning From Accidents: Case Studies from Transport." *Safety Science* 49 (2011): 98-105.

Hughes,D , Saeed A,. "Hazard Management." *Hazard Management, System Safety-Critical Systems: Problem, Process, and Practice, Springer, Proceedings of the 17th Safety Critical Systems Symposium, Brighton, UK.* London: Springer , The Safety-Critial Systems Club, 2009. pp. 23-37.

Human Factors Group, Health and Safety Laboratory . *Review of Railway Safety's Safety Risk Model, HSL/ 2002/06.* Sheffield: The UK Health and Safety Executive, 2002.

Hume, David. *A treatise of Human Nature.* 1984. London: Penguin, 1739/1984.

I.Mitrani. *Simulation technqiues for discrete event systems.* Cambridge: The Press Syndicate of the Cambrigde University Press, 1982.

IEC. *61508: Funtional Safety of electrical/electronic/programmable electronic safety- related systems.* Brussels: International Electro-technical Commission, 2001.

IEEE Computer Society. *IEEE 1220 Standard for Applicationn and Management of the Systems Engineering Process.* The Institute of Electrical and Electronics Engineers, 1998.

ISO/IEC. *System Engineering- System life cycle ISO/IEC 15288.* International Standard, Brussels: The International Elecctro-Technical Commission, 2002.

J.Fabrcky, Wolter, and Benjamin S.Blanchard. *Systems Engineering and Analysis.* Prentice Hall, 2005.

J.M.Coulson, and .J.F.Richardson. *Chemical Engineering.* Oxford: Pergamon, 1965.

J.Reason;E.Hollangel; J Paires. *Revisting the « Swiss Cheese » Model of Accidents.* Accident Model discussions, BRUXELLES: Eurocontrol Agency, 2006.

Joao Batista Camargo Junior, Jorge Rady de Almeida Junior. "The Safety Analysis Case in the Sao Paulo Metro." *Towards System Safety* . London: Springer, Safety-Critial Systems Club , 1999. 110.

Johnson, C.W. "Reasons for the Failure of Incident Reporting in the Healthcare and Rail Industries." *Proceedings of the Tenth Safety-critical Systems Symposium.* London: Springer-Verlag, 2002. 31-57.

Johnson, Chris. *What are Emergent Properties and How Do They Affect Engineering of Complex Systems.* Exploration, Glasgow: Glasgow University, 2006, 14.

Johnson.W.G. *Management Oversight and Risk Tree SAN 821-2.* Washington D.C: US Atomic Energy Agency, 1974.

Kelly, Tim. "Are Safety Cases Working?" January 2008: 5.

Kingston, John., Robert Nertney, Rudolf Frei, Philippe Schalllier, and FloorKoornef. "Barrier Analysis Analysed from MORT Perspective." *PSAM7/ESREL '04 International Conference on Probabilistic Safety Assessment and Management.* Berlin: Springer-Verlag, London , 2004.

Kletz, T. "Accident Investigations-Missed Oppurtunties." *Components of System Safety.* London: Springer-Verlag London Limited , 2002.

Kletz, T, D Mansfield, and L Poulter. *Improving Inherent Safety, OTH 96521.* Sheffiled: The UK Health & Safety Executive, 1996.

L.Derby, Stephen, and Ralph Keeny. "How Safe is Safe Enough." *Risk Analysiss* (Wiley & Sons) 1, no. 3 (1981): 21-24.

L.Janis, Irving, and Leon Mann. *Decision Making.* New York: The Free Press, 1977.

Ladkin, Peter. *Why Because Analysis* . 1 January 1995. http://www.rvs.uni-bielefeld.de/ (accessed Septmeber 13, 2011).

Leveson, N. "The Need for New Paradigms in Safety Engineering ." *Proceedings of the Seventeenth Safety- Critical Systems Symposium.* Brighton,UK: Springer-Verlag , 2009. 3-20.

Leveson, Nancy. "A New Accident Model for Engineering Safer Systems." *Safety Science 42,* 2003: 237-230.

Leveson, Nancy. "Applying Systems Thinking to Analyse and Learn from Events." *Safety Science* (Elsevier), 2011: 55-64.

—. "The Need for New Paradigms in Safety Engineering." *Safety-Critical Systems: Problems, Processes and Practice.* London: Springer-Verlag London Limited, 2009. pp. 3-20.

Lovallo, Dan, and Daniel Kahneman. "How Optimism Undermines Executives' Decisions." *Harvard Business Review* , July 2003: 8.

M.Copi, Irving, and Carl Cohen. *Introduction to Logic.* New Delhi: Pearson Education, 1998.

Maidment, David. "System Safety: challenges and pitfalls of Intervention." *New Technologies and Work* . Oxford : Elsevier Science, 2002.

Manu, Sage. *Laws of Manu, Manu Smrithi.* Translated by Wendy Donhier with Brian Smith. London: Penguin Classics, unknown/1951.

McCormick, Matt. *Immanuel Kant: Metaphysics.* 30 June 2005. http://www.iep.utm.edu/kantmeta/ (accessed October 2011, 2011).

NEP. *National ERTMS Programme Report.* London : National ERTMS Programme Team , 2003-2004.

Nicholson.M, and Rae.A. "IRSE Guidance on the Application of Safety Assurance Processes in the Signalling Industry (May 2010)." *Safety Critial Systems Club Newsletter* (Safety-Critical Systems Club) 20, no. 1 (Septembe 2010): 14-19.

Nikhilananda, Swami. *The Bhagavad Gita.* New York: Ramakrishna-Vivekananda Center, 1944.

Noordwijk Risk Foundation. *the NRI Foundation website.* 1998. www.nri.eu.com (accessed September 29, 2011).

Office of Rail Regulator. *HMRI's Risk Profile Topic Strategy for Level Crossings.* 2008-09 to 2009-10. http://www.rail-reg.gov.uk/upload/pdf/RPT-levxings.pdf (accessed October 04, 2011).

Office of Rail Regulator. *Level Crossings, A guide for managers, designers and operators, Railway Safety Publication 7.* London: Office of Rail Regulator, Aug 2011.

Penrose, Roger. *The Road to Reality.* London: Jonathan Cape, 2004.

Perrow, Charles. *Normal Accidents.* 1999. New Jersey: Princeton University Press, 1984.

Plato. *The Republic.* Translated by Desmond Lee. London: Penguin Books, 375 BC/1995.

Popkin H. R, Stroll,A. *Philosophy.* Oxford: Butterworth Heinemann, 1969.

Quayzi, X. "Are Best Practises Really Best Practises." *The Sixth International System Safety Conference* . Birmingham : The Instituion of Engieering and Technology , 2011. 4.

R.L.Maguire, and C.J. Brain. "History and Perception of the Language Used in the Safety Domain." *The 1st IET International Conference on System Safety.* London: Institution of Engineering and Technology, 2006. 196-01.

Rae, Andrew. *Safety Decision Making Using Cost-benefit Analysis* . Newcastle,UK: Safety Critical Systems Club Newsletter, 2010.

RAIB. *Fatal accident at Moreton-on-Lugg, near Hereford, 16 January 2010.* Accident Report 04/2011, The RAIB, UK Department for Transport, 2011.

Railway, Board British. *Group Standard GH/CZ0002* . Derby: The British Railway Board, 1993.

Rasmussen, J. *Information Processing and Human-Machine Interaction.* New York: Elsevier Science, 1986.

Rasmussen, J, A M Pejtersen, and L P Goodstein. *Cognitive Systems Engineering.* New York: John Wiley& Sons, Inc, 1994.

Rassmussen. "Risk Management in Dynamic System: A Modelling Problem." *Safety Science* (Elsevier Science Limited) 2/3 (1997): 183-213.

Rassmussen, J. "Risk Management in a dynamic society: a modelling problem." *Safety Science* (Elsevier Science Limited) 27, no. 2/3 (1997): 183-213.

Rassmussen, M.L. "A Systens View of The Self." *The 48th Annual Meeting of the International Society for the Systems Sciences.* The International Society for the Systems Science, 2004.

Reason, J. *Human Contribution: Unsafe Acts, Accidents and Heroic Recoveries.* Surrey: Ashgate Publishing , 2008.

—. *Human Error.* 17th. New York: Cambridge University Press, 1990.

Reason, James. *Recurrent Patterns in Transport Accidents: Conditions and Causes.* London: 18 th Parliamentary Advisory Council for Transport Safety, 2007.

Ring, Jack. "Towrds an Ontology of Systems Engineering ." *Insight* , April 2002: 19-23.

Risk Solutions. *Development and Calibration of Model for Guaging Societal Concern for the Railway Industry.* London: The RSSB , 2006.

Roberts, C, F P G Márquez, and A M Tobias. "A Pragmatic Approach to the Condition monitoring of Hydraulic Level Crossing Barriers." *Proceedings of the Institution of Mechanical Engineers, Part F: Journal of Rail and Rapid Transit,*. Birmingham: The Sage Publications , 2010. 605-10.

Robinson, D & Garrat. *Introducing Ethics.* London: Icon Books Limited, 2008.

Roland, Harold E., and Brian Moriarty. *System safety engineering and management* . Danvers. M.A: John Wiley& Sons, 1990.

RSSB. *Annual Safety Performance Report 2010/11.* London: RSSB, UK, 2010/11.

—. *Requirements for Level Crossings.* 06 Feb 2010. http://www.rgsonline.co.uk/Railway_Group_Standards/Control%20Command%20and%20Signalling/Railway%20Group%20Standards/GIRT7012%20Iss%201.pdf (accessed October 04, 2011).

Rupert Woodfin, Oscar Zarate. *Introducing Marxism.* Royston: Icon Books Ltd, 2004.

S.Baker, Gary. *Human Capital.* 3rd. Chicago : The University of Chicago Press, 1964.

S.Blanchard, Benjamin, and Wolter J.Fabrcky. *Systems Engineering and Analysis.* Prentice Hall , 2006.

Schopenhaeur, A. *On the Principle of Sufficient Reason.* Translated by Karl Hillebrand. New York: Prometheus Books, 1820/2006.

Singh, Simon. *The cracking codebook.* London: Harper Collins, 2001.

Skogstad, Øystein. *Experiences with Safety Case Documentation According to CENELEC Railway Safety Norms.* London: The Safety-Critical Systems Club, Springer, 1999.

Smith, Adam. *Wealth of Nations, The Nature and Causes of the Wealth of Nations.* New York: Modern Library, 1776/1937.

T, Taig., and Elliot.C. *Ethical Basis of Rai Safety Decisions.* London: RSSB, 2003.

Taleb, Nassim N., Daneil G. Goldstein, and and Mark W.Spitznel. "Six Mistakes Execuives Make In Risk Managment." *Harvard Business Review,* 2009.

The BBC. "Engineers can learn from slime." *The BBC.* 22 January 2010. http://news.bbc.co.uk/1/hi/8473316.stm (accessed October 03, 2011).

The BBC News. *China train crash: Design flaws to blame - safety chief.* News, Unknown: The BBC News, 12 August 2011.

The BBC News. *'Rats' cause misery for London King's Cross commuters.* unknown: The BBC News, 27 Septmber 2011.

The Economist. "Difference Engine: Disaster waiting to happen." 16th September 2011. http://www.economist.com/blogs/babbage/2011/09/reliability-grid (accessed September 17, 2011).

—. "The Deepwater Horizon Report." *Economist,* 13th January 2011.

The Economist. *Goodness has nothing to do with it.* Unknown: The Economist, 24 September 2011.

—. "So long, and thanks for all the quarks, Science and Technology section." 1-7th October 2011.

—. "All power tends to corrupt." *The Economist,* 1st-7th October 2011: 87-.

—. "Left to their own devices,Medtronic and the woes of America's medical-technology industry." *The Economist*, 10th September 2011.

The NEL Consortium. *Train Protection - Technical review of the ERTMS Programme Team report, The UK HSE Reserach Report 067.* Norwhich: The UK Health and Safety Executive, 2003, 54.

The Office of Rail Regulator . *ORR, Railway Safety Publication 7, Guide for level crossing managers, designers and operators.* The UK Office of Rail Regulator , 2011.

The Office of Rail Regulator. *Facts & figures on Level Crossings.* 25 January 2008. http://www.rail-reg.gov.uk/server/show/nav.1568 (accessed October 04, 2011).

The RAIB . *Investigation into an incident at Llanbadarn level crossing, 19 June 2011.* 19 June 2011. http://www.raib.gov.uk/publications/current_investigations_register/110619_llanbadarn.cfm (accessed September 29, 2011).

The RAIB. *Fatal Accident involving a Train Driver Deal, 29 July 2006.* Derby: The RAIB, 2006.

The RAIB. "RAIB Accident Report 04/2011, Fatal accident at Moreton-on-Lugg, near Hereford, 16 January 2010, Accessed on 15th June 2011." Derby, 2011.

The Royal Academy of Engineering. *Engineering Ethics in Practice: Short Version.* August 2011.
http://www.raeng.org.uk/societygov/engineeringethics/pdf/Engineering_ethics_in_practice_short.pdf (accessed October 07, 2011).

The UK Health and Safety. "Assessing compliance with the law in individual cases and the use of good practice." *The UK Health and Safety Executive.* 2003.
http://www.hse.gov.uk (accessed 10 04, 2010).

The UK Health and Safety Executive. *Why Control Systems Go out of Failure.* The UK Health and Safety Executive, 2003.

Tolstoy, Leo. *My Confession and the Spirit of Christ's Teaching.* London: Walter Scott, Limited, 1887/1930.

Turner, Barry A. "The organisational and Inter-organisational Development of Disasters." *Adminstrative Science Quaterly* (Johnston School of Management, Cornell University) 21, no. 3 (1976): 378-97.

Turner, S.D. "Jublilee Line Upgrade From Cross Acceptance to Revenue Service." *The 6th IET International System Safety Conference.* Birmingham: Institution of Engineering and Technology, 2011. 7.

Tversky, Amor, and Kahneman Daniel. "Judgement Under Uncertainty: heuristics and biases." *Science*, 1974: 1124-31.

Valerie, Roebuck. *The Upanisads.* London: Penguin Books, 2003.

W.Maier, Mark, David Emery, and Rcih Hillard. "ANSI/IEE 1471 and Systems Engineering." *Systems Engineering* (INCOSE and Wiley Subscription Services) 7, no. 3 (2004): 257-70.

Wainwright, Martin. *Two arrested over railway cable theft on east coast mainline.* The Guardian, 21 September 2011.

Wei, Choo Chun. "Organisational Disasters, Why they happen and how they may be prevented." *Management Division* 46, no. 1 (2008): 32-45.

Weyman, A & Pigdeon, N & Jeffcott, S & Walls, J. *Organisational Dynamics and Safety Culture in UK Train Operating Companies.* Norwich: UK Health and Safety Executive, 2006.

Whitehead, Alfred North. *Process and Reality.* 1985. New York: The First Free Press, 1927/1978.

Whittingham, R.B. *The Blame Machine.* Burlignton: Elsevier Butterworth-Heinmann, 2004.

Wicks, R. *The Stanford Encyclopedia of Philosophy.* Edited by Edward N. Zalta. 17 Nov 2007. http://plato.stanford.edu/archives/win2010/entries/schopenhauer (accessed September 05, 2011).

Wikipedia. *Earth potential rise.* 4 May 2011. http://en.wikipedia.org/wiki/Earth_potential_rise (accessed September 18, 2011).

—. *System safety.* 21 June 2011. http://en.wikipedia.org/wiki/System_safety (accessed September 25, 2011).

Winter, Peter. *Compendium on ERTMS.* Hamburg: DVV Media Group GmbH, 2009.

Wkipedia. "Swiss Cheese Model." *Wkipedia.* 15 September 2011. http://en.wikipedia.org/wiki/Swiss_cheese_model (accessed September 25, 2011).

Wolff, J. *T230b Railway Safety and Ethics of Tolerability of Risk.* London: RSSB, 2002.

9

The Compatibility and Preparation of the Key Components for Cement and Asphalt Mortar in High-Speed Railway

Fazhou Wang and Yunpeng Liu
Wuhan University of Technology
China

1. Introduction

1.1 Introduction of CA mortar for high-speed railway

Slab track is a non-ballast track form that has found wide applications in high speed railways in countries like Japan, Germany, Spain and Italy due to its advantages of reduction in structure height, lower maintenance requirements, increased service life and high lateral track resistance which allows future speed increases (Harada Y TS, Itai N, 1976, 1983; Esveld C, 2003; Miura S et al, 1998). Also, China is witnessing a rapid expansion in developing high-speed railway to alleviate the increasing pressure of transferring people around such a big country (SH Jin 2006; JQ Zuo et al, 2005).

Nowadays, two main forms of Slab track (CRTS I, CRTS II) are used in China which evolve from Shinkansen slab track of Japan and Max-BÖgl slab track of Germany, respectively (Harada Y TS, Itai N, 1976; Esveld C,2003; Vogel W, 1995). There are some differences in slab structure as shown in Fig 1.

Cement and asphalt mortar (short for CA mortar, sometimes abbreviated to CAM) is an interlayer injected in the spaces between the track slab and the concrete roadbed (or hydraulically stabilized base), which is the key component in the structure of slab track. It mainly consists of cement matrix, asphalt emulsion, fine aggregates and a variety of admixtures (SG Hu et al, 2009a, 2009b; FZ Wang 2008, 2009, 2010). CA mortar used in CRTS I and CRTS II slab track are divided into I CAM and II CAM, respectively, whose physical and mechanical properties are shown in Table 1.

Due to the differences of slab track structure, the inject depth and properties of CAM (mainly are compressive strength and elastic modulus) are different. As for I-CAM used in CRTS-I slab track, the inject depth is about 50-70mm, the 28d compressive strength is over 1.8MPa and elastic modulus is about 100-300Mpa; however, as for II CAM used in CRTS-II slab track, the inject depth, 28d compressive strength and elastic modulus are 20-40mm, over 15Mpa, 7000-10000MPa, respectively. The mechanical properties, elastic modulus, durability of CA mortar determine the service life, safety and comfort of high speed railway. CA mortar is a new type of semi-rigid high performance composite material and possesses tremendous potential of applications in high speed railway which supports the track and train, adjusts the track precision and plays as shock absorber. Figures 2--5 showed the different status of CA mortar under construction.

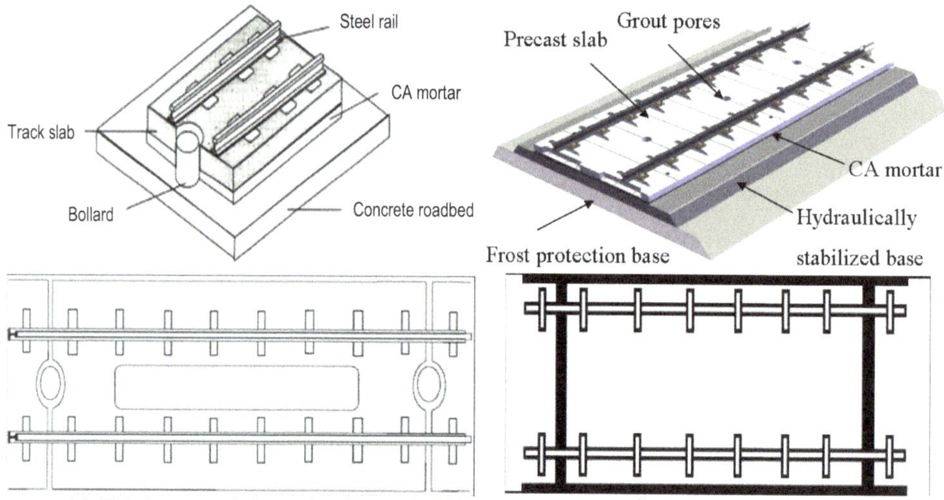

(a) Shinkansen slab track of Japan (b) Max-BÖgl slab track of Germany

Fig. 1. Structure of slab track

	Composition	Performance
I-CAM	asphalt emulsion is cationic A/C* is 1.4-1.6	Compressive strength >1.8MPa elastic modulus 100-300MPa
II-CAM	asphalt emulsion is anionic A/C is 0.3-0.4	compressive strength >15MPa elastic modulus 7-10GPa

Table 1. Comparison of I CAM and II CAM

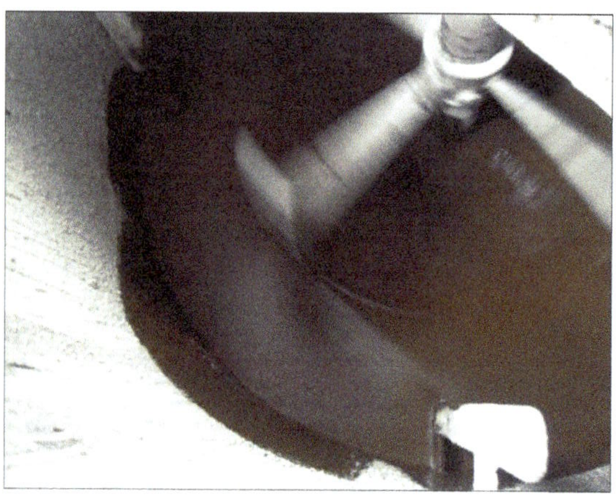

Fig. 2. Mix Process

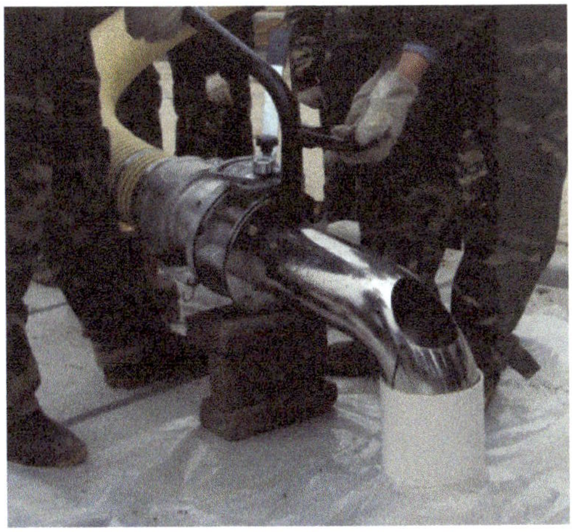

Fig. 3. Grouting Process

Fig. 4. Harden status

Fig. 5. Mortar Section

1.2 Hardening mechanisms of CAM

The hardening process of CAM was the process that CA mortar changed from flow paste to hardening state, which included flow state, plastic state and hardening state. The experimental methods that investigated the different states were proposed. For example, the methods to evaluate the compatibility between cement and asphalt emulsion was proposed to investigate the early interaction between cement and emulsion; the particle size analyze technology was employed to characterize the particle size variation of cement-emulsion paste thus to study the adsorption behaviour of asphalt droplets to the cement grains. A filtration method was used to investigate the adsorption ratio of asphalt to the cement grains (SG Hu et al, 2009).

1.2.1 The particle size analyze technology and filtration method

The laser diffraction technique has been proven well applicable in the analysis of cement and polymer particles (Masood et al, 1994; Su,Z et al,1993) .A GSL-101 laser particle size analyzer with an analysis range of 0.15–400 μm was used to test the particle size of the cement-asphalt emulsion (short for CAE) system. The given amounts of cement and asphalt emulsion were weighed and thoroughly mixed, then a representative sample was taken for test. The testing procedure was completed within 150 s. The variation of particle size D90 in the CAE system (droplet diameter for the 90th cumulative mass percentile) was obtained. The results were shown in Table 2.

Time/particle size	Cement: Asphalt emulsion (C:A)			
	0:1	0.4:1	0.625:1	1:1
0min/μm	2.61	9.86	13.14	14.84
15min/μm	2.61	13.26	18.31	20.16
30min/μm	2.61	16.95	24.85	28.64
60min/μm	2.61	18.84	29.48	33.73

Table 2. Particle size variation of CAE system with different A/C ratio

As can be seen from Table2, the cement content played a significant influence on the particle variations of CAE system. When there was on cement in the paste, the initial emulsion particle size was 2.61μm and was constant with time. With the increase of cement content, the particle size of CAE system increased with time. When A/C=0.41, the particle size of CAE paste increased by 34.5%, 71.9% and 91.1% , respectively, compared to the initial particle size, within the periods of 15, 30 and 60 min; with A/C=0.625 the percentages were 35.5%, 89.1% and 124.4%, respectively and with A/C=1, were 35.8%, 93% and 127.3%, respectively. The results indicated that the presence of cement in the CAE system accelerated the breaking of the asphalt emulsion, or the adsorption of the asphalt droplets to cement grains, both of which could be responsible for the particle size increase.

On the other hand, the increasing magnitudes within the corresponding periods increased with the presence of cement. Within the same time periods, the particle size increased more prominently with a larger C/AE. The particle size variation of CAE system with C/AE of 0.625 and 1.0 exhibited an analogous increasing magnitude within the periods of 30 and 60 min, and were both greater than that with C/AE of 0.4, which indicated that when C/ AE was larger than 0.625, the cement content had only a slight influence on the increasing magnitude of particle size of CAE system after 30 min.

The filtration method to evaluate the adsorption ratio of asphalt droplets to cement grains was as follows: first, the cement and asphalt emulsion were filtrated through the 45μm sieve; the cement retained on the sieve and asphalt emulsion passed the sieve were used for the tests; the cement and asphalt emulsion were mixed thoroughly and then the testing sample was diluted with distilled water and then filtered through the sieve, the residue was put in an oven until the weight of residue was constant. The adsorption ratio that asphalt droplets to cement grains was calculated as equation1 and the results were shown in Table 3.

$$v = \frac{m_2 - m_0}{m_1} \times 100\% \tag{1}$$

Where v is the adsorption ratio(%); m_0 is the cement content before mixing (%); m_1 is the asphalt content in the emulsion before mixing (g); and m_2 is the residues of cement and asphalt retained on 45μm sieve after mixing (g).

Time	Cement: Asphalt emulsion(C:A)		
	0.4:1	0.625: 1	1:1
0min/ Adsorption ratio	7.8	10.5	14.3
15min/ Adsorption ratio	12.0	16.9	19.6
30min/ Adsorption ratio	19.1	24.4	27.5
60min/ Adsorption ratio	39.9	50.1	56.8

Table 3. Adsorption ratios in CAE system with different A/C ratio

As can be seen from Table 3, the cement presented a significant adsorption effect on the asphalt droplets to cement grains within a period of 60min and the adsorption ratio increased with the time. When C/AE was 0.4, the adsorption ratios were increased by 4.2, 11.3 and 32.1%, respectively, within the periods of 15, 30 and 60 min, compared to the initial adsorption ratio. When C/AE was 1.0, the corresponding increasing percentages were 5.3, 13.2 and 42.5%, respectively. The adsorption ratio varied with the A/C ratio and increased with the cement contents. The increasing trend in adsorption ratio was analogous to that in particle size, which indicated that the particle size increase in the CAE system was mainly attributed to the adsorption behaviour of asphalt droplets to cement grains.

1.2.2 Non-contact resistivity

The resistivity of hardening-CAM with time was measured with CCR2 non-contact resistivity measuring instrument to investigate the evolution mechanism of CA mortar from flow state to hardening structure. CAM was prepared according to a certain ratio, then the fresh paste was poured into the mold and the resistivity was recorded every minute to get the resistivity-time curve. The ambient temperature was 20±1°C and the results were shown in Fig6.

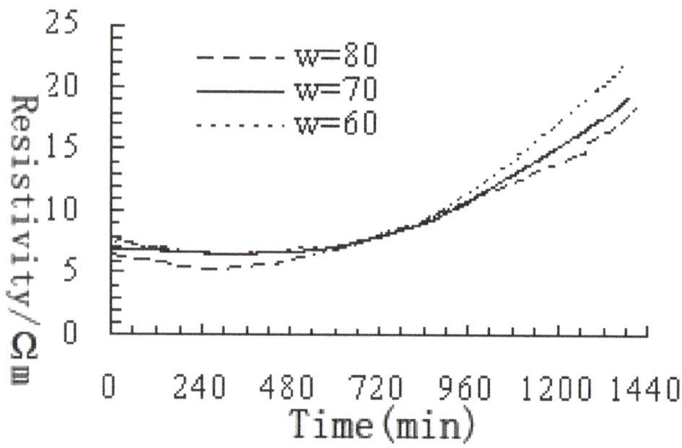

Fig. 6. Resistivity-time curve of CAM with different A/C ratio

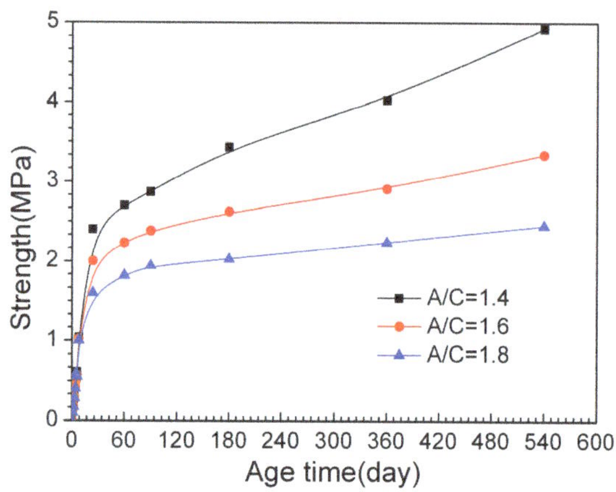

Fig. 7. CAM Strength development with age time

As can be seen from Fig6, the CAM resistivity was stable when CAM was under flow state and plastic state; also CAM with different water content presented the same trend. The resistivity decreased with the increase of water content, which indicated the ionic channel in the structure increased and the density decreased.

1.2.3 Strength development of hardening CAM

In the structure development of CAM, the asphalt droplets formed an asphalt net film in the CAM after emulsion breaking. However, the cement grains hydrated out-in and the cement grains would be coated with the hydration product formed before, thus the hydration of cement grains inside would be retarded. When asphalt emulsion broke and formed an asphalt film, the cement continued to hydrate which consumed the water inside and increased the solid contents. The density of CAM increased and lead to a continue increase of strength. The mix proportion and experimental results are shown in Table4 and Fig7, respectively.

Number	A/(C+UEA)	A/kg·m^{-3}	C/ kg·m^{-3}	W/ kg·m^{-3}
Y1	1.5	450	255	80
L2	1.6	480	255	70
Y2	1.7	510	255	60

Table 4. Mix proportion of CA mortar

As shown in Fig7, the CAM strength increased with the age time. The increase magnitude varied with different A/C ratio and increased with cement content. When the A/C ratio was 1.4, 1.6 and 1.8, the increase percentages of CAM strength within 180days compared to that within 28days were 41%, 30% and 25%, respectively; and that within 540days were 93%, 65%, 47%, respectively. This indicated that the cement in CAM continued to hydrate with age time, but the hydrate degree of later stage decreased, also were the increase percentages of CAM strength.

The elastic modulus of Ⅰ-CAM and Ⅱ-CAM were about 100-300Mpa and 7000-10000Mpa, respectively, according to the standard requirements. However, the CAM strength of later stage was not considered in the standard. The mechanic properties (strength and elastic modulus) variations with age time were investigated as shown in Table 5. As can be seen, the average increase percentages of CAM strength and elastic modulus between 20days and 120days were 6.8%~17.7% and 7.0~8.5%, respectively. Thus, the increase of elastic modulus should be considered in the CAM design to ensure the long-term mechanic properties.

Dry blend	Density/kg·m^{-3}	1d strength /MPa	7d strength /MPa	28d strength/ MPa	120d strength /MPa	28d elastic modulus /MPa	120d elastic modulus /MPa
JS	1940	3.1	14.0	18.1	21.3	8875	9500
DM	1900	3.3	12.2	20.1	22.4	8072	9044
MK	1950	3.3	15.2	19.1	21.8	9251	9953
ZY	2010	3.2	15.5	22.0	23.5	9075	9845

Table 5. Mechanic properties of CAM with different dry blend

1.3 Structure development of CAM

As mentioned before, the CAM hardening process was the process that CAM changed from flow paste to harden mortar, which included flow state, plastic state and hardening state. Moreover, the process could be divided into four states which included the dispersion state, the interaction between cement and asphalt emulsion, the formation of asphalt net structure and structure development of harden mortar.

1.3.1 Dispersion state

The cement grains and emulsion droplets dispersed isolately in the fresh cement-asphalt emulsion paste as shown in Fig 8.

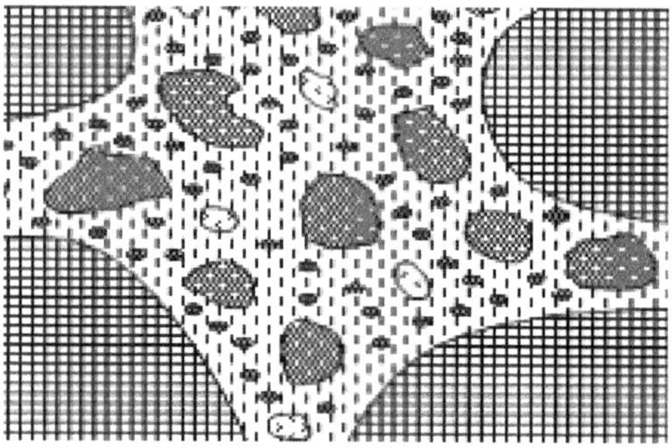

Fig. 8. Dispersion state

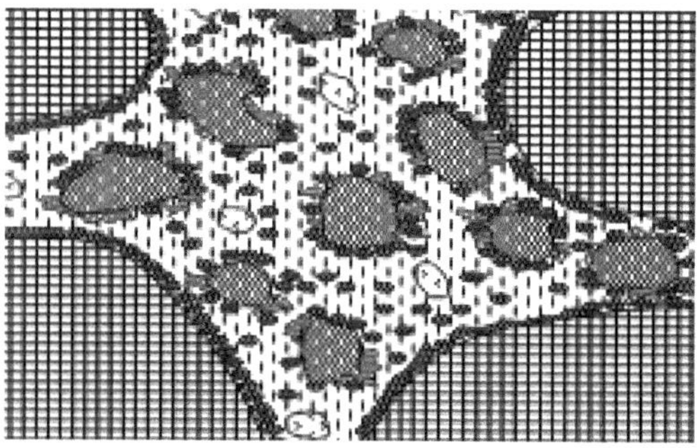

Fig. 9. Interaction between cement and emulsion

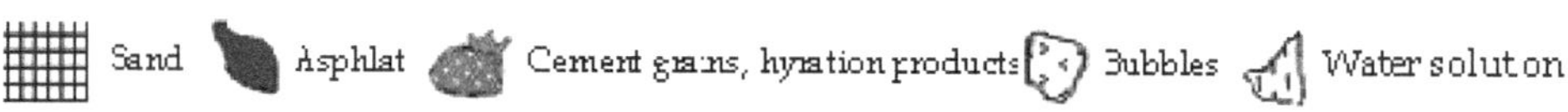

1.3.2 Interaction between cement hydration and emulsion breaking

The hydration product gradually formed when the $Ca(OH)_2$ concentration was saturated after the $Ca(OH)_2$ induced by cement hydration neutralized the acid in asphalt emulsion. Meanwhile, asphalt droplets would absorb to the cement grains and cement hydration product. The dispersion state turned to the interaction between cement hydration and asphalt emulsion breaking. (Fig9)

1.3.3 Formation of asphalt net structure

The cement hydration consumed much free water and decreased the distance between asphalt droplets which made droplets easy to contact and finally formed a continuous asphalt film. This process was within 20h and the asphalt net film contacted with the cement hydration products to form the harden mortar structure. (Fig10)

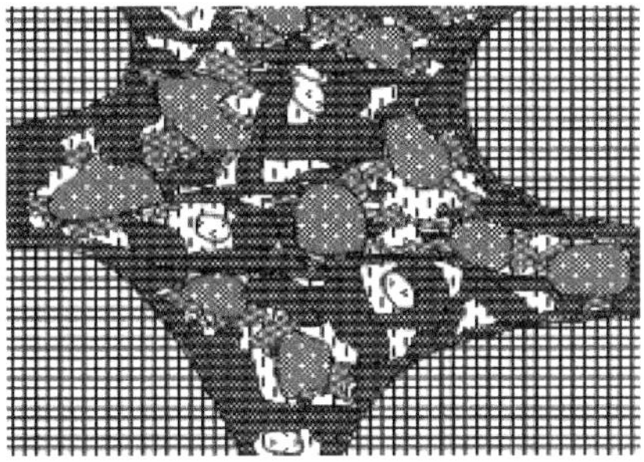

Fig. 10. Formation of asphalt net structure

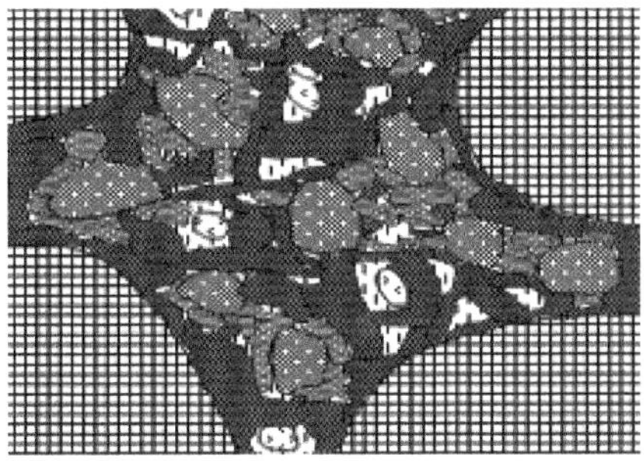

Fig. 11. Harden mortar structure development

1.3.4 Harden mortar structure development

After the emulsion broke and the asphalt net film formed, much water and cement grains were contained in the film, which would influence the mechanic properties of CAM. The cement grains surface would absorb the asphalt droplets selectively, some parts of the surface contacted with asphalt while other parts contacted with water, thus promoted the cement hydration and increased the solid contents. The density of CAM increased gradually and the strength increased with age time. However, the CAM strength of later stage increased slowly because of the retard effect of asphalt film on the contact between cement grains and water (Fig11).

2. Compatibility between raw materials

CAM is mainly consisted of dry blend which is mainly composed of cement, expansive agent, sand and some minor components for properties adjustment, asphalt emulsion, water, superplasticizer and anti-foaming agents. Of which cement and asphalt emulsion are the main components of CAM. However, the cement hydration and emulsion breaking are affected by each other and influenced the CAM's properties. The cement content determines the CAM's strength and elasticity, also the cement hydration influences the emulsion breaking, thus the organic and inorganic materials should have a proper ratio which means the cement hydration rate should match the emulsion breaking rate. These tough requirements for the match of proportion, hydration rate and breaking speed are named compatibility.

The setting process of cement and asphalt emulsion paste (CAE) is of great importance on the properties of CA mortar, which requires a proper match between the two materials, namely an optimum cement/asphalt emulsion ratio (C/AE ratio) and the coordinated and compatible rate between the hydration of cement and the breaking of asphalt emulsion, which is vital to the homogeneity of CA mortar. In this chapter, the setting process of CAE was characterized by that of the paste prepared with cement and asphalt emulsion. The influences of cement types (two ordinary Portland cements and one sulfoaluminate cement), C/AE ratios (0.6, 1.0, 1.6, and 2.0) and blended cement on the setting process of the paste and on the properties of CA mortar were investigated.

A new method to evaluate the setting process of cement and asphalt emulsion in CA mortar was proposed (FZ Wang, 2008). The method was developed from the standard specification for the test of setting time of cement paste (GB/T 1346-2001). The cement-asphalt emulsion paste was prepared in a mould, then every 30 min, a steel needle penetrated the grout from the upper surface and after 30 s a depth was read. The setting process could be characterized by plotting the depth against time. The time of first catastrophe point was defined as the initial setting time and the time when the depth plotted was stable was defined as the final setting time.

2.1 Influence of cement types on the setting process of CAE in CA mortar

Cement: three types of cements (C1, C2 and C3) were used, among which C1 and C2 were ordinary Portland cements and C3 was rapid hardening sulfoaluminate cement. The properties of the three cements were listed in Table6.

Cement	Surface density g/cm^3	Setting time /min		Compressive strength /MPa		
		Initial set	Final set	3d	7d	28d
C1	3.13	138	190	17.7	22.9	47.3
C2	3.18	131	170	13.0	21.3	46.5
C3	2.79	29	50	38.4	46.7	50.3

Table 6. Physical properties of cements

Sand: River sands S2, S3 and S4 with fineness modulus of 1.4, 1.7 and 2.0, respectively. The properties were shown in Table7.

Number	Properties	Results	Number	Properties	Results
1	Bulk density (kg/m^3)	1380	4	Mud content (%)	0.3
2	Packing density (kg/m^3)	1480	5	Mica content (%)	0.6
3	Apparent density (kg/m^3)	2630	6	Organic content (%)	Qualified

Table 7. Properties of river sand

Expansion agent: Aluminum powder (AP) was Flaky and its properties were given in Table 8. UEA expansive agent for cement, the properties are shown in Table9.

Appearance	Silvery white
Purity (%)	≥99.0
Soluble organic substance (%)	≤4.0
Volatiles (105°C %)	≤10.0
Particle size (m)	50–75
Solubility	Soluble in acid, alkali, not in water

Table 8. Specifications of aluminum powder

Num ber	Properties		Result s	Num ber	Properties		Result s
1	Fineness	Material retained on sieve 0.08mm (%)	2.0	3	Density (kg/m3)		2750
		Material retained on sieve 1.25mm (%)	0	4	Limited expan-sion ratio	In water 14d (%)	0.03
2	MgO contents (%)		1.7			In air 28d (%)	-0.013

Table 9. Properties of UEA expansive agent

Polycarboxylate-based superplasticizer (SP)

Defoamer: silicone type.

Asphalt emulsion: Independent development and manufacture.

The influences of cements C1, C2 and C3 on the cement-asphalt emulsion paste setting time were shown in Fig12.

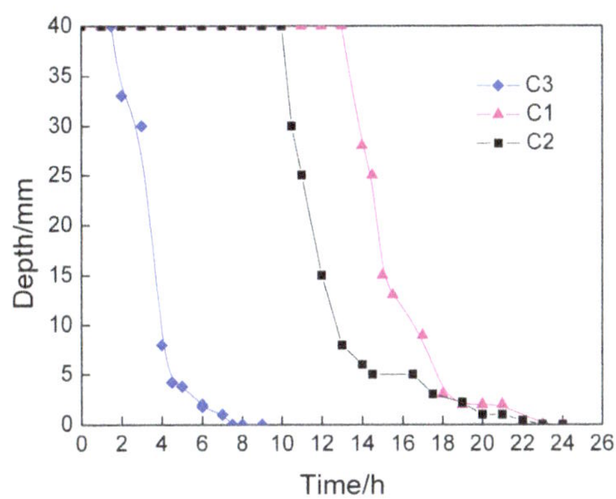

Fig. 12. Influences of cements on the setting process of CAE paste

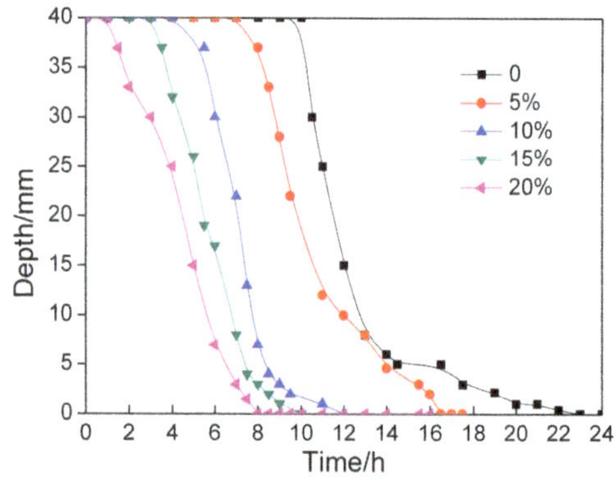

Fig. 13. Influences of blending cements on the setting process of CAE paste

As indicated in Fig.13, the setting process of CAE3 (C3 was used) was the fastest with the initial setting time of less than 2h and final setting time of 9h. The CAE1 (C1 was used) and CAE2 (C2 was used) had a slower setting time, with the initial time of 11, 13h and final setting time of 21, 23h, respectively.

No.	Cements	Separation rate /%	Fluidity /s	Workable time /min	Compressive strength /MPa		
					1d	7d	28d
CAM1	C1	0.6	20	70	0.14	0.88	2.0
CAM2	C2	0.7	19.5	85	0.13	0.85	2.1
CAM3	C3	0.5	19.5	35	0.82	1.8	2.0

Table 10. Properties of CA mortar prepared with different cements

As can be seen from Table10, the cements played a significant effect on the CAM's workable time and strength. The workable time of CAM3 was only 35min while that of CAM1 and CAM2 were as long as 70 and 85min, respectively. However, the 1d and 7d strength of CAM3 were 0.82 and 1.8Mpa, which was quite higher than that of CAM1 and CAM2. The properties of CAM 1 and CAM2 were quite similar except the difference in early strength.

The influences of blending cements on the CAM properties were also investigated. In this test, blended cements with different combinations of C1 and C3 were used, and C1 was replaced by C3 with different replacement ratios (0%, 5%, 10%, 15%, and 20%). Results were given in Fig.13.

As indicated in Fig.13, the setting process of a blended cement-asphalt emulsion system was markedly accelerated compared to that using C1 only, and the accelerating effect was more obvious with the higher the replacement ratio. When the replacement ratio exceeded 20%, the paste had an early initial setting time of less than 2h and final setting time of 9h compared to 11h and 21h of the paste using C1. So the emulsion breaking and CAE paste setting could be well controlled with the blending cement to have a proper harden time and prevent the early bleeding of CA mortar.

2.2 Influence of C/AE ratios on the setting process of CAE in CA mortar

The C/AE ratio is the key factor that influences the setting process of CAM. The influence of C/AE ratio on the setting process was studied with four C/AE ratios (0.6, 1.0, 1.6, and 2.0) and cement C1 used. The results were shown in Fig. 14.

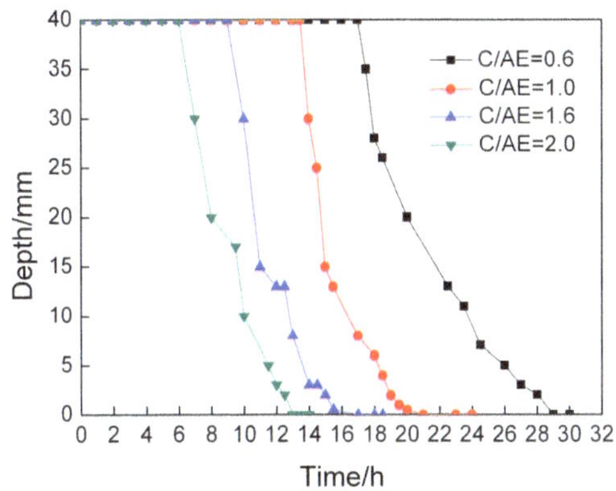

Fig. 14. Influence of C/AE ratio on the setting process of CAE paste

As can be seen from Fig14, the setting process of CAE was slowed with the decrease of C/AE ratio. The higher C/AE ratio with more cements lead to a quicker setting process. When C/AE ratio was 2.0, the final setting time of CAE was about 15 h, while that of C/AE ratio 0.6 was as long as 26 h. So, the setting process of CAE was directly proportional to the cement content. The cement hydration promoted the emulsion breaking and was benefit to the paste hardening.

No.	A/C	Fluidity/s	Workable time /min	Separation rate/%	Compressive strength /MPa		
					1d	7d	28d
CAM1	1.4	19	95	0.8	0.16	0.92	2.6
CAM2	1.5	19	95	0.8	0.13	0.87	2.3
CAM3	1.6	19.5	85	0.7	0.13	0.85	2.1
CAM4	1.7	20	75	0.6	0.11	0.82	1.9
CAM5	1.8	20	70	0.6	0.09	0.80	1.7

Table 11. Influence of A/C ratio on the properties of CAM

As can be seen from Table11, the separation rate decreased with the increase of A/C ratio also did the fluidity and compressive strength. The proper ratio of A/C was 0.8~ 1.7 and it would be better when the ratio was around 1.6.

3. Preparation of asphalt emulsion for CAM

3.1 Introduction of asphalt emulsion for CAM

An Asphalt emulsion is asphalt dispersed through water and chemically stabilized for the emulsifier, whose preparation process is shown in Fig 15. There are two main types of emulsions with different affinity for aggregates, cationic and anionic, based on salts of fatty long chain molecules. Asphalt emulsions are used in cold processes for road construction and maintenance and have a wide variety of applications .Asphalt emulsion is an important part of CA mortar which possesses 10-30% of the CA mortar and its properties have greatly influenced the quality of CA mortar (FZ Wang 2009, 2010; JQ Zuo, 2005)

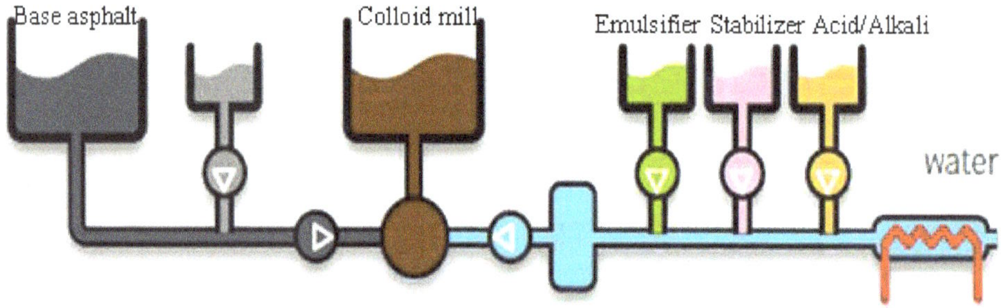

Fig. 15. Preparation process of asphalt emulsion

As shown in Table 1, the asphalt emulsion is different for different CAM and slab structure. As for I-CAM, the asphalt emulsion is cationic to make a tight combination between asphalt and inorganic material because the charges of sand when absorbing water and cement grains during hydration are both negative. However, the asphalt emulsion for II-CAM is anionic to ensure the nice cement compatibility because of the large cement content of II-CAM.

Compared to the asphalt emulsion used in highway, asphalt emulsion for CA mortar needs both nice storage stability and excellent cement compatibility. The storage stability is quite important to guarantee the properties of emulsion after long-term storage and long-distance transportation. On the other hand, the C/A (mass ratio of cement and asphalt emulsion) of CAM is quite high (C/A=0.60~2.04) , this is quite different from the situation when asphalt emulsion is used just as modifier in cement mortar (Pouliot N,2003; G Li 1998; H Song 2006) . The asphalt emulsion needs a good cement compatibility (or chemical stability) to keep stable when mixed with so much cement to avoid breaking thus to guarantee the workable time of CA mortar. Generally speaking, asphalt emulsion for CA mortar should have the following characteristics.

3.1.1 Nice cement compatibility

In the fresh cement-asphalt emulsion paste, the cement grains and asphalt droplets are dispersed isolately. The hydrophilic group of emulsifiers is directed in the aqueous phase and cement grains are intrinsically hydrophilic, also the particle sizes of cement are about 10 times as large as asphalt droplets, all of which facilitate the adsorption of asphalt droplets to cement grains. Moreover, the cement hydration consumes free water in emulsified asphalt, the spaces between the micelles in the emulsified asphalt are reduced, the electrical attraction and probability of collision between micelles increases (Song H, 2006); the Ca^{2+}, Al^{3+} induced by cement hydration will also influence the double electrical layer and asphalt emulsion stability(Kulmyrzaev A,2000). Thus, asphalt emulsion needs quite nice cement compatibility to guarantee the workable properties of CAM.

3.1.2 Good storage stability

Asphalt emulsion is a thermodynamically unstable system with asphalt droplets uniformly dispersed in a liquid of water and various surfactants. The droplets tend to coalescence together to decrease the surface areas and make the emulsion more stable (Ivanova R, 1999; Al-Sabagh AM 2002, Bury .M.1995). Besides that, the droplets tend to sediment due to the gravitational forces (A.S. Dukhin SSD, P.J. Goetz, 2007). So, its storage stability plays an important role in guaranteeing long-term storage and long-distance transportation (FZ Wang 2012; Ivanov IB, 1997; Ostberg GB, 1994), which is crucial in the preparation of CA mortar.

3.1.3 Low temperature sensitivity

When used in CAM, asphalt emulsion would finally breaking and work as the base asphalt. However, the asphalt has a high temperature sensitivity which softens at high temperature and brittles at low temperature due to its molecular structure. The climate in China is varied in different regions and the construction of high-speed railway often covers several different districts, which proposes tough requests for the temperature sensitivity of asphalt used in CAM.

3.1.4 Environment-friendly

As mentioned before, asphalt emulsion is the key component of CAM. But quite few emulsifiers used in asphalt emulsion are biodegradable which is harmful to the environment. Thus, the asphalt emulsion used for CAM and high-speed railway should be environment-friendly to reduce pollutions.

3.2 Technical methods to prepare the asphalt emulsion with high performance for CAM

To meet the requirements of asphalt emulsion for CA mortar, the technical methods were proposed as shown in Fig16, according to colloid chemistry, surface and interfacial chemistry and preparation techniques. This part would like to introduce these technique methods, respectively.

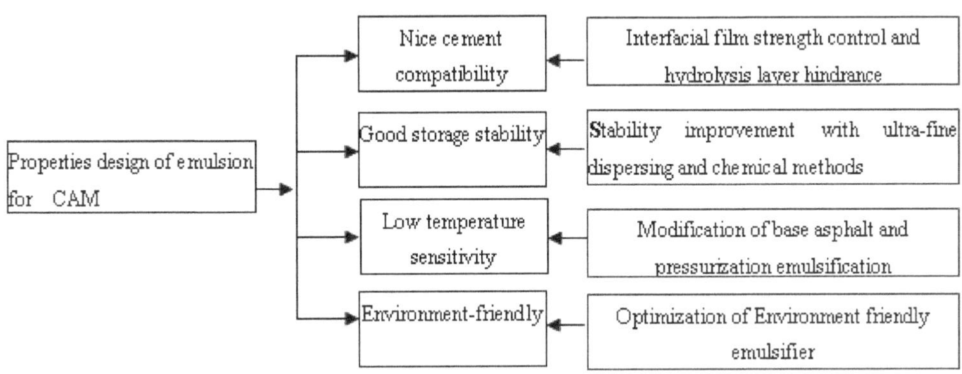

Fig. 16. Technical methods to prepare the asphalt emulsion for CAM

3.3 Interfacial film strength control and hydrolysis layer hindrance

Interfacial film is the hydrolysis shell formed by the emulsifier, ionic and counter-ionic around the asphalt droplets. The tight interfacial film structure and high film strength are the key factors for emulsion stability. However, the film strength needs to be well controlled in the cement-asphalt emulsion system. When the cement absorbs the asphalt droplet or the droplets tend to coalescence, the interfacial film should have enough strength to provide hindrance to guarantee the workability of CA mortar; on the other hand, the interfacial film strength couldn't be too strong which would retard the cement hydration and influence the early strength development of CA mortar. Moreover, the temperature has a significant accelerate effect on cement hydration and droplets coalescence, the film strength needs to be controlled and adjusted according to the temperature. Thus the methods to control the interfacial film strength (FZ Wang, 2012) and hydration layer hindrance were proposed.

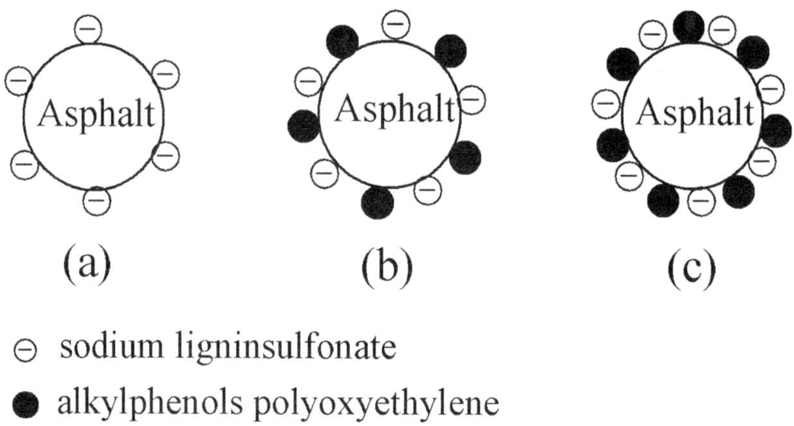

Fig. 17. Simulation of interfacial film

Fig17 showed the simulation structure of interfacial film in anionic asphalt emulsion. As shown in Fig17-(a), the emulsifier had a loose distribution around the droplets due to the electrostatic repulsion caused by the same charges of hydrophilic molecular, so the cationic ionic induced by cement hydration were easy to damage the thin and weak interfacial film and double electric layer of droplets thus lead to emulsion breaking.

However, as for cationic asphalt emulsion, the hydrophilic molecular is mainly quaternary ammonium, of which, N atom has a strong adsorb ability and affinity to both cement grains and aggregates, When mixed with cements and aggregates, the droplets would soon absorb to the surface of aggregates and cement grains thus make the emulsion break soon.

The nonionic is free of charges, so it is little affected by the salts in the solution and can act like a protective colloid around the main emulsifier layer at the asphalt oil-water interface to increase the interfacial film strength and prevent the coalescence of oil droplets thus improve the emulsion stability (Gullapalli RP, Sheth BB, 1996). On the other hand, the length of nonionic emulsifier molecular is varied which provides different steric hindrance. The interfacial film can be well controlled by the optimization of nonionic emulsifiers.

The interfacial film structures controlled by this technique can be simulated as Fig17-(b) and Fig17-(c). The composite emulsifiers were combined by the cationic emulsifiers or anionic emulsifiers with the nonionic-emulsifiers and inorganic salts. The emulsions prepared with the composite emulsifiers can be applied in the CAM constructions under different temperatures. The workable properties and early strength of CA mortar were both considered. The compatibility results with cement were shown in Table12 and Table13.

Emulsifier	Cement compatibility results
A1	40°C,stable for 4h, 78ml flow out in 20s
A2	40°C,stable for 4h, 92ml flow out in 20s
D301	40°C,stable for 4h, 238ml flow out in 9s
D301-H	50°C,stable for 4h, 238ml flow out in 10s

Table 12. Preparation of anionic composite emulsifiers

Emulsifier	Results
C1	Breaking after mix
C2	25°C, sieve residue 1.9%
R201	25°C, sieve residue 0.4%
R204	50°C, sieve residue0.8%

Table 13. Preparation of cationic composite emulsifiers

Figures 18 and 19 showed the viscosity variation of cement-asphalt emulsion paste to indicate the interaction between cement and emulsion (YP Liu, FZ Wang, 2011). Compared to the ordinary emulsion ASS-1, the ASS-2 prepared by the interfacial film strength control technique had nice compatibility with cement under 25°C, 45°C.

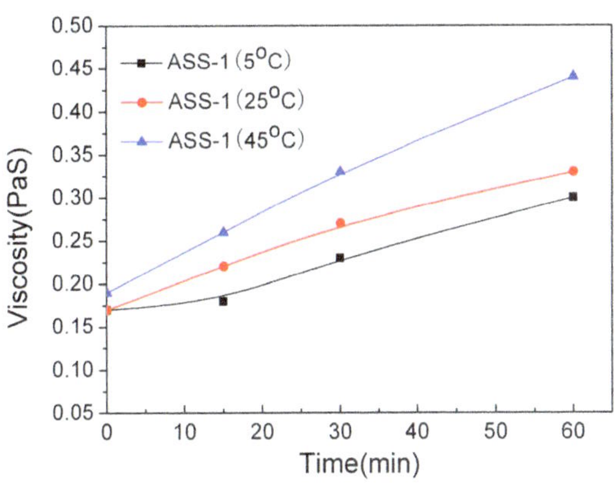

Fig. 18. Viscosity variations of CAE-1 system (ASS-1 used)

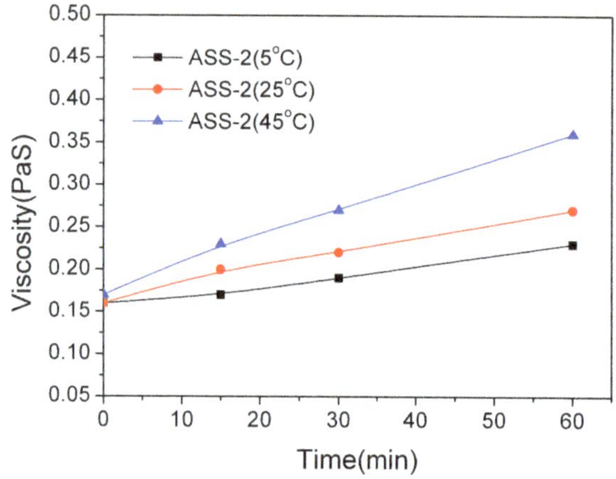

Fig. 19. Viscosity variation of CAE-2 system (ASS-2 used)

Figures20 and 21 showed the IR spectrum of R301 and R301-H in Table 12, respectively. Figures22 and 23 showed the IR spectrum of R201 and R204 in Table 13, respectively. The practical IR spectrums were almost the same with the theoretic results, which indicated that there were no chemical reactions when the main emulsifier mixed with the nonionic emulsifier. The improvement of emulsion's cement compatibility could be attributed to the addition of nonionic emulsifier. This was mainly because the nonionic emulsifier used had a larger EO numbers (EO number=40), water was easy to be absorbed to the ether group due to the hydrogen bond. One ether group usually absorbs 20~30 H_2O molecular. Thus a big hydrolysis layer was formed around the asphalt droplets which prevented the influences of cement hydration on emulsion breaking and improved the emulsion stability. So both the increase of interfacial film strength and thick hydrolysis layer could contribute to the improvement of emulsion's cement compatibility.

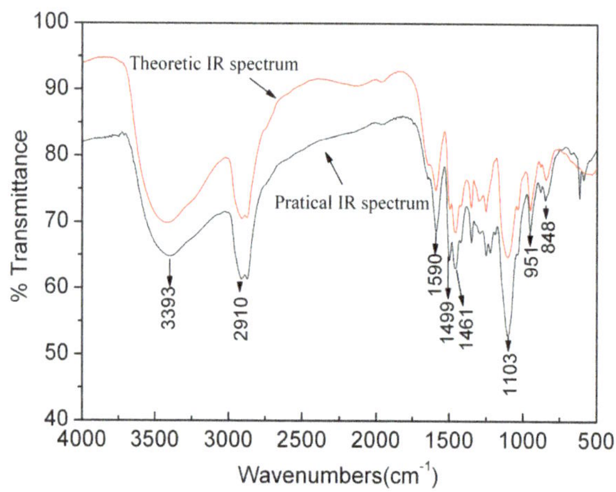

Fig. 20. Theoretic and practical IR spectrums of D301

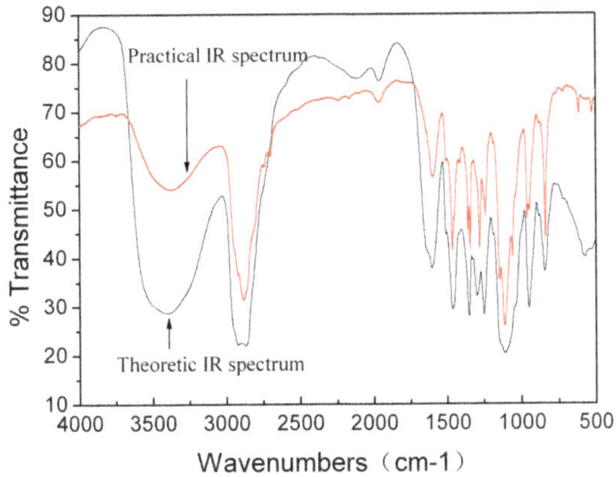

Fig. 21. Theoretic and practical IR spectrums of D301-H

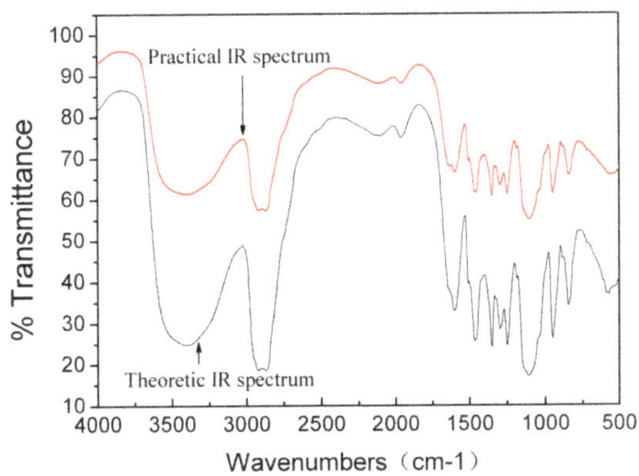

Fig. 22. Theoretic and practical IR spectrums of R201

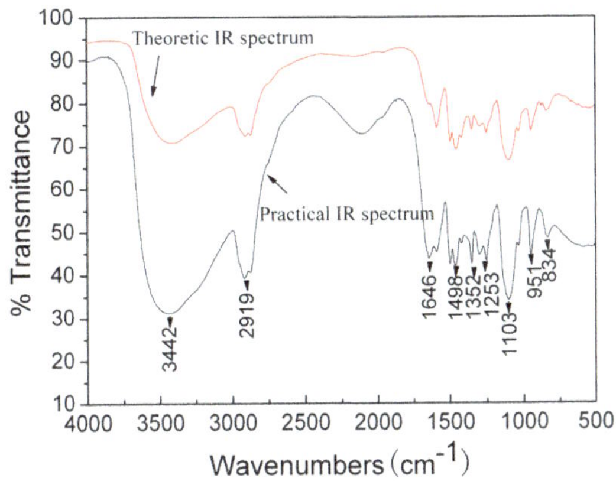

Fig. 23. Theoretic and practical IR spectrums of R204

Figure 24 and Table 13 showed the influences of different emulsifier's solution on the early cement hydration. As can be seen, the cement hydration rate was retarded when mixed with emulsifier solution compared to the cement only mixed with water. The heat flow peak of the cement mixed with water happened after 8.8 hours; however, the time of heat flow peaks of D4 and C1 were 13.2 and 14.6 hours. This was mainly because the nonionic emulsifier and lignin-amine absorbed to the cement grains and retarded the cement hydration for steric hindrance. The 1d heat of cement mixed with water was higher than that of cement with emulsifier solution, but the results after 3days were on the contrast.

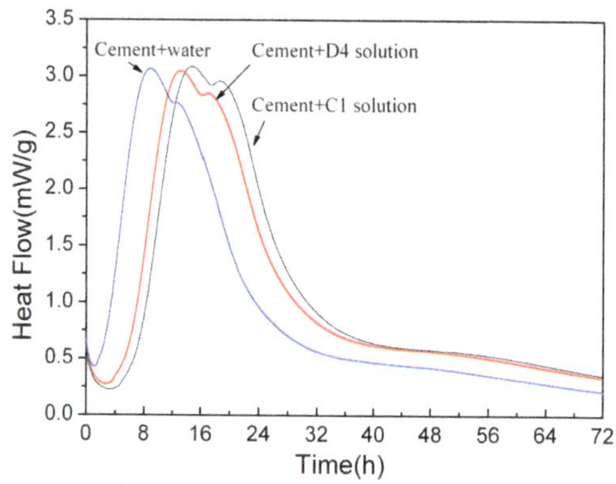

(a) Effect of emulsifier solution on heat flow of cement hydration

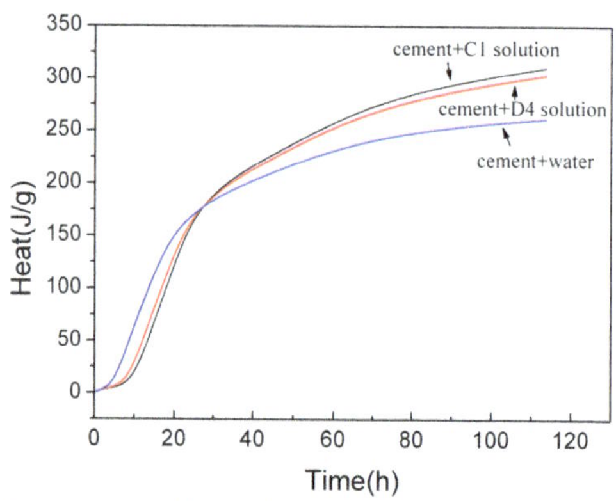

(b) Effect of emulsifier solution on heat of cement hydration

Fig. 24. Effect of emulsifier solution on cement hydration

	Cement with water	Cement with C1 solution	Cement with D4 solution
Peak (mW/g)	3.06	3.08	3.04
Time of Peak(h)	8.8	14.6	13.2
3days heat(J/g)	241.1	274.1	267.6

Table 13. Influence of emulsifier solution on cement hydration

3.4 Stability improvement with ultra-fine dispersing and chemical methods

The asphalt emulsion was stabled by the double electric layer induced by the emulsifier. The stability is easy to be damaged when under storage and transportation. The droplets tend to sediment and coalescence together especially under stress condition. The emulsion stability mostly can be explained by the Stoke's law.

According to Stoke's law (Equation 2), the resulting sedimentation rate v_0 of a single droplet is proportional to the particle size:

$$v_0 = \frac{2r^2(\rho - \rho')}{9\eta} g \quad (2)$$

Where r is the hydrodynamic radius of the droplet, ρ is the density of the external phase, ρ' is the density of the internal phase, η is the macroscopic shear viscosity of the external phase, and g is the gravitation constant.

In order to improve the emulsion stability, the particle sizes need to be decreased to slow down the sedimentation rate and the thickeners are used to increase the viscosity to improve the emulsion stability.

Fig25 showed the effects of particle sizes on the storage stability. As can be seen, the smaller particles lead to a stable emulsion. The smaller the particles were, the slower the sediment rate was and the emulsions were more stable. But when the particle was less than 3μm, the stability was not obviously changed.

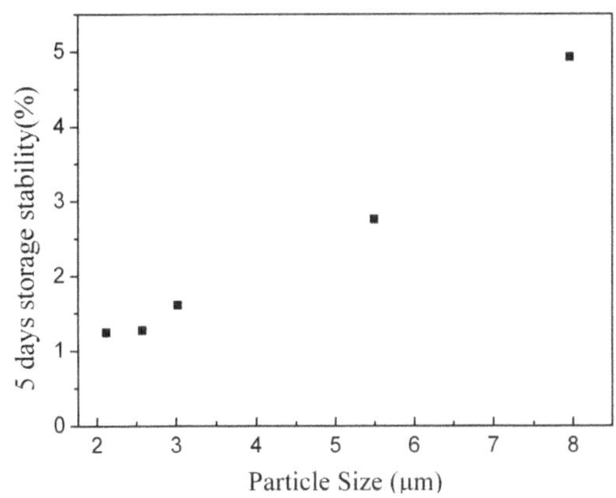

Fig. 25. Effect of particle sizes on emulsion stability

The relation between emulsification process (flow rate, flow velocity and temperatures of asphalt and emulsifier solution) and particle size was studied to develop the ultra-fine dispersing technique. The results were shown in Fig26, and as can be seen, the particle sizes of droplets were greatly reduced. The composite stabilizers combined with thickeners, inorganic salts and nonionic emulsifiers were prepared to improve the emulsion stability. As shown in Fig27, the emulsion stability was significantly improved, even under higher temperature (FZ Wang, 2012).

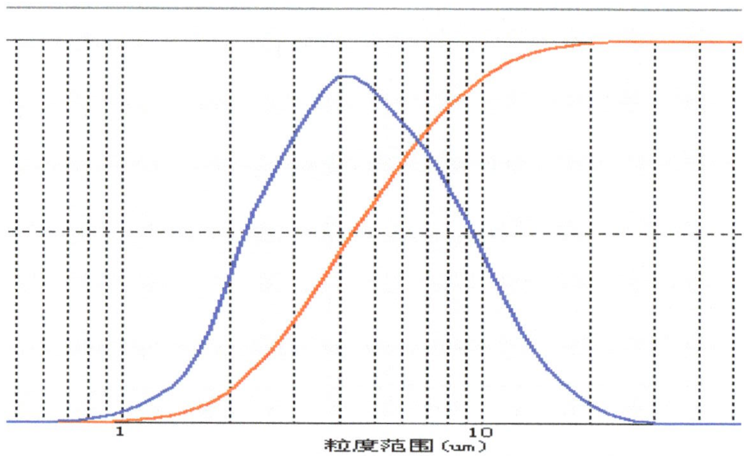

(a) Common technique (Average particle size 4.99μm)

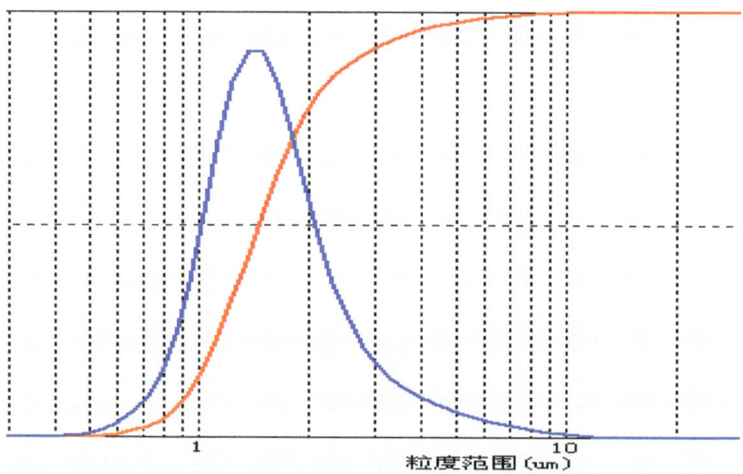

(b) Ultrafine dispersing technique (Average particle size 1.37μm)

Fig. 26. Effect of the ultrafine dispersing technique on particle size of asphalt emulsion

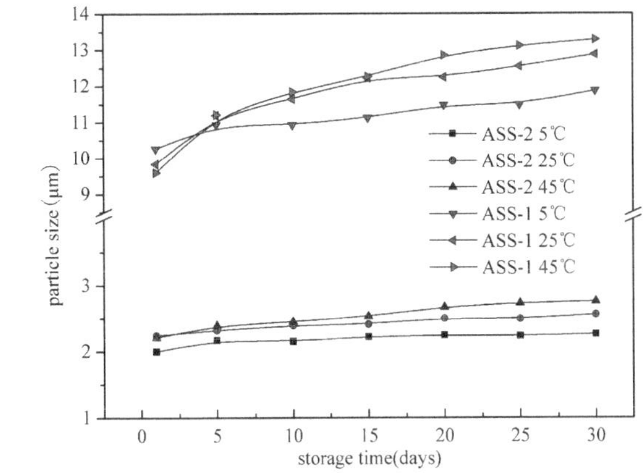

(a) Particle size variations

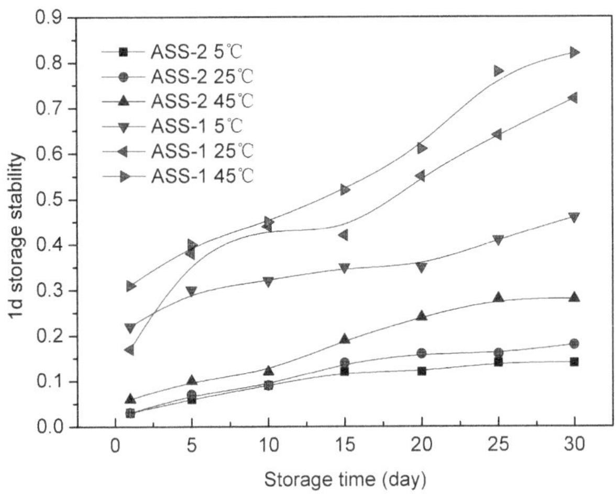

(b) 1d storage stability variations

Fig. 27. Emulsion storage stability variations under different temperatures

Table14 showed the influences of different types thickeners including polyvinyl alcohol (PVA), corn starch (CS), cellulose ether with low molecular weight (K1), cellulose ether with large molecular weight (K4) on the storage stability and Engler viscosity.

Thickeners type	1d Storage stability (%)	Engler viscosity (25℃)
-	13.6	4.17
PVA(0.03%)	8.0	4.43
C S (0.03%)	0.3	10.36
K1(0.03%)	2.4	8.2
K4(0.03%)	4.1	6.9

Table 14. Influence of thickeners on the emulsion stability

As can be seen from Table 14, the thickeners significantly increased the Engler viscosity of emulsion, even with the dosage 0.03%. Of which, the corn starch played the best thickening effects with the Engler viscosity 10.36 and the best emulsion storage stability 0.3%. The effects of PVA were not obvious with the dosage 0.03% for the Engler viscosity was slightly increased.

Fig28 showed the influences of K4 dosages on the Engler viscosity and emulsion stability. As can be seen, the Engler viscosity was opposite with the storage stability. The Engler viscosity increased and the emulsion stability decreased with the increase of K4 dosage, which indicated the emulsion more stable. The emulsion stability had a slight change when the K4 dosage was over 0.06%.

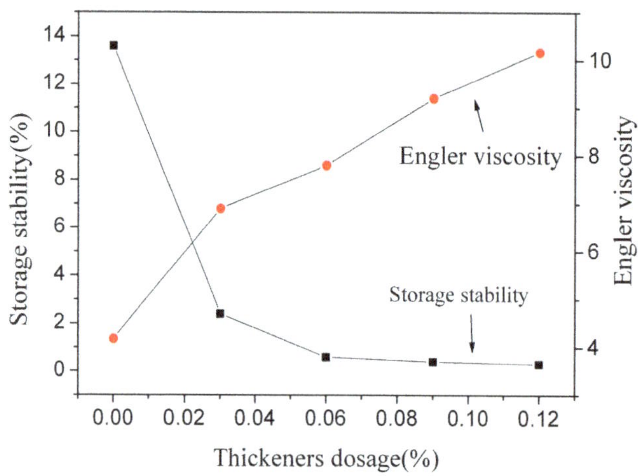

Fig. 28. Effect of thickeners dosage on emulsion stability

3.5 Modification of base asphalt and pressurization emulsification

The base asphalt used for asphalt emulsion are usually 90#、70# heavy traffic paving asphalt without modification. However, the asphalt emulsion evaporation residue prepared with these asphalt had a smaller softening point and low temperature ductility which can't stand the complicated climate, especially in China. To improve the temperature sensitivity of emulsion, styrene-bu-tadiene block copolymer (SBS), styrene-butadiene rubbers (SBR) are often added into the asphalt emulsion to prepare modified asphalt emulsion. However, emulsions prepared with the "first emulsification, then modification" method is usually not homogeneity because it's hard to form crosslink structure between asphalt and modifier. The softening point of the emulsion evaporation residue is hard to over 70°C and the 5°C ductility is less than 20cm, which could not meet the requirements of hot and cold regions.

There is also another modification method namely "first modification, then emulsification" method. This technique is to modify the base asphalt with SBS first and then prepare the emulsion. The temperature sensitivity is improved with the increase of SBS dosages, but the emulsification is more difficult at the same time.

When prepared the asphalt emulsion, the base asphalt modified with SBS usually need to be heated to be over 170°C, but the temperature of emulsifier solution is usually 50~60°C, which make the temperature difference between these two parts over 100°C, thus induce serious water gasification.

To solve this problem, the pressurization emulsification method was proposed combined with high shearing speed and pressurization in emulsion system to prevent the water gasification and guarantee the emulsification process.

Table15 showed the different emulsification process parameters for modified asphalt with different SBS dosages. As can be seen, both the colloid mill speed and the emulsification system pressure increased with the increase of SBS dosage.

Number	SBS dosages (%)	Emulsification process parameter	
		Sheer speed of colloid mill /r/min	Emulsification system pressure/MPa
1	3	9000	0.3
2	4	10000	0.4
3	5	11000	0.6

Table 15. Emulsification process parameter of asphalt modified with different SBS dosages

Table16 showed the properties of modified emulsion's evaporation residue with different SBS dosages. As can be seen, the temperature sensitivity was improved with the increase of SBS dosage. When the SBS dosage was over 3%, SBS formed an interpenetrating structure through the asphalt. The formation of net structure increased the viscosity and elasticity of modified asphalt thus increased the softening point.

Number	SBS dosage (%)	Properties of emulsion evaporation residue	
		Softening point/°C	Ductility /5°C,cm
1	0	48	/
2	3	62	20
3	4	75	22
4	5	81	26

Table 16. Properties of resin asphalt modified with different SBS dosage

Considering the costs and emulsion stability prepared with modified asphalt with higher SBS dosage, the properties of asphalt emulsion modified both with SBS and SBR latex were studied. The base asphalt was first modified with SBS and then the emulsion was prepared, after this, SBR latex was used to modify the emulsion. Table 17 showed that 5°C ductility was greatly improved with this method, and the softening point also reached 64°C which indicated the asphalt emulsion also had low temperature sensitivity.

Number	Modification Method	Properties of emulsion evaporation residue	
		Softening point /°C	Ductility /5°C,cm
1	/	48	/
2	3%SBS+2%SBR	64	52

Table 17. Properties of asphalt emulsion prepared with different modification methods

3.6 Properties of asphalt emulsion

The asphalt emulsion with excellent cement compatibility, nice storage stability and low temperature sensitivity was prepared with the technology mentioned before which included interfacial film strength control and hydrolysis layer hindrance, stability improvement with ultra-fine dispersing and chemical methods, modification of base asphalt and pressurization emulsification. The properties of cationic emulsion and anionic emulsion were listed in Tables18 and 19, respectively.

Number	Technical index		Unit	Standard requirements	CAE-1	CAE-2	CAE-3
1	Particle charge		/	Cationic	Cationic	Cationic	Cationic
2	Engler viscosity(25°C)		/	5~15	9	10	9
3	Sieve tests (1.18mm)		%	< 0.1	0	0	0
4	Storage stability (1d, 25°C)		%	<1.0	0.6	0.3	0.1
5	Storage stability (5d, 25°C)		%	<5.0	2.2	1.6	1.2
6	Storage stability(-5°C)		/	No coarse grains	Qualified	Qualified	Qualified
7	Cement compatibility		%	<1.0	0	0	0
8	Residue form distillation	Solid contents	%	58~63	59.2	60.5	59.8
		Penetration(25°C, 100g)	0.1mm	60~120	64	51	74
		Softening point (℃)	°C	/	70.4	76.5	48.1
		Ductility (5℃)	cm	≥20	33.5	20.8	55.6

Table 18. Properties of cationic asphalt emulsion for CA mortar

Number	Technical index		Unit	Standard requirements	AAE-1	AAE-2	AAE-3
1	Sieve tests (1.18mm)		%	<0.1	0.03	0.02	0.05
2	Particle charge		/	anionic	anionic	anionic	anionic
3	Particle size		μm	Average particle size≤7 ;	Average particle size=2.3;	Average particle size=3.8;	Average particle size=1.8;
4	Cement compatibility		ml	≥70	240	240	240
5	Storage stability (1d, 25°C)		%	<1.0	0.6	0.4	0.8
6	Storage stability (5d, 25°C)		%	<5.0	2.4	2.3	2.4
7	Storage stability (-5 °C)		/	No coarse grain	Qualified	Qualified	Qualified
8	Residue form distillation	Solid contents	%	≥60	61.8	61.0	62.1
9		Penetration (25°C, 100g)	0.1mm	40~120	64	48	73
10		Softening point (°C)	°C	≥42	70.6	75.8	46.0
11		Ductility (5°C)	cm	≥20	31.8	22.1	53.4

Table 19. Properties of anionic asphalt emulsion

4. Preparation of dry blend for CAM

The properties design of CAM must meet the requirements of construction. The structure of slab track in high-speed railway determines the unique construction method- grouting method. CAM was directly grouted in the narrow spaces between the concrete roadbed and slab track with its gravitational force. CAM must have both nice mechanical properties and workability to adjust to this method. This part mainly introduced the main techniques that improved and controlled the rheological properties of CAM by the dry blend composition and mix proportion designs, which was also the core technology of preparation for dry blend and CAM.

4.1 Design of rheological properties of CAM

4.1.1 Construction procedure of slab track

As for CRTS Ⅰ slab track, CAM would be grouted into the inject bag that was preset under the slab track. There were three types of different dimensions of slab track which were 4962x2400mm, 4856x2400mm and 3685x2400mm. However, as for CRTS Ⅱ slab track, the CAM would be directly grouted into the narrow space between the slab track and concrete roadbed (6450×2550×30mm). The grouting process was influenced by the interface property between CAM and concrete because the CA mortar contacted with the roadbed and slab track directly.

4.1.2 Design of CAM flow state

Fig29 showed the schematic diagrams of grouting process for CRTS I and II slab tracks. The grouting and filling processes of CAM under the slab track were accompanied with the gas exhaust process. The I CAM were injected in the filling bag that was squeezed vacuum, CAM filled the filling bag with extrusion which could prevent the bubbles to be brought in the CAM that would influence the interfacial quality between CAM and slab track. However, this was quite different for II CAM. When the II CAM were grouted, the mortar directly contacted with the roadbed concrete and slab track. The filling process indeed was the process that replacement of gas in the space by CAM. Thus, the flowability of CAM directly influenced the replacement effects.

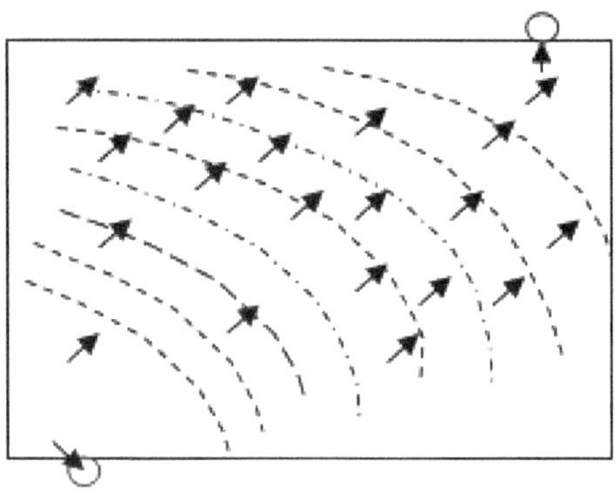

(a) CRTS I CAM

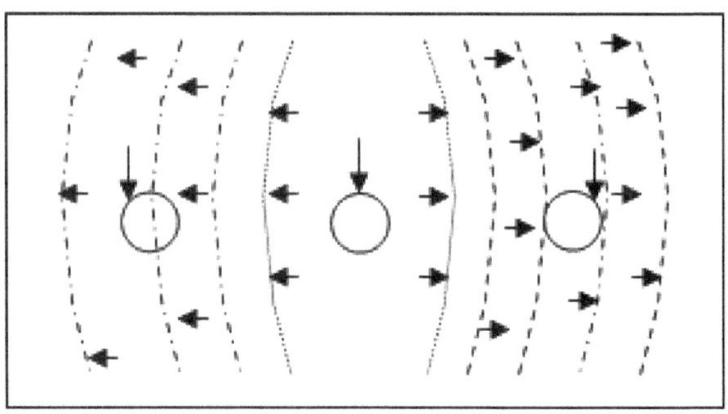

(b) CRTS-II CAM

Fig. 29. Schematic diagram of grouting process for CRTS I and II slab tracks

Thus, the rheological properties of CAM were the key factors to guarantee the filling effect of CAM under the slab track. The basic flow pattern of fluid could be divided into Newtonian fluid and non-Newtonian fluid. Newtonian fluids like water and air are the fluids with the constant viscosity. However, the Bingham model is usually to study the rheological properties of fresh concrete and cement paste. The Bingham liquid usually had a critical shear stress (or "yield stress"), once the yield stress is exceeded, the fluid began to flow. The equation is shown as

$$\tau = \tau_0 + \eta_0 \frac{dv}{dx} \quad (3)$$

Where: τ-shear stress(Pa); τ_0-yield stress(Pa); η_0-viscosityPa S); dv/dx—shear rate(S^{-1}).

According to our study (Wang FZ, Wang T, 2008), CAM also belonged to Bingham fluid. The shear stress and shear rate followed the equation 3.

CAM needs to have both nice flowability and good homogeneity without separation, to fill the huge narrow space thoroughly with the help of rheological agents. As can be seen from the equation 3, the yield stress τ_0 and the viscosity η_0 determines the rheological properties. CAM would have an increased flow resistance with increased viscosity and a bad homogeneity with small viscosity; on the other hand, CAM needs a proper shear stress τ_0. When the shear stress τ_0 was too small, the rheological properties of CAM would be similar with the Newtonian fluid, which had a nice flowability but was easy to separate; when the shear stress τ_0 was too large, it would be hard for CAM to fill the space fully because CAM would not flow unless with a larger shear rate. In short, the purpose to use the rheological agents is to coordinate the balance between the flowability and homogeneity; also to control the flow state of CAM thus to guarantee the grouting process.

4.2 Preparation of rheological agents

The rheological agents are usually divided into inorganic and organic agents. Fig 30 showed the influences of common inorganic rheological agents on the CAM's rheological properties and CMC were prepared to control the flow state of CAM (yield stress τ_0 and viscosity η_0). The organic agents that were commonly used included Polyacrylamide, sodium polyacrylate, polyvinyl alcohol and cellulose ether, which mainly increased the paste viscosity. The organic rheological agents VMA were prepared which could keep the homogeneity of paste but have little effect on the flowability.

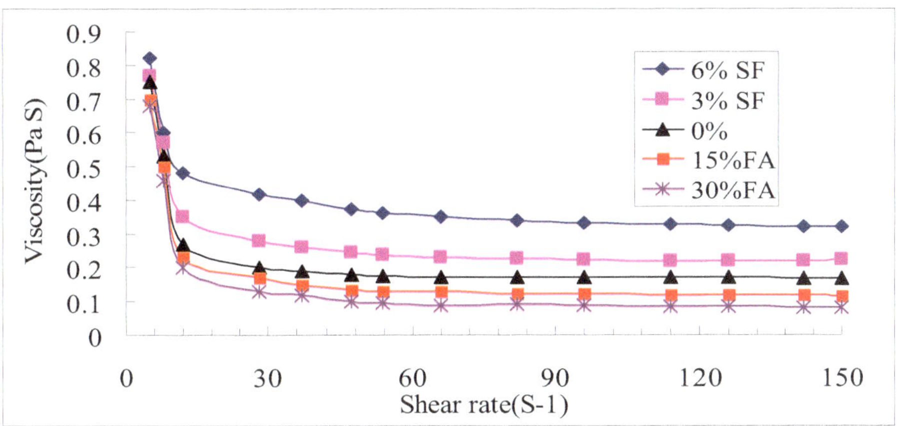

Fig. 30. Influence of inorganic rheological agents on the CAM rheological properties

The flow state of CAM can be evaluated by indexes like flowability, spreading and spreading speed. The influences of CMC and VMA on the rheological properties of CAM were shown in Table20.

Number	VMA	CMC	Flowability/s	Spreading/mm	D280/s
CAM-1	0	0	90	350	5
CAM-2	√	0	118	340	6
CAM-3	√	√	115	320	8

Note: CAM here referred to II CAM

Table 20. Influence of CMC and VMA on the rheological properties of CAM

4.3 The key technology to prepare the dry blend for CAM

As mentioned before, the CAM for high-speed railway was usually consisted of five parts which were asphalt emulsion, dry blend, superplasticizer, anti-foaming agent and water. Of which, dry blend was consisted of cement, sand and other powder material that adjusted the CAM properties. The powder materials that adjusted the CAM properties were usually expansion agent, aluminum powder and rheological agents. The key technology to prepare the dry blend for CAM was to control the stability of raw materials, grading of sand and uniform addition of adjust materials with small amount.

4.3.1 Dynamic controlling of sand grading

Sand is the raw material with maximum demand for CAM preparation whose properties are hard to control. The grading and mud contents are quite important for the CAM quality. The sand grading influenced its fineness. At a constant weight, the surface area increased with the decrease of fineness, which increased the paste that coated the sand and the water to make the CAM have the same flowability. At the same time, the grading and surface morphology of sands influenced the packing tightness of aggregates and the flow state of CAM. The influences of sand fineness on the CAM properties were shown in Table 21.

Number	Sand	Fineness modulus	Flowability /s	Workable time/min	Separation rate/%	Compressive strength /MPa		
						1d	7d	28d
S2	River sand	1.4	20.5	65	0.6	0.13	0.83	2.1
S3	River sand	1.7	19.5	85	0.7	0.13	0.85	2.1
S4	River sand	2.3	19	95	3.1	0.11	0.75	1.75
S5	Silica sand	1.4	19.5	90	0.9	0.12	0.8	2.05

Table 21. Influences of sand fineness on the CAM properties

As shown in Table21, the homogeneity of CAM decreased with the increase of sand fineness modulus. The proper fineness modulus should be controlled within 1.4~1.7.

The powders with the particle sizes less than 0.15mm in dry blend were cements and mineral powders. Of which, cement is the key component to guarantee the compressive strength of CAM and inert material is the component to increase the viscosity of dry blend to prevent the CAM separation. The experimental results indicated that, CAM had a nice rheological properties without separation when the contents of powders with particle sizes less than 0.15mm were about 40~50%; when the content was less than 40%, CAM was easy to separate and bleed; when the content was larger than 50%, more water was needed and the viscosity and flowability of CAM increased which made CAM hard to fill the spaces.

4.3.2 Uniform addition of adjustment materials with small amount

The contents of adjustment materials were quite small in the dry blend of CAM. For example, the weight of aluminum powder in the dry blend per ton was about 10~30g and the weight of rheological agents were 3~8kg. Although the amount was quite small, the uniform distribution of these materials in dry blend was quite important for the rheological properties and durability of CAM. Thus, the uniform addition technology is quite important. So the adjustment materials were pre-mixed with carrier like sand or cement, then the sand was added by several times to make the carrier with the adjust materials dispersed more evenly.

5. Application of CAM on Beijing-Shanghai high-speed railway

Beijing-Shanghai high speed railway is the longest high-speed railway with the most strict quality requirements in the world. We have taken part in the railway construction by providing the CAM prepared the technology mentioned in the chapter. Fig31-41 showed the key construction process during construction of CRTS II CAM in Jing-Hu high-speed railway.

Fig. 31. Preparation before grouting

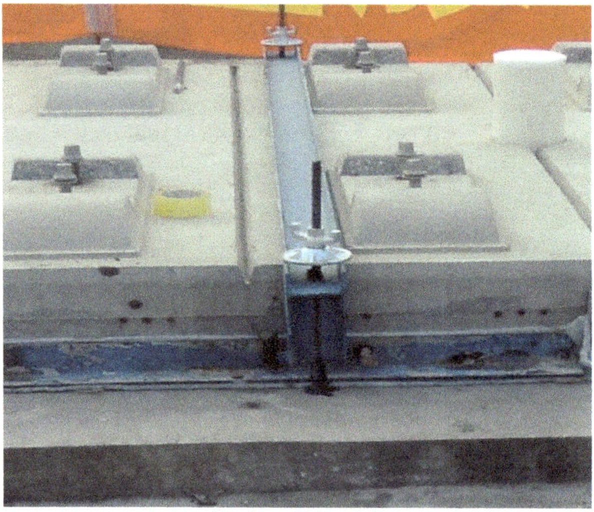

Sealing Edge (a)

Sealing Edge (b)

Fig. 32. Sealing Edge

Fig. 33. Tight the slab track

Fig. 34. Wetting

Fig. 35. Preparation of CA mortar

(a)Transportation of CA mortar

(b)Transportation of CA mortar

Fig. 36. Transportation of CA mortar

Fig. 37. Grouting

Fig. 38. Grouting finish

Fig. 39. Harden state of CA mortar

Fig. 40. Railway track laying

Fig. 41. Construction finish

6. Acknowledgment

The author would like to thank the National Natural Science Foundation of China (No. 50602033) and the Fundamental Research Funds for the Central Universities (No. 2010-II-005) for the financial supports.

7. References

Al-Sabagh AM. (2002).The relevance HLB of surfactants on the stability of asphalt emulsion. *Colloids and Surfaces A-Physicochemical and Engineering Aspects*. Vol. 204, No.1-3, pp. 73-83.

A.S. Dukhin SSD, P.J. Goetz. (2007). Gravity as a factor of aggregative stability and coagulation. *Advances in Colloid and Interface Science*. No.133-135, pp.35-71.

Bury .M. Gerhards JE, W. Stamm, A. (1995).Application of a new method based on conductivity measurements to determine the creaming stability of o/w emulsions. *International Journal of Pharmaceutics*. Vol.124, No.2, pp: 183-194.

Esveld C. (2003) .Recent development in slab track. *European Railway Review*, No. 2, pp: 81–85.

Fazhou Wang, Yunpeng Liu, Yunhua Zhang, Shuguang Hu. (2012). Experimental study on the stability of asphalt emulsion for CA mortar by laser diffraction technique. *Construction and Building Materials*, No.28, pp.117-121.

Gullapalli RP, Sheth BB. (1996). Effect of methylcellulose on the stability of oil-in-water emulsions. *International Journal of Pharmaceutics*. Vol.140, No.1, pp. 97-109.

Harada Y TS, Itai N. (1983). Development of cement asphalt mortar for slab tracks in cold climate. *Quart Rep RTRI (Railway Technical Research Institute)*, Vol.15, No.1, pp.62-67.

Harada Y TS, Itai N. (1976). Development of ultrarapid-hardening cement-asphalt mortar for grouted-ballast track structure. *Quart Rep RTRI (Railway Technical Research Institute)*, Vol.17, No.1, pp.6-11.

Hu SG, Wang T, Wang FZ, Liu ZC, Gao T. (2009) .Adsorption behavior between cement and asphalt emulsion in CA mortar. *Advanced Cement Research*, Vol.21, No.1, pp. 11-14.

Hu SG, Wang T, Wang FZ, Liu ZC, etc. (2009). Freezing and thawing resistance of cement asphalt mortar. *Key Engineering Materials*, No.400-402, pp.163-167.

Ivanov IB, Kralchevsky PA. (1997).Stability of emulsions under equilibrium and dynamic conditions. *Colloids and Surfaces A: Physicochemical and Engineering Aspects*. Vol. 128, No.1-3, pp. 155-175.

Ivanova R, Balinov B, Sedev R, Exerowa D. (1999). Formation of a stable, highly concentrated O/W emulsion modeled by means of foam films. *Colloids and Surfaces A: Physicochemical and Engineering Aspects*. Vol.149, No.1-3, pp. 23-28.

Jin SH, C XF, Yang J. (2006). Key technologies of CA mortar for slab track. *China Railway Science*, Vol.27, No.2, pp.20-25. (In Chinese)

Kulmyrzaev A, Chanamai R, McClements DJ. (2000). Influence of pH and $CaCl_2$ on the stability of dilute whey protein stabilized emulsions. *Food Research International*. Vol. 33, No.1, pp.15-20.

Li G., Zhao Y., Pang S.-S, Huang W. (1998).Experimental study of cement-asphalt emulsion composite. *Cement Concrete Research*, Vol.28, No.05, pp.635–641.

Masood, Irshad, Agarwal, S.K., Sinha, U.N. (1994).Effect of various admixtures on the particle size distribution of cement determined with the aid of laser particle analyzer. *Cement Concrete Research*, Vol.24, No.3, pp.527-532.

Miura S, Takai H, Uchida M. and Fukada Y. (1983) .The mechanism of railway tracks. *Japan Railway & Transport Review*, No. 3, pp.38–45.

Ostberg GB, Bjorn; Hulden, Margareta. (1994). Mechanical Stability of Alkyd Emulsions. *The Journal of Coatings Technology*, No. 66, pp.37-46.

Pouliot N., Marchand J.,Pigeon M.(2003).Hydration mechanisms, microstructure, and mechanical properties of mortars prepared with mixed binder cement slurry-asphalt emulsion. *J Mater Civil Engng*, No.1–2, pp. 54–59.

Song H., Do J., Soh Y. (2006).Feasibility study of asphalt-modified mortars using asphalt emulsion. *Construction and Building Materials*, No.20, pp. 332–337.

Supratim Ghosh JNC.(2008).Factors affecting the freeze-thaw stability of emulsions. *Food Hydrocolloids*, No.22,pp.105-111.

Su,Z., Bijen,J.M.J.M, Fraaij,A.L.A. (1993). Interaction of polymer dispersions with Portland cement paste. *Mater Res Soc Symp Proc*, No.289, pp. 199-204.

Vogel W. (1995). Earthwork structures for new railway lines slab track-Principles and suggestions for realization. *Railway Technical Review*, No.1, pp. 324~430.

Wang FZ, Wang T, Hu SG, Liu ZC, Gaotao, Chen Liang. (2008). Rheological behaviour of cement asphalt mortar. *Engineering Journal of Wuhan University*, Vol.41, No.4, pp.06~27.(In Chinese)

Wang FZ, Liu ZC, Wang T, Hu SG. (2008) .A novel method to evaluate the setting process of cement and asphalt emulsion in CA mortar. *Materials and Structures*, Vol.41, No.4, pp.643-647.

Wang FZ, Liu ZC, Wang T, Hu SG.(2010).Temperature stability of compressive strength of cement asphalt mortar. *ACI Mater J*, Vol.107, No.1, pp.27-30.

Wang FZ, Liu ZC, Hu SG.(2009). Early age volume change of cement asphalt mortar in the presence of aluminum powder. *Materials and Structures*, Vol. 43, No.4, pp.493-498.

Wang FZ, Zhang YH, Liu YP, Tao T, Zou JZ, Chen L .(2009).Preliminary study on asphalt emulsion used in CA mortar. *J Test Evaluation*, Vol.37, No.5, pp.483-5.

Yunpeng Liu, Fazhou Wang.(2011).Influence of temperature on the absorption behaviour between cement and asphalt emulsion in CA mortar. *Advanced Materials Research*, No. 295-297, pp. 939-944.

Zuo JQ, Cai BF. (2005). Experimental research on the asphalt emulsion special for CA mortar in slab track. *Railway Construction*, No.2, pp.68-70. (In Chinese)

Zuo JQ, Fu DZ. (2005). Research on CA mortar in slab track. *Railway Construction*, No. 20, pp. 96-98. (In Chinese)

10

Reliability and Lifetime Prediction for IGBT Modules in Railway Traction Chains

X. Perpiñà[1], L. Navarro[1], X. Jordà[1], M. Vellvehi[1],
Jean-François Serviere[2] and M. Mermet-Guyennet[2]
[1]*IMB CNM CSIC, Barcelona,*
[2]*Alstom Transport, Tarbes*
[1]*Spain*
[2]*France*

1. Introduction

Electric railway traction chains (ERTCs) emerged at the end of the nineteenth century. Over the years, they have been consolidated as a better solution than their counterparts, i.e., the traction systems with generating power on board (e.g., diesel or steam-based systems). In terms of performances, ERTCs show the highest power-to-weight ratio, fastest acceleration and highest traction effort on steep gradients of the railway traction scenario. They also offer other advantages; such as: less noise, lower maintenance requirements of the traction units, and a higher rational use of energy respecting and preserving the environment (e.g., energy harvesting systems as regenerative brakes or no greenhouse gasses' emissions). However, their drawback is the electrification line cost, which always entails an economical viability study according to a trade-off between line distance and volume traffic. When this investment is not economically profitable, the optimum solution is a hybrid approach combining generating power on board (based on fossil fuels) and electric engines.

Such figures of performance for ERTCs would not be possible without power electronics development. As a matter of fact, ERTCs are based on power conversion strategies (power converters), which use semiconductor power devices. Power converters are electrical circuits which supply the energy from high voltage electrified lines (DC voltage from 1.5 to 3 kV or AC voltage of 25 kV at 50 Hz) to railway motors (2800-8800 kW) by using semiconductor power devices, passive elements (e.g., DC link capacitors), and mechanical actuators (e.g., circuit breaker). In this framework, semiconductor power devices are responsible for diverting and switching the electrical current (rectifiers and power switches, respectively) among the different branches of the power converter, supporting high current and voltage ratings under working conditions. Such extreme working conditions fix their reliability problems, since they usually arise either from the ruggedness of such devices to non-desired electrothermal events (Perpiñà et al., 2007a, 2010a) (e.g., short-circuit, overvoltage, overcurrent, or overtemperature) or from the ageing of one of the power converter constituting elements (Ciappa, 2002; Malagoni-Buiatti et al., 2010) (e.g., capacitors, bus bar or power device packages). Both phenomena have a negative effect on power devices: their working conditions become more adverse (overload conditions in terms of current, voltage or temperature), eventually causing their explosion. Thus, a complete knowledge of the physical signature of their failure not only could provide precious information about its cause, but also could reveal the origin of the overloading condition. For this reason, a good comprehension of this topic is desired by ERTC manufacturers. Several efforts have been invested in investigating (e.g., projects LESIT, PORTES, etc.) which is the impact on the converter long-term reliability when the packaging of power devices (multi-chip power module) wears out and/or when such power devices fail. In fact, this is one of the main hot topics in the current research in the fields of railway traction and power devices design for reliability.

This chapter presents and revises this problematic issue, eventually proposing an approach for the determination of the most thermo-electrically stressed devices of a power converter. For this purpose, this chapter will be organised as follows: Section 2 provides a description of ERTC parts, also introducing their reliability problems. Section 3 presents the most used power devices for railway applications and their packaging technology, highlighting the packaging wear-out problems. The lifetime prediction methodology currently used in railway traction is critically reviewed in Section 4, stressing their main limitations and challenges. In the last two sections, an experimental approach based on determining local thermal stresses within the package is proposed. This work has been performed in the framework of a European funded program called PORTES (*POwer Reliability for Traction ElectronicS*) (Portes, 2004), mainly devoted to explore and understand the physical failure mechanisms for the first generation of IGBT power modules.

2. ERTC description and reliability problems

Fig. 1 depicts an ERTC schematic, in which the energy delivery process to the motor can be followed. By using a line contactor (pantograph or floaters), the energy is extracted from an external source (electrified lines or catenary) and it is adapted by several power conversion strategies until arrive to the electrical traction motors. The final conversion stage is performed by the motor drives or inverters (see Fig. 1), which consist in a DC-AC converter (Mohan et al., 1995). They drive the train electrical engine (train speed control and braking) (Mermet-Guyennet et al., 2007) using IGBT (Insulated Gate Bipolar Transistors) multi-chip power modules, in which IGBT devices and freewheeling diodes are packaged together (multi-chip packaging strategy). In this application, the IGBTs are switched according to a pulse-width-modulation pattern (i.e., between 200 Hz and 2 kHz, indicatively) (Mohan et al., 1995) to supply the required sinusoidal input current to the motor (inductive load); whereas the freewheeling diodes ensure a continuous path for the current coming from the motor (Mohan et al., 1995). Fig. 1 also shows a schematic of a 3-phase power inverter for driving traction motors: it is formed by three branches or legs, which contain two IGBT modules each one (M_x in Fig. 1). The motor is represented by three inductive loads interconnected to the nodes between IGBT modules. Aside from IGBT modules, they also include a cooling system (based on air-forced convection, heat-pipes or liquid pump technologies) (Bouscayrol et al., 2006; Baumann et al., 2001), bus-bar high power connections, a close control command or IGBT drivers for switching them (Steimel, 2004;), low voltage connections, capacitors (decoupling or filtering) (Malagoni-Buiatti et al., 2010), mechanical parts for support and connectors, and a control board and current sensors integrated in the IGBT module (Bose, 2006).

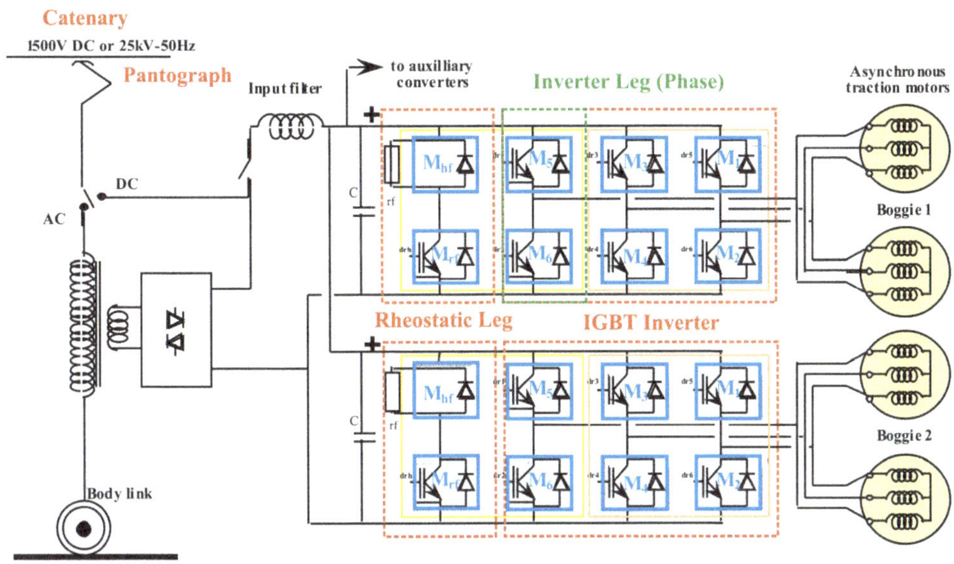

Fig. 1. Example of an ERTC showing the external supply line, a circuit breaker which connects the required conversion scheme, a first power conversion stage when an AC supply line is employed (AC-DC converter in this case), the body link, two 3-phase inverters, two choppers, and electrical motors. Special emphasis is put on the electrical connections within the 3-phase inverters.

Within ERTCs, the power inverter represents the most stressful application scenario for power devices, since it has an inductive load and this impose severe working conditions during the IGBT turn-off or diode reverse recovery (coexistence of high voltage and current levels in both cases). Therefore, the key points to achieve a satisfactory design for reliability of power inverters are (Mohan,1995):

- *Safe operating area*: the voltage and current of IGBT modules must remain within the limits given by their manufacturer under operating conditions.
- *Dielectric requirements*: the high voltage operation imposes minimum distances among the materials within the package to avoid dielectric breakdown and partial discharges.
- *Thermal limits*: the designer verifies that for a train duty cycle representative of a service line (mission profile), the working temperature and its variations (temperature swing) does not exceed certain maximum values.

This last aspect requires an accurate estimation of the train duty cycles or mission profiles, the IGBT and diode junction temperatures, and the behaviour of IGBT power modules under thermal cycles. Such thermal cycles come from the actual working conditions of the packaged devices: their selfheating (Arnold et al., 1994) induce a temperature field inside the IGBT module, which evolves depending on the train speed along a railway service line. Consequently, the power module experiences several local thermal cycles defined by the train acceleration and braking processes (Coquery et al., 2003) that, in turn, provokes local stresses among such materials because they show different thermal expansion coefficients (CTE). This leads to the apparition of crack propagation or interface delamination of the solder joint of the package assembly, which will locally increase the thermal resistance of the power module (thermal performance degradation) (Ciappa, 2002). During this process, there is another aspect that has an important role: the thermal contact between the power module and the cooling system. A perfect thermal contact between an IGBT power module and its cooling system is not always assured all over the module backside (uneven thermal resistance along the module backside), as evidenced in (Hamidi, 1998a; Perpiñà et al., 2010b). Then, the addition of both effects can provoke that some components reach temperature values, which lead to their destruction by burn-out (Sankaran et al., 1997; Ciappa, 2002; Perpiñà et al., 2010a).

The packaging wear-out due to thermo-mechanical effects was extensively studied in the nineties, particularly through the research programs called LESIT (*LeistungElektronik Systemtechnik InformationsTechnologie*, 1994-1997) and RAPSDRA (*Reliability of Advanced high Power Semiconductor Devices for Railway traction Applications*, 1996-1998). They led to the definition of Power Cycling Tests (PCTs) and lifetime prediction methods for railway traction IGBT modules (Hamidi et al., 1998b), that will be presented further on. Up to then, thermal cycling was the most common approach for accelerating the IGBT modules' ageing: the whole power module was submitted to the same temperature swings following square waveforms with different thermal profiles (duty cycles) in a temperature-controlled chamber (Bouarroudj et al., 2008; Lhommeau et al., 2007). The problem of this test was that the local thermal stress induced by power devices was not taken into account. Therefore, this approach was not suitable for wear-out studies of power modules and lifetime prediction from the final application viewpoint. However, power thermal cycling presents some limitations, as will be further discussed in Subsection 4.2:

- it deals with a macroscopic approach for lifetime prediction (average temperature values of the module) and do not determine the most thermally stressed devices for a given cooling system,
- it does not reproduce the power module failure under a real working condition, i.e., high current and voltage ratings.

Therefore, promoting thermal studies at chip level is fundamental for a better understanding of ERTC failures and for increasing its lifetime (Coquery et al., 2001). In fact, the key point is to monitor the temperature inside a power module under a thermal working condition representative of the final application. In these studies, one of their main goals should be verifying whether reliable operating and uniform temperatures are assured within a power module. With these tests, dysfunctions on cooling systems behaviour or non-reliable thermal designs of multi-chip modules would be evidenced. Besides, this information will contribute to the knowledge of the local influence of the temperature distribution on the devices assembled within the module, since there is a lack of experimental results reported in the literature.

3. Multichip power modules for railway traction

The current generation of power inverters for railway traction is based on the use of high voltage IGBT (*Insulated Gate Bipolar Transistor*) power modules (Rahimo et al., 2004). They consist in multi-chip devices (IGBTs and diodes) integrated in the same package, which cover a large range of voltage (1700V, 3300V, 6500V) and nominal currents (1600A-2400A,

1200A, 600A). This package, electrically interconnects several power semiconductor devices, extracts the dissipated heat through a thermal path separated (or not) from the electric path, isolates the high voltages from other critical parts of the system, and provides a mechanical support to the devices. In this framework, power electronics packaging plays an important technical and economical role, because it constitutes the interface between the raw semiconductor device and the circuit application. In this section, its main parts will be explained in Subsection 3.1, the semiconductor devices usually employed in railway applications will be briefly presented in Subsection 3.2, and its ageing mechanisms will be revised in Subsection 3.3.

3.1 Power electronics package and IGBT multi-chip power module structure

Fig. 2 shows the structure of a typical power electronics package for a single (discrete) or multiple (multi-chip module) dies, whose main constitutive elements are (Sheng & Colino, 2005):

- *Power semiconductor devices* (IGBT, FRED). They provide the current flow control function previously described.
- *Die-attach technology*. It contacts the die to the substrate mechanically, electrically and thermally.
- *Top-side interconnections*. They perform the electrical contact on the device top-side. In some cases, they can also provide top heat extraction (e.g., bump interconnection technology).
- An *electrically insulating and thermally conductive substrate*. It is the mechanical support of the devices, tracks and terminals, as well as it permits the electrical insulation between some of them. It also provides an efficient heat extraction path.
- An *encapsulating material*. Typically it consists on a conformal coating or casting for environmental and mechanical protection.
- *A base plate*. It provides a mechanical support, as well as it favors both the spreading and conduction of heat towards the cooling system.
- *Case and cover*. It is a housing structure that protects all devices and interconnections.

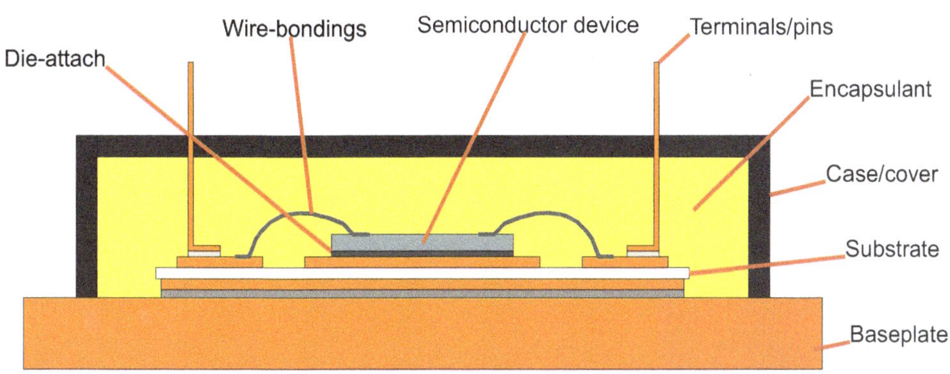

Fig. 2. Typical power electronics package structure.

Each one of these constitutive elements is made of a certain material depending on its functionality and position within the package. The material selection is performed according to its physical properties (electrical resistivity, thermal conductivity, or mechanical properties) to meet the package requirements. Typical materials used in power packaging are:

- *Power semiconductor devices.* Mainly silicon-based devices, although other semiconductors have been also introduced (SiC JFETs and rectifiers, AlGaN/GaN HEMTs, AsGa diodes).
- *Electrically insulating/thermally conductive substrate.* Usually ceramic substrates are (Al_2O_3, AlN, Si_3N_4, BeO) with copper layers on top and backside. The typical technology in railway power modules is the DCB (Direct Copper Bonded). For low and medium power ratings, metal (typically aluminum) substrates with a ceramic-filled polymer (typically epoxy) layer are used instead of the ceramic ones (e.g., Insulated Metal Substrate –IMS- technology).
- *Base plate.* Typically, they are based on nickel plated copper slabs. Other materials are metal matrix composites such as copper matrix composites reinforced with diamond, aluminum matrix reinforced with SiC, carbon-reinforced composites, etc.
- *Die-attach.* The usual solder joint material is the PbSnAg alloy, but now other lead-free alloys are used or new materials, such as Silver nano-sintering, have also been introduced.
- *Top-side interconnections.* The most used technique is large aluminum wire-bonding. Other solutions include pressure-type contact and metal bumps.
- *Case and cover.* Thermoset and thermoplastic materials including silicones and epoxies.

In order of importance, the package starts at the interface of the chip itself (first level packaging) and extends to higher levels of integration; such as substrate-level and system-level. The first level consists in the attachment of one or more bare chips to a substrate, the interconnection from these chips to the package leads, and the encapsulant (cover). The first level interconnection has a major role, because it directly interfaces with the chips, not only electrically, but also thermally and mechanically. For reducing the thermal stresses in the chips, it is necessary a thermal-friendly interconnection design. And lastly, the reliability of the first level interconnection is vital to ensure an extended lifetime of the electronic assemblies. At this point, a good thermo-mechanical matching property (mainly the CTE) between neighboring materials is crucial, as will be discussed in Subsection 3.3. Other aspects taken into account for their design are the optimization of the electrical circuit of the package (parasitic inductance and resistance reduction), its electromagnetic compatibility (low parasitic radiation and conduction), the mechanical roughness, and the package thermal management.

In the case of IGBT power modules, the chips are typically connected by wire-bonding technology. In such modules, a DCB substrate is commonly used as a ceramic substrate for the power device. DCB provides an excellent electrical insulation as well as good thermal conduction due to the direct bonding of copper on ceramic materials; such as alumina and aluminum nitride. The materials used for device and DCB attachment are usually solder alloys. Fig. 3 (a) shows a commercial 3.3kV IGBT module, and Fig. 3 (b) shows the chip interconnection within the modules. Their internal structure consists of three elementary phases (A, B, and C) according to the distribution of the anode (collector) and cathode (emitter) power terminals, as depicted in Fig. 3 (a). Each elementary phase includes eight IGBTs (T_x in Fig. 3 (a)) and four free-wheeling diodes (D_x in Fig. 3 (a)), electrically interconnected as shown in Fig. 3 (b). Such devices are equally shared between two direct copper bonded (DCB) substrates (DCB 1 and DCB 2 in Fig. 3 (a)). The DCB substrates are soldered on a common AlSiC base plate. Metal-bars, also visible in Fig. 3 (a), implement a pairwise connection between the DCB substrates.

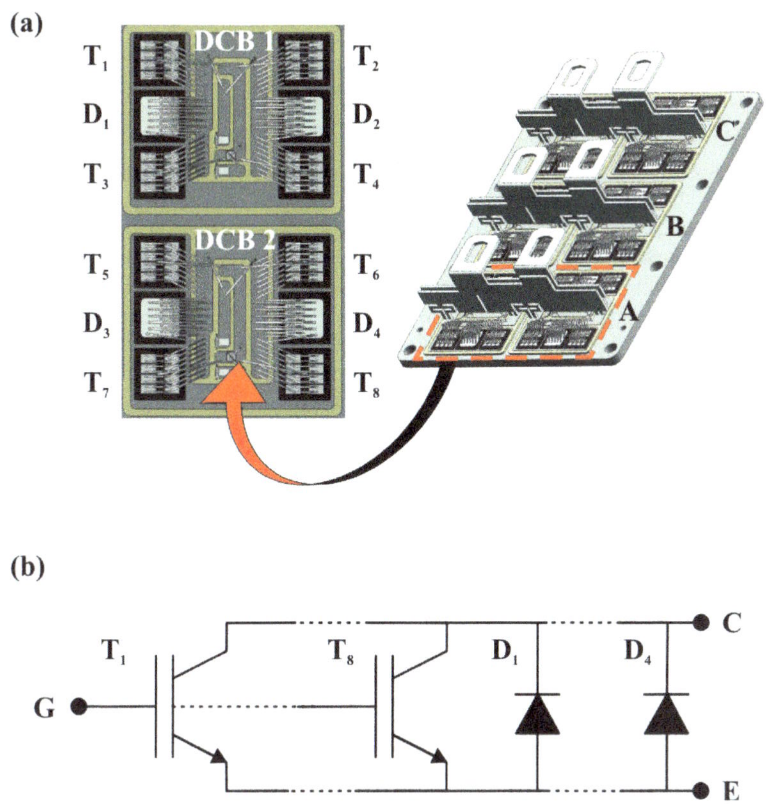

Fig. 3. (a) Power module internal view, showing all the elementary phases and providing a major detail in one of them. (b) Schematic showing the devices connection in one phase and highlighting the terminals: Gate (G), collector (C) and emitter (E).

3.2 Failures of semiconductor power devices in railway traction scenario

The power semiconductor devices most commonly used in railway traction are IGBTs (acting as switches) and diodes (acting as rectifiers). The main difference between them is that the former are switched-off and -on by an external control signal (low power driver), whereas the latter depend on circuit variables. Their role in ERTCs is to control the current flow within the power converter (forward state) and sustain the voltage when the current is diverted to another place (blocking state), e.g. another converter leg (Mohan et al., 1995).

In a railway power inverter, such power devices are always working in switching operation with an inductive load. This can give rise to very stressful processes in terms of instantaneous dissipated power during both the diodes reverse recovery and the IGBTs turn-off, i.e., ruggedness failures when high current and high voltage levels coexist (Perpiñà et al., 2010a). Besides, IGBT modules are also the most sensitive elements to other system failures. For instance, dysfunctions on the IGBT drivers (Bouscayrol et al, 2006; Steimel, 2004), on the sensing elements to monitor the critical electrical and thermal variables of the inverter (Bose, 2006), on the cooling system (Baumann et al., 2001), and on the capacitors (decoupling or filter) (Malagoni-Buiatti et al., 2010) undoubtedly lead to their destruction (induced-failures). Furthermore, their working conditions can be more adverse when the IGBT power module wears out (Lhommeau et al., 2007), a non-optimum thermal management design has been carried out (Perpiñà et al., 2007a, 2010b), and external electrical dysfunctions occur (Malagoni-Buiatti et al., 2010). As mentioned before, the package degradation limits the power converter lifetime since the working temperature inside the module increases due to the solder fatigue among the material stack used for the power module manufacturing (Ciappa, 2002; Mermet-Gunyennet et al., 2007). This ageing process can be enormously accelerated as a consequence of a non-optimum thermal management design (e.g., cooling system design or thermal interface selection) (Perpiñà et al., 2007a). On the other hand, it has been observed that several failures depend on the switching strategy of the power device. Sometimes, the dysfunctions on the driver induce abnormal events (short-circuit, over-current, or over-voltage), which severely stress the components (Khargekar, 1994). However, such events not only result from driving anomalies, but also can be linked to environmental or load conditions (Perpiñà et al., 2007a). In this failure scenario, it is interesting to determine "a posteriori" the device failure signature so as to derive precious information about the device failure origin. Over the years, this analysis has been tackled in both diodes and IGBTs, providing a physical insight into the failures occurred during the diode reverse recovery and the IGBT turn-off under inductive loads (Perpiñà et al., 2010a). Obviously, the working conditions for such components can be more adverse when the power module wears out by thermal fatigue, as can be inferred from their derived failure mechanism explained in the next subsection.

3.3 Failure mechanisms in multi-chip modules aged by thermal fatigue

ERTCs manufacturers should fix at least, a long-term warranty of 30 years. During this long period of time, the main problems present in railway power modules are related to their wear-out. Although several efforts have been addressed to improve the power device ruggedness under overloading conditions, the package wear-out will always be the limiting factor for long-term reliability. For this reason, they follow several accelerated ageing tests not only targeted to the devices or materials performance degradation (e.g., leakage current, gate dielectric, humidity), but also for determining the thermomechanical impact on the material stack conforming the power module (long-term failure mechanisms). Concerning the thermomechanical effect on the power module wear-out, it can be divided into two contributing effects: power devices local dissipation and external temperature variation. Both will exert thermal cycles on the material stack of the package assembly producing the so-called thermal fatigue. The term fatigue comes from the material mechanics field and refers to a cyclic load on a material stack (Pecht, 1991). There are two different tests to explore the thermomechanical induced wear-out of power modules: thermal and power cycling. As mentioned, the former submits the whole module to uniform temperature swings, whereas the latter considers the local overheating performed by the devices. For the packaging technology previously outlined, PCTs have put in evidence the following failure mechanisms (Ciappa, 2002): reconstruction of metallization, wire-bonding fatigue, and solder alloys fatigue. They are briefly revised in the next subsection, leaving the concept of PCTs to be treated further on.

3.3.1 Reconstruction of metallization

Power cycles induce important periodical compressive and tensile stresses on the device upper metallization because of its large thermo-mechanical mismatch with silicon. As a result, such stresses can undergo far beyond the elastic limit, and their relaxation can occur by mechanical processes (e.g., diffusion creep, grain boundary sliding, or by plastic deformation through dislocation glide) (Ciappa, 2002). This leads either to the extrusion of the aluminium grains or to cavitation effects (empty cavities formation) at the grain boundaries depending on the texture of the metallization, which give rise to the aluminium reconstruction and results into an increase of its sheet resistance with time. Then, it can be monitored by measuring the voltage drop across the device on-state.

3.3.2 Wire-bonding fatigue

Under working conditions (high current ratings and switching operation), wire-bondings are exposed to almost the full temperature swings imposed by both the power dissipation in the silicon and the wire-bonding itself. Moreover, the current density distribution across its section is strongly inhomogeneous due to the skin effect (Ciappa, 2002). In general, the failure of a wire-bonding occurs predominantly as a result of a fatigue caused either by shear stresses generated between the bond pad and the wire-bonding. As a result, they are gradually disconnected from the power devices until they reach the open circuit condition. Two phenomena are observed: crack propagation at the wire-bonding heel (heel cracking propagation) or the lift-off of the wire-bonding (wire-bonding lift-off). In the first case, this problem comes from a non-optimum wire-bonding process which mechanically damages the wire-bonding heel (crack creation). In the latter, the wire-bonding is aged, eventually inducing its lift-off because of the high CTE mismatching between materials. This failure starts at a crack at the tail of the wire-bonding, and propagates through the interface defined between the wire bonding and the chip upper metallization until it completely lifts off (see Fig. 4). The failure evolution can be externally monitored by sensing a change either in the contact resistance or in the internal distribution of the current, such that it can be traced by monitoring the device voltage drop during on state (Cova et al., 1997).

3.3.3 Solder joint fatigue

The main failure mechanism of railway multi-chip power modules is associated with the thermo-mechanical fatigue of the solder joint. The most critical interfaces are the solder joints between the die attach-ceramic substrate and the ceramic substrate-base plate. At such locations one finds the worst CTE mismatching (in both interfaces), the maximum temperature swing combined with the largest lateral dimensions (only for ceramic substrate-base plate). The most frequently used materials as solder joints in multi-chip power modules are based on tin–silver, indium, or tin–lead alloys (Lutz et al., 2011). They have excellent electrical properties and as soft solders, they exhibit good flow characteristics. Often, solder joints are considered as a single homogeneous phase, but this is not true as their phase evolves with time. For instance, when a material with a copper metallization is soldered with a standard lead–tin alloy, the bond is mainly provided through the formation of a Cu5Sn6 intermetallic phase located close to the copper layer (Ciappa, 2002). Two additional distinct phases, one tin rich and one lead rich, are formed in the central part of the solder joint. During accelerated ageing tests, these phases coarsen rapidly due to the high homologous temperature at which the alloy is operated, experiencing a modification of its thermo-mechanical properties: since the copper phase is much more brittle than the tin–lead phases, thermomechanical fatigue cracks often propagate within the copper rich intermetallic. Due to the larger CTE mismatch and to the higher temperature, fatigue cracks are found preferably in the vicinity of the intermetallic layer immediately below the ceramic substrate. Metallographic preparations have shown that cracks initiate at the border of the solder joint, where the shear stress reaches its maximum (Ciappa, 2002). Additionally, crack formation is highly promoted by the presence of sharp angles at the edges of the ceramic substrate (Ciappa, 2002).

Fig. 4. Wire-bonding lift-off after PCTs

Fig. 5 shows an acoustic microscope image which depicts this behaviour: higher solder delamination (bright areas) is observed at the ceramic substrate edges. The discontinuity at the edges is responsible for a stress peak at this location and especially at the corners. This is due to the fact that attached materials can freely expand with temperature along the unbounded directions, but, at the material attachment interface, they are bonded and their thermo-mechanical properties, mostly the Young modulus and thermal expansion coefficients, will fix the dynamics of the assembly. Therefore, for higher (lower) Young modulus and very different thermal expansion coefficients and material thickness, it is expected to obtain more localized (distributed) strains close to the discontinuity present at edges of the assembly, which will finally induce a higher and more local (lower and more distributed) stress (Ciappa, 2002; Lutz et al., 2011). Therefore, the fractures start at the outside corners and edges and propagate towards the center of the soldered materials, thus absorbing this stored mechanical energy. When a non uniform temperature is observed in the case of the modules and high temperature gradients are present, the degradation of the solder joint starts at the corners pointing towards the position of the module at which the higher temperature is reached and moves outward. Obviously, several parameters will fix the solder fatigue pattern: the thermal interface between the power module and cooling device, the cooling device design (e.g., water or heat-pipe based), and the base plate backside flatness (Perpiñà et al., 2007b).

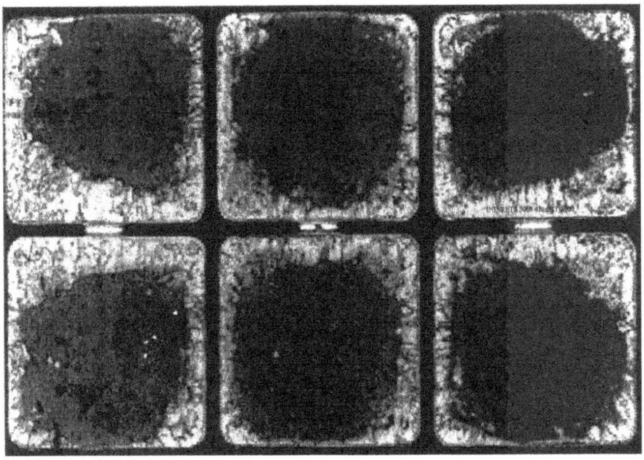

Fig. 5. Solder delamination inside IGBT 3300V 1200A with copper base-plate after 40 Kcycles.

4. Critical revision of lifetime prediction approach in traction modules

In actual IGBT power modules, its lifetime calculation is based on deriving a reference curve obtained when the modules are submitted to local thermal stresses. This curve depicts the number of cycles to failure versus temperature swing (thermal load) at an indicative temperature of the thermal cycle (maximum, average or minimum value), as performed in mechanical fatigue of materials (number of duty cycles to failure versus mechanical load) (Ciappa, 2002). This curve is derived by following accelerated ageing tests representative of the final application (PCTs), in which the die provokes local thermal cycles within the module structure (see Subsection 4.2). For lifetime calculation, a virtual experiment is carried out to extract the thermal cycles (temperature swings) responsible for the thermal fatigue experienced by each constitutive element of the package. They should be computed considering a mission profile representative of a real service line and should be properly counted following an adequate algorithm. Therefore, the current flow and voltage drop across the module are modelled for a given motor and service line. Finally, it is supposed that only an average of all thermal cycles experienced by the module along a service line will contribute on the module ageing to compare such a result with the reference curve. The key points related to its lifetime calculation, the followed procedures, and their limitations are critically discussed in the following subsections. First of all, Subsection 4.1 presents the lifetime methodology, and after Subsections 4.2, 4.3, and 4.4 critically revises the limitations present in PCTs, local temperature measurements in power modules, and the counting and superimposition of thermal cycles, respectively.

4.1 Current followed methodology

The starting point for lifetime prediction in railway power modules is determining a duty cycle based on a train service line (mission profile). Several approaches are possible:

- for trains with a well known mission profiles, like metros or tramways, it is considered the total duty cycle on a given service line assuming maximum load conditions;
- for trains without well defined mission profiles, it is defined a reference duty cycle corresponding to the most stressed course for the train;
- for high speed trains and locomotives, it is very difficult to define a reference cycle.

The key aspect in the above approach is that we only consider thermal cycles in the second timescale (package timescale), which will be referred as traction-braking cycles. Long thermal cycles, like day-night or season cycles, are not considered. In the same way, very short cycles (below second scale) are not accounted for, as they will consider the device or IGBT module-only thermal dynamics without taking into account the effect of the cooling system (Berg & Wolfgang, 1998). For instance, this is the case of statoric cycles at very low speed in locomotives.

In order to determine the thermal cycles experienced by the IGBT modules, several calculations are required. Starting from the cinematic of the train, currents in the motors are estimated. Subsequently, in accordance to the control strategy applied on the converter, the current and voltage ratings seen by the IGBT modules are inferred. At this stage, with the static and dynamic characteristics of IGBT modules, losses at die level (both, IGBTs and Diodes) are estimated. Such losses are the input data for temperature cycle estimation, but another key parameter of the power converter is required: its equivalent thermal model.

Generally, this equivalent model is extrapolated from thermal measurements and is presented under the form of a compact thermal model based on a RC stacking (Foster or Cauer representation) (Ciappa, 2005). These equivalent models are used to calculate the virtual junction temperature (T_{vj}) of the whole IGBT module for diodes and IGBTs. In fact, T_{vj} is an average temperature of the module among all IGBTs or diodes. In a 3.3 kV-1200A IGBT power module (see Fig. 3 (a)), T_{vj} for IGBT (diodes) is the average temperature for 24 chips (12 chips). At this point, we obtain the curve T_{vj} versus time for both IGBT and diode dies (see Fig. 6 (a)), T_c versus time for the power module, and a criterion (detailed in 4.4) to translate these curves into a histogram of cycles, as shown in Fig. 6 (b) (Ciappa, 2000).

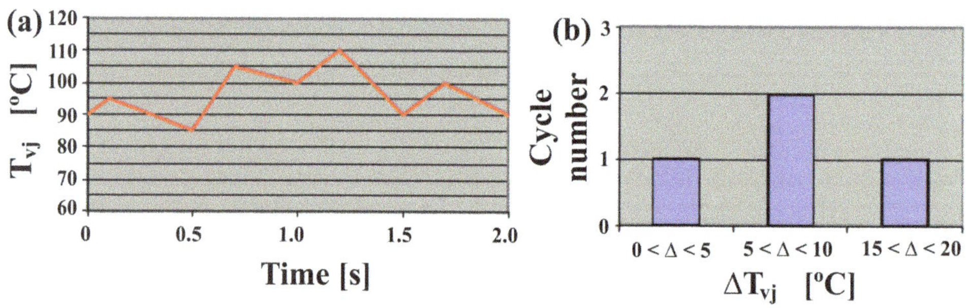

Fig. 6. (a) Thermal cycles versus time and (b) histogram of the thermal cycles.

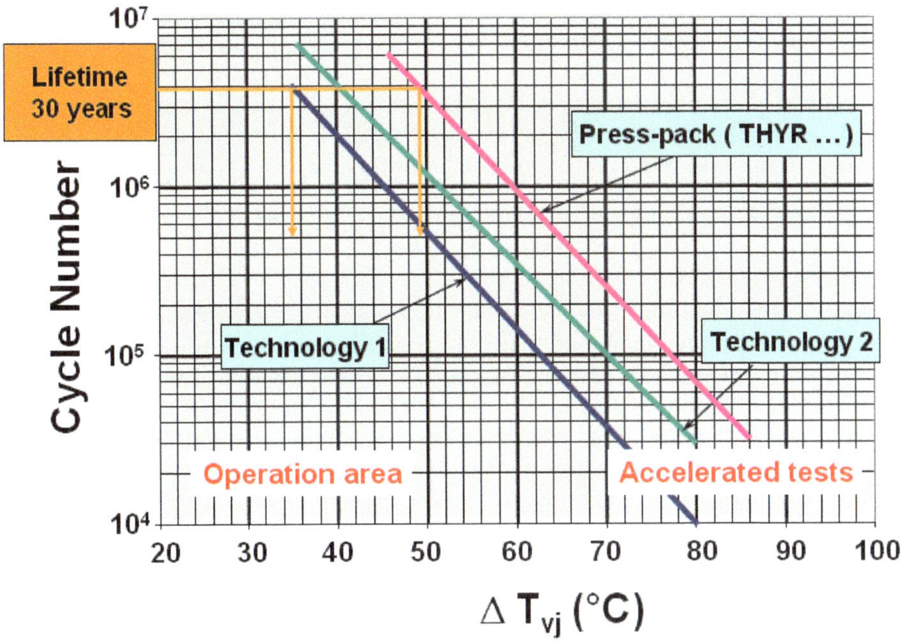

Fig. 7. Reference curves N cycles to failure versus virtual junction temperature cycles for various technologies

The final step is to compare the histogram of thermal cycles obtained for the whole life of the train (example shown in Fig. 6 (b)) with the reference curves for lifetime prediction. Such reference curves correspond to the dependence of the number of cycles to failure N_f versus the amplitude of thermal cycles (temperature swing) experienced in average at die level (ΔT_{vj}), usually by IGBTs. For instance, Fig. 7 shows one of these reference curves: N_f versus ΔT_{vj}. Both reference curves are obtained from PCTs and an extrapolation law based on plastic deformation fatigue (Coffin-Manson-like approximation).

This law assumes that when ΔT_{vj} ranges from 5 to 80°C, the number of cycles until failure N_f may be expressed as:

$$N_f = \lambda \times (\Delta T_{vj})^{-\alpha_{vj}} \times \exp[E_{a,vj}/(k_B T_{vj,m})] \quad (1)$$

where λ and α_{vj} are fitting constants depending on the material mechanical properties, α_{vj} ranging from 5 to 6. Concerning the exponential term (Arrhenius effect), $E_{a,vj}$, k_B, and $T_{vj,m}$ are the activation energy, the Boltzmann constant and the average temperature of the thermal cycle, respectively. It has also been observed in other scenarios that there are other parameters which fix these kind of curves when performing PCTs, such as: heating up time, current and voltage ratings, and wire-bonding diameter. Incorporating such new variables in a new expression inspired in Eq. (1) has led to definition of a more complex law (CIPS model) which allows performing more accurate lifetime predictions. However, this modified expression is no longer valid in traction (Lutz et al., 2011), since it has been settled for Al_2O_3 substrates and not for AlN and AlSiC (as it is the case of traction modules).

In order to compare multiple thermal cycles determined along a given mission profile N_{mp} (see Fig. 6) with the reference curves (N_f versus ΔT_{vj}, se Fig. 7), an equivalent point of the mission profile test should be determined. First of all, an equivalent temperature increase ΔT_e is defined and corresponds to ΔT_{vj} of the histogram at which more cycles have been detected. After, a cumulative cycle to failure N_e at ΔT_e is computed assuming that every thermal cycle consumes a fraction of power module lifetime given by reference curve N_f, i.e.:

$$N_e\,(\Delta T_e) = N_f\,(\Delta T_e) \times \sum_{i=1}^{n} \left(N_{mp}(\Delta T_{vj,i}) / N_f\,(\Delta T_{vj,i})\right) \quad (2)$$

where i is a subindex which varies within the range of temperatures of the histogram and $\Delta T_{vj,i}$ represents the temperature swing for each one of the considered points in the histogram. In such a calculation, $N_{mp}(\Delta T_{vj,i})$ corresponds to the thermal cycles experienced by the module during 30 years, considering the number of mission profiles performed along this time. The criterion to know if we meet the lifetime requirement is made by placing the derived equivalent point (N_e, ΔT_e) with respect to the considered reference curve and by checking if N_e is lower than the number of cycles of the reference curve at ΔT_e.

4.2 Role of power cycling tests in traction lifetime prediction

As previously mentioned, RAPSDRA program proposed the sketch of reliability tests called active Power Cycling Tests or PCTs (Berg & Wolfgang, 1998). The advantage of these tests in comparison with the followed until then (thermal cycling) was that they considered the local thermomechanical stresses induced by dies dissipating power. These tests are run with IGBT modules mounted on a water-based cooling system (water plate). In order to simplify the test set up, the power module is locally heated by only IGBT devices, which are switched on with a low voltage power supply. The temperature swing is adjusted with the device conduction time (heating up time) and the current level. This means that during PCT, the stress conditions are completely different from real conditions: no switching, no high voltage, no dynamic losses are considered.

In order to determine the end of life of aged modules, a statistical treatment of the experimental results is carried out according to the Weibull statistics (Lutz et al., 2011). Such a statistical criterion gives the number of cycles after which 10% of the modules have reached the end of life. The main criterion for the end of life is the evolution of saturation voltage of the IGBT module $V_{CE,sat}$ (or forward voltage for diodes). A commonly accepted limit is 5%. In the case of monitoring the junction to case thermal resistance, the main criterion is defined as 20% increase. However, such accelerated ageing tests do not provoke the same failure mechanisms than in real conditions and we only observe indicators of the following failure mechanisms:

- collector-emitter saturation voltage for IGBT,
- forward voltage for diode,
- collector-emitter leakage current,
- loss of control of the gate,
- loss of power connections.

There is no clear relation between the evolution of these indicators and the failure mechanisms in the field. A first conclusion on PCT is that they are well adapted to compare durability of wire-bondings and soldering technology, but the extrapolation of PCT tests for lifetime estimation is not so straight-forward. For this reason, the PCT should be redesigned on the basis of the following principles, more oriented to the final application:

- accelerated ageing tests must induce the same failure mechanisms than in real service,
- ageing of packaging structure induces thermal failure mechanisms linked with high voltage and/or switching,
- ageing of packaging is mainly activated by active and passive thermal cycles, and
- distribution of temperature between chips inside the power module is a key factor for inducing a catastrophic failure.

Therefore, not only a thermal stress should be induced, but also the IGBT module should be electrically operated under real working conditions (high voltage and switching) to be representative of the failure observed in the final application. Besides, the cooling conditions for the IGBT power module must be the same than in real service operation.

This is the reason why new tests covering these misleading points should be carried out. They should consist in mounting power modules with inverter legs in a back to back configuration (high current and voltage) and driven by control boards (PWM pattern) modulating the package heating time in order to adjust ΔT_{vj} at 80°C and 60°C. Obviously, the cooling conditions will be the same than in real operation. In order to exactly know the context of the induced failures, current and voltage tracking should be made. These tests should be carried out with at least 10 IGBT modules in order to get a good estimation of the Weibull shape factor of the failure distribution at a given ΔT_{vj}. Then, acceleration factor should be calculated between different ΔT_{vj} under the condition that the same range of Weibull shape factor is maintained whatever is ΔT_{vj}. Unfortunately, this kind of studies have not been already reported in the literature for IGBT power modules for railway traction, only finding a pioneering work in automobile traction (Smet et al., 2011). This work has performed ageing tests with 600 V-200 A IGBT power modules under high voltage and PWM driving conditions. Their parameters have been set in a reference temperature of 60°C with a heating up time of 30 s long. They have observed that wire-bonding and upper metallization degradations are the predominant ageing mechanisms in the case of high thermal stresses (ΔT_{vj}=100°C reaching a maximum junction temperature of 160°C), whereas in the case of less stress-inducing protocols (ΔT_{vj}=60°C reaching a maximum junction temperature of 120°C), the aforementioned failure modes (i.e., wire-bonding fatigue, metallization reconstruction and solder delamination) occur simultaneously. This difference is due to the fact that such stressing conditions generate local thermal stresses at a higher frequency than in PCT conditions, which locally acts on the upper part of the device and reveal different effects.

In any case, PCTs does not allow accessing to a thermal mapping of the IGBT module in thermal working conditions (just monitoring points as detailed in such works). This is a key point for a better understanding of the thermal distribution inside the module and the local thermal cycles generated by the devices and understanding their failures.

4.3 Module thermal distribution measurements

Temperature monitoring at chip level is mandatory to understand what occurs when power devices fail, since it is a key parameter in all processes previously explained. For an individual die, such issue can be easily tackled by means of several approaches based on (Bouscayrol et al., 2006): device thermo-sensitive electrical parameters (TSPs), temperature sensor monolithic integration in the device, thermo-optical techniques. Depending on the spatial resolution, one may define the junction temperature concept, which consists in an average temperature inside the device. However, multi-chip packaging introduces an apparent limitation on temperature local monitoring, since multiple devices are connected in parallel leading to temperature spatial resolution problems. Consequently, an average temperature measurement of all devices is always performed (Bose, 2006). Since any sensing method inside the package has not been envisaged by the manufacturers, usually TSPs are used for power cycling purposes, such as the $V_{CE,sat}$ at low current level (Coquery, 2003).

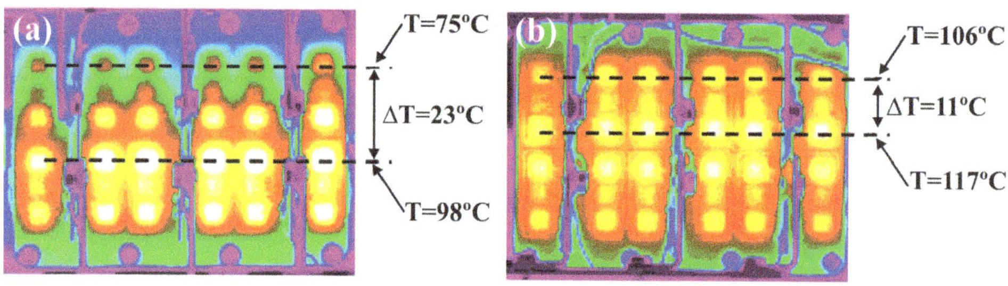

Fig. 8. Infrared thermal mappings of the same module when cooled down by a heatpipe-based thermosyphon (a) or a water cooled-based system (b). They have been performed by measuring first thermal mappings biasing IGBTs and diodes separately (1.8 kW in IGBTs and 0.6 kW in diodes) and after applying the superposition principle.

It has been observed that an uneven temperature distribution inside the module is induced by the module-cooling system interface, the chips placement inside the module, and the module mechanical mounting procedure (Ciappa, 2005; Perpinya et al., 2007c). This can induce a huge temperature dispersion, which leads to more stressed devices depending on their location inside the module. Therefore, determining such temperature distribution becomes essential when facing reliability studies. Fig. 8 depicts this dependence on the heat exchange system employed for cooling down the IGBT module. In this case, a non-optimal forced-convection heat pipe-based thermosyphon (a) and a water-based cooling system (b) are compared, in which the air and inlet water temperatures are 25°C and 60°C, respectively. In both measurements, the thermal interaction among other modules is also taken into account. The thermal mappings have been derived in two steps. The thermal mappings referred to each device type are firstly acquired by using 60 s long power pulses (1.8 kW in IGBTs and 0.6 kW in diodes), and afterwards, they are synchronised and added, according to the sources superposition principle (Mohan et al., 1995a). From these thermal mappings, it is clearly evidenced that the commonly followed way to tackle the thermal design of a converter only considering the virtual junction is not the most reliable approach.

4.4 Counting and superimposition of thermal cycles

Commonly, the Rainflow Method (Bannatine et al., 1990) is the most employed, easy to use and practical criterion for stress cycle counting. It consists in an algorithm that transforms the stress (or temperature) curves versus time into a frequency histogram of cycles number versus stress amplitude (or temperature) cycles and stress average value (or temperature) for the cycle. It is also used to measure the likely impact of the most damaging stress cycles in fatigue studies. On the contrary, to the observed thermal cycles in automotive applications (Ciappa et al., 2003), the rainflow method is applicable in railway scenario; because their mission profiles present simple waveforms and their acceleration-braking cycles can be well distinguished. In addition, this method is fairly easy to implement in a computer code, is capable to filter noisy data, and determines the minimum, average, and maximum temperature values of the cycle. Usually, this data is stacked in groups of 5°C, since it has been observed that this small amount of temperature has not an important effect on lifetime prediction and serves as a filtering criterion to make groups of cycles for the derived histogram. When the equivalent point N_e and ΔT_e is extracted for its comparison with the reference curve N_f-ΔT_{vj} obtained from PCT tests, there are some misleading points in all this procedure:

- large cycles are not taken into account (day-night, season cycles),
- temperature swing of the analysed cycles have a linear cumulative effect on the module ageing, i.e., two cycles with a temperature swing of 20 °C have the same effect on the packaging ageing than one cycle of 40°C,
- reference curve should be based on physics based models rather than statistical approaches without fixing a standard procedure to perform PCT tests (e.g., fixing a heat up time or using a cooling system representative of the final application)

5. Mission profile-based approach for local thermal cycles measurement

On the basis of the limitations of the lifetime prediction method explained in section 4, the last part of this chapter proposes an experimental approach for determining local thermal cycles within a power module. This is faced by means of a thermal test bench, in which the actual thermal working conditions of multi-chip power modules are emulated. The main objective is determining how the cooling system and the thermal interface locally affect the temperature distribution inside the module, eventually inferring the thermal cycles at chip level (both diodes and IGBTs). This approach contrasts with the most commonly adopted solutions based on either averaged temperature measurements on the power module by thermo-sensitive parameters (TSPs) (Su et al., 2002) or locally monitoring the temperature at the interface between the set module-cooling system with thermocouples (Coquery et al., 2003). Moreover, this solution is less complex and more straightforward than those explained in 4.2, becomes complementary to accelerated ageing tests, and also evidences the impact of non-uniform thermal distributions inside the package on the long-term reliability of the power inverter. The proposed procedure is carried out on new designs of cooling systems (forced-convection heat pipe-based thermosyphon) to study their suitability for railway applications. First of all, Subsection 5.1 introduces the thermosyphon cooling systems. Next, the details about the used experimental set-up and mission profile computation are outlined in Subsections 5.2 and 5.3, respectively.

5.1 Thermosyphon cooling system description

Generally speaking, a thermosyphon cooling system refers to a method of passive heat exchange based on natural convection, in which circulates a liquid (coolant) in a closed-loop circuit without requiring an external pump. This strategy simplifies the coolant pumping and the heat transfer by avoiding the cost, operation reliability, and complexity problems linked to conventional liquid-pumping systems (Baumann et al., 2001). One particular case of these cooling systems is the convection-forced heat pipe-based thermosyphon investigated in this work (see Fig. 9). It is formed by several sealed tubes which contain a certain amount of coolant, usually methanol pure or mixed with water (heat pipes). The heat pipes are partially inserted in a copper block (evaporator), and the rest are covered by thin fins to facilitate the thermal exchange by air-forced convection (condenser), as depicted in Fig. 9. The IGBT modules are screwed on both sides of the evaporator (see Fig. 9), simultaneously cooling down two legs of one power inverter (two modules per side). This cooling solution is used in railway inverters belonging to the power range from 500 kW to 1000 kW (medium power range), in which the power dissipation requires higher cooling performances than dry panels cooled by air-forced convection, but lower than water pumping-based cooling systems (Baumann et al., 2001).

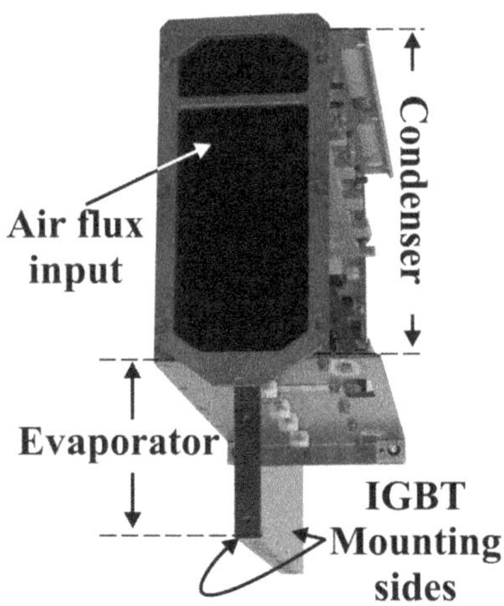

Fig. 9. Photograph of the inspected heat pipe-based cooling system detailing its main parts.

The proper operation of this cooling system relies on the fact that the coolant is located at the evaporator and the condenser is cooled down by air-forced convection. In these systems, the heat removal is produced as follows: the coolant absorbs the thermal energy dissipated by the power module until inducing its evaporation (first phase change). Subsequently, the coolant migrates to the condenser due to the presence of a thermal gradient. At the condenser, the coolant condenses (second phase change) releasing the thermal energy to the air flux. Afterwards, the coolant flows back to the evaporator through the internal wicks present in the walls of the heat pipe cavity (film evaporation process), mainly assisted by capillary forces (coolant pumping effect) (Dunn & Reay, 1983; Romestant, 2000).

5.2 Experimental set-up and thermal test conditions

The thermal interaction between the module and the cooling system has been studied on 3.3 kV-1.2 kA IGBT modules shown in Fig. 3 (a). The naming criterion for the devices and locations also obey to that depicted in this figure; i.e: elementary phases (A, B, and C), DCB substrates (DCB 1 and DCB 2 in Fig. 3 (a)), and devices (IGBTs and diodes T_x and D_x, respectively).

The thermal behaviour of the set module-cooling system is analysed with the set-up detailed in Fig. 10. In this set-up, the thermal interaction between two legs of one inverter is characterised. The approach consisted in inspecting two opened modules (M_3 and M_4) screwed on one thermosyphon side (one inverter leg) by an infrared camera, while the other two modules (M_1 and M_2) are also operating. The opening process of the modules M_3 and M_4 has been carefully performed maintaining their initial static electrical characteristics. No modifications have been introduced on the other two modules M_1 and M_2. The modules have been interconnected in series to a current source to bias them with a given current waveform I_{bias}, thus dissipating any kind of power profile in the analysed thermal system (see Fig. 10 (a)). With this interconnection scheme, the gate-to-emitter voltage in IGBT devices is maintained all the time at 15 V. In this approach, the selected power profile has a time resolution in the hundreds of millisecond range, neglecting the possible effects on the set module-cooling system due to the instantaneous power dissipated under switching operation. In fact, this approximation perfectly allows analysing the inverter long-term reliability, since the thermal performance degradation occurs within the heating up time-scale (second range) of the set module-cooling system (Smet et al., 2011; Yun et al., 2001). The modules have been thermally excited by current instead of power waveforms, as the voltage evolution with the temperature must be ensured to be much closer to the actual working conditions.

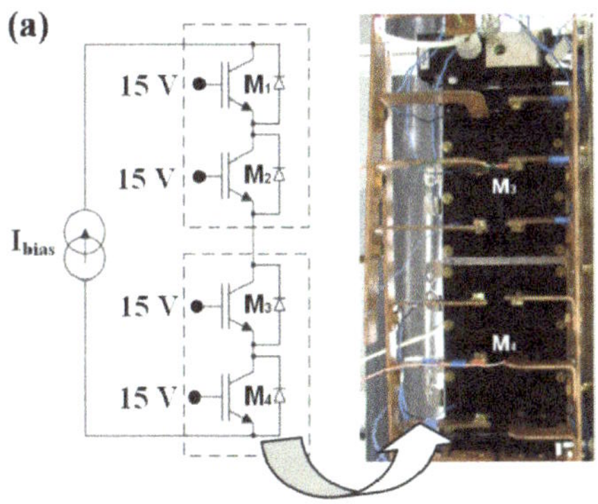

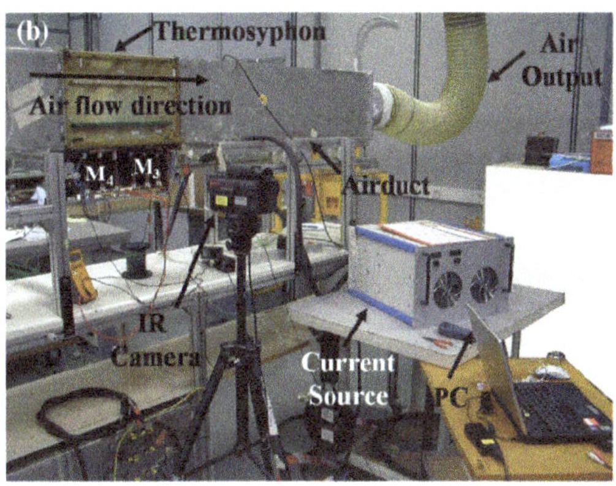

Fig. 10. (a) Circuit schematic showing the connections between the modules. The opened modules are also highlighted. (b) Experimental set-up used for testing the thermosyphon, in which the infrared camera, the airduct, the current source, and the monitoring PC are highlighted.

Fig. 10 (b) presents a photograph of the experimental set-up, highlighting the infrared camera, the current source driven by a computer, and the thermosyphon with the inspected modules. The modules have been screwed on the evaporator in such a way that for each elementary phase, IGBTs and diodes are identified according to their position in the evaporator as Fig. 3 (a) depicts; i.e., the lowest subscript shown in Fig. 3 (a) corresponds to the upper position in the evaporator. The thermal coupling between diodes and IGBTs within the same package can also be inferred from the experimental results. The thermal mappings referred to each type of device are firstly acquired, and afterwards, they are synchronised and added, according to the sources' superposition principle mentioned in 4.3 (Franke et al., 1999; Necati, 1993). Obviously, it is a first approximation to the real thermal mapping under actual working conditions. However, this approach allows us to perform a very good assessment of the whole system local thermal performances, and to predict the base plate-DCB solder delamination, as shown further on. These phenomena have been often analysed assuming an average virtual junction temperature inside the module, resulting in a non interesting approach when performing reliability prediction studies (Ciappa et al., 2003; Khatir & Lefebvre, 2004; Yun et al., 2001).

5.3 Railway service line mission profile determination

In order to obtain a representative mission profile for the power dissipation, the conduction and switching losses are numerically inferred in diodes and IGBTs separately by means of a traction drive design tool called CITHEL (CInematic THermal ELectric) (CITHEL, n.d.; Kreuawan, 2008). Given the route profile, train speed, and motor torque, CITHEL calculates the kinematics and electric aspects at inverter level for a given train characteristic. Thus, the traction chain performances as a function of different electrical motors, power components, and available inverters can be analysed and compared in a very early design stage. CITHEL is a fast solver capable to perform easy iterative calculations in electrical steady state, but it is not an electrical simulator. Unfortunately, this is the only way to estimate P_{total}, since no experimental data are currently available due to the complexity of measuring such a parameter in a real service line. With this software, the power module losses are determined on the basis of IGBT and diode parametric models extracted from experiments at 125°C (component temperature limit). The total power losses (P_{total}) are determined at each calculation time considering both conduction (P_{cond}) and switching ($P_{switching}$) contributions. Exactly, both electrical parameters are averaged over one statoric period, which is a function of the train instantaneous speed (Mohan et al., 1995a). P_{cond} is extracted from (Baliga, 1996):

$$P_{cond} = V_0 \times I_m + R_d \times I_{ef}^2 \tag{3}$$

where V_0 makes reference to the knee voltage and R_d is the on-resistance of the device. As a first approximation, the on-state of both devices, IGBT and diode, is modelled by a static characteristic defined by two straight lines that join at V_0, whose slopes are zero and R_d, respectively. I_m and I_{ef} refer to the average and effective current supplied to the motor for one statoric period, respectively. On the other hand, by assuming a sinusoidal output current resulting from the considered PWM scheme (Pulse Width Modulation) (Holmes et al., 2003), $P_{switching}$ is calculated as (Mohan et al., 1995a):

$$P_{switching} = F_s \times \sum_{i=1}^{n} E_i \tag{4}$$

where E_i corresponds to the energy loss due to the turn-on or -off, i counts each event indistinctly, and n represents the total number of turn-on and -off events during one statoric period. The diode turn-off losses are neglected in CITHEL computations. E_i has been experimentally related to the instantaneous current when a turn-on or -off event occurs (I_i), which can be written as (Mohan et al., 1995a):

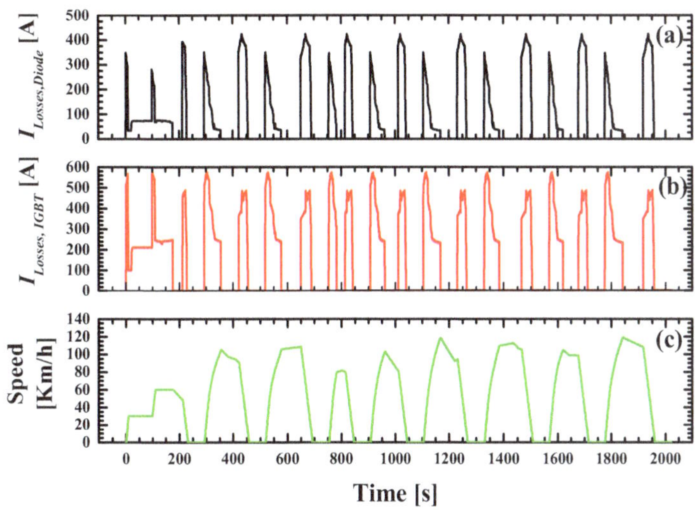

Fig. 11. I_{Losses} used to reproduce the dissipated power corresponding to switching and conduction losses for IGBTs (a) and diodes (b) extrapolated from behavioural models, as well as the speed required along a half of the service line (c).

$$E_i = V^{\alpha} \times \left(a\, I_i + b\, I_i^2 + c\, I_i^3 \right) \tag{5}$$

in which V is the line voltage (input voltage); α, a, b, and c are fitting parameters extracted from experiments. I_i can be easily determined by the following expression (Mohan et al., 1995a):

$$I_i = I_{Amp} \times \sin\left(2\pi\, i \frac{F_{stator}}{F_{switching}} \right) \tag{6}$$

where I_{Amp} is the amplitude of the resulting sinusoidal current waveform, F_{stator} denotes the motor statoric frequency, and $F_{switching}$ gives the modulation frequency associated with the followed PWM scheme (Mohan et al., 1995a). Regardless of the several limitations dealing with the performed time average, CITHEL is capable to calculate the P_{total} evolution at the second time-scale. Thereby, the contribution of the thermal effects due to the package and cooling system will be accounted for. The obtained mission profile can be used for reliability estimations, in spite of the simplicity of the used models, especially to infer P_{cond}.

Finally, the required current profile is analytically inferred from CITHEL results by using the measured I-V characteristic of the considered modules at 125°C. On the basis of a real route profile, Fig. 11 illustrates for a half of the service line, the derived current mission profile for both devices, IGBTs ((a), $I_{Losses,IGBT}$) and diodes ((b), $I_{Losses,Diode}$), and the speed profile (c). The total service line duration is 90 min (5400 s.). In this case, the speed profile has been selected to follow the theoretical maximum speed permitted along the service line.

6. Experimental results

Section 6 presents the main results derived from the approach detailed in the previous section. Subsection 6.1 shows how the proposed test bench allows the determination of the device with the highest thermal stress during real acceleration-braking cycles. Subsection 6.2 depicts that the most delaminated areas in the DBC-solder interface correspond to both the location of the most stressed components and the presence of a higher thickness of thermal interface material (non uniform distribution). Besides, this subsection correlates these results to those obtained with IGBT modules with a copper base plate aged with endurance cycling tests. Such tests have consisted in connecting two inverters in a back to back configuration (high voltage and current), which reproduces the real operation of power modules in PWM driving conditions.

6.1 Local thermal cycles measured on devices

From the followed experiments, several interesting results have been obtained at package and cooling system level. Figs. 12 (a) and (b) illustrate the inferred temperature mappings for each device between two consecutive stations at the end of the train acceleration and braking processes, respectively. This temperature distribution is representative of the real thermal mapping of the module for a time-scale in the second range. Therefore, a first experimental estimation of the thermal interaction of the module with its cooling system is extracted in the worst case (acceleration or braking cycles), which clearly evidences the temperature dispersion among the dies. Under these conditions, the maximum temperature differences are reached. Fig. 12 also evidences a clear vertical temperature gradient due to the chosen cooling system. At first sight, the temperature in the module M_3 is higher than in M_4 (around 10°C). This fact is a direct consequence of the air heating by M_4, when the heat exchange is produced within the thermosyphon condenser region. Moreover, the DCB 2 of phase B within module M_3 suffers the highest thermal stress during real acceleration-braking cycles.

Fig. 13 illustrates the inferred temperature evolution T_{IGBTs} (see Fig. 13 (a)) of four representative IGBTs in M_3 (see Fig. 13 (b)) during a part of the mission profile, considering the thermal interference with the diodes. Fig. 13 (a) shows 20°C temperature differences between dies from M_3. This fact could be attributed to two possible effects related to the cooling process. First, once the liquid has been heated up and the evaporation process starts, a pressure difference inside the heat pipe (at the evaporator) is induced, which in turn, introduces a temperature gradient within the heat pipe (Dunn & Reay, 1983). In the second case, the turning liquid (i.e., coolant) in the wick coming back from the condenser has a higher efficiency in cooling down the chips T_1 and T_3 (upper position) than T_5 and T_7, because of the liquid level inside the tube. As a consequence, chips T_1 and T_3 are cooled down by film evaporation process (the real heat pipe working principle); whereas the other ones follow a pool-boiling heat exchange mechanism (Dunn & Reay, 1983; Romestant, 2000).

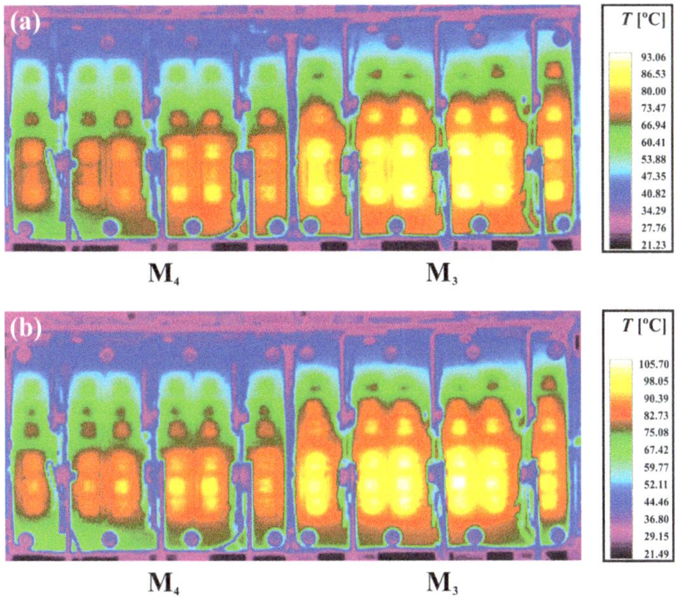

Fig. 12. Infrared thermal mappings of the power modules at the end of an acceleration (a) and braking (b) process shown in Fig. 11, considering the thermal interaction between diodes and IGBTs (not at the same colour scale).

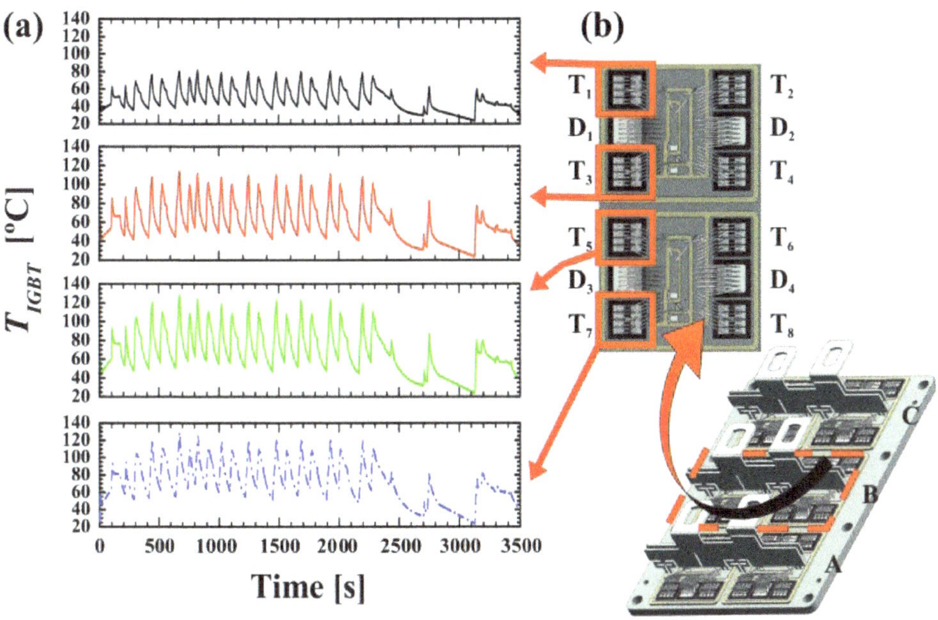

Fig. 13. (a) Junction temperature of four IGBTs (T_{IGBT}) in the same module taking into account the thermal interaction with diodes. (b) IGBTs under study.

Another remarkable fact observed in Fig. 13 (a) is the high temperature swing experienced by the chips (up to 65°C) due to the slow thermosyphon dynamic response. Such behaviour can be expected, since heat pipes must be heated up to be efficiently operative. This point is very important, since the module wear-out is very sensitive to the temperature swing (Ciappa, 2002).

In order to compare and summarise the temperature information shown in Fig. 13 (a), the thermal cycles experienced by each die have been counted by following the rainflow method and classified according to their temperature swing ΔT_{cycle} and the reached temperature peak T_{peak}. T_{peak} has been selected instead of T_m, because this variable provides us an idea of the maximum temperature reached by each IGBT. The ΔT_{cycle} class is divided into fourteen categories going from 5°C to >65°C in steps of 5°C, and T_{peak} is distributed into six sets going from 40°C to 140°C in steps of 20°C. Fig. 14 presents the thermal cycles experienced by the four selected IGBTs in M_3. The most (less) stressed set corresponds to the highest (lowest) T_{peak} and ΔT_{cycle} values. It can be observed from Fig. 14 that the thermal stress is distributed differently among chip T_1 and the other dies, which is in accordance to the remarks related to Fig. 13 (a). Fig. 14 also demonstrates that chip T_5 experiences the most stressing thermal cycles. Therefore, an asymmetrical wear-out effect on the module is expected to occur at long-term.

6.2 Effects of local thermal cycles on thermal grease and solder

Fig. 15 show the thermal grease distribution originated from the experienced thermal cycles, when the power modules M_3 and M_4 are removed. Fig. 15 (a) is obtained when the air flux goes from M_4 to M_3, whereas Fig. 15 (b) is originated from performing twice the mission profile changing at each time the air flux direction. Fig. 15 (a) presents that the coldest module (M_4) has attached more thermal grease than the hottest one (M_3). In the case of Fig. 15 (b), one can observe a similar grease distribution. This result highlights how from a certain temperature value, the thermal grease liquefies, improving the thermal contact of the module. By capillary effects, the thermal grease fills in the voids of the module-cooling system interface. However, this eventually leads to the thermal grease to be displaced observing that it has been partially or completely removed (thermal interface degradation). This indicates that a good measure for maintenance would be replacing the thermal grease after some working time.

Fig. 16 (a) depicts the DCB solder delamination observed in a module (M3 position), which comes from an endurance test performed on the analysed cooling system. In a back to back configuration by using two inverters, they have been operating to obtain the solder delamination observed in Fig. 16 (a). This process has been carried out on modules that have a copper base plate. The reason to select such modules is that a high degree of delamination in the solder DCB-base plate can be obtained with few thermal cycles (Ciappa, 2002). This result is compared with the solder delamination results obtained in a previous work (Khatir & Lebefvre, 2004), where classical power cycling tests were performed to induce the module wear-out (see Fig. 16 (b)). In that case, the module was cooled down by a water-pumping based cooling system. One may observe from both figures that the solder delamination patterns are different. Fig. 16 (a) presents a DCB substrate with a higher solder delamination in the zone where Fig. 12 shows higher temperature values during both acceleration and braking processes (DCB 2 corresponding to the phase B of M3). On the contrary, Fig. 16 (b) presents a higher solder delamination at the centre of the module, coinciding with the region where the module has the worst thermal contact to the cooling system (Perpiñà et al., 2007a).

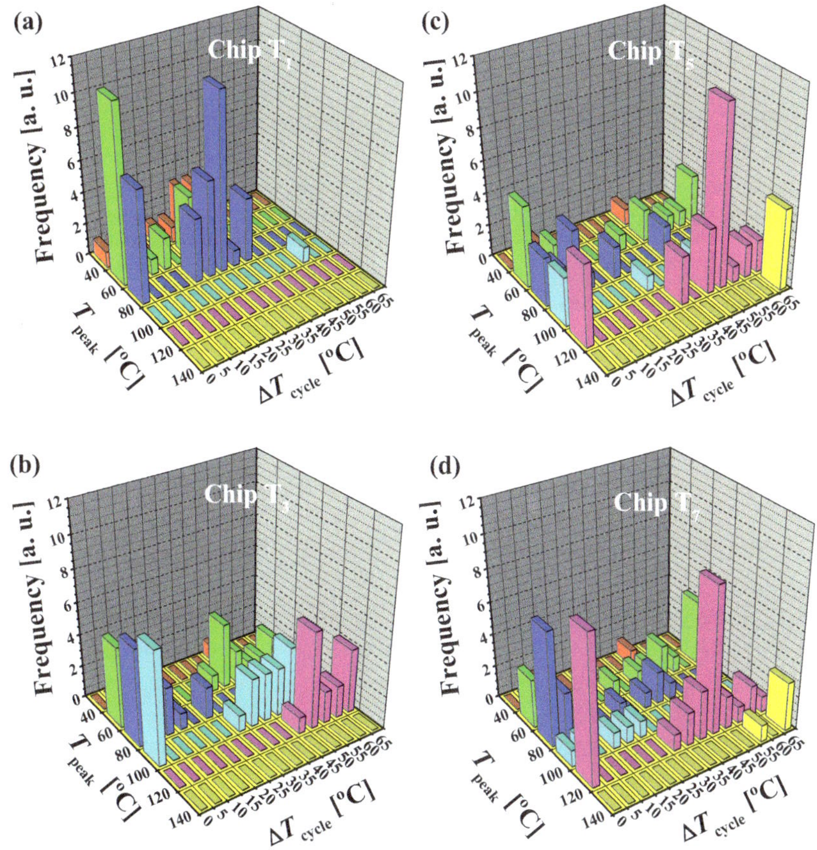

Fig. 14. Number of temperature cycles experienced by the four IGBTs shown in Fig. 13 (b), counted by means of the rainflow method: (a) Chip T_1, (b) Chip T_3, (c) Chip T_5, (d) Chip T_7.

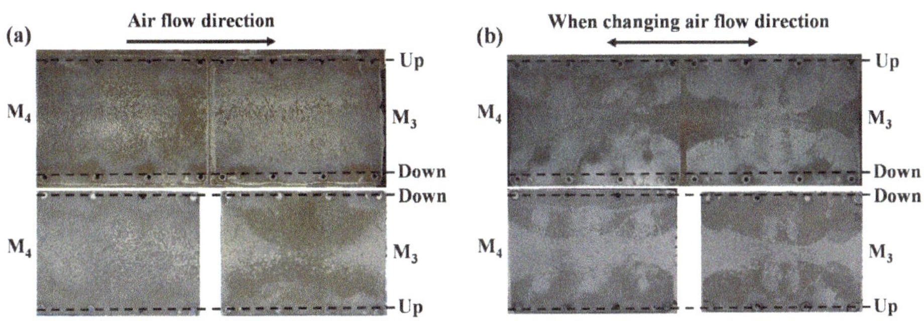

Fig. 15. Thermal grease distribution on both the modules and cooling system resulting from the local thermal cycles after a mission profile in two different conditions: (a) when a single mission profile is effectuated maintaining the same air flow direction, (b) when the mission profile is repeated twice and changing the air flow direction.

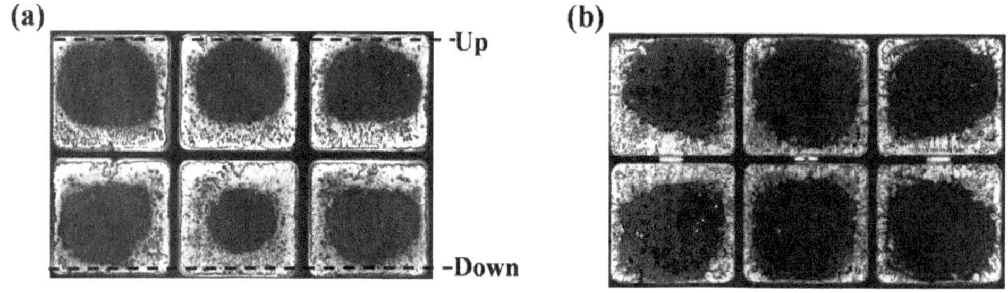

Fig. 16. Base plate-DCB solder delamination observed in IGBT modules with copper base plate, when cycled in a heat pipe (a) or water pumping-based (b) cooling systems.

7. Conclusions

Packaging wear-out of IGBT modules is one of the main limitating factors for ERTC long-term reliability. They are mainly due to the thermal cycles originated from the power devices working conditions across the package structure and the mismatch between its constitutive materials. This leads to crack initiation and propagation across the package interfaces or solder joints, which degrade the thermal performance of the package.

For this reason, the final user has developed a procedure to determine the IGBT module lifetime. After reviewing this methodology, it is evidenced that the results obtained are not representative of the failures observed from the field. In fact, they are representative for the manufacturer interest to improve its product, but not for the final user. Several times, the harsh environmental conditions or the real working conditions, both electrical and thermal, could induce a device failure before to the predictions given by only considering an ageing process. Therefore, it is crucial to link the failure criteria of Power Cycling Tests ($V_{CE,sat}$, $R_{th,j\text{-}c}$, leakage current) to real electrical failure mechanism. For this reason, new tests covering these misleading points should be carried out. They should consist in mounting power modules in inverter legs in a back to back configuration (high current and voltage) and driven by control boards (PWM pattern) modulating the package heating time in order to adjust ΔT_{vj} at 80°C and 60°C and considering the cooling conditions in real operation. Some comments have also been addressed to some misleading points on the methodology followed for thermal cycles superimposition. In this procedure, some approximations are taken into account: large cycles are not considered (day-night, season cycles) and temperature swing of the analysed cycles have a linear cumulative effect on the module ageing. Moreover, lifetime prediction reference curves should be based on physics based models rather than statistical approaches without fixing a standard procedure to perform PCT tests (e.g., fixing a heat up time or using a cooling system representative of the final application).

At the end of this chapter, the impact of the interaction between the power module and its cooling system on the inverter reliability is analysed. This study is mainly based on an air-forced convection heat pipe-based thermosyphon. The thermal mapping is experimentally determined and the non uniform temperature distribution inside the power module is justified by the special characteristics of the cooling system. Moreover, a comparison of the temperature dispersion between devices within the power module with a water-based cooling system is also provided. The local thermal cycles obtained from a real mission profile are measured, and their effects on solder delamination and thermal grease degradation are indicated. Finally, these results are compared with some features observed on failed power modules coming from the field, both being in agreement.

8. Acknowledgment

This work has been partially supported by the "Consejo Superior de Investigaciones Científicas" (CSIC) (under contract "Junta para la Ampliación de Estudios", JAE-Doc), the Spanish Ministry of Science and Innovation (Research Programs: THERMOS TEC2008-05577, RAMON Y CAJAL RyC-2010-07434, TRENCH-SIC TEC2011-22607) and the European Community with the project PORTES (POwer Reliability for Traction ElectronicS, MTKI-CT-2004-517224). Authors would like to thank M. Piton, M. Ciappa, O. Garonne, J.-P. Rochet, and P. Jalby for their enriching comments and support in the performed measurements.

9. References

Arnold, E.; Pein, H. & Herko, S. (1994). Comparison of self-heating effects in bulk-silicon and SOI high voltage devices, *Proceedings of International Electron Device Material (IEDM)*, San Francisco (USA), pp. 813-816, 1994.

Baliga, B.J. (1996). *Power Semiconductor Devices*, PWS Publishing Company, Boston, MA, 1996, ISBN: 0-534-94098-6.

Bannantine, J.; Comer, J. & Handrock, J. (1990). *Fundamentals of Metal Fatigue Analysis*, New Jersey, Prentice Hall, 1990, ISBN: 0-133-40191-X.

Baumann, H.; Heinemeyer, P.; Staiger, W.; Töpfer, M.; Unger, K. & Müller, D. (2001). Optimized cooling systems for high-power semiconductor devices, *IEEE Transactions on Industrial Electronics*, vol. 48, no. 2, pp. 298-306, April 2001, ISSN: 0278-0046.

Berg, H. & Wolfgang, E. (1998). Advanced IGBT Modules for Railway Traction Applications : Reliability Testing. *Microelectronics Reliability*, Vol. 38, pp. 1319-1323, 1998, ISSN: 0026-2714.

Bose, B.K. (2006). Power electronics and motor drives recent progress and perspective, *IEEE Transactions on Industrial Electronics*, vol. 56, no.2, pp. 581-588, April 2006, ISSN: 0278-0046.

Bouarroudj, M.; Z. Khatir, J-P Ousten, S. Lefebvre (2008). Temperature-level effect on solder lifetime during thermal cycling of power modules, *IEEE Transactions on Device and Materials Reliability*, vol. 8, no. 3, September 2008, ISSN: 1530-4388.

Bouscayrol, A.; Pietrzak-David, M.; Delarue, P.; Peña-Eguiluz, R.; Vidal, P-E. & Kestlyn, X. (2006). Weighted control of traction drives with parallel-connected AC machines, *IEEE Transactions on Industrial Electronics*, vol.53, no. 6, pp. 1799-1806, December 2006, ISSN: 0278-0046.

Ciappa, M. & Fichtner, W. (2000). Life-time Prediction of IGBT Modules for Traction Applications. *Proceedings of International Reliability Physics Symposium (IRPS)*, San Jose (California, USA), 2000.

Ciappa, M. (2002). Selected failure mechanisms of modern power modules, *Microelectronics Reliability*, vol. 42, no. 4-5, pp. 653-667, April/May 2002, ISSN: 0026-2714.

Ciappa, M.; Carbognani, F.; Cova, P. & Fichtner, W. (2003). Lifetime prediction and design of reliability tests for high-power devices in automotive applications", *Proceedings of International Reliability Physics Symposium (IRPS)*, Dallas (USA), pp. 523-528, 2003.

Ciappa, M.; Fichtner, W.; Kojima, T.; Yamada, Y. & Nishibe, Y. (2005). Extraction of Accurate Thermal Compact Models for Fast Electro-Thermal Simulation of IGBT Modules in Hybrid Electric Vehicles, *Microelectronics Reliability*, Vol. 4, issues 9-11, 2005, ISSN: 0026-2714.

CITHEL (n.d), User Manual.

Coquery, G.; Carubelli, S.; Ousten, J.P. & Lallemand, R. (2001). Power module lifetime estimation from chip temperature direct measurement in an automotive traction inverter, *Microelectronics Reliability*, vol. 41, no.9-10, pp. 1695-1700, September/October 2001, ISSN: 0026-2714.

Coquery, G.; Piton, M.; Lallemand, M.R.; Pagiusco, S. & Jeunesse, A. (2003). Thermal stresses on railways traction inverter IGBT modules: concept, methodology, results on sub-urban mass transit application to predictive maintenance, *Proceedings of European Power Electronics and Drives (EPE)*, Toulouse (France), 2003.

Cova, P.; Ciappa, M.; Franceschini, G., Malberti, P. & Fantini, F (1997). Thermal characterization of IGBT power modules. *Proceeding of Microelectronics and Reliability*, Vol. 37, Issues 10-11, pp. 1731–1734, 1997

Dunn, P.D. & Reay, D.A. (1983), *Heat Pipes*, Third Edition, Pergamon Press, Oxford, 1983. ISBN: 0-08041903-8.

Franke, T.; Zaiser, G.; Otto, J.; Honsberg-Riedl, M. & Sommer, R. (1999). Current and temperature distribution in multi-chip modules under inverter operation, *Proceedings of European Power Electronics and Drives (EPE)*, Lausanne (Switzerland), 1999.

Hamidi, A. (1998a). *Contribution à l'étude des phénomènes de fatigue thermique des modules IGBT de forte puissance destinés aux applications de traction*, PhD. Thesis, Institut National Polytechnique de Lorraine (Lorraine, France), 1998.

Hamidi, A.; Coquery, G.; Lallemand, R.; Vales, P. & Dorkel, J.M. (1998b). Temperature measurements and thermal modeling of high power IGBT multichip modules for reliability investigations in traction applications, *Microelectronics Reliability*, vol. 38, no. 6-8, pp. 1353-1359, June/August 1998, ISSN: 0026-2714.

Holmes, D.G. & Lipo, T.A. (2003). *Pulse Width Modulation for Power Converters: Principles and Practice*, IEEE press Series on Power Engineering. Piscataway, NJ:IEEE press, October 2003, ISBN: 978-0-471-20814-3.

Khargekar, A.K. & Kumar, P.P. (1994). A novel scheme for protection of power semiconductor devices against short circuit faults, *IEEE Transactions on Industrial Electronics*, vol. 41, no. 3, pp. 344-351, June 1994, ISSN: 0278-0046.

Khatir, Z. & Lefebvre, S. (2004). Boundary element analysis of thermal fatigue effects on high power IGBT modules, *Microelectronics Reliability*, vol. 44, no. 6, pp. 929-938, June 2004, ISSN: 0026-2714.

Kreuawan, S. (2008). *Modelling and optimal design in railway applications*, PhD thesis, École Centrale de Lille, Lille (France), 2008.

Lhommeau, T.; Perpiñà, X.; Martin, C.; Meuret, R.; Mermet-Guyennet, M. & Karama, M. (2007). Thermal Fatigue Effects on the Temperature Distribution inside IGBT

Modules for Zone Engine Aeronautical Applications, *Microelectronics Reliability*, vol. 47, no. 9-11, pp. 1779-1783, September/November 2007, ISSN: 0026-2714.

Lutz, J.; Schlangenotto, H.; Scheuermann, U. & De Doncker, R. (2011). *Semiconductor Power Devices*, Springer, Heidelberg (Germany), 2011, ISBN 978-3-642-11124-2.

Malagoni-Buiatti, G.; Martín-Ramos, J.A.; Rojas-García, C.H.; Amaral, A.M.R. & Marques Cardoso A.J. (2010). An on-line and non-invasive technique for the condition monitoring of capacitors in boost converters, *IEEE Transactions on Instrumentation and Measurement*, vol. 59, no. 8, August 2010, ISSN: 0018-9456.

Mermet-Guyennet, M.; Perpiñà, X. & Piton, M. (2007). Revisiting power cycling test for better lifetime prediction in traction, *Microelectronics Reliability*, vol. 47, no. 9-11, pp. 1690-1695, September/November 2007, ISSN: 0026-2714.

Mohan, N.; Undeland, T.M. & Robbins, W.P. (1995a). *Power Electronics: Converters, Applications and Design*, Second Edition, J. Wiley & Sons, Inc., April 1995, ISBN: 0-471-22693-1

Necati, M (1993). *Heat conduction*, 2nd edition, John Wiley and Sons, New York, USA, 1993, ISBN: 0-47153256-8.

Pecht, M. (1991). *Handbook of electronic package design*, Marcel Dekker, New York, 1991, ISBN: 0-8247-7921-5.

Perpinya, X.; Garonne, O.; Rochet, J.P.; Jalby, P.; Mermet-Guyennet, M. & Rebollo, J. (2007c). Experimental Analysis of Temperature Distribution within Traction IGBT Modules", *Proceedings of European Power Electronics and Drives (EPE)*, Aalborg (EUA), 2007.

Perpiñà, X.; Castellazzi, A.; Piton, M.; Mermet-Guyennet, M & Millán, J. (2007a). Failure-relevant abnormal events in power inverters considering measured IGBT module temperature inhomogeneities, *Microelectronics Reliability*, vol. 47, no. 9-11, pp. 1784-1785, September/November 2007, ISSN: 0026-2714.

Perpiñà, X.; Serviere, J.F.; Urresti-Ibañez, J.; Cortés, I.; Jordà, X.; Hidalgo, S.; Rebollo, J. & Mermet-Guyennet, M. (2010a). Analysis of Clamped Inductive Turn-off Failure in Railway Traction IGBT Power Modules under Overload Conditions, accepted in *IEEE Transactions on Industrial Electronics*, 2010, ISSN: 0278-0046.

Perpiñà, X., Mermet-Guyennet, M.; Jordà, X.; Vellvehi, M. & Rebollo, J. (2010b). Long-term reliability of railway power inverters cooled by heat pipe-based systems, submitted to *IEEE Transactions on Industrial Electronics*, 2010, ISSN: 0278-0046.

PORTES (2004), *Power Reliability for Traction Electronics* (PORTES), Marie Curie Transfer of Knowledge, MTKI-CT-2004-517224, EU funded project.

Rahimo, M.; Kopta, A., Schnell, R.; Schlapbach, U.; Zehringer, R. & Linder, S. (2004). 2.5kV-6.5kV Industry Standard IGBT Modules Setting a New Benchmark in SOA Capability, *Proceedings of Power Electronics/Intelligent Motion/Power Quality (PCIM)*, Nuremberg (Germany), 2004.

Romestant, C. (2000). *Etudes théoriques et expérimentales de caloducs et de thermosiphons soumis à de fortes accelerations*, PhD Thesis, Université de Poitiers, Poitiers (France), 2000.

Sankaran, V.A.; Chen, C.; Avant, C.S. & Xu, X. (1997). Power Cycling Reliability of IGBT Power Modules, *Proceedings of IAS*. New Orleans (USA), pp 1222-1227, 1997.

Sheng, W.W. & Colino, R.P. (2005). *Power electronic Modules, design and manufacture*. CRC press, ISBN: 0-203-50730-4.

Smet, V.; Forest, F.; Huselstein, J-J.; Richardeau, F.; Khatir, Z.; Lefebvre, S. & Berkani, M. (2011). Ageing and failure modes of IGBT modules in high-temperature power cycling, *IEEE Transactions on Industrial Electronics*, Vol. 58, no. 10, October 2011, ISSN: 0278-0046.

Steimel, A. (2004). Direct self-control and synchronous pulse techniques for high-power traction inverters in comparison, *IEEE Transactions on Industrial Electronics*, vol.51, no. 4, pp. 810-820, August 2004, ISSN: 0278-0046.

Xu, D.; Lu, H.; Huang, L.; Azuma, S., Kimata, M. & Uchida, R. (2002). Research on the power loss and junction temperature of power semiconductor devices, *IEEE Transactions on Industrial Applications*, vol. 38, no. 5, pp. 1426-1431, September/October 2002, ISSN: 0093-9994.

Yun, C-S.; Ciappa, M.; Malberti, P. & Fichtner, W. (2001). Thermal Component Model for Electrothermal Analysis of IGBT Module Systems", *IEEE Transactions of Advanced Packaging*, vol. 24, no. 3, pp. 401-406, August 2001, ISSN: 1521-3323.

Part 3

Parameters Monitoring in Railway Safety and Reliability

11

Lateral Resistance of Railway Track

Jabbar Ali Zakeri
Iran University of Science and Technology, School of Railway Engineering, Iran

1. Introduction

Increasing the speed limit of railway tracks is applicable by welding rail joints and employing Continues Welded Rails (CWR). Unfortunately, there are large numbers of curves with the radius less than 400 m in the most conventional railway tracks. Field investigations show that the lateral resistance of the railway track is not adequate for welding the rail joints in the mentioned railway tracks. In fact, elimination of rail joints causes huge longitudinal forces in the rails leading to the track lateral movement. There are several methods to increase the lateral resistance of the railway track such as employing winged sleeper, dual block sleepers, sleeper anchoring, frictional sleeper, Xi-track method and large sleeper [3, 18]. Various studies have been conducted to measure ballasted railway track characteristics [1, 10, 13, 14]. The main purpose of these studies is examining the possibility of track stability against longitudinal forces due to the changes in temperature. Former studies were usually conducted on ballasted tracks with conventional sleepers, and they often showed that enough lateral stability for the straight CWR tracks is applicable with reasonable dimensions in the ballast section. However, more arrangement is needed in curves according to these studies [11, 15].

The resistance between a simple mono-block concrete sleeper and the ballast bed mainly consists of three components: friction on both sides of the sleeper, passive pressure at both ends of the sleeper (shoulders) and friction at the bottom (figure 1). Components of the frictional resistance at the bottom of the sleeper are created with sliding the ballast particles on the relatively uniform surface of the concrete sleeper. The friction coefficient is measured to be around 0.5, while the internal friction of ballast particle is in the range of 0.9 to 1.4. [16]

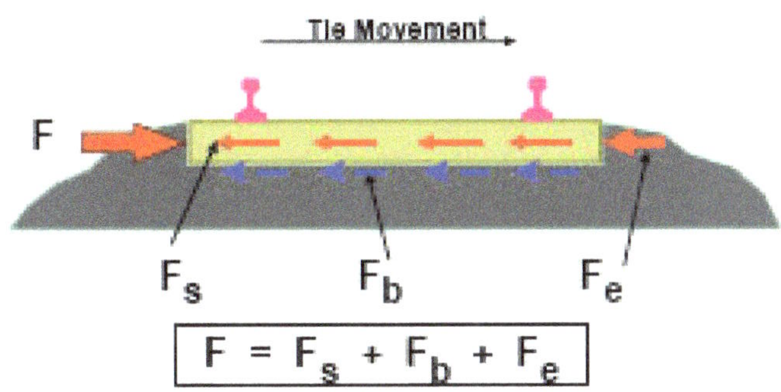

Fig. 1. Resistance between a simple mono-block concrete sleeper and the ballast bed

Near the rail joints of conventional railway tracks, different kinds of damages may occur: plastic deformation of head and sides of the rails, failure in sleepers and fasteners, ballast bed damage, lateral movement of railway track and etc. To reduce the enormous cost of maintenance and also take the advantage of the use of CWR, the track lateral resistant of must be increased.

2. Measuring methods of lateral resistance

Determination of lateral resistance is one of the key points for safety and stability of railway tracks, which is influenced by: sleeper type, weight and dimensions of the sleeper, intervals, ballast gradation, ballast stone quality, ballast depth in the crib and the shoulder height from the bottom of sleeper, ballast compaction, rail and fasteners type.

The resistance to lateral displacement can be measured by the following methods [13]:

- Single sleeper (Tie) Push Test (so called STPT)
- Panel displacement test
- Mechanical track displacement test
- Continuous dynamic measurements of lateral resistance

These methods measure the force versus sleeper or track panel displacement.

Generally, the STPT is a laboratory method to determine the lateral resistance of a sleeper. In this method, the displacement of sleeper is measured proportional to the applied force by applying the force to the sleeper. Often, this displacement is recorded up to 0.2 mm.

In order to implement this test, the four fasteners of a sleeper must firstly be removed and then a hydraulic jack must be installed to the shoulder fasteners. Consequently, the sleeper is pushed against the rail with the help of a hydraulic device by applying the force to the shoulder fastener. On the other hand, a LVDT should be mounted on the sleeper (such as a hydraulic cylinder connection) to measure the amount of sleeper displacement by moving the sleeper and gauge returning and to record the lateral displacement of a sleeper by the processor.

In Panel displacement Test method, a frame of 4-6 meters is placed on the foundation, and then after applying the force to the panel, the displacement is measured. Thus, in this test, the lateral resistance of track panel is measured.

In Mechanical track displacement Test method, additional equipments are attached to the tamping machine, in which the track lining and lifting cylinders provide the lateral and vertical forces, respectively. The lining value transducers also measure the displacements in tamping machines.

The track lateral resistance can be measured by employing The Dynamic Track Stabilizer (DTS) with the additional equipments during the operation [11].

In figure (2) test results are presented by Plasser and Theurer [1] for different speeds and frequencies. As can be seen, the lateral resistance of the track has been increased by employing the dynamic track stabilizer after the tamping.

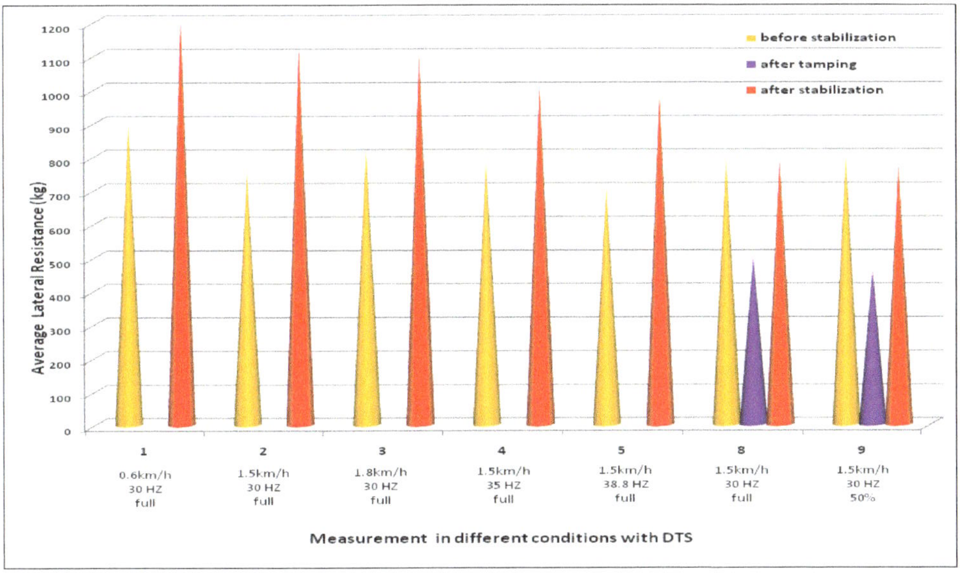

Fig. 2. Track lateral resistance, results measured by the DTS

Several studies have been carried out to measure the lateral resistance of the sleeper and railway track; among these studies the researches done by Plasser and Theurer [1] and the U.S department of transportation of federal railroad administration can be noted. The effect of maintenance involves the use of the following Tamping Machine, Dynamic Track Stabilizer and Ballast Regulator. [6, 7]

In this chapter, the frictional concrete sleepers and wing concrete sleepers will be introduced. It is important to note that the design and construction of wing concrete sleepers has been implemented before in China, but due to some executive problems it had no pervasive use in the actual railway tracks.

In this chapter, the results of tests conducted on the simple and frictional concrete sleepers [3, 9] in the actual railway track curve with radius of 250 meters have been presented and the impact of employing the frictional concrete sleeper has also been described.

This chapter also deals with the effect of vertical loads and running rolling stocks on track lateral resistance and on the change of stability in both frictional sleeper tracks (with ribbed bottom sleepers) and conventional sleeper tracks (with flat bottom sleepers). The results of the studies have indicated considerable increase in lateral resistance of track with frictional sleepers as well as track bearing under vertical loads.

3. Lateral resistance of ballast

Lateral resistance of the ballast is one of the most important factors to prevent the expansion and track buckling. According to the tests results, from the summation of the track lateral resistance it is known that the proportion of the ballast, rails and fastenings from the total lateral resistance are 65%, 35% and 10%, respectively.

Ballast section used in CWR tracks should be well constructed and compacted in accordance with standards.

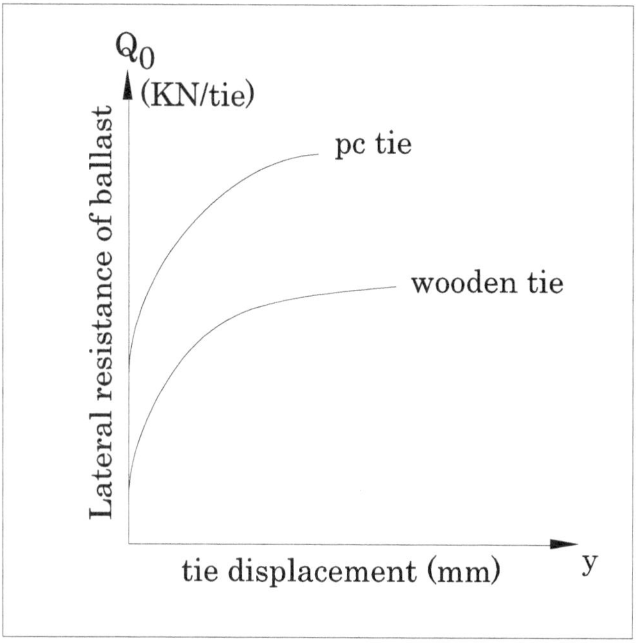

Fig. 3. The relationship between lateral resistances and displacement

In the lateral resistance of ballast tests, the lateral resistance of one sleeper is initially measured. The ballast resistance against the movement of a sleeper is shown by $Q_\circ$. Experiments show that the $Q_\circ$ has a nonlinear relationship with the lateral displacement of the sleeper.

As shown in figure (3), $Q_\circ$ initially increases with the increase of y. But when the lateral displacement reaches to a certain level, the resistance of ballast will not increase and remains at a constant level. If more displacement occurs, the ballast bed will cause to failure.

The amounts of $Q_\circ$ depends on the type, weight, size, shapes and the cross section dimensions of sleeper, the aggregation of ballast and annual passing traffic. The resistance of the ballast is usually expressed in terms of unit length of the track. If it represented by q then:

$$q = \frac{Q_0}{a}(N / cm) \quad (1)$$

where a is the sleeper spacing.

As it has been mentioned, the value of q could be determined by the lateral displacements of sleepers, and the relationship between them is as follows:

$$q = q_0 - c_1 y + c_2 y^n \tag{2}$$

where y is the lateral displacement of the track and $q_0, c_1, c_2, n < 1$ are parameters obtained from experiments[18]. For instance, after implementation of several tests in China, the following relationships were presented for the lateral resistance of the ballast:

- For wooden sleepers, crushed stone ballast:

Ballast shoulder, 30 cm

$$q = 12.2 - 201.9y + 255.8y^{2/3} \tag{3}$$

Ballast shoulder, 40 cm

$$q = 12.2 - 201.7y + 290.1y^{2/3} \tag{4}$$

- For concrete sleepers, crushed stone ballast:

Ballast shoulder, 30 cm

$$q = 13.7 - 388.1y + 511.6y^{\frac{2}{3}} \tag{5}$$

Ballast shoulder, 40 cm

$$q = 14.7 - 435.1y + 571.3y^{\frac{2}{3}} \tag{6}$$

It should be noted that the lateral resistance provided by sleeper is different in respect to types of sleepers and the dynamic effects resulted from the passing vehicles. This is due to existence of positive curvature in the region ahead of the train wheels on rails while there are simultaneously negative curvatures of the rails between two bogies. Therefore, the lateral resistance decreases because of the loss of contact between the sleeper and the ballast. Also the rigidity of track (EI) is a parameter which reduces the deformation of the flexible track which is necessary for preventing the lateral track buckling. This resistance comes from two sources: one is the lateral bending rigidity of rails and the other is the torsion rigidity of fasteners. Fastener torsion resistance is depending on the kind of sleepers and fasteners, toe load, and the relative rotation between the sleeper and the rails. The lateral rigidity of track can be obtained as follows:

$$EI = \beta EI_y \tag{7}$$

in which EI_y is the lateral rigidity of each rail.

In this function, $ß$ is the conversion ratio. In wooden sleepers, because of the high relative rotation between the sleeper and the rail, the torsion resistance of fasteners is considered to be zero. Thus, the value of $ß$ is considered as 2.

3.1 The resistance of ballast between the sleepers (Crib)

The Resistance, which prevents the displacement of sleepers in the longitudinal direction, is provided by the ballast and can be calculated based on each sleeper resistance or resistance per unit length of the track. The resistance of ballast depends on ballast gradation, ballast quality, tamping quality and the weight of track panel. The resistance of ballast will increase as the displacement of sleepers increases. But after a certain value of displacement, the ballast between sleeper yields and the resistance of ballast will not increase. Usually in the calculations, the resistance of ballast is taken into account when the displacement of sleeper to ballast is equal to 2 mm.

It should be noted that the resistance of ballast is not the same in all tracks. Therefore, field investigations should be done for the calculation of ballast resistance, if necessary. The track maintenance operation disrupts the arrangement of the particles of ballast and decreases the resistance of ballast, which will be gradually compensated by the load of passing vehicles.

In real conditions, the resistance of ballast is caused by the relative motion between the sleeper and ballast.

3.2 The resistance of ballast shoulders

The upper part of the ballast that is between the end of sleepers and beginning of ballast slope is known as the ballast shoulder. The obvious solution to prevent the lateral movement of sleepers during the buckling, is the increase of track lateral stiffness. A method to increase the track lateral stiffness is increasing the width of the ballast shoulder and the ballast between the sleepers. Another method is to increase the lateral stiffness of the rail-sleeper structure. Germany and Austria railways attempted to increase the ballast shoulder width to 0.35 m, as can be seen in figure (4). In the former Soviet Union railways, in freight tracks with wood and concrete sleepers, the shoulder width of 0.45 m have been proposed like figure (5). [19]

Fig. 4. Section of the different types of ballasted railway tracks in German (dimensions are in meters)

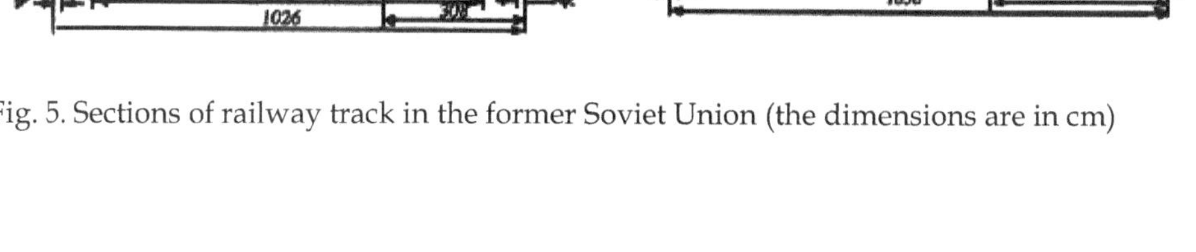

Fig. 5. Sections of railway track in the former Soviet Union (the dimensions are in cm)

In Japan, the width of ballast shoulders should be chosen at least 40 cm. In addition, ICE railways in Germany also use 50 cm wide shoulders. In 1970, the North American railways has been recommended to take the advantage of shoulders wider than 15.3 cm in straight tracks. In addition to reducing the risk of the track buckling, these shoulders reduce the deterioration of the track and maintenance costs. The reason refers to the movement of ballast particles under the load from the seat of rail to the areas of low stress. (Figure 6)

Fig. 6. Changing the location of ballast under the load of passing train

In this case, considering the wider shoulders create more resistance against the movement of particles. This theory was tested on the FAST tested track in the Colorado state. The results of the tests confirmed this theory. The main findings of the experiments are shown in Figure (7).

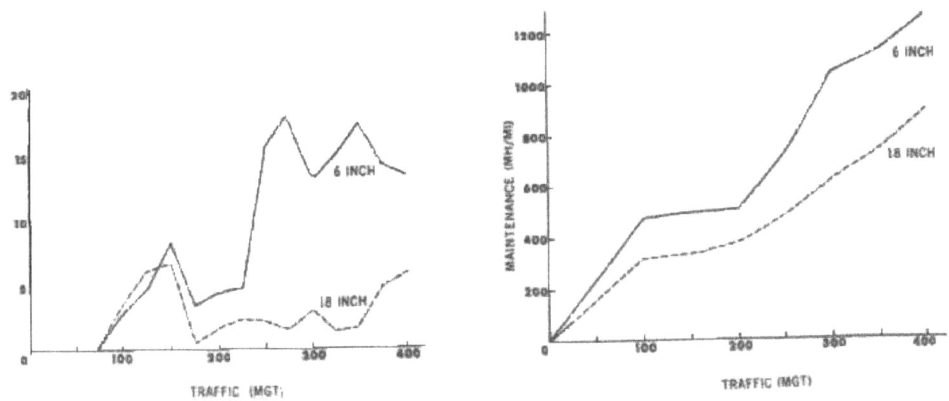

a) Trying the maintenance of the track against the cumulative traffic passing (MGT)
b) Change in the rail profiles to cumulative traffic passing (MGT)

Fig. 7. The results of the width of ballast shoulder on the FAST tested track

The results of these tests show that in addition to improve of the track safety against the buckling, the increase of ballast shoulder width is also effective in reducing the geometric deterioration of track and therefore, reduces the maintenance cost.

At the aim of reducing the buckling, German and Russian railways were recommended to use at least 35 cm ballast shoulder width in straight and curve tracks, but so far no action has been performed about shoulder width optimization to reduce the Superstructure deterioration.

Kind of sleeper	Ballast specification		Resistance per sleeper (KN)					
	Size (mm)	Density	Ballast shoulder :150mm		Ballast shoulder :230mm		Ballast shoulder :300mm	
			Total	Only ballast shoulder	Total	Only ballast shoulder	Total	Only ballast shoulder
Wood	38	Loose	2.12	0.1	2.46	0.44	2.71	0.69
	38	Dense	3.45	0.74	3.52	0.80	3.59	0.88
Concrete	38	Loose	3.41	0.15	3.72	0.46	4.39	1.13
	38	Dense	-	-	5.06	0.51	6.31	1.76
Concrete	75	Loose	-	-	3.24	0.13	3.48	0.10
	75	Dense	-	-	8.50	1.30	8.80	1.65

Table 1. The effect of increasing the ballast shoulder width on the lateral resistance of sleeper

4. The effect of operation and maintenance on the track stability

Maintenance operations, such as track laying, welding and other operations which are rising the rail neutral temperature will help to develop the potential of track buckling. However, stabilization of track after the maintenance operation increases the track safety factor against the buckling.

Tamping is a maintenance method that has an adversely affect on the lateral stability of the track because it distribute the ballast under the sleepers. The distribution of ballast leads to track vertical sinking and the lateral track defects. After tamping, often dynamic track stabilization method is used to re-dense the ballast.

The test was conducted using the Single Tie Push Test (STPT) to measure the variation in lateral resistance during a common track tamping operation. Lateral resistance was measured prior to tamping, after Tamping/before stabilization, and after stabilization. A sample of this device is shown in Figure (8).

Fig. 8. STPT Test Setup

The pre-tamping STPT measurements were made to characterize the condition of the track in operational condition [6]. Common characteristics of the pre-tamping tests are a steep initial slope, distinct peak, and a post-peak which decreases to an approximately constant value of lateral resistance for large deflections. (Figure 9)

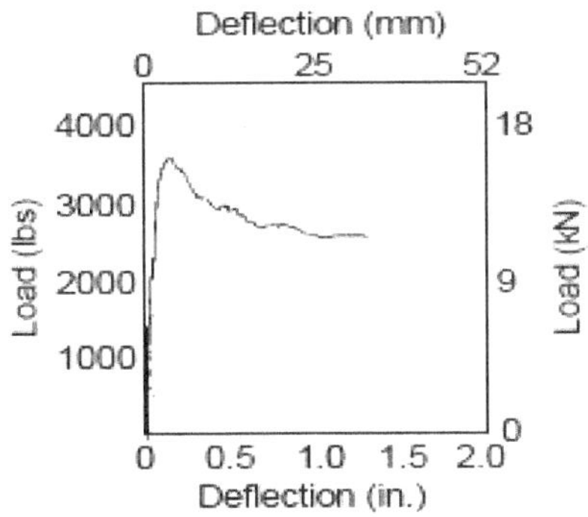

Fig. 9. Pre-tamping tests results

Figure (10) describes the measurements of STPT method after tamping and before stabilization. Specifications of these measurements are a gradual increase to a constant peak value. In this case, the lateral resistance of the conditions before tamping has been reduced by 43%.

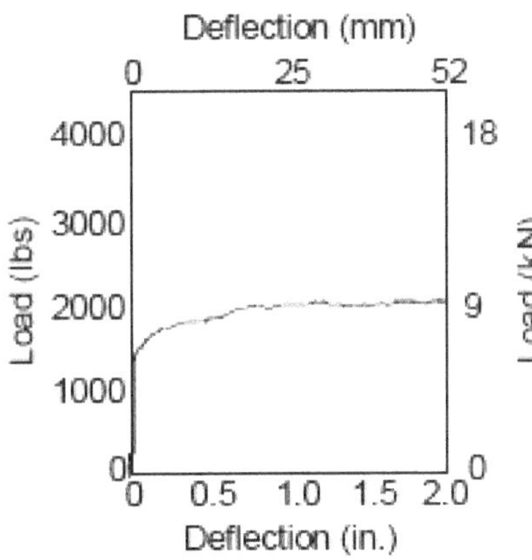

Fig. 10. Post-tamping and before stabilization tests results

The STPTs which has been conducted in the post-stabilization stage of the test, characterized the stability of the track after Tamping maintenance followed by the dynamic track stabilizer. This stage is just before subjecting the track to the revenue traffic. Common characteristics of the post-stabilization STPTs are a steeper initial increase than the pre-stabilization stage as shown in Figure 11. Stabilization produces an initial peak value similar to the trend of the data from the pre-tamping stage, but the peak value is significantly lower and less well defined than the pre-tamping stage peak. The increase in the initial slope (although less discernable) and thereafter, observance of a peak in load-deflection curve is consistent with behavior associated with the more dense, stronger and stiffer ballast. This represents an average increase in lateral resistance of approximately 31% over the post-tamping condition.

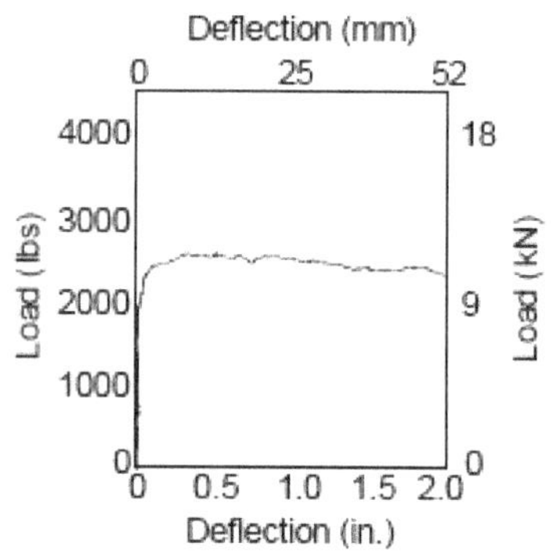

Fig. 11. Post-stabilization and before passing traffic results

5. The effect of elastic pads under sleepers on the lateral resistance of track

Now we describe the practical results of the effect of using under sleeper pads on the lateral resistance of track in Germany and Switzerland.[17]

5.1 Germany

One of the features of a modern ballast track is increasing the efficiency of track with improvements while introducing new components. Using elastic elements can be the best way to increase the track elasticity and damping. In the recent years, Germany Railways has tested the sleepers that are equipped with elastic pads under them. The results of field measurements, laboratory tests and theoretical evaluations have been very satisfactory. For example the maintenance period (tamping) has been increased.

Indeed, the substructure of ballasted railway tracks has two major problems: one is the difficulty contact of concrete sleeper with ballast, and the other is its rigid structure because of the new construction methods. These factors causing the deterioration of ballast, the speed limitation, and increasing the maintenance costs [20].

Using the elastic pads under sleepers has a satisfactory influence on the ballasted tracks, stability of tracks with longer durability, and lower life-cycle cost.

In recent years, in order to study the effect of elastic pads under sleepers on the lateral resistance of the track, field tests have been conducted on these type of sleepers in Germany [21].

After two years of track operation, lateral resistance of track has been measured for the conditions of with and without USP. The results are given in Table 2.

Under sleeper pads	non used	Stiff	SOFT
Stiffness (KN.mm)	-	70	30
Lateral resistance per rail (KN)	9.7	10.3	9.4

Table 2. Amount of the lateral resistant

It is concluded that the lateral resistance of both tracks with and without USP is the same, and does not change [17].

5.2 Switzerland

In early 2005, five types of USP in the Kysn test line, that were obtained from different plants, have been installed by the Switzerland Federal Railways headed by Mr. Schneider to assess the effects of elastic pads under sleepers on the railway tracks. USP specifications used are as follows:

USP	Brand	Manufacturer	measured Modulus (N/mm3)	
			Field	Laboratory
USP 1	SLB 3007G	Getzner/A	0.30	0.15
USP 2	M 01	Muller-RST/D	0.30	0.17
USP 3	S 01	Spreepolymer	0.35	0.38
USP 4	PRA	Sateba/F	0.30	0.14
USP 5	TR1-85M	Tiflex/UK	0.25	0.08

Table 3. USP specifications used in the Kysn test line

The length of each five track sections, where the USP has been installed, was 216m long. And the length of the sixth control section, where USP has not been installed, was 283m long. The test line was determined based on the UIC Classification system, with type of D4, with a maximum axial load of 22.5 ton. The objective of this test was evaluating the quality of tracks and increasing the tamping intervals.[22]

To measure the lateral resistance of each section, first, the fasteners were loosened to allow the concrete sleepers to have the lateral displacement and then replaced the rail pads with greased steel plate [14]. Second, the concrete sleepers were subjected to lateral load with a specially developed hydraulic device which is shown in Figure 12. Third, the force between the hydraulic cylinder and the rail was measured with minimal friction losses. Then, the amount of rail displacement has been recorded towards a reference fixed point on the front rail. In each section, four sleepers were measured in each part: with and without the USP.

Fig. 12. The Device that applied the lateral force to the track

As shown in Figure (13), the measurement of lateral resistance indicates that the lateral resistance against the displacement of sleepers with USP is less than those recorded from sleepers without USP. This means that the track lateral resistance with sleepers which are equipped with USP is less than the sleepers that are not equipped with USP. The Softest USP (USP 5), the lowest lateral resistance.[20]

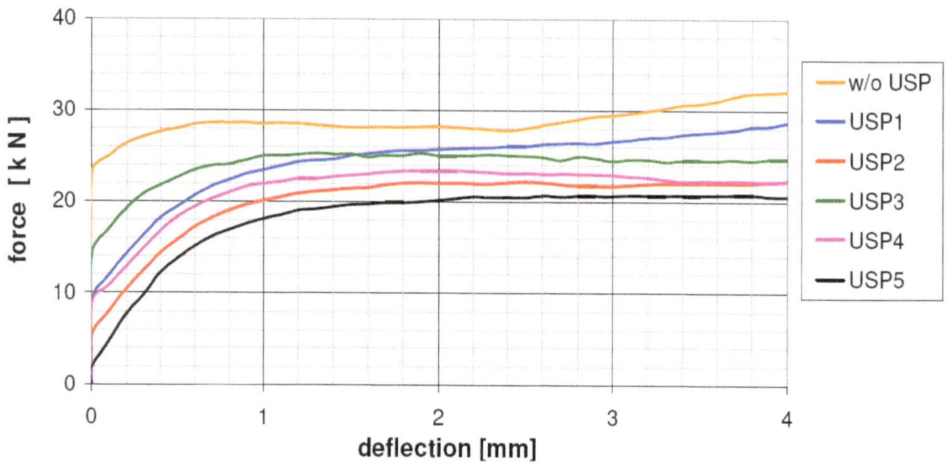

Fig. 13. Results of track lateral resistance at different sections

It should be noted that these results can be unrealistic, since the vertical axial load on the system has not been taken into account and also because of the test method. It is obvious that when a train passes, the vertical component of the load will increase, and consequently the amount of friction force will differ. Therefore, the role of pad below the sleeper in lateral resistance become important. This does not match with the resistance measured by STPT method.

Therefore, it can be said that the lateral resistance of the track equipped with USP is the same as the track without USP.

The results of the field investigations in SATEBA company under the supervision of International Union of Railways in 2006 indicates that using the elastic pads under the sleepers increases the lateral resistance of track approximately by 9%.

Laboratory investigations on the models which were built at the University of Zagreb in Croatia indicates that using the elastic pads under the sleepers increases the lateral resistance of track approximately by 9% (same as SATEBA results).

Further researches with the aim of studying the impact of these components on the lateral resistance of track have not been conducted yet. Other researches have implicitly reported the results in this aspect, and because of using various USP, the results of studies are different. In many cases, using the elastic pads under the sleepers increases the lateral resistance and in some other cases decreases the lateral resistance or has no impact.

Based on investigations conducted by the International Union of Railways and other researches on the track lateral resistance, it is demonstrated that the less hard pads under the sleepers, the less lateral resistance of railway track.

In all experiments, only STPT (Single Tie Push Test) has been carried out without any loading which may affect the results of these experiments.

The environmental temperature has a direct impact on the vertical hardness of the plate under the sleeper. Therefore, the selection of plate will be directly influenced by environmental conditions.

With regard to the positive effect of using USP in the reduction of environmental vibrations, it can be concluded that the existence of these pads (with the appropriate choice) will not create a problem for the track lateral stability.

6. Frictional concrete sleeper

Ballast is a layer of crushed stone particles with a diameter in range of 20 to 60 mm where the track panel including the set of sleepers, fasteners and rails are layed over. A frictional concrete sleeper has been designed to increase the involvement of ballast stone particles and sleeper base, the coefficient of friction beneath the base of sleeper, and to increase the lateral resistance of the track. As it shown in Figures (14) and (3), all parts of the frictional concrete sleeper is like a special mono-block concrete sleeper (B70) except that its beneath surface is trapezoidally- shaped. The dimensions of the height and the two small and large bases of the trapezoidal are 2.5cm, 6cm and 10 cm, respectively. It should be noted that the trapezoidal shape of the sleeper base has no effect on the maintenance operations [23].

Fig. 14. The bottom of frictional concrete sleeper

6.1 Field tests to determine the lateral resistance of railway track

Field tests were conducted on the curve of an actual railway track with a radius of 250m. First, the lateral resistance was measured on a track with B70 mono-block concrete sleepers. Then, the B70 mono-block concrete sleepers were replaced by the frictional concrete sleepers (B70-F) and the lateral resistance of the track was measured again. In the test process, the following conditions were maintained the same: 1) Weather conditions, and 2) Physical conditions of the railway track.

In order to maintain the same weather conditions, both tests (with simple concrete sleeper and frictional concrete sleeper) were conducted in equal atmospheric situations.

In order to maintain the same physical conditions, first the track have been tampd and then the Dynamic Track Stabilizer (DTS) has stabilized the track in both tests. Furthermore, other same track characteristics were: the ballast shoulder width, the slope of ballast layer on both sides, the depth of ballast in the crib zone, and the space between sleepers.

To carry out the field investigation, the panel displacement method was employed using a hydraulic jack. In this test, the lateral displacement was measured by the LVDT which was installed on pedestals along with the applied lateral force. As shown in Figure (15), for more establishments, the hydraulic jack has been installed with the buffer (to the inner rail).

Fig. 15. Installation of the hydraulic jack

Also, the jack's tip was perpendicular to the rail to insure a perpendicular lateral load. These procedures were the preliminary preparations to conduct the test. The test has been started by applying the lateral force to the inner rail at t=0. The time also has been recorded during the test. Simultaneously, the LVDT has recorded the displacements due to lateral force applied by the hydraulic jack. Therefore, displacements could be saved versus time.

In the second stage, the tests were repeated after the replacing simple concrete sleepers with frictional concrete sleepers. Finally, the data obtained from both test were compared.

Note that the effect of the vertical load was disregarded in this test.

The procedure of replacing the conventional railway track (with simple concrete sleepers) with the new railway track (with frictional concrete sleepers) is clearly shown in Figures (16) and (17).

Investigating the effect of frictional concrete sleeper on the track lateral resistance in curves, the results obtained from each stage are explained, interpreted and compared in the next subsection.

Fig. 16. Installation of railway track prepared with B70-F sleeper by employing 120- tons railway crane

Fig. 17. Tamping and stabilizing operations after the installation of the railway track with B70-F sleepers (before the second stage of test)

6.2 Results

In conventional railway tracks, there are many sharp curves with radius less than 400m. Therefore, it is impossible to remove all rail joints by welding due to the lack of lateral resistance. Investigations show that lots of damages occur near rail joints such as failure in fasteners, plastic deformation of head and sides of the rails, failure in sleepers, damage of ballast bed, lateral movement of railway track etc.

In this research, regarding the effect of using frictional sleepers on the lateral resistance of actual curves with radius of 250m, the use of the frictional sleepers instead of conventional sleepers has been recommended.

In summary, the test described in this research was conducted on a curve of an actual railway track with radius of 250m. In the first stage, the lateral resistance of a track with B70 mono-block concrete sleepers was first measured. In the second stage, the B70 mono-block concrete sleepers were replaced by the frictional concrete sleepers and the lateral resistance of the track was measured again.

In general, the test results show that the amount of the force which was necessary to displace the track with the frictional concrete sleepers at a specified value, is about 1.67 times greater than the force applied in a conventional track.

In conclusion, by employing the frictional concrete sleepers, the lateral resistance of a railway track will increase approximately by 67%. Thus, to insure the lateral stability of sharp curves (curve radius between 250m and 400m) of railway tracks, the use of frictional sleeper is recommended.

7. References

[1] Plasser & Theurer Publications, *The Lateral Resistance of the Track*, Technical Report, 2007.

[2] Nordal, S.R & Løhren, A.H.: Concrete Friction Sleeper for Increased Lateral Track Resistance, International Conferences on railway engineering, London, Uk, 2003. .

[3] Fakhari, M. (2009) "Laboratory investigation of Frictional sleepers effects on Lateral Resistance of Railway Track" BSc. Thesis, Iran University of Science and Technology, School of Railway Engineering, 2009.

[4] Esveld, C. (2001) Modern Railway Track, Second edition, TU-Delft.

[5] Selig, T.E &. Waters, J.M.(1994) "Track Geotechnology And Substructure Management", Thomas Relford Publications,1994.

[6] U.S. Department Of Transportation Federal Railroad Adminstration, The Influence Of Track Maintenance On The Lateral Resistance Of Concrete Tie Track, 2003.

[7] Zakeri J. A. & Rezazadeh M.(2007) "Railway Track Maintenance Methods" Iran university of science and Technology press, 2007. (in persian)

[8] KS-625N"Sleeper Holding Test Recorder" Instruction Manual, KANEKO Corporation

[9] Zakeri, J.A. Mir fattahi, B. Fakhari, M. Field and laboratory Investigation on the lateral resistance of sleepers by employing STPT test. CD proceding. First International Conferance on Road and Rail Infrastructure, Croatia, 2010

[10] "Measurement of lateral resistance characteristics for ballast track." (1998) Report ERRI D 202/DT 361, Utrecht, Netherlands.

[11] Miyai, T., et al. (1983). "Lateral resistance of operating line". Quarterly Report of RTRI, No. 4.

[12] Nam-Hyoung, Lim, Nam-Hoi, Park, Young-Joung, Kang (2003). "stability of continuous welded rail track." Computers and Structures, Vol. 81.

[13] Lichtberger, B. (2007). " The lateral resistance of the track." European Railway Review, Issue 3 & 4.

[14] "Measurements of lateral resistance, longitudinal resistance and change of neutral temperature (NRT) of ballasted track." (1997) Final Report, Vol. 2, ERRI D 202/DT 361, Utrecht, Netherlands.

[15] " Theory of CWR track stability." (1995) Report ERRI, D 202/PR3, Utrecht, Netherlands.

[16] Barati, M. (2010) " Effect of vertical loads on increasing of lateral resistance of the track." M.Sc. Seminar, Iran University of Science and Technology.

[17] Shahrokhi-nasab, E (2011) "study on effects of under sleeper pads on lateral resistance of sleepers" M.Sc. Seminar, Iran University of Science and Technology.

[18] Fattollahzadeh, M. (2004) " stability of CWR track" B.Sc. Thesis, Iran University of Science and Technology.

[19] Kerr , Arnold (2003) fundamentals of railway track Engineering, Simmons Boardman Pub Co

[20] International Union of Railways, (۲۰۰۵) "UIC Project, Under Ballast Mats" , UIC Project

[21] Bolmsvik, R. (2005)"Influence of USP on track response – a literature survey",Abetong Teknik ABVäxjö, Sweden, pp. ۱۳

[22] Sateba Company, "Sateba Expreince in the Field of Under Sleeper Pads",Presentation to STC (۲۰۰٤)

[23] Zakeri J. A., Mirfattahi B., Fakhari M. (2012) "Lateral Resistance of Railway Track with Frictional Sleepers" *Proceedings of the Institution of Civil Engineers - Transport* journal (under press).

[24] Mirfattahi, B. 2009. Field Investigation on Lateral Resistance of railway track with frictional sleepers. M.Sc. Thesis, School of Railway Engineering, Iran University of Science and Technology, Tehran.

12

Multifunction Portals for Train Monitoring: Recent Advances and Innovative Optoelectronic Instrumentation

Luca Fumagalli[1], Paolo Tomassini[1], Marco Zanatta[1], Giorgio Libretti[1], Marco Trebeschi[1], Giovanna Sansoni[2] and Franco Docchio[2]
[1]*Q-Tech Srl, Rezzato (BS),*
[2]*Laboratory of Optoelectronics, University of Brescia, Brescia, Italy*

1. Introduction

Railway transportation is the ideal infrastructure for remote control and high-tech monitoring solutions. After the completion of the remote control operation of the italian railway system, in fact, the need arose of monitoring equipment that could detect the status of the trains travelling along the tracks, and could transfer the status information to the control centres, for a prompt response in case abnormalities were reported (train blocking, signalling). This is of primary importance for Italy, full of mountains and therefore of tunnels crossing them. The need to control train integrity well before the tunnel input is motivated by the need of arresting the convoy on time in presence of (i) misalignments of carriage loads, and (ii) abnormal temperature rises in any part of the engine or of the carriage. This need has motivated the national railway Company *Ferrovie dello Stato* to launch a project aimed at installing a number of train monitoring multifunction portals at suitable distances from the tunnel entrances. The concept of multifunction portal is related to the presence of a number of concurrent sensors, each performing a separate task, and integrated in a common software platform that elaborates the sensor information and issues the train status information and possible alarms.

The first monitoring portal put in exercise in the middle 2000's is the portal of Exilles, on the Turin-Modane Railway line. This portal was implemented after the tragic Montblanc tunnel motorway accident. The portal was equipped with thermal cameras to monitor temperature abnormalities. In 2005 the Company launched a first experimental multifunction portal, that was supposed to combine 3D shape, thermal distribution and visual profile, to carriage length and type information. The portal had to be located to monitor one of the tracks at the entrance of the Station of Rastignano, along the Firenze - Prato (Tuscany) Railway line. Sirti S.p.A. was encharged of the realization of the portal. A partnership was then formed, which included our company, Q-Tech s.r.l., and the companion Laboratory of Optoelectronics of the University of Brescia. Q-Tech s.r.l. was in charge of the development of the 3D-profile, visual and thermal sensors for the monitoring portal, together with all the software for the control and synchronization of the sensors and the setting of the alarms to be sent along the line. To this aim, it collaborated with the parent Laboratory of Optoelectronics, which focused on the development of the software for data acquisition, elaboration and alarm generation. The portal was set in operation in 2007.

In the same year a second project was launched by the Ferrovie dello Stato, for the development of two second-generation, two-track monitoring, multifunction portals, to be installed at the two sides of a three-tunnel series on the high-velocity Rome-Formia-Neaples Railway line. One of the two portals has been assigned to the Sirti-Q-Tech partnership (the one Located at Minturno, south side of the monitored line), the other to Ansaldo S.p.A. (located in Sezze Romano, north side). This was a more ambitious project, since the high-velocity nature of the line required more stringent performance specifications of the sensors and the data collection and elaboration procedure. Both assignees completed their work in 2010, and the two portals are now operative.

The two portals developed by Sirti and Q-Tech are significantly different, due to the above mentioned requirements. The major difference is in the 3D shape sensors which, in the first portal, have been purchased from a German Research Institute, and in the second have been completely in-house designed, developed and tested by Q-Tech. This is due to the fact that the previous sensors had a number of drawbacks that were inherently due to the measurement principle (phase difference telemetry). Q-Tech's solution for the second portal is based on pulsed time-of-flight telemetry, which allows a better signal-to-noise ratio in the visible, and results in much better defined 3D profiles. The avaliability of high repetition rate lasers allowed, in addition, to obtain four 3D shape measurement sensors with only one laser source, multiplexed in emission and detection, and making use of fibre optics to reach the scanning mirrors. This resulted in much lower sensor costs, better maintainability, and safety of operation.

This chapter is devoted to the description of the work done by Q-Tech in the Sirti/Q-Tech partnership, with reference to (i) the optoelectronic instrumentation in multifunction portals, and (ii) the innovative optoelectronic 3D scanner based on time-of-flight telemetry, for the high speed monitoring of the train profile in three dimensions and with a multiplexed geometry.

2. Optoelectronic instrumentation in multifunction portals

2.1 Architecture of a portal

In a Railway Multifunction Portal a number of elaboration systems acquire, combine and elaborate data coming from a number of sensors, and produce suitable alarms in case of non-conformities of the train. A photograph of a Railway Multifunction Portal is shown in Fig. 1.

Fig. 2 shows the overall system architecture of a portal. The project and the development of the systems shown on the left side of this figure is the core of Q-Tech know-how; these systems will be described in the next paragraphs of this chapter. In particular we recognize:

- 3D shape monitoring systems, making use of the innovative telemeters described in the next section: they acquire the train profiles and compare them to reference profiles, issuing alarms when the measured profiles exceed the reference ones;

Fig. 1. Picture of a train-monitoring Portal. Sensors in positions A and C monitor the left track, whereas sensors in positions B and D monitor the right track; sensors in positions E and F perform the monitoring of the lower part of the train.

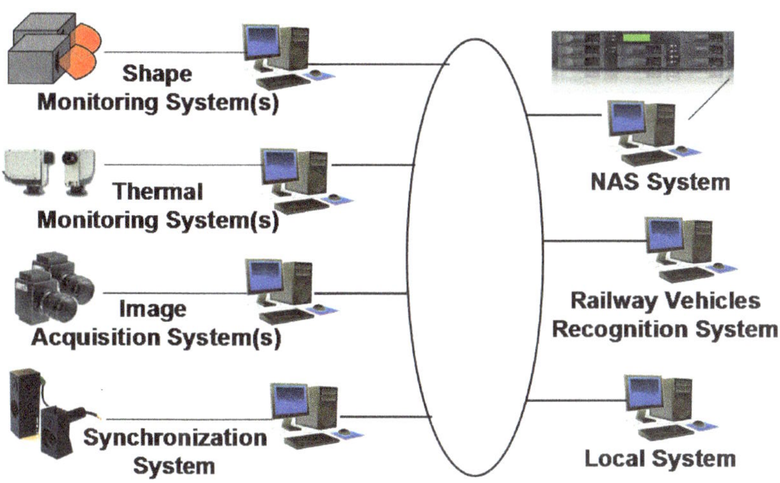

Fig. 2. Overall architecture of a Railway Multifunction Portal

- thermal monitoring systems: they acquire the thermal maps of the trains and compare them to nominal operating temperatures, which are different according to the part of the carriage considered; the aim is to issue alarms in case of risk of fire caused by abnormally high temperatures;
- image acquisition systems, providing visual images of the train by means of high-speed linear cameras;
- a synchronization system: its purpose is to set a reference to all the data acquired by the other systems to the actual configuration of the train in transit; this is achieved by detecting: (i) the transit of the axles under the portal as detected by suitable sensors, (ii) the discontinuities between different carriages, and (iii) the presence of a train on the opposte track.

These systems interact with the high-level systems (not developed by Q-Tech) as depicted in the right part of Fig. 2:

- a NAS system is a high-capability storage unit intended to save the data resulting from the elaboration of all measurements;
- a railway vehicle recognition system, which allows the identification of the train by recognizing the number and the type of carriages that compose it;
- a local system, hosting the user interface which allows the operator to view data and alarms and to configure the parameters for each system.

The measurement and elaboration units are industrial PCs contained in an isolated shelter situated in proximity to the portal (see the right side of Fig. 1); the elaboration units are connected to their measuring sensors.

Issues concerning fast data elaboration and data handling were particularly critical in this project, given the huge amount of data acquired from all the sensors: in fact the systems are required to monitor trains running up to 300 Km/h, and any alarm must be raised within 90 seconds after the end of the transit.

2.2 The image acquisition system

The train image acquisition is performed by using line-scan cameras operating in the visible. Their purpose is to visualise the train in transit. More systems can be installed on a Portal, in order to acquire both the upper part of the trains (e.g. positions A, B, C, D in Fig. 1) and their lower part (e.g. positions E, F in Fig. 1). The frequency of the scan is calculated according to the train speed, to keep the spacing (in mm) between two successive scans to a constant value.

Two illuminators, driven by crepuscular sensors, start operating at sunset. The exposure time is set according to the track illumination conditions in order to avoid over- and underexposures.

Fig. 3 (a) and (b) show details of the visual images of the left side of a train in transit, acquired by the upper and the lower image acquisition system respectively.

2.3 The thermal monitoring system

Thermal monitoring is performed by using commercial pyroelectric line cameras. More systems can be installed on a Portal, in order to acquire both the upper part of the trains (e.g. positions A, B, C, D in Fig. 1) and their lower part (e.g. positions E, F in Fig. 1). In particular, to monitor the lower part of the train, the cameras can be installed either on both sides of the portal, or within a hollow crosspiece situated under the track.

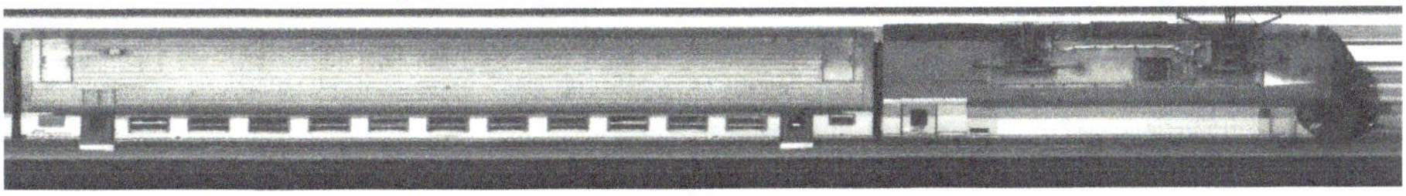

(a)

(b)

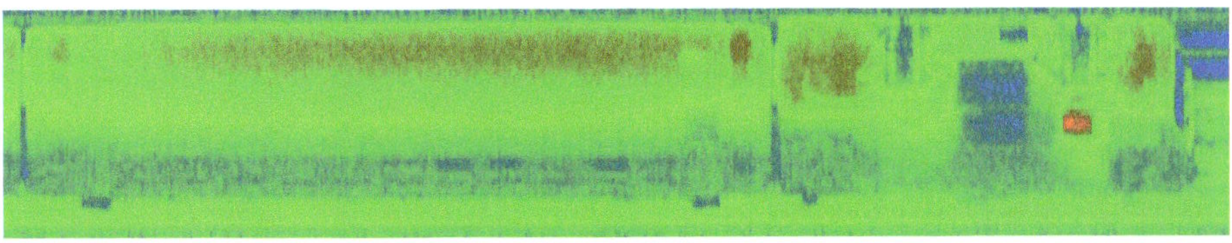

(c)

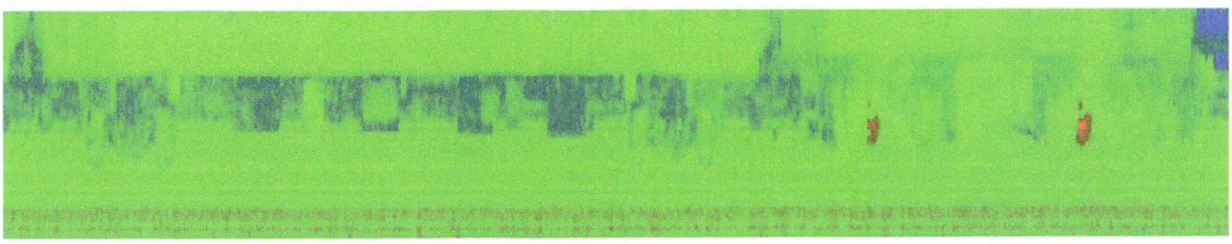

(d)

Fig. 3. Sample acquisition of the left side of a train (only the locomotive and the first passenger car are shown): (a) visual image, upper part; (b) visual image, lower part; (c) thermal map, upper part; (d) thermal map, lower part.

Each sensor acquires a thermal map of the train in transit; starting from the acquired data, up to four different thermal inspections can be performed:

1. "Fire on board" inspection: comparison with a maximum reference temperature;
2. "Maximum carriage temperature" inspection: comparison with a reference temperature depending on the type of carriage;
3. "Regions of Interest (ROI) grid inspection": each railway vehicle is split into a grid of rectangular regions: each region is characterized by a specific reference temperature; this inspection has the purpose to detect over-threshold situations in different parts of each carriage, each characterized by its own operating temperature and alarm threshold;
4. "Axle boxes" inspection: the position of each axle box is detected, to launch a warning in case of overheat.

Fig. 3 (c) and (d) show a detail of the thermal maps of the left side of a train in transit, acquired by the upper and the lower thermal acquisition system respectively.

Fig. 4 shows the thermal acquisition of the lower part of a locomotive (left side) and a passenger car (right side) of a train at the speed of 153 Km/h: the carriage wheels are evident in the map as circles. The lower part of Fig. 4 shows a detail of the corresponding visual image of the wheels acquired by the lower image acquisition system. The situation of the four axles of the passenger car (right side of Fig. 4) is normal, whereas the green-coloured portions in correspondence to the four axles of the locomotive (left side of the Figure) indicate clearly that the wheels are hotter, presumably because the carriage brakes are active.

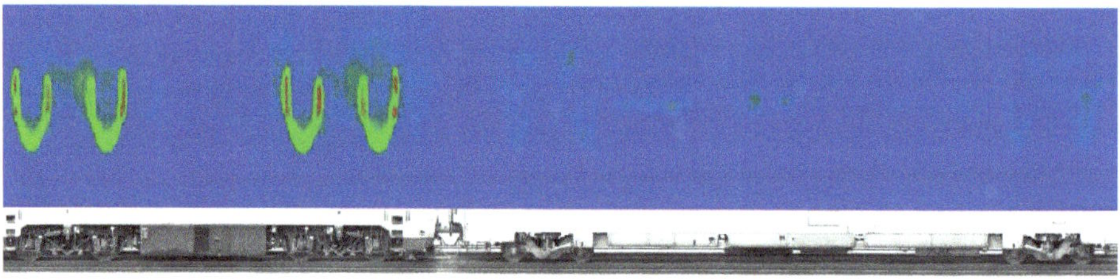

Fig. 4. Example of thermal acquisition of the lower part of a train (thermal map, upper, and visual image, lower).

2.4 The synchronization system

The multifunction portal is based on the cooperation among different acquisition and elaboration units (Fig. 5), which operate separately but must (i) refer their data to a common time base, and (ii) trace the elaboration flux and outline potential malfunction situations. In fact, the purpose of each system is to acquire all the data pertaining to a transit from all the portal sensors, setting its own internal resolution and synchronizing its information through control signals that allow the establishment of a common time base for all the systems. Moreover, each system must provide the data according to suitable formats, useful for the analysis of the convoy at the local level (verification of the presence of alarms) and at the remote level (visualization of the overall train situation).

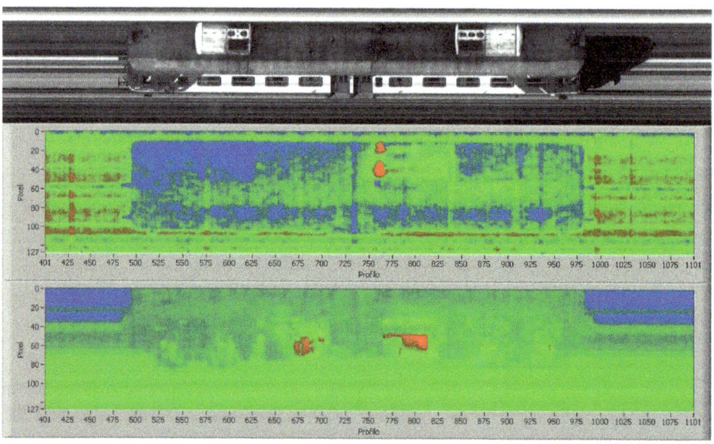

Fig. 5. Example of the acquisition of the left side of a locomotive concurrently performed by the upper image acquisition system, and the upper and lower thermal monitoring systems.

The common time base to the units is provided by a synchronization system, developed by Q-Tech; this system performs the acquisition and the elaboration of signals coming from sensors that detect the transit of each wheel (the so-called "pedals") at known distances. Moreover the synchronization system makes use of a discontinuity sensor to detect the instant of beginning and end of each carriage. The synchronization system has also the task of handling the presence of two trains concurrently under the portal, as shown in Fig. 6.

Fig. 6. Example of a concurrent transit.

3. New optoelectronic developments: The 400-kHz distributed time-of-flight telemeter

After the first portal in Rastignano it was decided to carry out an in-house development of a time-of-flight distributed telemeter unit, also due to the fact that four scanning heads were required, to scan two railway tracks. This distributed telemeter uses a single transmitter/detector unit. This unit is well shielded from the trains and from the environment inside the control shelter of the portal, and feeds, in a multiplexed way, the required number of scanning stations through transmission/detection optical fibre pairs.

3.1 Description of the telemeter

The system operates at 400 kHz, with an extended dynamic range to solve the problem of measuring distance of targets in a variable range of distances and angles. The use of a sub-nanosecond risetime pulse allows to obtain accuracies well below the centimeter at distances of some meters. The use of fibre pairs for transmission and detection allows to place the remote measuring stations at distances of 50-100 m from the main unit.

Fig. 7 shows the overall architecture of the system.

The distributed telemeter is composed of four measuring stations, namely 1l, 1r, 2l and 2r. The architecture of the system is rather flexible, to adapt to different measuring requirements (higher number of measuring stations, different timing, etc.). A single control unit contains the transmitter laser source with all the optical components for the beam shaping, and the receiver detectors and electronics for the measurement of the optical signals from the targets. Each remote station is connected to the central control unit by means of (i) a transmitter fiber carrying the signal from the transmitter, (ii) a receiver fiber carrying the signal from the target, and (iii) an adequate number of control signals.

Fig. 8 shows the details of the transmitter and receiver optical units of the telemeter setup.

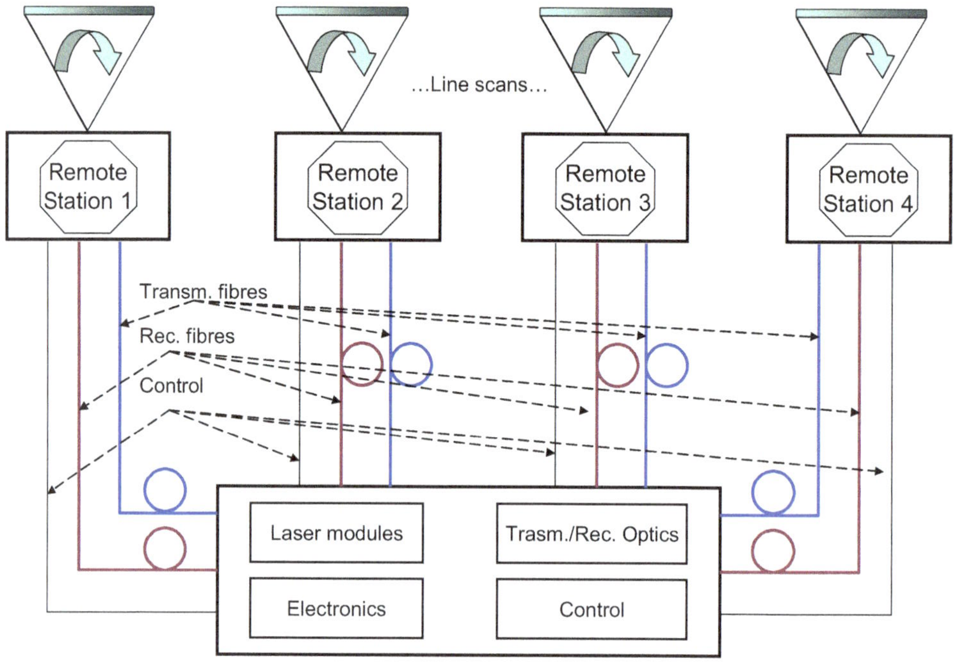

Fig. 7. Overall layout of the telemeter.

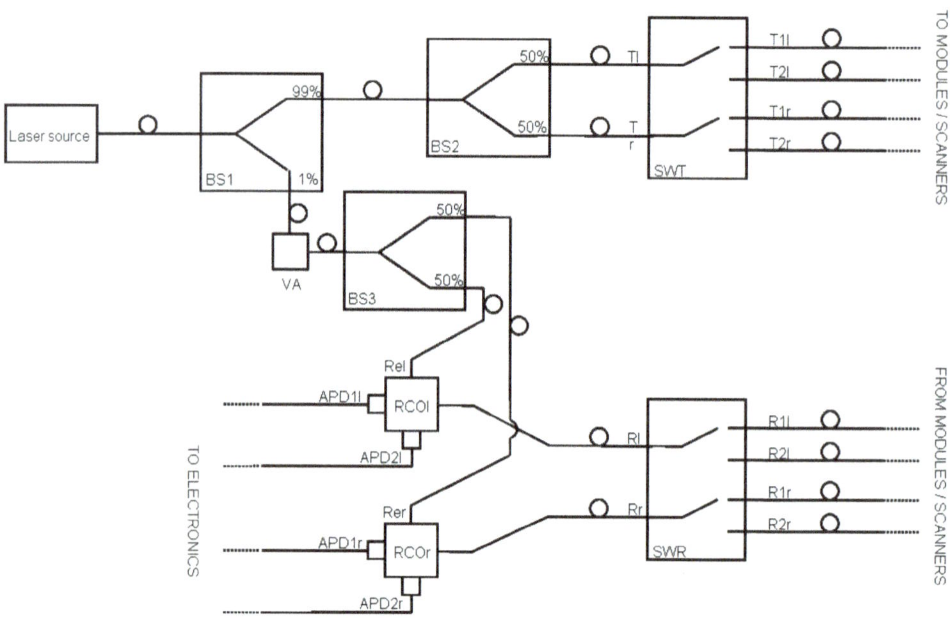

Fig. 8. Optical layout of the central system unit, with transmitter and receiver sections.

3.1.1 Transmitter section

The transmitter section of the system is depicted in the upper part of Fig. 8. The transmitter source is a fast-risetime, short pulse emitter. This can be, according to the application requirements, a high-frequency fibre laser with an output average power of 1.5W, or a nanosecond diode laser. In the application developed for the railway monitoring, the source was a fibre laser emitting pulses having an overall duration of 2.1 ns, and a subnanosecond risetime. The laser fibre has 100μm/150μm core/cladding dimensions. The repetition rate of the emitted pulses is 400kHz.

The output of either the fibre laser or the diode laser is coupled to a first beam splitter, BS1, that spills 1% out of the transmitter laser power from the main beam, and is sent directly to the receiver unit. Its purpose is to act as the reference, (Re) or "start" signal, as it will be clear later on. The main output of BS1 splitter is then sent to a second, 50%/50% fibre optics splitter BS2, that equally divides the transmitter beam. The two output fibres at the splitter exit deliver the transmitter pulses for the scanning of the left part (Tl) and the right part (Tr) of the train respectively (here the prefix T stands for "transmitter"). The two fibres are directly sent to a 2x4 switch, SWT, that feeds four outputs, two at the same time. These outputs are, respectively, T1l, T1r, and T2l and T2r, and are connected to 100μm core fibres that feed the four remote stations.

3.1.2 Receiver section

The 400μm receiver fibres from the four scanning stations are here called R1l, R1r, R2l and R2r. They enter the receiver section and are coupled to a second 2x4 switch, SWR. The output of the switch, which is driven by the same control signal that controls the transmitter switch, is composed of two fibres that carry the receiver left signal Rl and right signal Rr respectively. Each fibre is sent to a Receiver collimation optics (RCOl and RCOr). The 1% part of the Transmitter signal (Re), carried by the fibre at the exit of the first splitter of the transmitter section, enters the receiver section, is attenuated by a variable attenuator and further split in two parts by means of a 50%/50% splitter, BS3. Each of these fibres (Rel and Rer) enters the Receiver Collimation Optics (RCO).

A 40dB dynamic range, difficult to achieve at 400 KHz with standard electronics, has been split in two 20dB dynamic range portions by means of a beam splitter in the RCO. (Fig. 9). Here the Rl (the same holds for Rr) fibre enters the RCO, the beam is collimated and reaches a 10% - 90% beam splitter. The Rel fibre enters the RCO at 90° with respect to the Rl, is also collimated and combined to the Rl signal through the splitter. Two beams are generated: a first one with Rl attenuated 90% and Rel attenuated 10%, and a second beam with Rl attenuated 10% and Rel attenuated 90%.

The purpose of this splitting is clear, since in the presence of very weak optical signals from the object they need to feed a high-gain detector/amplifier system, and in this case most of the signal has to be collected, together with a strongly attenuated reference beam. The opposite holds in the case of strong signals, that feed lower gain detector-amplifier systems.

Each RCO is equipped with two high-speed avalanche photodiodes, APD1 and APD2 [12,13]. The signal from the two photodiodes, as seen in the figure, is composed of a train of two pulses, i.e. the attenuated transmitted pulse, acting as a "start" signal, and the received pulse, acting as a "stop" signal.

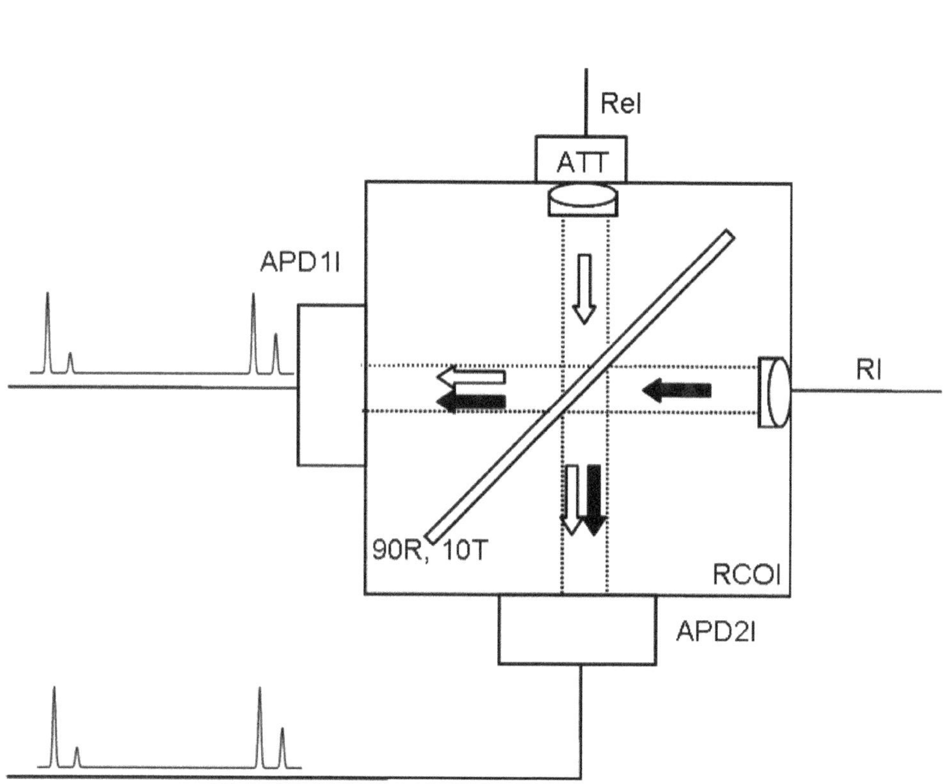

Fig. 9. Detail of the receiver collimation optics (RCO).

3.1.3 Remote stations

Fig. 10 shows the detail of each remote station (as an example, the station "left" of Track 1). The optical fibre T1l enters the module. The beam is collimated and sent to the polygonal scanner rotating at 70 r.p.s. angular speed, driven by the electronic control unit equipped with a position encoder. The polygonal mirror has the purpose of directing the transmitted beam over the required angular range (in our case ±36°). The beginning and the end of this angle represents the time interval where collected signals are valid.

The Transmitter fibre T1l enters the station and is coupled to the Transmitter/Receiver combination blocks. The fibre output is collimated by the 25mm lens L1, and sent, through a 45° mirror, to the polygonal mirror, and hence to the target. The light diffused by the target reaches, through the polygonal mirror, the combiner: 75% of it passes around the 45° mirror and reaches the 50mm focusing lens L2, that injects it into the receiver fibre R1l.

Electrical signals connected to the Remote Stations are the motor start/stop signal, the valid data signal from the encoder, etc.

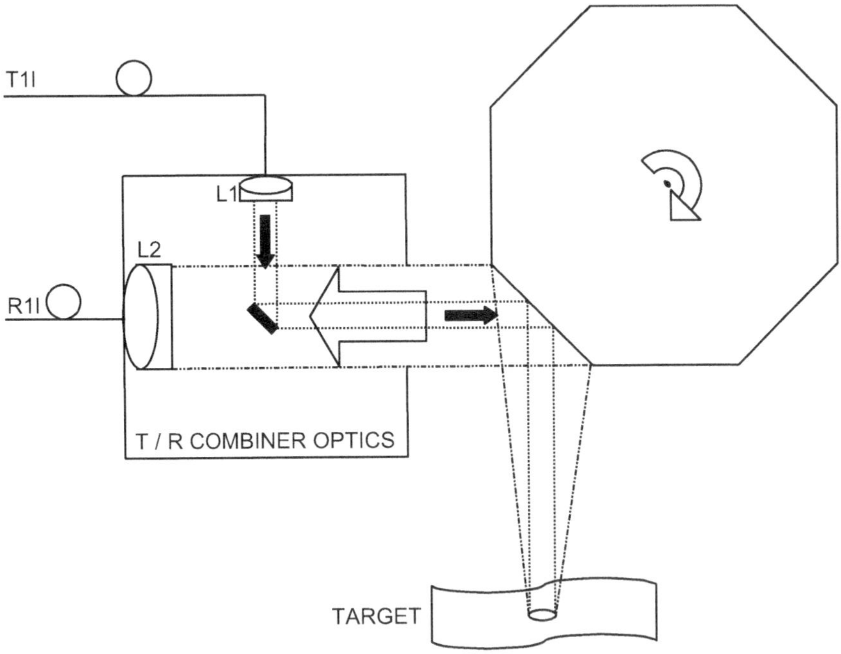

Fig. 10. Optical layout of the remote station, with polygonal mirror and T/R combiner optics.

3.1.4 The system electronics

The output signals from the four APDs described in the Receiver section are sent to the system electronics. They enter the front-end electronics section, with transimpedance amplifiers and conditioning electronics [14-16]. The output of the Front-end Electronics feeds a four-channel Constant Fraction Discriminator (CFD) for the accurate measurement of the arrival time of the train of two pulses for each channel. The digital outputs of the CFD are in turn sent to a Time-to-Distance converter section, which yields the measured distance. Depending on the level of the signal from the receiver, only one of the two signals from the photodiodes should be processed for each couple (l or r). The two signals are sent to two peak detectors, each of them followed by a comparator. If the result of the comparison is positive, i.e., if the APD1l/r can be considered valid, then the result of the time-to-distance conversion is valid. Otherwise, the result of the conversion of the other signal in the couple is valid. All the controls of the electronic unit are made by a control unit based on a FPGA.

3.2 Experimental results

3.2.1 System characterization in house

The laser telemeter has been fully characterized in house to test its performances. A first test consisted in measuring the overall Type B uncertainty with the use of a fixed target of known diffusion coefficient (85% reflectance). The target was accurately placed at a nominal 3m distance from the exit window of the instrument. The nominal distance was in turn measured by means of a high accuracy laser telemeter. At this distance, the Signal-to-noise Ratio (SNR) of the detected signal was 100.

Fig. 11 shows a typical distribution of the readings obtained from a sequence of 1 Million single-shot samples, carried out at the full frequency of 400 kHz at this position. The analysis of the distribution reveals a standard error of 5mm and an overall accuracy of 2.4mm.

Tests were repeated at various nominal distances, ranging from the minimum distance of 0.5m to the maximum of 20m. The results are summarised in Fig. 12. In the figure the measured distance (left scale) together with the single shot accuracy (right scale) are plotted vs the nominal distance. One notes the excellent linearity of the instrument, with errors well contained in an interval of -5mm - +5mm throughout the nominal distance.

In summary, from the in-house characterization testing it is possible to draw the conclusion that the laser, telemeter, in the version that mounts the fibre laser is well in line with the state-of-art laser telemeters, although operating at much higher frequencies with respect to the existing literature.

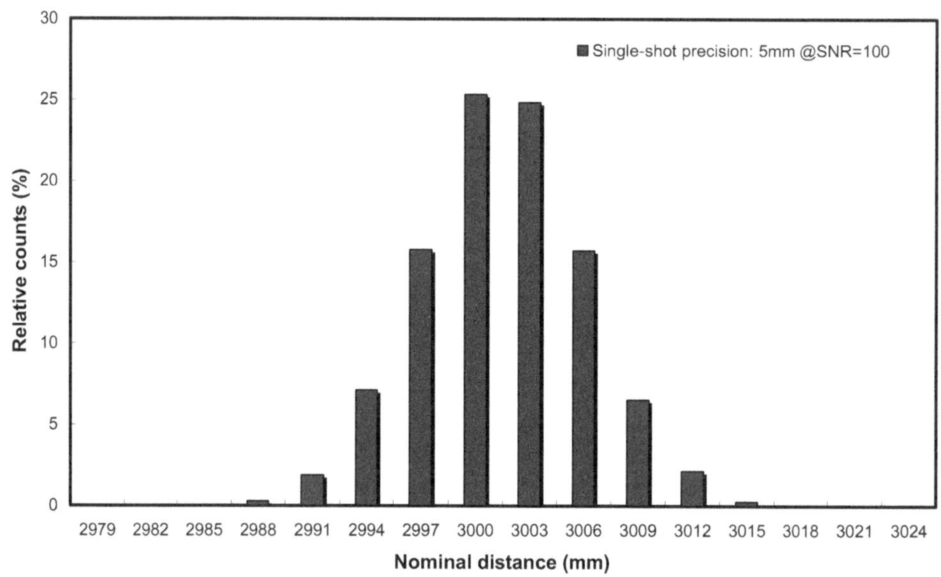

Fig. 11. Distribution of single-shot measurements for a target placed at a nominal distance of 3m from the instrument. SNR: 100. Frequency of acquisition: 400 kHz

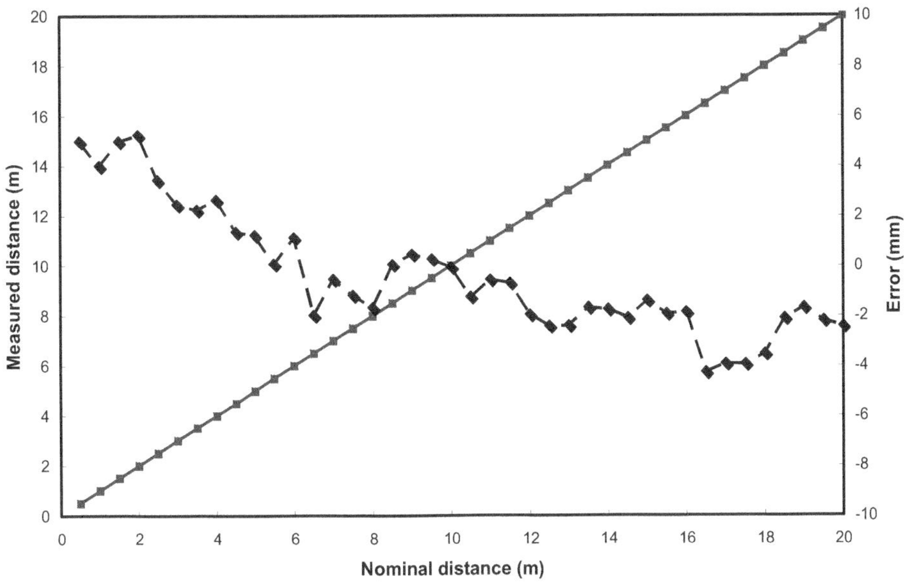

Fig. 12. Plot of the measured single-shot, static distance values (squares, left scale) and of the errors (rhombs, right scale) in a range of nominal distances 0.5 to 20m, at 400 kHz

3.2.2 In-field system testing: The shape monitoring system

The distributed telemeter has been installed on the train-monitoring portal. Here, the four stations have been mounted in such a way as to scan the whole half-carriage under an angle of ±36° (positions A, B, C, D of Fig. 1).

The central unit, containing all Transmitter and Receiver Optics and electronics, is placed into a shelter that contains also all the other electronic and computing equipment, well shielded from the vibrations induced by the trains passing through the portal at about 200km/h. The length of the fibres range from 50 to 100m depending on the location of the remote unit to be fed.

The operating laser frequency has been set to 400kHz, thus allowing a complete profile of the left and right side of the carriage to be obtained every 1.4ms.

This results in a longitudinal resolution of 8cm for a train passing under the portal at a speed of 200km/h. A complete convoy having a length of 500m contains about 4 million measure points.

Considering e.g. the locomotive seen in Fig. 5, Fig. 13 shows a typical result of a shape monitoring process. Fig. 13(a) shows the acquisition of the cross-section of the locomotive highlighted by the dotted line in Fig. 13(b). The point cloud of the obtained by combining together all profiles acquired by the Q-Tech time-of-flight telemeter is shown in Fig. 13(c). In this figure, the 3D shape data are combined to the information acquired by the visible cameras.

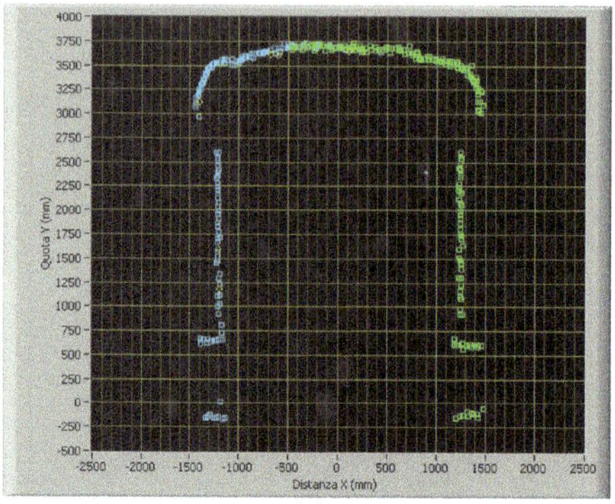

(a)

(b)

(c)

Fig. 13. (a) Acquisition of a single profile of the locomotive shown in Fig. 5; (b) detail of the visual image of the locomotive, highlighting the acquired profile; (c) 3D profile the entire locomotive as obtained by the elaboration of the telemeter measurements.

The point cloud is shown without filtering or digital processing of the points. Nevertheless, in the figure only very few points are evidently "false", due to particularly unfavourable target conditions (glass, dirt, etc.).

Fig. 14(a) shows a point cloud of a passenger train as obtained by the acquisition and the elaboration of the telemeter measurements; Fig. 14(b) shows the presence of an open door.

(a)

(b)

Fig. 14. (a) 3D profile of a passenger train with an open door; (b) detail of the open door.

Fig. 15 shows an example of detection of a shape alarm, i.e. when the measured shape profiles are larger than pre-defined reference shape profiles: Fig. 15(a) shows a profile with an alarm condition; the alarm is caused by a man stretching out his arm, as the visual image acquired by the upper visible camera shows in Fig. 15(b).

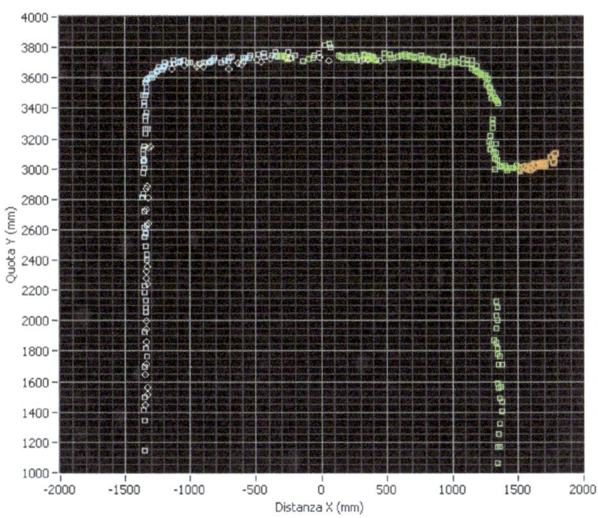

(a)

(b)

Fig. 15. (a) Profile with a shape alarm; (b) detail of the visual image of the train, highlighting the acquired profile.

The results obtained in the monitoring process are extremely positive for the detection of shape alarms for a train before entering a tunnel.

3.3 Conclusions

The distributed telemeter has proved to be a substantial breakthrough with respect to the previous version based on the phase-shift commercial telemeter. The use of a single transmitter-receiver unit allowed cost reduction of the overall system, increased portability and ease-of-maintenance of the remote stations, protection of the central unit in the shelter. The remote stations resulted to be extremely simplified and lightweight with comparison to the previous version. Considerable increase of the measurement accuracies with respect to the previous version could be obtained without exceeding the emitter power level prescribed by the safety class.

A distributed-architecture concept such as the one described in this paper opens perspectives for the use of the telemeter in a number of multipoint measuring situations without the need of purchasing several telemeters. Robotics, mechatronics and production line monitoring could all be industrial domains where this approach could be beneficial.

For these reasons we plan to implement new versions of the instrument, conceived in a modular way, with all modules connected to standard industrial buses, for laboratory, as well as for industrial applications.

4. References

[1] H. Höfler, C. Baulig, A. Blug, M. Dambacher, N. Dimopoulos, and H. Wölfelschneider, "Optical high-speed 3D metrology in harsh environments: Recording structural data of railway lines", in: Optical Measurement Systems for Industrial Inspection IV Proc. SPIE 5856 pp. 296-306, 2005.

[2] H. Höfler and G. Schmidtke (1994): "Three-dimensional contouring by an optical radar system (ORAS)". In: Industrial Applications of Laser Radar, Proc. SPIE Vol. 2271, p. 116-123, 1994.

[3] A. Blug, C. Baulig, M. Dambacher, H. Wölfenschneider and H. Höfler: Novel laser scanners for 3D mapping of railway tracks or roads. Conference on Optical 3-D Measurement Techniques 8, 2007

[4] H. Höfler, C. Baulig, A. Blug, H. Wölfelschneider, O. Fleischhauer, H. Wirth, J. Meier, C. Lehmkühler, and H. Lenz, "High speed clearance profiling with integrated sensors". World Congress on Railway Research (WCRR), July 2006, Montreal.

[5] J. Kalisz, "Review of methods for time interval measurements with picosecond resolution", Metrologia, vol. 41, pp. 17-32, 2004.

[6] T. Bosch, and M. Lecure, Selected Papers on Laser Distance Measurements, SPIE Milestone Series, vol. 115, pp. xi - xiii, 1995.

[7] A. Kilpela, "Pulsed time-of-flight laser range finder techniques for fast, high precision measurement applications", PhD Dissertation, Dept. Electronics, Univ. Oulu university press.

[8] K. Määttä, J. Kostamovaara, and R. Myllylä, "Profiling of hot surfaces by pulsed time-of-flight laser range finder techniques", Appl. Opt. Vol. 32, pp.5334-5347, Sept. 1993.

[9] I. Kaisto, J. Kostamovaara, M. Manninen, and R. Myllylä, "Optical range finder for 1.5-10 m distances", Appl. Opt. Vol. 22, pp. 3528-3264, 1983.

[10] L. Fumagalli, P. Tomassini, M. Zanatta and F. Docchio: A 400kHz, high-accuracy laser telemeter for distributed measurements of 3D profiles. Submitted to IEEE Transaction on Instrumentation and Measurements (2009).

[11] L. Fumagalli, P. Tomassini, M. Zanatta, and F. Docchio, Laser Telemeter for Distributed measurements of object profiles. Italian patent (submitted)

[12] High-performance emitters & detectors for the most demanding applications, available online at http://optoelectronics.perkinelmer.com.

[13] Avalanche Photodiode - A User's Guide, available online at http://optoelectronics.perkinelmer.com.

[14] P. Battacharya Semiconductor Optoelectronic Devices, Prentice-Hall Inc., 1997.

[15] P. Horowitz, and W. Hill: The art of electronics, Cambridge university press, Cambridge, 1989.

[16] P. Palojärvi, T. Ruotsalainen, and J. Kostamovaara, "A Variable Gain Transimpedance Amplifier with a Timing Discriminator for a Time-of-Flight Laser Radar". Proc. ESSCIRC'97 Conference, Southampton, UK, Sept. 16-18 1997, pp. 384-387, 1997.

13

Condition Monitoring of Railway Track Using In-Service Vehicle

Hitoshi Tsunashima[1], Yasukuni Naganuma[2], Akira Matsumoto[3], Takeshi Mizuma[3] and Hirotaka Mori[3]
[1]*Nihon University*
[2]*Central Japan Railway Company*
[3]*National Traffic Safety and Environment Laboratory*
Japan

1. Introduction

Condition monitoring of railway tracks, vehicles are essential in ensuring the safety of railways (Goodall et al., 2006, Buruni et al., 2007). In the field of road traffic, research is proceeding to acquire detailed traffic flow information and reflect it in traffic control by using cars that are regarded as "probes" with an information-obtaining function and having them transmit real-time traffic information such as traffic jams and travel times to traffic control station.

Monitoring of such parameters are not necessary in railways that are operated according to time tables. However, in-service vehicles equipped with simple sensors and GPS may serve as probes to detect and analyze real-time vehicle vibration and signaling systems while running. So called "probe vehicles" (see Fig. 1) (Kojima et al., 2005, 2006) may also dramatically change the current style of rail maintenance and thus contribute to establishing safe transport systems.

The probe vehicles can change the current maintenance style to focus on locations regarded as essential maintenance areas, utilizing data acquired by real-time monitoring of actual vibration together with positional information obtained by GPS. Monitoring based on information obtained by in-service vehicles may enable the detection of maintenance problem at an early stage (Hayashi et al., 2006), thus contributing to the revitalization of local railways by making maintenance tasks more efficient.

The aim of this chapter is to summarize the track-condition-monitoring system based on vehicle measurements for conventional and high speed railway. Section 2 describes the track-condition-monitoring system for conventional railway. In this application, track irregularities are estimated from the vertical and lateral acceleration of the car body. The roll angle of the car body, calculated using a rate gyroscope, is used to distinguish line irregularities from level irregularities. Rail corrugation is detected from cabin noise with spectral peak calculation. A GPS system and a map-matching algorithm are used to pinpoint the location of faults on tracks. Field test using a in-service vehicle was carried out to evaluate the developed system. In section 3, track-condition-monitoring system for high speed railway, shinkansen, called RAIDARSS 3 is introduced. Finally, conclusions are given in Section 4.

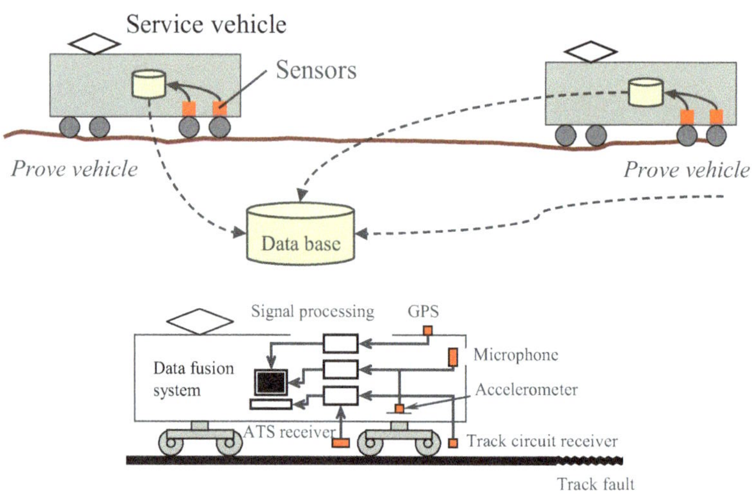

Fig. 1. Condition monitoring of railway by probe vehicle system

2. Condition monitoring of conventional railway track

2.1 Detection of track faults from cabin vibration

2.1.1 Cabin vibration due to track faults

Several kinds of track faults can be detected by measuring the acceleration of bogies (Waston et al., 2006, 2007, 2007). However, if track faults can be detected in-cabin, condition monitoring of track irregularities will be much easier. As the distinctive signal of track faults are hidden in natural frequency of car-body vibration, signal processing is necessary for the acceleration measured in-cabin to detect track faults.

Track faults include corrugation, that is a phenomenon in which cyclic wear patterns are formed on rail heads with wavelengths of a few centimeters to 10 to 20cm as shown in Fig.2 (Matsumoto et al. 2002). Corrugation in tight curves poses particularly serious problems. Corrugation growth causes considerable noise and vibration and leads to rail damages, so it has become an important issue in track maintenance.

Figure 3 shows measurement result from a curved section of track with significant corrugation using sensors on a in-service vehicle. This is the measurement result for travelling a curve with a radius of 202m at a constant speed of 38km/h. The vertical acceleration of the left axle-box, i.e. the inner-rail side, is shown in Figure 3(a). This is a classic characteristic of corrugation in tight curves and confirms the occurrence of corrugation on an inner-side rail.

Figure 3(b) shows the vertical accelerations of the vehicle body measured on the floor of the cabin. Vertical acceleration of a car body measured on the cabin floor is greatly influenced by low-frequency vibration of the car body, and no significant difference is observed in measurement signal by the presence or absence of corrugation in tracks. Thus, it is difficult to detect corrugation by methods using measurement signals directly such as threshold processing, and therefore signal processing is required for detecting corrugation from the acceleration of car bodies.

Fig. 2. Example of rail corrugation

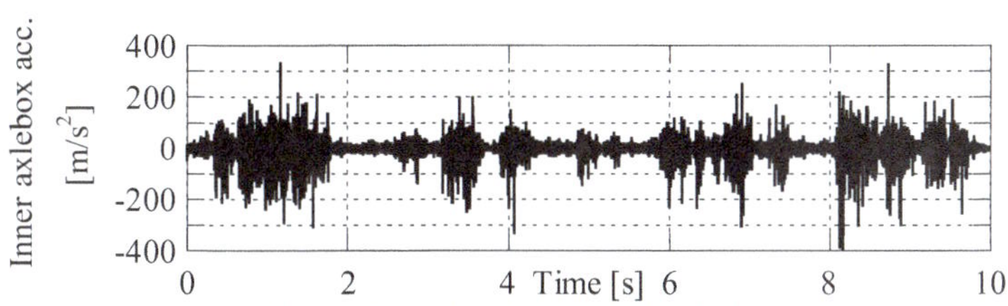

(a) Vertical acceleration of left axle-box

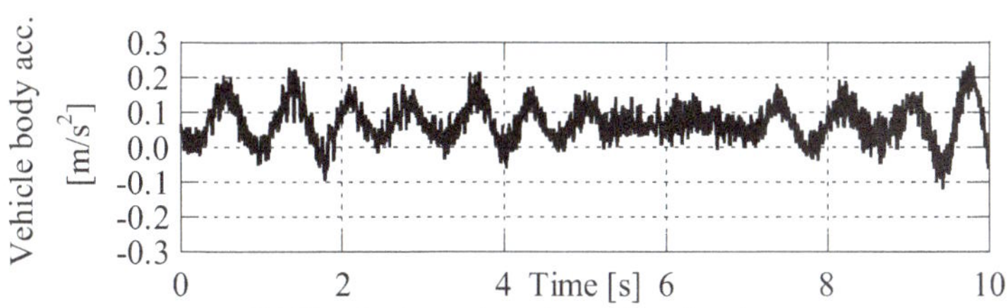

(b) Vertical acceleration of car body

Fig. 3. Measurement results of curved section with corrugation

2.1.2 Detection of track faults by multi-resolution analysis (MRA)

Fourier analysis is a technique for converting time domain data to frequency domain data, but it loses time information. Short-time Fourier transform (windowed Fourier transform) is a technique for time-frequency analysis of signals, but the results depend on the window size. Knowledge of the objects to be analyzed and the ability to be estimated are needed for fault detection, as a certain size of window may not necessarily detect a fault.

In contrast, a wavelet transform that changes window size automatically according to frequency is considered to be suitable for analyzing unknown signals. Therefore, a method was developed to detect track faults from accelerations measured at the car body, using discrete wavelet transforms. This method detects faults by decomposing the measurement signal into an approximation component of low frequency and a detailed component of high frequency (Kojima et al., 2005, 2006).

Wavelet based multi-resolution analysis (MRA) (Daubechies, 1992) decomposes a signal into a number of components at different resolution by using the discrete wavelet transform. A signal is decomposed into some detailed (high-frequency) components and an approximated (low-frequency) component as shown in Fig. 4.

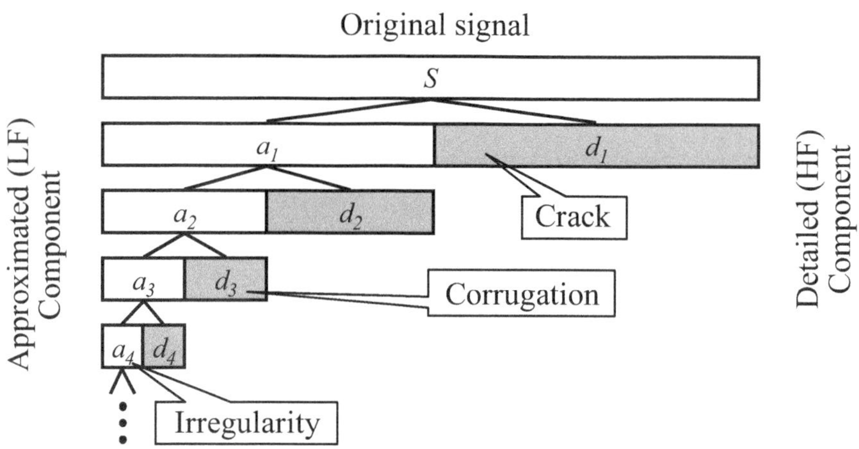

$$S = d_1 + d_2 + d_3 + d_4 + ... + d_n + a_n$$

Fig. 4. Multi-resolution analysis (MRA)

The results of the MRA of the vertical accelerations of car body with significant corrugation, Fig. 3(b), are shown in Fig. 5. Due to the sampling frequency being 2kHz, d_1, d_2, d_3, d_4, and a_4 correspond approximately to 500–1000Hz, 250–500Hz, 125–250Hz, 62.5–125Hz, and frequencies not greater than 62.5Hz, respectively.

It should be noted that the component d_3, in particular, which includes the frequency, 160Hz, for corrugation, shows conspicuous acceleration, which is of a waveform similar to that for the axle-box shown in Fig. 6. This result indicates that the vibration component due to corrugation can be extracted from the acceleration of a vehicle body by using MRA.

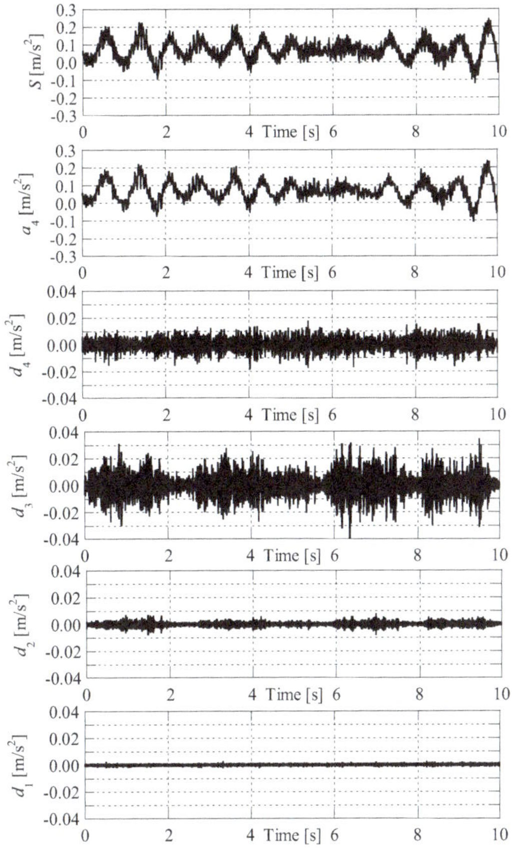

Fig. 5. MRA of acceleration of vehicle body with corrugation

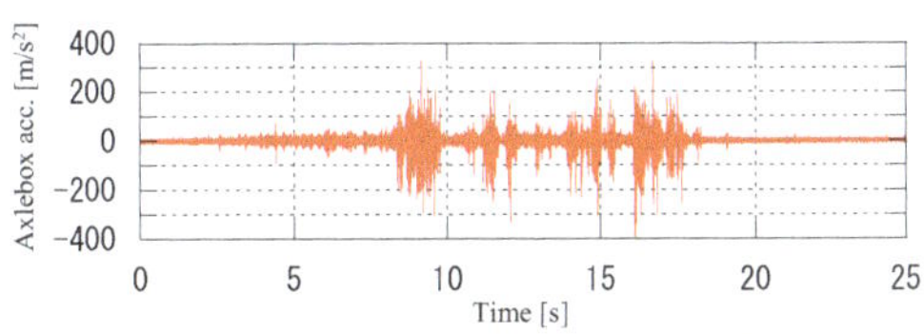

(a) Vertical acceleration of axle-box

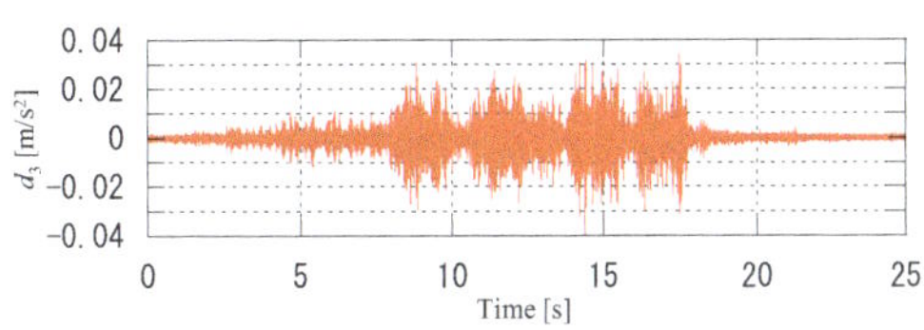

(b) Signal d_3 extracted from vertical acceleration of car body

Fig. 6. Extraction of signal due to corrugation

Figure 7 shows the calculation results of vertical acceleration of car-body and bogie. In this calculation, vertical acceleration of car body and bogie are calculated using 5 degree-of-freedom vehicle model as shown in the left side of Fig. 7. Three types of track faults (corrugation, 1-2s, irregularity, 4-6s, crack, 8s) are considered. It should be noted that only the effect of irregularity can be seen on the vertical acceleration of car-body.

Figure 8 shows the calculated results of MRA. We can see that the corrugation can be detected in 1-2s in the d_3 component which include the frequency of 167Hz due to the corrugation. It can be seen that a crack of a rail can be detected as a impulsive signal in d_1, d_2, d_3, d_4, particular in d_2. The irregularities of track appear in the lower frequency ranges, i.e. d_5-d_{10}. These results show that condition monitoring of track irregularities from car body accelerations is possible using MRA.

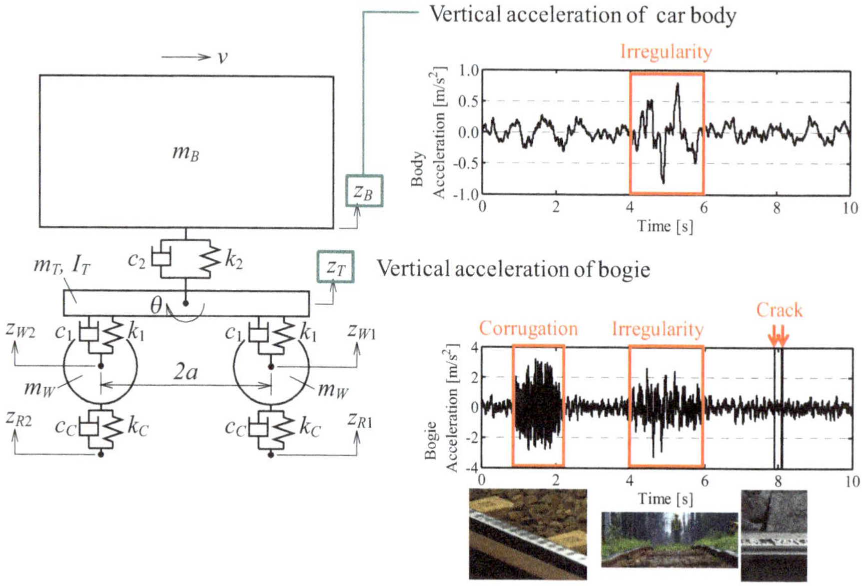

Fig. 7. Vertical acceleration of car body and bogie due to track faults

2.1.3 Detection of track irregularities from car body acceleration

The previous section shows that the MRA is effective for detecting track faults from car body accelerations. However, irregularities of track can be detected from the acceleration of car body directly without using MRA. Simulation studies were carried out using multi-body dynamics code, SIMPACK, to find the possibility of detecting track irregularities from car body vibration directly.

Figure 9 shows the SIMPACK model used for simulation study. The right side of Figure 10 shows vertical acceleration, lateral acceleration and roll angle of car body while the vehicle is travelling with 72km/h on track with irregularities in the vertical direction, the lateral direction and the roll direction, respectively. It can be seen that the car body acceleration and roll angle can be used for detection of track faults.

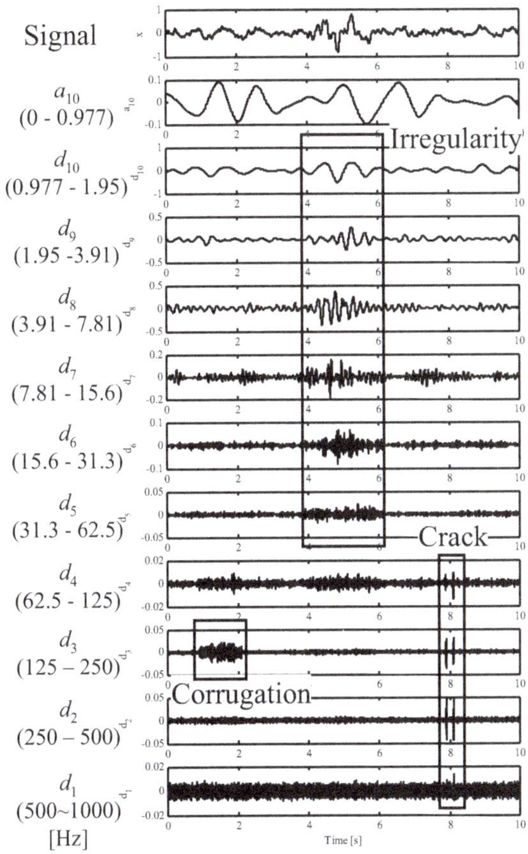

Fig. 8. Decomposition of vertical acceleration of car body by MRA

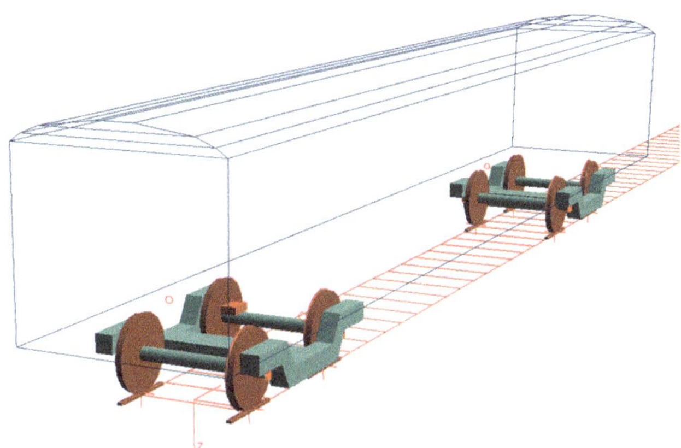

Fig. 9. Full vehicle model

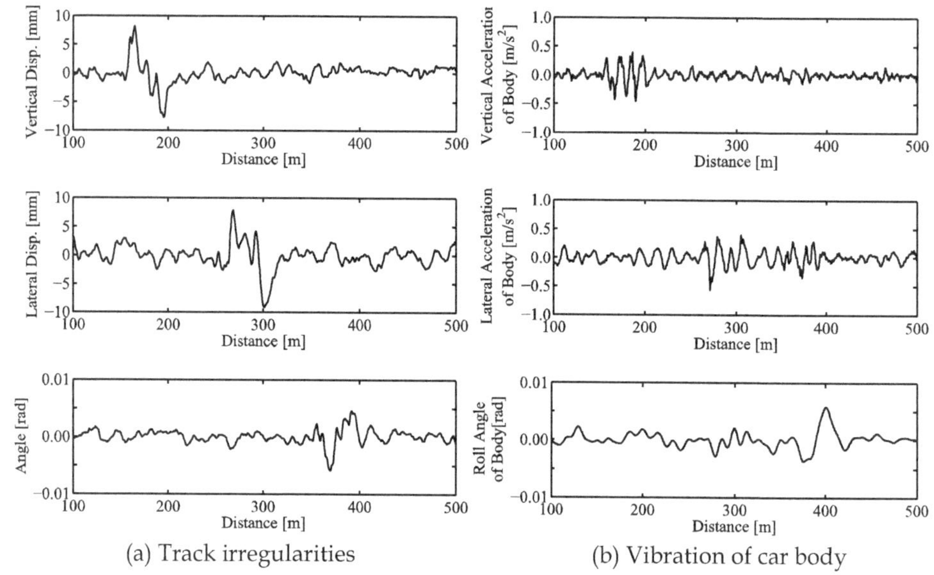

Fig. 10. Vibration of car body due to track irregularities

2.1.4 Detection of corrugation from cabin noise

Some measures should be taken to ensure accurate measurement of high-frequency vibration components using an accelerometer, e. g., it should be attached tightly to the cabin floor. A method was therefore invented to detect corrugation using cabin noise that is uniquely generated when trains run on rails with corrugation.

In this method, spectra are obtained using a short-time Fourier transform of cabin noise data. Peak heights of specific frequencies in the spectra together with the corresponding frequencies are calculated in real-time, and their time-related changes are evaluated as shown in Fig. 11.

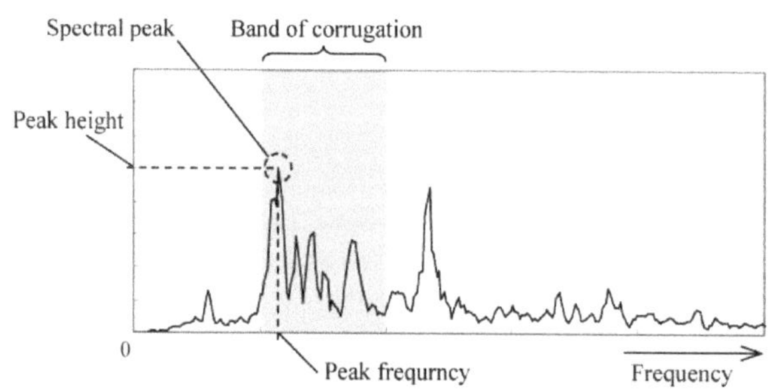

Fig. 11. Detection of corrugation from cabin noise

Corrugation can be detected by simpler measurement with this method using a microphone in the cabin. It was also confirmed that the extent of corrugation can also be diagnosed by this method, in an experiment using a commercial railway line.

Figure 12 presents the results of corrugation detection by cabin noise. Figure 12(a) depicts the noise level of a relatively small corrugation section (hatched part) with a wave height of 0.1 to 0.2mm, indicating that the corrugation section could not be specified by cabin noise. In contrast, spectral peak values (Fig. 12(b)) were elevated in the corrugation section, suggesting that early stage corrugation could not be detected by noise level alone but could be detected successfully by spectral peak.

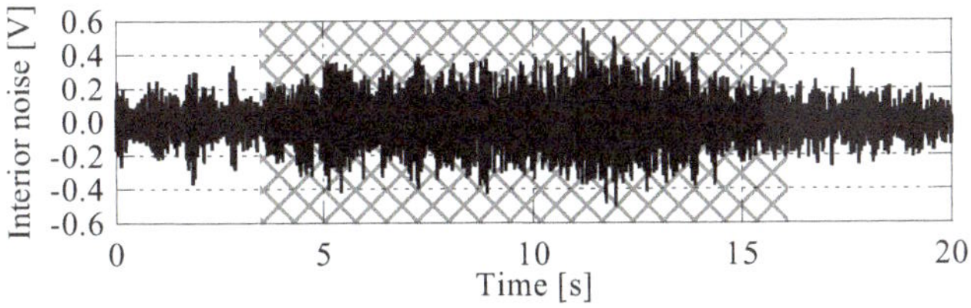

(a) Cabin noise level

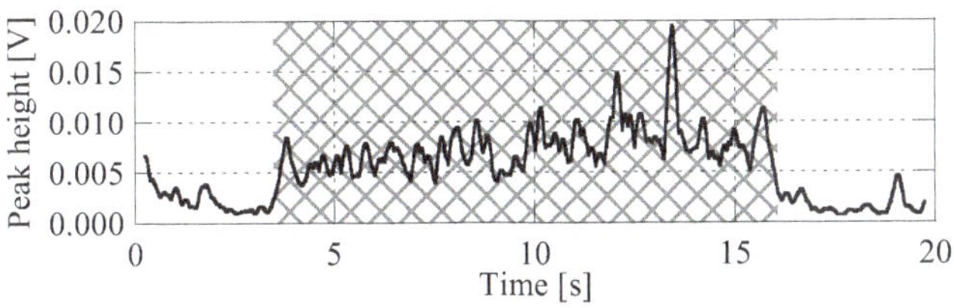

(b) Spectral peak of cabin noise

Fig. 12. Cabin noise measured at in-service vehicle and its spectral peak

2.2 Onboard sensing system

A portable onboard sensing system was developed for an in-service vehicle to enable simple diagnosis of tracks on a commercial line (Tsunashima et al., 2008). Figure 13 depicts components of the sensing system developed. It consists of a microphone for detecting corrugation, accelerometers for detecting track irregularity, a rate gyroscope, a GPS receiver for detecting position, a computer for analysis, and an analog input terminal for inputting signals from each sensor to the computer. The signal output from each sensor is converted into a digital signal by the analog input terminal and input into the computer.

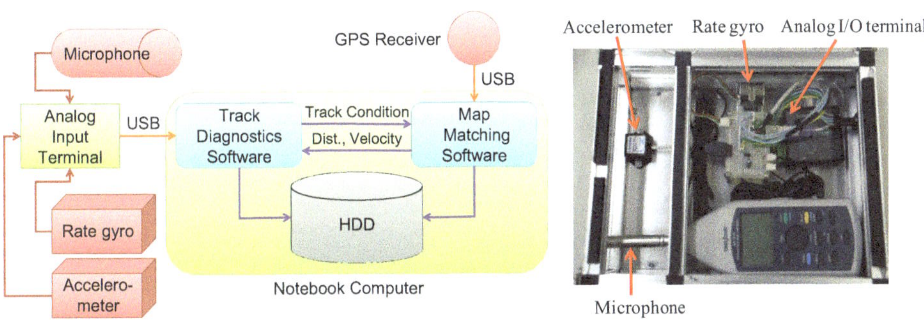

Fig. 13. Portable onboard sensing system

Position information acquired by a GPS receiver is also input into the computer. The computer not only estimates current position and velocity based on the position information from the GPS receiver and acceleration signal from the acceleration sensor, but it also estimates track condition by processing the signal from each sensor and displays it in time sequence in the present position on a screen. Data obtained by signal processing is also recorded on a hard disk drive and utilized for detailed diagnosis of track status by off-line analysis.

2.3 Field test

A method for estimating car body vibration from track displacements has been created as an index for controlling track irregularities. This method estimates riding comfort by calculating car-body vibration and evaluates track condition more effectively by obtaining response characteristics of the car body from field tests (Fig. 14). Figure 15 shows the real time monitoring of track condition in the field test.

Fig. 14. Field test

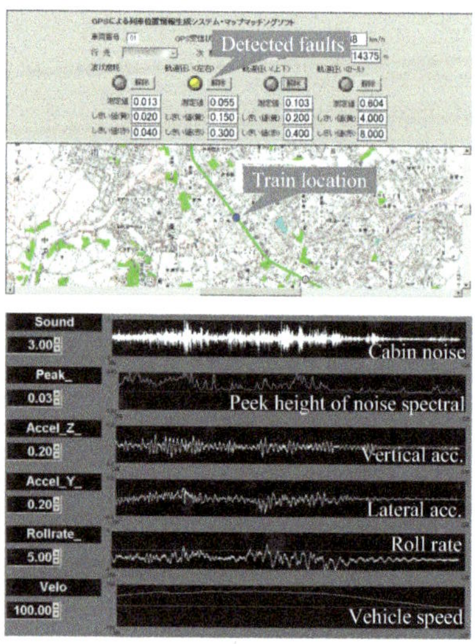

Fig. 15. Real time condition monitoring of track

The response characteristics of a car body to track irregularities may vary depending on the conditions, such as characteristics specific to vehicles, running velocity, and loading; however, it can be roughly estimated by evaluating the RMS value of the car body acceleration with time. Irregularities in cross level can be also distinguished from that in the pure vertical and lateral using a rate gyroscope.

Figures 16 and 17 present the results of measurement when a vehicle is run at 75km/h on a straight section. The horizontal axis indicates distance from the origin, obtained by GPS. Figure 16 depicts the relationship among lateral acceleration, its RMS value, and irregularities in line alignment. Irregularities in line alignment were demonstrated by expressing displacement toward the left as travelling direction. The RMS value of lateral acceleration (Fig. 16(b)) was relatively high around 11.1km.

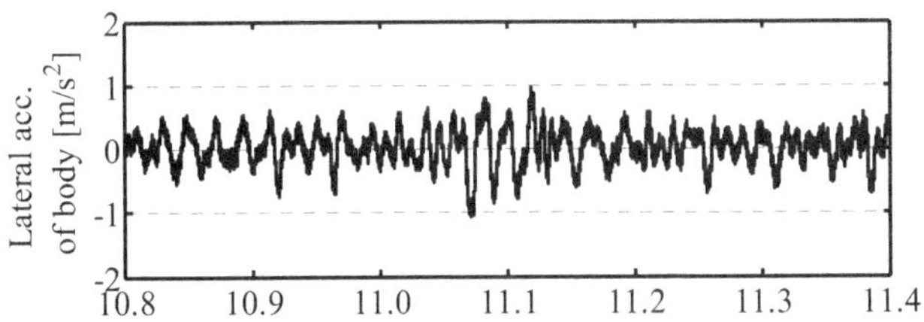

(a) Lateral acceleration of car body

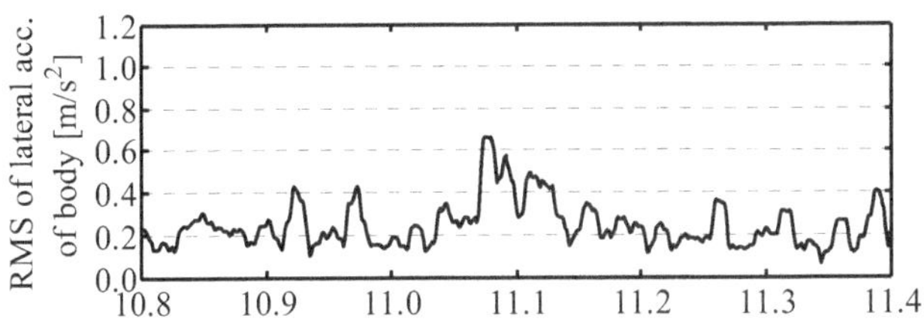

(b) RMS of lateral acceleration of car body

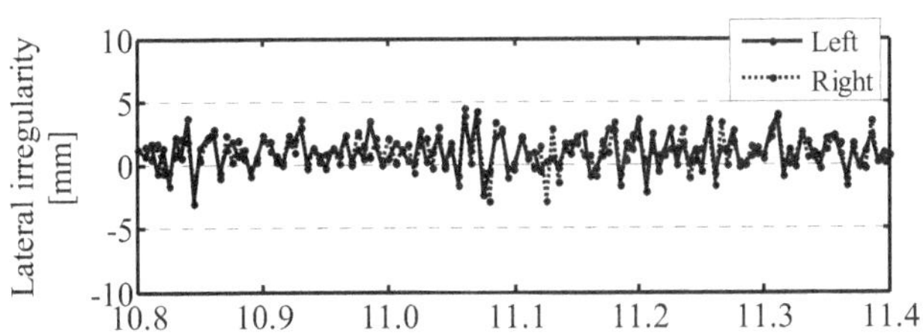

(c) Lateral irregularity of track

Fig. 16. Lateral acceleration of car body and lateral irregularity of track (traveling direction from right to left)

Figure 17 depicts the relationship among roll angle, its RMS value, and irregularities in alignment of level. The roll angle is obtained by integrating the roll rate measured by a rate gyroscope. The RMS value of the roll angle was high around 11.1km (Fig. 17(b)), which corresponded to a location with great irregularity in the alignment of level (Fig.17(c)). It is also near the peak of lateral acceleration RMS (Fig. 17(b)). Based on these findings, it is considered that track irregularities and positions can be detected by using the RMS of normal track as standard and setting a threshold.

Figure 18 show the result of corrugation detection. Gray parts of the figure indicate the areas where the corrugation is observed. It is shown that the corrugations are successfully detected by the proposed method from cabin noise.

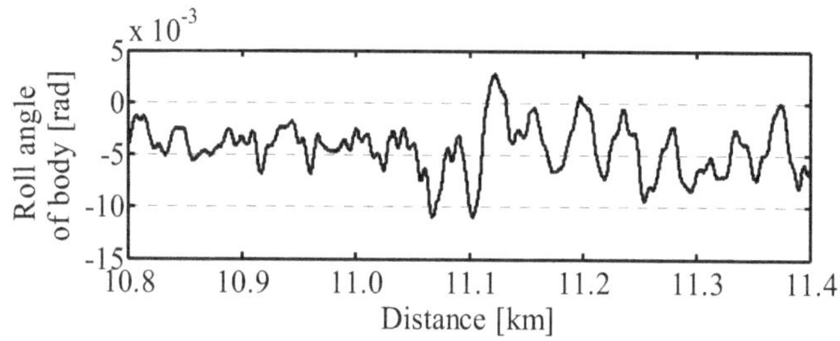

(a) Roll angle of car body

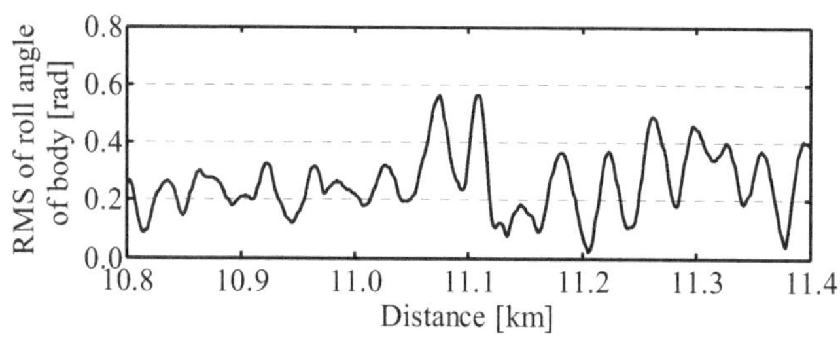

(b) RMS of roll angle of car body

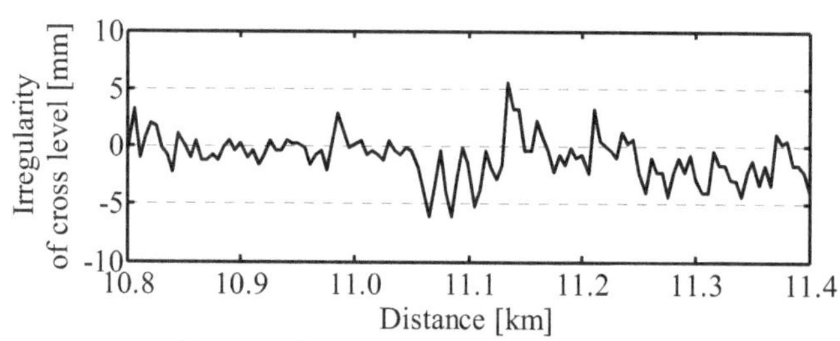

(c) Irregularity of cross level of track

Fig. 17. Roll angle of car body and irregularity of cross level of track (traveling direction from right to left)

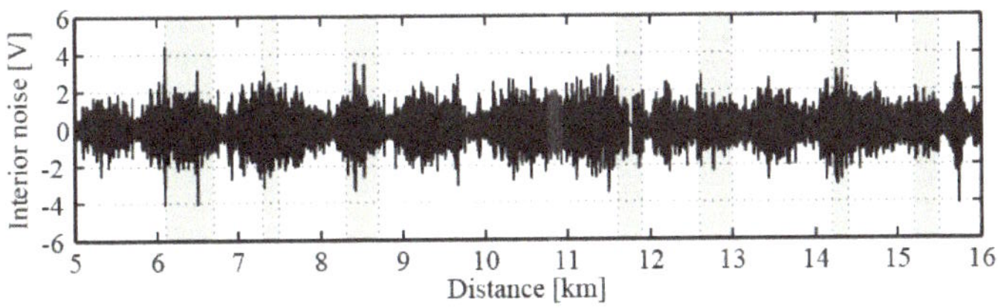

(a) Cabin noise

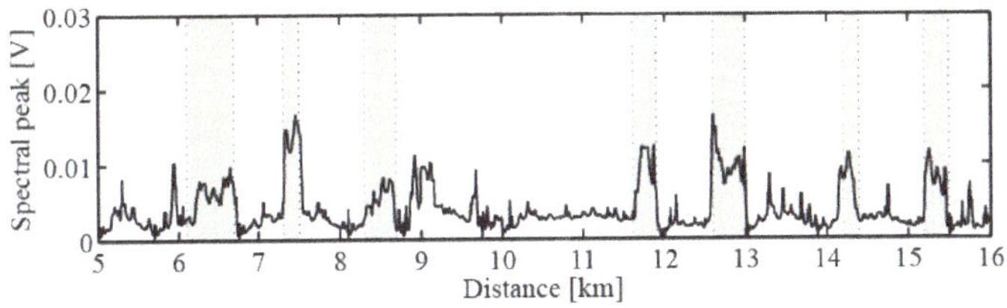

(b) Spectral peak of cabin noise

Fig. 18. Detection of corrugation

2.4 Condition monitoring of track in local lines

Long term condition monitoring of railway tracks is carried out on local lines where the irregularity or corrugation is significant.

In the first step, evaluation was made of the reliability of measurement results. Figure 19 shows a comparison result of two measurements with a time interval of 18 days. It should be noted that the two measured wave forms are almost the same and the significant peeks due to the rail irregularities can be seen at the same location.

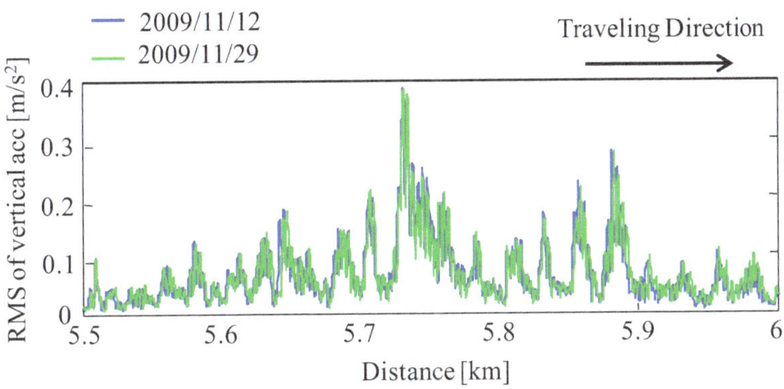

Fig. 19. Comparison of RMS of vertical acceleration in first and second runs

A railway company made their repair works for the track where the significant peek were observed. Figure 20 depicts changes of RMS of vertical acceleration before and after the repair work. It can be seen from the figure that the track condition before and after the repair works were clearly observed using the portable onboard sensing system.

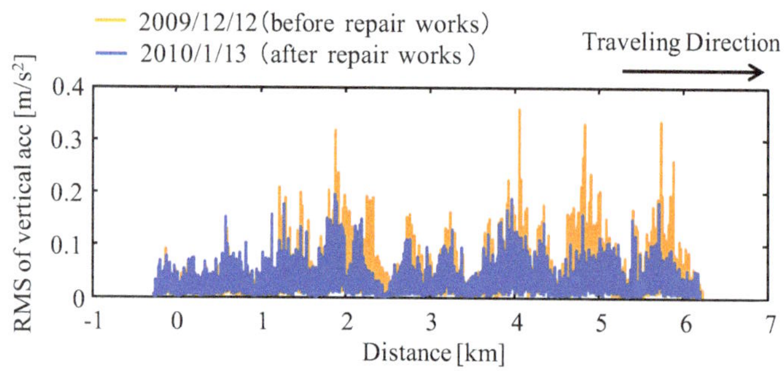

Fig. 20. Changes of RMS of vertical acceleration before and after repair work

Corrugation condition monitoring on another commercial line was also carried out. Eight measurements in two years were collected and evaluated. Figure 21 and 22 show the change of corrugation condition evaluated by the spectral peek from cabin noise.

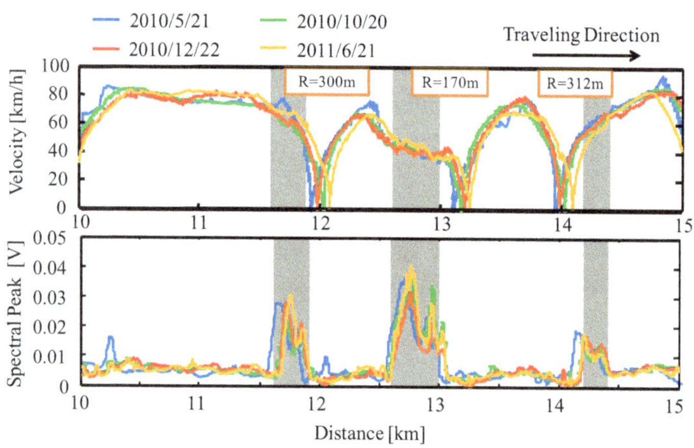

Fig. 21. Spectral peak of cabin noise

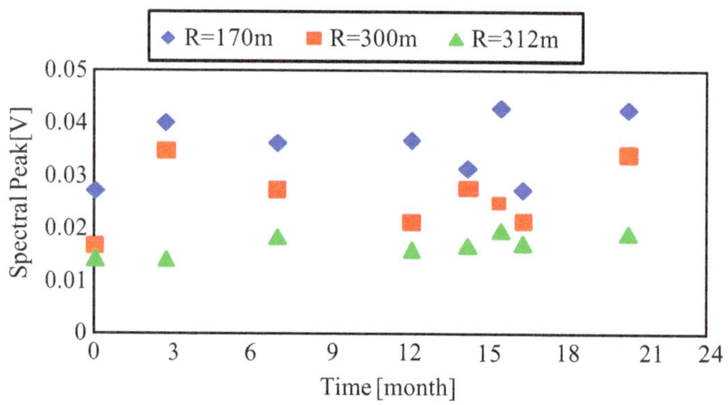

Fig. 22. Change of maximum spectral peak in two years operation

3. Condition monitoring of high-speed railway track

3.1 Overview

To enhance the safety and reliability of high-speed transportation, one of the most important tasks is to examine infrastructure frequently and accurately. On the shinkansen lines, catenary and track infrastructure is examined by multiple inspection trains in business hours in the daytime. The multiple inspection train called 'Dr. Yellow' runs on the Tokaido & Sanyo shinkansen, and 'East-i' runs Tōhoku, Jōetsu, Nagano, Yamagata & Akita shinkansen. The newest Kyushu Shinkansen carries a track recording device on some operating trains for cost reduction purpose.

In this way, every shinkansen track is examined every 10 days. In order to confirm the safety in the period of the interval of track inspection, car body acceleration of commercial trains has been measured every day since the inauguration of service on the Tokaido shinkansen line in 1964. Track condition monitoring that only references car body accelerations become difficult because the recent shinkansen vehicle is equipped with high-performance suspension and is therefore not as responsive to track irregularity. To solve this problem, a new device which is able to measure track irregularity was developed.

3.2 RAIDARSS-3: Track condition monitoring system on Tokaido shinkansen

In 2009, a new track condition monitoring device that is able to measure vertical track irregularity using double integration of the axle-box acceleration, was developed. The new devices called RAIDARSS-3 (see Fig. 23,) are now installed on six N700 series shinkansen train sets in order to check the track condition several times a day. If the measured accelerations or track irregularities exceed the predetermined target values, these measured values and locations are automatically reported to the train control centre and track maintenance depots. Table 1 shows the main features of RAIDARSS-3.

The inertial measurement is more suitable for track condition monitoring by commercial trains than the 3-points method because it doesn't require a special car body structure or bogies.

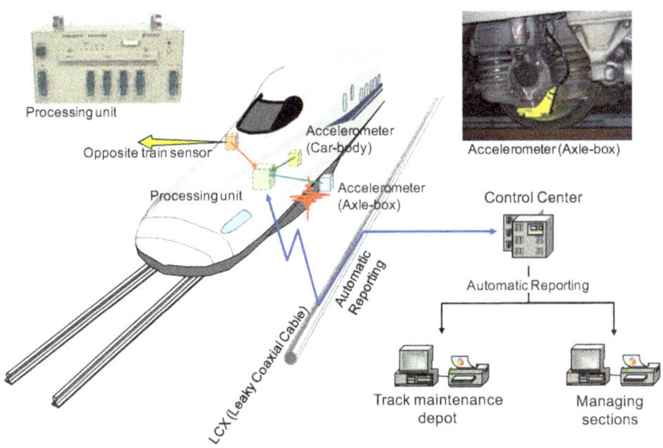

Fig. 23. RAIDARSS-3: Track condition monitoring system for Tokaido shinkansen

Automatic reporting	Exceeded values and locations are automatically reported to the train control center and track maintenance depots.
Wheel diameter adjustment	Wheel diameters are automatically revised by wheel pulse and position information from train.
Opposite train sensing	Opposite trains are detected by optical sensors
Data acquisition	All data are automatically transmitted to a server via LCX(Leaky Coaxial) Cable. 50 runs of operation data are stored on 2GB memory card.
Settings	Setting change and data transmission are available via LCX cable.

Table 1. Main features of RAIDARSS-3

3.3 Former inertial measurement system

Inertial measurement is based on a simple law where double integration of the acceleration indicates a position on an accelerometer. For example, the vertical position of a wheel can be found by using double integration of the axle-box acceleration. The result provides the longitudinal level due to the wheel being continuously in contact with the rail (see Fig. 24). On the other hand, for the measurement of the track alignment, the change of the clearance laterally existing between the wheel flange and the rail needs to be taken into account and thus needs to be measured by means of sensors.

The conventional inertial system uses an analogue integral circuit. If an input signal has a slight offset, the output of an analogue integrator is completely saturated in the vicinity of the power supply voltage, and therefore cannot function as an integrator. To avoid saturation, a high-pass filter is added before the integrator. The cut-off frequency of the high-pass filter varies with vehicle speed to maintain the cut-off wavelength at a fixed value (ex. 120 m) over the distance domain.

Unfortunately, the high-pass filter distorts the output waveform so that the output signal does not agree with the track profile on the ground. It is caused by a nonlinear phase shift of the analogue high-pass filter. Another issue is that the measured waveforms of alternate directions are largely different. The distortion can be corrected by reversing the phase of the output signal. But a smarter solution to wave distortion is to introduce digital processing.

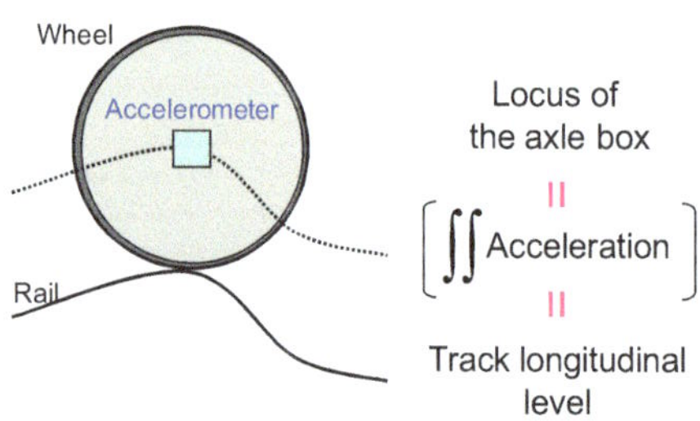

Fig. 24. Inertial track measurement in longitudinal level

3.4 Digital processing for RAIDARSS-3

As with saturation in an analogue integrator, an offset of the input data gains in number of bits by digital integral calculus, the bit width of the processor will shorten immediately. To avoid this problem, as in analogue devices, a digital high-pass filter is necessary before the integrator. But it is difficult to solve by changing the many coefficients of the high-order digital HPF according to the vehicle speed in real time.

To overcome the difficulty, RAIDARSS-3 uses the 10 m versine characteristic to stabilize the double integration (see Fig. 25).

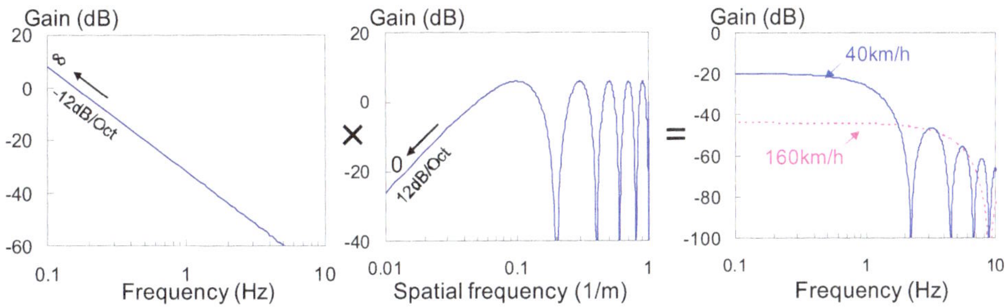

(a) Double integration (b) Mid-chord offset (10 m versine) (c) Integration and mid-chord offset composite calculation

Fig. 25. Digital inertial processing for RAIDARSS-3

The 10 m versine method is expressed by

$$y(\xi)=x(\xi)-\frac{x(\xi-5)+x(\xi+5)}{2}, \tag{1}$$

where $y(\xi)$ is measured 10m versine signal and $x(\xi)$ is original track profile on the ground.

From Eq. (1), a transfer function for a 10m versine measurement on the z-plane yields Eq. (2).

$$H_{10}(z)=-\frac{1}{2}+z^{-5}-\frac{1}{2}z^{-10}, \tag{2}$$

In this equation, the sampling distance is 1.0m and an output delay of 5m to satisfy the law of causality.

Furthermore,

$$H_{10}(z)=-\frac{1}{2}\left(1-2z^{-5}+z^{-10}\right)=-\frac{1}{2}\left(1-z^{-5}\right)^2, \tag{3}$$

Equation 3 shows that a characteristic of the 10m versine consists of two difference filters and one multiplier. Figure 26 shows a block diagram of the 10m versine method, and Fig. 27 shows the characteristics of a difference filter.

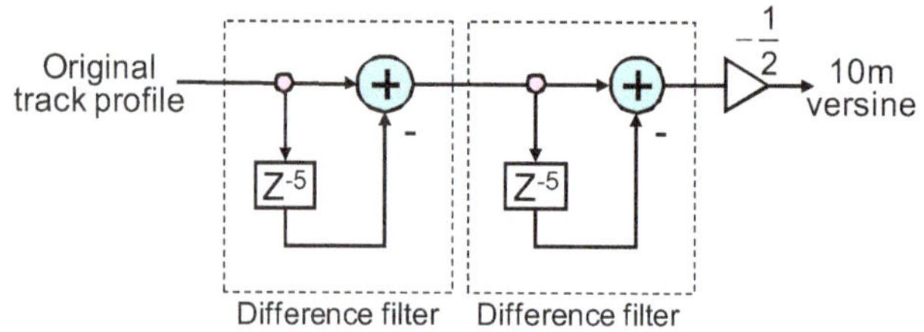

Fig. 26. Brock diagram of 10 m versine method

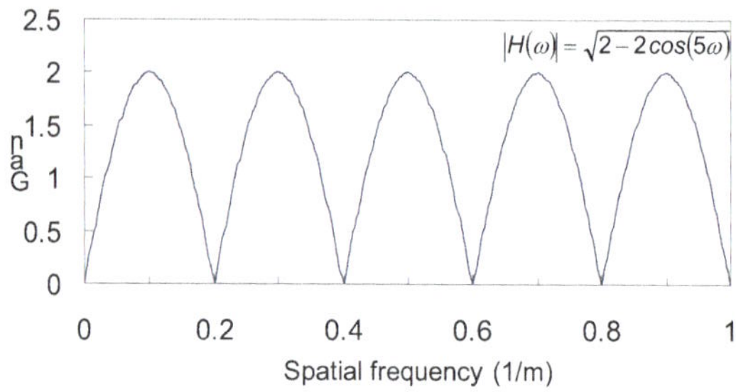

Fig. 27. Frequency responce of difference filter

As shown in Fig. 27, because the difference filter, which is divided from a 10m versine characteristic, exhibits a high-pass characteristic and the DC (0Hz) gain is zero, allowing this filter to utilize integration stabilization. Since the difference filter operates by taking the difference between the input signal and its delayed signal, the processing load is very light. Furthermore, a variable frequency filter can easily be used by changing the delay corresponding to the vehicle speed. With a sampling frequency of 1.0Hz, the characteristics of the 10m versine method in the time domain yields Eq. (4).

$$H_{10}(z) = -\frac{1}{2}\left(1 - z^{-L}\right)^2 \tag{4}$$

where L = 5/v, and v is the vehicle speed (m/s).

The transfer function for the simplest digital integrator on the z-plane is

$$H_I(z) = \frac{1}{1 - z^{-1}}, \tag{5}$$

Using Eqs.(4) and (5), the digital inertial processing is expressed as

$$H_{DI}(z) = H_{10} \times H_I = -\frac{1}{2}\left(\frac{1 - z^{-L}}{1 - z^{-1}}\right)^2, \tag{6}$$

This digital inertial measurement technique is called the "Frequency variable difference filter". Figure 28 shows a block diagram of this processing technique. This system, mainly composed of adders with a single multiplier, can maintain quite low CPU loading condition.

10m versine longitudinal level obtained by RAIDARSS-3 is shown in Fig. 29. For comparison, measurement signals from an existing track geometry car in Dr. Yellow are shown in the figure. There are the good correspondences between the signals.

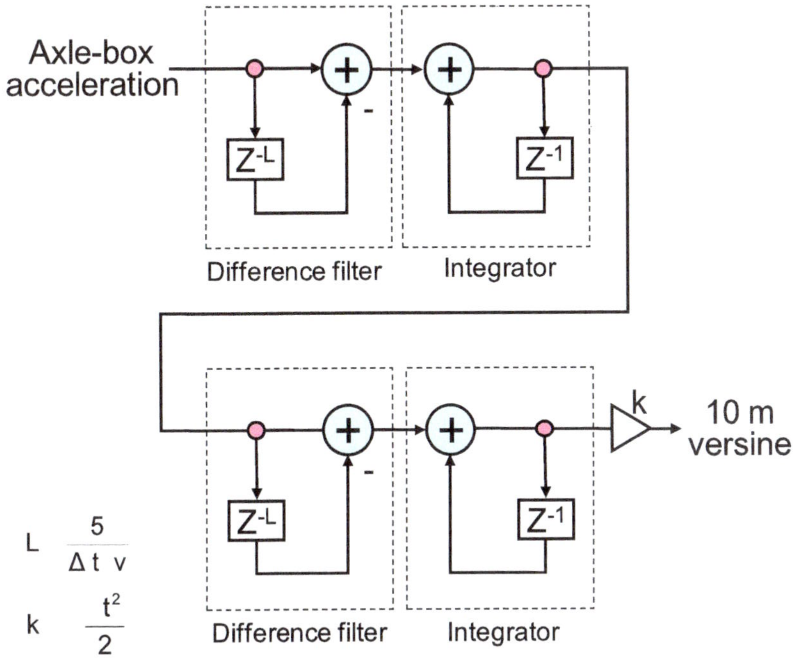

Fig. 28. Block diagram of frequency variable difference filter

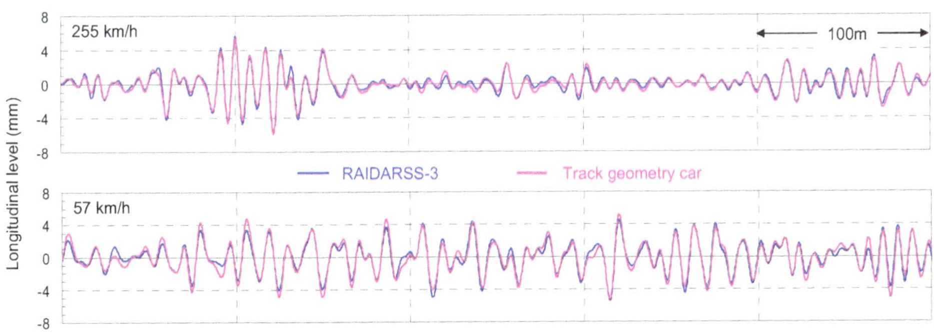

Fig. 29. Comparison between RAIDARSS-3 and Dr. Yellow

3.5 Cant and twist

Track cant is calculated from the vertical track profile of right and left rail. This method is suitable for track monitoring by a commercial vehicle because an expensive gyro is not necessary. Figure 30 shows an operation flow to calculate cant.

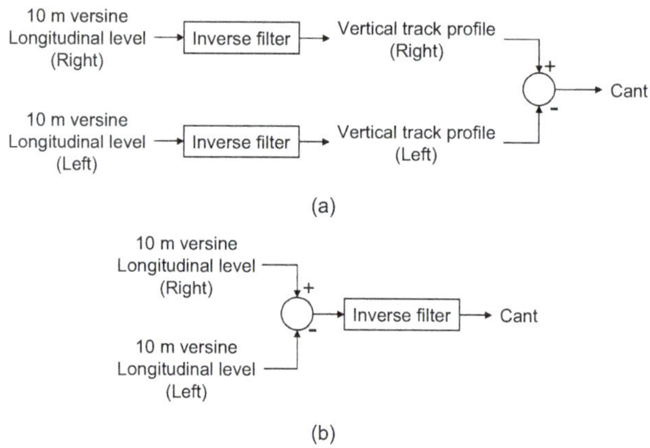

Fig. 30. Calculation flow for cant

Since RAIDARSS calculates a 10m versine of longitudinal level directly, an inverse filter shown in the Fig. 31 is used to get vertical track profile from 10m versine. Fig. 30(b), that is equivalent to Fig. 30(a), is used for effective calculation.

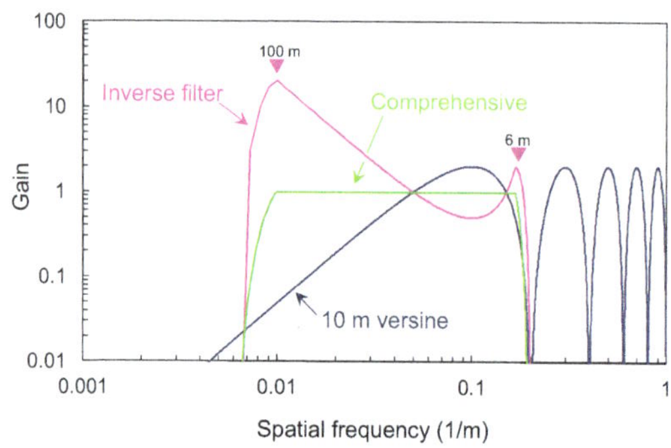

Fig. 31. Inverse filter for track profile restoreation

Cant calculated from RAIDARSS is compared with the signals from an existing Dr. Yellow in Fig. 32. There are the good correspondences between the signals. Both agree well, and the standard deviation of the difference of both is 0.28mm. Track twist shown in Fig. 33 is calculated by means of a difference of cant between 2.5 m, and the standard deviation of the difference with the Doctor Yellow is 0.14 mm. Even if an expensive gyro is not used, track cant and twist are obtained with high precision.

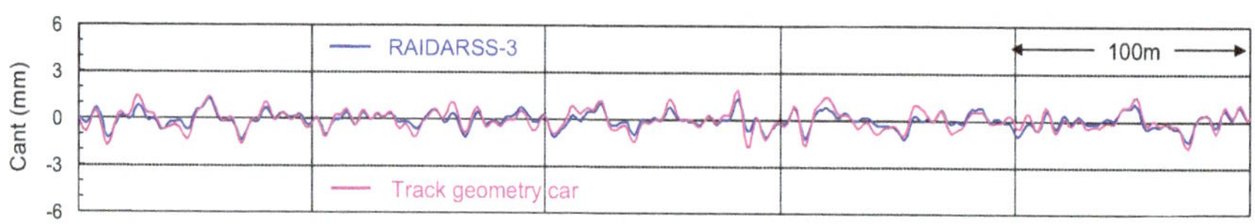

Fig. 32. Cant calculated from RAIDARSS-3

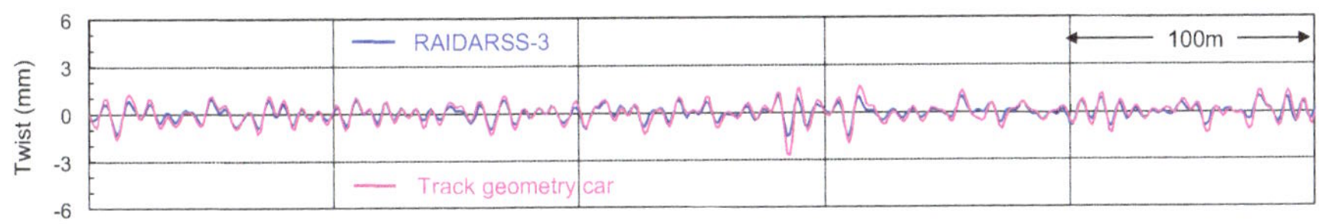

Fig. 33. Twist calculated from RAIDARSS-3

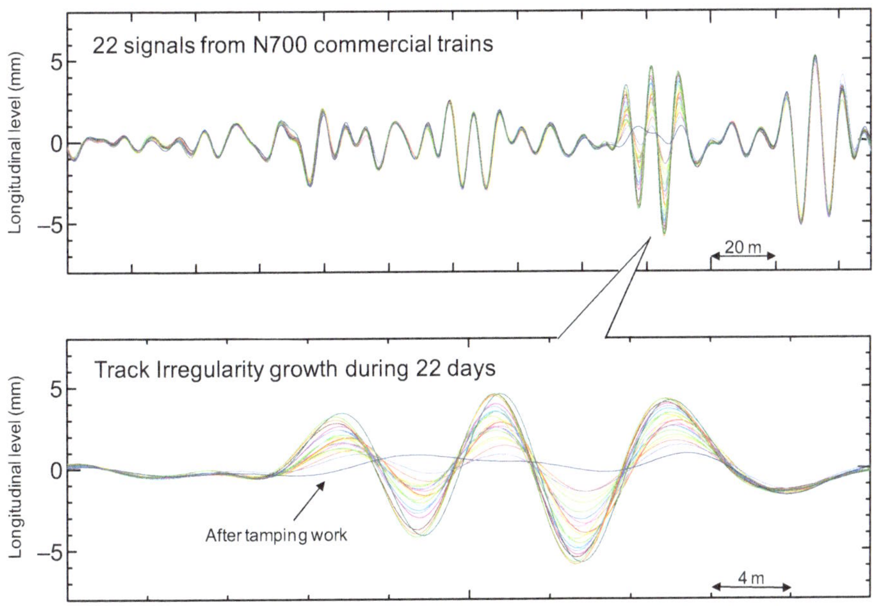

Fig. 34. Repeatability of RAIDARSS-3 for 3 weeks

3.6 Practical use of RAIDARSS-3

RAIDARSS-3 is installed in six N700 train sets already working on Tokaido shinkansen, and car-body accelerations and longitudinal levels are checked several times every day. The repeatability for 3 weeks is shown in Fig. 34.

In spite of the varied train sets and running speed, good correspondence was obtained among 22 different signals, which indicates good repeatability of the device. It can be seen that rapid track geometry degradation occurs in this section. Although the growth rate of the track irregularity is small every day, the good repeatability of the device can identify even the slight change of the track. RAIDARSS-3 will contribute to the track safety of the Tokaido Shinkansen in the future.

4. Conclusions

Two types of track-condition-monitoring system for conventional and high-speed railway are summarized in this chapter. The development of a portable condition monitoring system for track which is easily integrated on in-service vehicles is introduced. In this system, irregularities of the rail are estimated from vertical and lateral acceleration of car body. A roll angle of car body, which is calculated using a rate gyroscope, is used to distinguish irregularity of line from irregularity of level. Rail corrugation can be detected from cabin noise with spectral peek calculation. A GPS system and map matching algorithm localizes the fault on track.

A field test was conducted using a commercial line in cooperation with a railway operating company. Track irregularity was detected by vertical and lateral acceleration measured while the vehicle was running, and corrugation was detected by spectral peaks of cabin noise. Track condition was displayed on a route map in real time together with information of the location based on the position information obtained by GPS. The field results in local lines showed that the long term condition monitoring of railway track using the developed probe system gives us useful information for condition-based-maintenance.

For condition monitoring of high-speed railway track, a new device for measuring vertical track irregularity using double integration of the axle-box acceleration, RAIDARSS-3, is introduced. RAIDARSS-3 is installed in six N700 train sets already working on Tokaido shinkansen, and car-body accelerations and longitudinal levels are checked several times every day. It will contribute to the track safety of the Tokaido Shinkansen in the future.

5. Acknowledgment

This study was supported by the Program for Promoting Fundamental Transport Technology Research from the Japan Railway Construction, Transport and Technology Agency (JRTT) and Adaptable and Seamless Technology Transfer Program through Target-driven R&D, Japan Science and Technology Agency.

6. References

Buruni, S., Goodall, R. M., Mei, T. X. and Tsunashima, H. (2007). Control and monitoring for railway vehicle dynamics, *Vehicle System Dynamics*, Vol. 45, No. 7-8, , pp.765-771

Daubechies, I. (1992), Ten Lectures on Wavelets, CBMS-NSF Regional Conference Series in Applied Mathematics, *Society for Industrial and Applied Mathematics*, (61)

Goodall, R. M. and Roberts, C. (2006). Concept and Techniques for Railway Condition Monitoring, *The International Conference on Railway Condition Monitoring 2006*, pp. 90-95

Hayashi, H., Kojima, T., Tsunashima, H. and Marumo, Y. (2006), Real Time Fault Detection of Railway Vehicles and Tracks, *Railway Condition Monitoring 2006*, pp. 20-25

Kojima, T., Tsunashima, H. and Matsumoto, A. (2005), Fault detection of railway track by multi-resolution analysis, *International Conference on Wavelet Analysis and Its Applications*

Kojima, T., Tsunashima, H. and Matsumoto, A. (2006), Fault detection of railway track by multi-resolution analysis, *Computer in Railway X*, WIT Press, pp. 955-964

Matsumoto, A., Sato, Y., Ono, H., Tanimoto, M., Oka, Y., and Miyauchi E. (2002), Formation mechanism and countermeasures of rail corrugation on curved track, *Wear*, 253(1), pp. 178-184

Naganuma, Y. & Sato, Y. (1999). Practical Use of TRASC on Track State Confirming Cars, *Proceedings of World Congress on Railway Research 1999 (WCRR '99)*

Naganuma, Y. & Sato, Y. (2000). Track state control with use of real time digital data processing, *International Journal of Heavy Vehicle Systems 2000*, Vol. 7, No. 1, pp. 82-95.

Naganuma, Y., Kobayashi, M., Nakagawa, M.and Okumura, T. (2008), Condition Monitoring of Shinkansen Tracks using Commercial Trains, *International Conference on Railway Condition Monitoring 2008*

Naganuma, Y.; Kobayashi, M.; Nakagawa, M. and Okumura, T. (2010). Inertial Measurement Processing Techniques foe Track Condition Monitoring on Shinkansen Commercial Trains, *Journal of Mechanical Systems foe Transportation and Logistics*, Vol. 3, No. 1

Otake, T.; Naganuma, Y. and Sato, Y. (1997). Rectification of Distorted Track Irregularity Record Obtained by Inertia Method, *Proceedings of the 6th International Conference of the IHHA*

Takeshita, K. (1997). Track Irregularity Inspection Method by Commercial Railway Vehicle, *QR of RTRI*, Vol. 38, No. 1, pp. 6-12.

Tsunashima, H., Kojima, T., Marumo, Y., Matsumoto, A. and Mizuma, T. (2008), Condition Monitoring of Railway Track and Driver Using In-Service Vehicle, *International Conference on Railway Condition Monitoring 2008*

Waston, P. F., Ling, C. S., Goodman, C. J., Roberts, C., Li, P. and Goodall, R. M. (2007), Monitoring lateral track irregularity from in-service railway vehicles, *Proceedings of the Institution of Mechanical Engineers, Part F, Journal of Rail and Rapid Transit*, Vol. 221, No. F1, pp. 89-100

Waston, P. F., Ling, C. S., Roberts, C., Goodman, C. J., Li, P. and Goodall, R. M. (2007), Monitoring vertical track irregularity from in-service railway vehicles, *Proceedings of the Institution of Mechanical Engineers, Part F, Journal of Rail and Rapid Transit*, Vol. 221, No. F1, pp. 75-88

Waston, P. F., Roberts, C., Goodman, C. J. and Ling, C. S. (2006), Condition monitoring of railway track using in-service trains, *Railway Condition Monitoring 2006*, pp. 26-31

14

New Ultrasonic Techniques for Detecting and Quantifying Railway Wheel-Flats

Jose Brizuela[1], Carlos Fritsch[1] and Alberto Ibáñez[2]
[1]*Consejo Superior de Investigaciones Científicas (CSIC)*
[2]*Centro de Acústica Aplicada y Evaluación No Destructiva - CAEND (CSIC-UPM)*
Spain

1. Introduction

The railway transport braking processes are likely to form surface defects on the tread if the wheel locks up and slides along the rail. This action can be produced by a defective, frozen or incorrectly tuned brake, as well as by a low rail-wheel adhesion caused by environmental conditions (rain, snow, leaves, etc.). The abrasive effect of skidding causes a high wear on the rolling surface (a wheel-flat), with lengths ranging typically from 20 to over 100 mm.

The rise in temperature caused by abrasion followed by a fast cooling may lead to the formation of brittle martensite beneath the wheel-flat. This can be associated to the beginning of further flaws like cracks and spalls with the loss of relatively large pieces of tread material. When the wheel rolls over a flat, high impact forces are developed and may cause a rapid deterioration of both, rolling and fixed railway structures. Moreover, the incidence of hot bearings, broken wheels and rail fractures are coincidental with the number of wheel-flats and out-of-round wheels (Kumagai et al., 1991; Snyder & Stone, 2003; Vyas & Gupta, 2006; Zakharov & Goryacheva, 2005).

Besides, when a critical speed is reached, a loss of rail-wheel contact occurs which produces high-levels of noise and vibrations that also affect passengers comfort. Finally in the worst-case scenario, surface defects on treads can cause a derailment (Wu & Thompson, 2002; Zerbst et al., 2005).

Therefore, there is a great interest in finding methods for an early detection and evaluation of surface defects without dismantling wheelsets due to their complex assembly (Pohl et al., 2004). Ideally, in order to reduce time and costs, inspection systems should be placed at the end of train-wash stations or at the entrance of maintenance shops, where trains pass frequently at low speed.

Nowadays, many automatic and on-line wheel tread surface defect detection systems have also been developed for the railway industry. Most of them can be classified into one of the following categories:

***Measurement of the impact forces in an instrumented rail*:** the Wheel Impact Load Detector (WILD) developed by Salient Systems, Inc. (2010) is the most popular system to detect wheel-flats; it consists of a large number of strain gauges mounted in the rail web, which are used to quantify the force applied to the rail through a mathematical relationship between the applied load and the deflection of the foot of the rail. As a result, the wheels health and the safe train operations can be ensured by monitoring these impact forces (Stratman et al., 2007). In other cases, accelerometers are used instead (Belotti et al., 2006; Madejski, 2006). These techniques analyse the impacts produced by flats or other kind of flaws, but give no indications about their size.

***Wheel radius variations measurement*:** in this case, the flange is considered perfectly round and wear-free, being used as a reference to estimate the variations in the wheel radius. Mechanical (Feng et al., 2000; He et al., 2005) and optical (Gutauskas, 1992) systems have been designed based on this idea. Nevertheless, the hostile railway environment involves some disadvantages for these methods, both are sensitive to vibrations and their resolution is limited. Furthermore, small irregularities or adherences in the flange will lead to false indications.

***Ultrasonic flaw detection and measurement*:** the Non Destructive Testing (NDT) techniques by ultrasound are often used for offline wheel examination; most of them require complex installations and/or machineries (Kappes et al., 2000). However, recent designs have been developed for online wheel tread inspection using ultrasonic methods. Such systems consist in sending an ultrasound pulse over the rolling surface to detect echoes produced by cracks. The interrogating Rayleigh wave is generated by a transducer (EMAT or piezoelectric) which is fired when the wheel passes over it; the same or other sensor can receive reflections originated on flaws (Fan & Jia, 2008; Ibáñez et al., 2005; Salzburger et al., 2008). Unfortunately, wheel-flats usually have smooth edges that do not generate echoes, so it is very difficult to detect them by these techniques. Moreover, acoustic coupling between transducers and wheels frequently give rise to unreliable measurements due to variability problems.

The aim of this chapter is to present an innovative ultrasound technique designed to detect and quantify wheel-flats that have been formed on the railway wheel tread. An extended theoretical framework supports the proposed method. It can be applied to trains moving at low speed (10-15 Km/h typical) and allows all the wheelsets mounted in a train to be inspected within a few seconds.

2. Alternative methodology

The proposed method differs from other conventional ultrasonic flaw detection approaches, which are based on the reflectivity of static flaws. In this case, surface waves are sent over a measuring rail instead of the rolling surface to interrogate the rail-wheel contact point position. Wheel-flats are then detected and sized by analysing the kinematic of the wheel-rail contact point echo.

The proposed technique comes up with many other advantages for railway industry, e.g., no moving parts are involved in the measuring system; the set of transducer and measuring-rail is invariant, so it can be fully characterized once; results are independent of wheel wear degree, no exploring pulses along rolling surface are sent; finally, an optimum ultrasonic coupling between transducer and rail can be achieved.

Following this methodology two alternatives have been considered to determine wheel-flats: the first one is based on Doppler techniques which are suitable to detect velocity changes of a moving reflector (in this case the rail-wheel contact point). The second consist in measuring

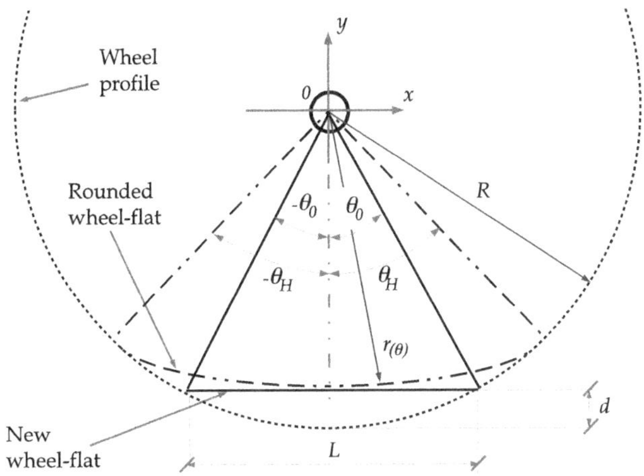

Fig. 1. Large-scale representation of a new wheel-flat and its wear stage.

the round trip time of flight (RTOF) of the echo produced by the rail-wheel contact point. The RTOF value variations allow detecting and quantifying wheel-flats with enough resolution.

2.1 Theoretical background

A wheel-flat of recent formation (*new or fresh wheel-flat stage*) can be represented as a circumference chord defined by two parameters: its length L and the loss of material d. The wheel-flat edges are singular points, where the curvature changes from the nominal wheel radius R to ∞ in the angular interval $[-\theta_0, \theta_0]$. The geometry shown in Fig. 1 lets to obtain the length L as a function of d as follows:

$$L = 2\sqrt{2Rd - d^2} \approx \sqrt{8Rd} \tag{1}$$

Nevertheless, as the wheel keeps in motion, the wheel-flat edges become worn progressively by plastic deformation (*partially rounded wheel-flat stage*). Finally, the irregularity profile eventually disappears due to the successive impacts over the rail (*degenerated or rounded wheel-flat*). At this stage, the wheel-flat becomes a continuous curve of length greater than L. The wheel wear may cause a further increase of the wheel-flat length but, since this process is rather slow and uniform, so the loss of material d remains unchanged until it is removed by a turning machine (Baeza et al., 2006). However, if the rounded wheel-flat length increases above four times from the original size, the defect makes the wheel be out-of-round (Snyder & Stone, 2003).

The wheel tread having a wheel-flat can be described by a function $r(\theta) < R$ in the irregularity and $r(\theta) = R$ outside of the angular interval which defines the defect. For degenerated wheel-flats, this function is continuous with no singular points, so that $\dot{r}(\theta)$ denotes a smooth unbroken curve as well. Although not strictly required, it will be also assumed for clearness a symmetric wheel-flat profile around the interval center located at the point of maximum loss of radius: $d = R - R_{min}$ (Fig. 1).

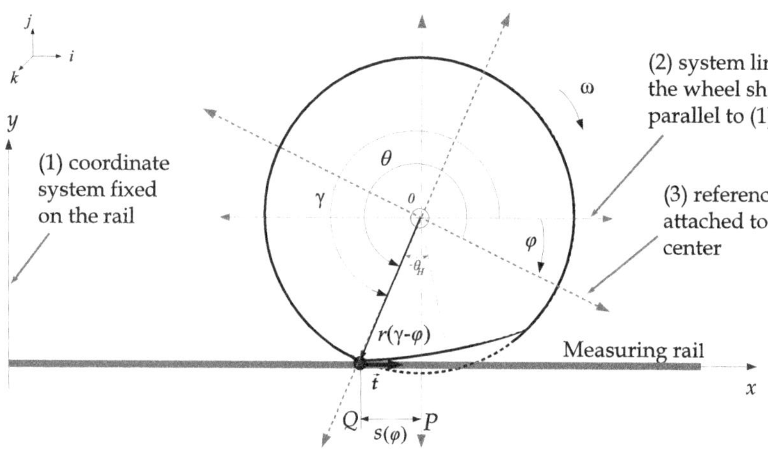

Fig. 2. Scheme used to determine the wheel-flat loss of material d. The wheel moves towards $+x$ direction on the reference system (1). Pay attention that $\varphi < 0$ and the displacement $s < 0$. It is assumed that effects such as creep, spin, and slip, are ignored, so the contact is not lost in the time domain. The contact point Q is defined in system (2) by its polar coordinates $(r(\theta), \theta) = (r(\gamma - \varphi), \gamma - \varphi)$, while in (3) they are $(q(\gamma, \varphi), \gamma)$. The rotation angle φ at each instant represents the difference between fixed and mobile systems, both mounted on the wheel center.

2.1.1 Rail-wheel contact point kinematic

The wheel rotation around its contact point Q can be described by following the scheme in Fig. 2, where can be find three reference systems:

1) A coordinate system fixed on the measuring rail.
2) A system linked to the wheel shaft and parallel to the system 1.
3) A coordinate system attached to the wheel center in motion with vector $\vec{Q}$.

Note that, the rail is always tangent to the wheel at the contact point. When the wheel rolls over a perfectly rounded region the wheel center projection P on the rail coincides with the wheel-rail contact point Q (that means $P = Q$) since, in a circunference, radius and unitary tangent vector $(\vec{t})$ at Q are perpendicular. However, when the wheel rolls over its non circular region $[-\theta_H, \theta_H]$ it rotates an angle φ, so vectors $\vec{Q}$ and $\vec{t}$ are not orthogonal. Moreover, the contact point and the projection are moved apart in a distance s which is a function of φ, reflecting an advance or delay of Q in relation to P. In other words:

a) when $-\theta_H < \varphi \leq 0$, P leads Q and $s \leq 0$;
b) for $0 < \varphi < \theta_H$, P lags Q and $s > 0$.

Fig. 2 shows the situation when $\varphi < 0$, so the contact point is behind the wheel center projection. On the other hand, each point of the wheel tread is described by $\vec{r}(\theta) = (r(\theta), \theta)$ on reference system 2, while on system 3 the coordinates are: $(r(\theta), \theta) = (q(\gamma, \varphi), \gamma)$. Then, the transition from one system to another is given by:

$$q(\gamma, \varphi) = r(\gamma - \varphi) \tag{2}$$

and

$$\frac{\partial\left(q(\gamma,\varphi)\right)}{\partial\gamma}=\frac{\partial\left(r(\gamma-\varphi)\right)}{\partial\gamma}=\dot{r}(\gamma-\varphi) \tag{3}$$

Assuming that contact is not lost while the wheel is moving and the train speed v is constant over the measuring rail. So, the rail-wheel contact point is defined for each rotated angle φ as $Q=\left(q(\gamma_Q,\varphi),\gamma_Q\right)$, such that its ordinate value $q(\gamma,\varphi)\sin(\gamma)$ is minimal on γ_Q, that is:

$$\left.\frac{\partial\left(q(\gamma,\varphi)\sin(\gamma)\right)}{\partial\gamma}\right|_{\gamma=\gamma_Q}=\left.\frac{\partial\left(q(\gamma,\varphi)\right)}{\partial\gamma}\right|_{\gamma=\gamma_Q}\sin(\gamma_Q)+q(\gamma_Q,\varphi)\cos(\gamma_Q)=0 \tag{4}$$

Replacing eqs. (2) and (3) on (4), it is possible to figure out the tangent value at Q, which is:

$$\tan(\gamma_Q)=-\frac{r(\gamma_Q-\varphi)}{\dot{r}(\gamma_Q-\varphi)} \tag{5}$$

The contact point Q belongs to the instantaneous axis of rotation, which remains steady for each φ value, and the wheel center movement can be described, at all time, as a pure rotation around Q. Additionally, as the angular velocity ω is an invariant and independent of any reference systems used, the velocity vector is defined by:

$$\vec{v}=-\vec{\omega}\times\vec{Q}=-\vec{\omega}\times\vec{r}(\gamma_Q-\varphi) \tag{6}$$

where

$$\vec{\omega}=\omega\vec{k} \tag{7}$$

$$\vec{r}(\gamma_Q-\varphi)=\left(q(\gamma_Q,\varphi)\sin(\gamma_Q)\right)\vec{j}+\left(q(\gamma_Q,\varphi)\cos(\gamma_Q)\right)\vec{i} \tag{8}$$

Solving the cross product (6), it yields to:

$$\vec{v}=\begin{vmatrix}\vec{i} & \vec{j} & \vec{k}\\ 0 & 0 & \omega\\ q(\gamma_Q,\varphi)\cos\gamma_Q & q(\gamma_Q,\varphi)\sin\gamma_Q & 0\end{vmatrix}=\omega q(\gamma_Q,\varphi)\sin(\gamma_Q)\,\vec{i}-\omega q(\gamma_Q,\varphi)\cos(\gamma_Q)\,\vec{j} \tag{9}$$

As a result, there are two velocity components on (9). The first on $\vec{i}$-direction, means the train speed. By taking into account (2):

$$v=\omega\, q(\gamma_Q,\varphi)\ \sin(\gamma_Q)=\omega\, r(\gamma_Q-\varphi)\ \sin(\gamma_Q) \tag{10}$$

On the other hand, regarding eqs. (5) and (10), the velocity component in the $\vec{j}$-direction is:

$$\vec{v_y}=-\omega\, q(\gamma_Q,\varphi)\ \cos(\gamma_Q)=\omega\,\dot{r}(\gamma_Q-\varphi)\ \sin(\gamma_Q)=-\frac{v}{\tan(\gamma_Q)} \tag{11}$$

As (11) expresses, for a perfect rounded wheel, the contact point Q will be located at $3\pi/2$, so $\vec{v_y}$ will be null. Note as well on Fig. 2, that the displacement distance between the contact point Q and the wheel center projection P denoted by $s(\varphi)$ is the x-coordinate on the reference system 2. That is:

$$s(\varphi)=-r(\gamma_Q-\varphi)\ \cos(\gamma_Q)=-q(\gamma,\varphi)\ \cos(\gamma_Q) \tag{12}$$

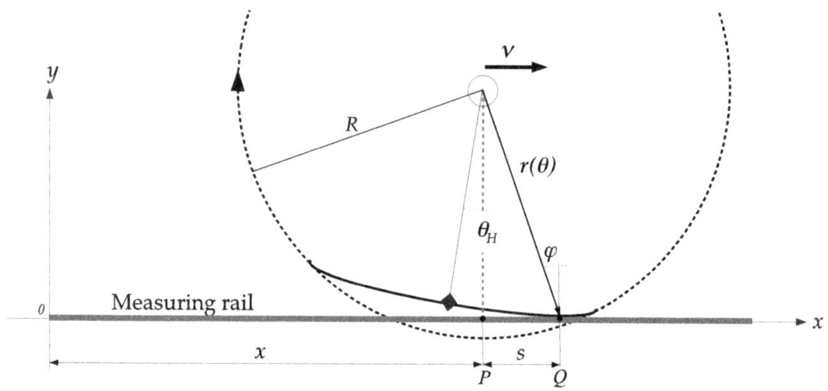

Fig. 3. Position of points P and Q on the reference system fixed to the rail, when the wheel passes over an irregularity.

Therefore, the wheel center vertical movement can be related with the displacement $s(\varphi)$ by replacing (12) on (11):

$$\vec{v_y} = \omega\, s(\varphi) \tag{13}$$

Thus, the wheel center vertical movement is zero for perfectly rounded wheels. However over a wheel-flat irregularity, the wheel center goes down a distance equal to the maximum loss of material d. Afterwards, it starts to rise again up to reach the radius nominal value just when the contact point comes out of the irregularity. As a result, the total wheel center displacement is $2d$ from the time instant t_1 to t_2, when the wheel rolls over the irregularity:

$$2d = \int_{t_1}^{t_2} |v_y|\, \mathrm{d}t = \int_{t_1}^{t_2} |\omega\, s(\varphi)|\, \mathrm{d}t = \int_{-\varphi_H}^{\varphi_H} |s(\varphi)|\, \mathrm{d}\varphi \tag{14}$$

where the limits of integration $-\varphi_H$ and φ_H correspond to the angular range that determines the irregularity. Hence, (14) gives the parameter d as a function of displacements $s(\varphi)$ which can be measured; then the original wheel-flat length is obtained by (1). On the other hand, as $s(\varphi)$ is a continuous function even if singular points are present, (14) is valid for any kind of wheel-flats (new, partially rounded or degenerated). This property can be generalized as:

"For small irregularities, whatever is the wheel-flat wear degree from the original one, the area below $|s(\varphi)|$ is two times the loss of material d".

As well, since $s(\varphi) = 0$ in the round part of the wheel ($|\varphi| > \varphi_H$), the limits of integration may be extended to any angle $\varphi_A \geq \varphi$ for single or isolated wheel-flats cases. Using an auxiliary parameter α in order to cover a full wheel revolution (14) can be written as:

$$d(\alpha) = \int_{\alpha}^{\alpha+\varphi_A} s(\varphi)\, \mathrm{d}\varphi \qquad \text{where} \quad 0 \leq \alpha \leq 2\pi - \varphi_A \quad \text{and} \quad \varphi_A \geq \varphi \tag{15}$$

2.1.2 Rail-wheel contact point velocity

When the wheel moves over the rail the instantaneous position of the wheel center projection is given by x, while the contact point Q is $x + s$. Note that the distance x is now measured

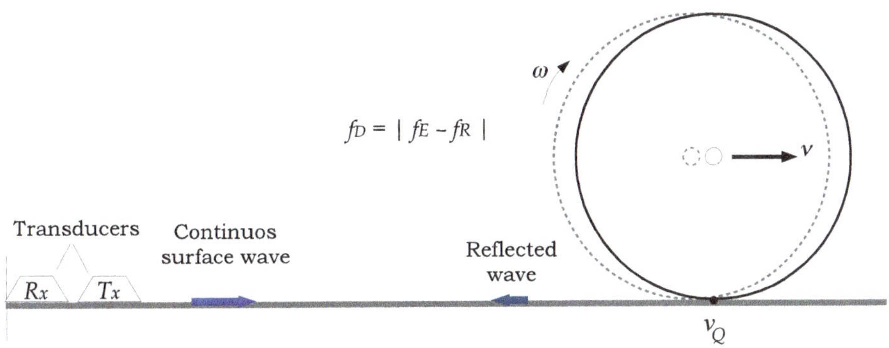

Fig. 4. Arrangement used for wheel-flat detection by Doppler effect.

from the origin located at reference system fixed to the rail (Fig. 3). The time derivative of the Q position gives the wheel-rail contact velocity as:

$$v_Q = \frac{\mathrm{d}x}{\mathrm{d}t} + \frac{\mathrm{d}s}{\mathrm{d}t} = v + \frac{\mathrm{d}s}{\mathrm{d}t} = v + \omega\,\frac{\mathrm{d}s}{\mathrm{d}\varphi} \tag{16}$$

where v is the train speed. As the wheel rolls over a perfectly rounded region, $s = 0$ and $v_Q = v$. However when the contact point reaches an irregularity, s takes negative values because the point Q is behind the wheel center projection P and $v_Q \leq v$. On the other hand, when Q is ahead of P, s becomes positive and $v_Q \geq v$. Therefore, the presence of an irregularity can be determined by measuring variations in the contact point velocity; or what is the same, the time variations of s, or s in relation to φ, since $\mathrm{d}\varphi = \omega \mathrm{d}t$.

3. Measuring techniques

Taking into account the arrangement shown on Fig. 3, the rail can be used as the physical support for Rayleigh waves in order to determine velocity changes while the rail-wheel contact point is in motion.

A conventional ultrasound transducer mounted on a plastic wedge, which has a propagation velocity c_w, can be used to generate Rayleigh waves by properly adjusting the incidence angle β of the emitted signal (Bray et al., 1973). Following the Snell's law for refraction at 90°:

$$\sin(\beta) = \frac{c_w}{c} \tag{17}$$

By this way, ultrasonic waves propagate over the measuring rail, eventually producing an echo when it arrives at the rail-wheel contact point. Two alternatives to perform the measurement will be described as follows.

3.1 Wheel-flat detection using Doppler effect

In this case, monochromatic surface waves of frequency f_E generated by an emitter transducer are propagated over the measuring rail. Another sensor receives the weak echo signal produced by the wheel-rail contact. The emitting transducer is placed ahead of the receiver to avoid reflections; both are fixed to the rail to achieve a better acoustic coupling (Fig. 4).
The received frequency f_R, once amplified, is compared with the emitted f_E by a quadrature

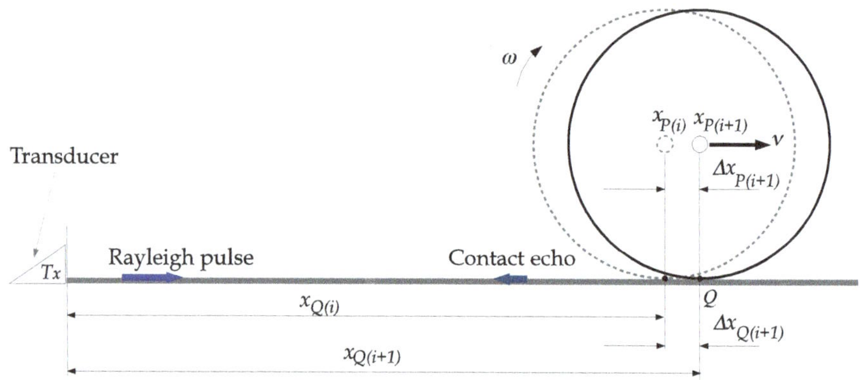

Fig. 5. Arrangement used for measuring the round trip time of flight (RTOF).

demodulator for recovering the Doppler frequency $f_D = |f_E - f_R|$. This frequency shift is proportional to the rail-wheel contact point velocity v_Q by:

$$f_D = 2\frac{v_Q}{c}f_E \tag{18}$$

where c is the propagation velocity for Rayleigh waves. Then by using (16), it is possible to reach the following expression:

$$f_D = \frac{2v}{c}f_E + \frac{2f_E}{c}\frac{\mathrm{d}s}{\mathrm{d}t} \tag{19}$$

The first term on (19) represents the nominal Doppler shift which is considered constant and it is proportional to the train speed. When the wheel rolls over a perfectly rounded region, the second term is zero ($s = 0$). However if the wheel passes over a wheel-flat, the displacement s varies over the time, shifting the nominal Doppler frequency. Then:

$$\frac{\mathrm{d}s}{\mathrm{d}t} = \frac{c}{2f_E}f_D - v \tag{20}$$

Moreover, (20) can be easily implemented for detecting flaws in a continuous way as:

$$s(t) = \left(\frac{c}{2f_E}f_D - v\right)t \tag{21}$$

since c, f_E are design data, and f_D, v are the measurements taken.
The practical implementation of (20) is restricted by the time-frequency uncertainty conflict which makes the resolution be reduced. This limitation must be addressed through compromise solutions to get reliable flaw indications. Nevertheless, wheel-flats longer than 30 mm can be detected although sizing is difficult. The method also provides a time-frequency analysis that gives a quick reference about the wheel tread state (Brizuela, Ibáñez, Nevado & Fritsch, 2010).

3.2 Wheel-flat detection by measuring the round trip time of flight (RTOF):

This alternative uses only one transducer located at one end of the rail (Fig. 5). The sensor generates pulses of Rayleigh waves with a period T_{PRF} and receives the echo produced by the moving rail-wheel contact point. The rail-wheel contact position obtained from measuring the RTOF ($T_{Q(i)}$) of a pulse i is:

$$x_{Q(i)} = \frac{c\, T_{Q(i)}}{2} \tag{22}$$

where c also means the propagation velocity for Rayleigh waves. Assuming that the train moves at a constant speed v; the wheel center projection over the rail for the same pulse i is:

$$x_{P(i)} = i\, T_{PRF}\, v \tag{23}$$

Therefore, the relative displacement between P and Q can be figured out as follows:

$$s(i) = x_{Q(i)} - x_{P(i)} = \frac{c\, T_{Q(i)}}{2} - i\, T_{PRF}\, v \tag{24}$$

On the other hand, a sampled system can approach the integration indicated on (15) as:

$$d_k(M) = \frac{v\, T_{PRF}}{R} \sum_{i=k}^{k+M} s(i) \tag{25}$$

where the differential on (15) has been replaced by $\Delta\varphi = \omega\, \Delta t = v\, T_{PRF}/R$ and M is the discrete version of the angle φ_A. The sum given by (25) is extended over M samples of $s(i)$ which is obtained from (24). Consequently, M should be chosen to cover at least half the largest irregularity. Since the sampling period of $s(i)$ is T_{PRF} and L_{max} is the length of the largest wheel-flat of interest,

$$M \geq \frac{L_{max}}{2\, v\, T_{PRF}} \tag{26}$$

The measuring process (25) is carried out as a convolution of a rectangular unity window of M samples with the values of $s(i)$. This way M measurements of d are taken: the sequence $d_k(M)$ represents the value of the loss of material d, which is estimated by convolving the k with a window of size M. Since the window must be wider than the irregularity, the peak value of this sequence corresponds to the best estimation of d for isolated wheel-flats. Fig. 6 shows graphically the measuring process. For a given value of M samples, the resulting sequence $d_k(M)$ has two peaks, one negative d_N followed by another positive d_P, which correspond to the lead-lag of P respect to Q on the signal displacement s. Their absolute values scaled by the factor $v\, T_{PRF}/R$, make the result be equal to d. However in real applications where noise is present, d can be much better estimated by the peak-to-peak average:

$$d_E(M) = \frac{v\, T_{PRF}}{R} \frac{|d_P(M)| + |d_N(M)|}{2} \tag{27}$$

Once the estimated value d_E has been found, the length corresponding to the original wheel-flat L_E can be obtained by application of (1).

This methodology also provides a suitable possibility of estimating the train speed required by (27). The contact point velocity is $v_Q = v$ while the wheel rolls over a defect-free region, otherwise during an irregularity $v_Q \neq v$. Nevertheless, the average speed along the wheel-flat is $\hat{v}_Q = v$. Consequently, by calculating the contact point mean-speed for a long enough time interval the train speed is obtained with a good accuracy.

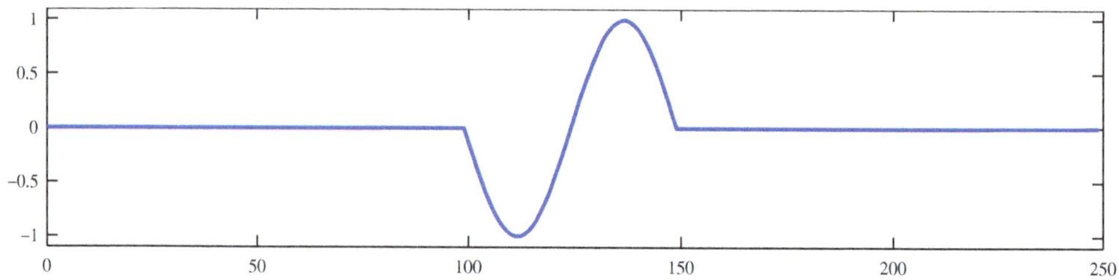

(a) Displacement signal s

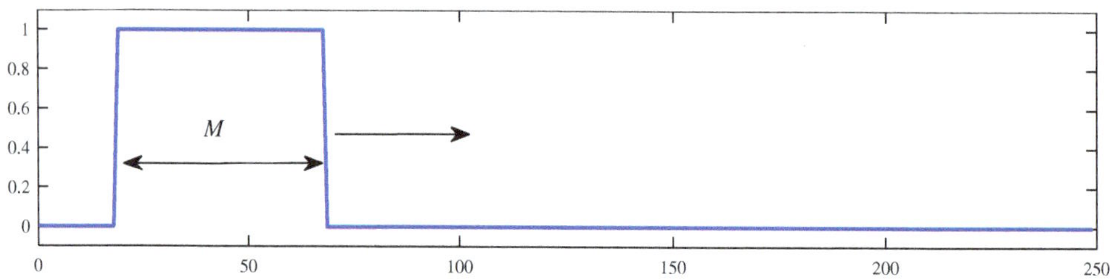

(b) Rectangular unity window $\mathbf{w}$

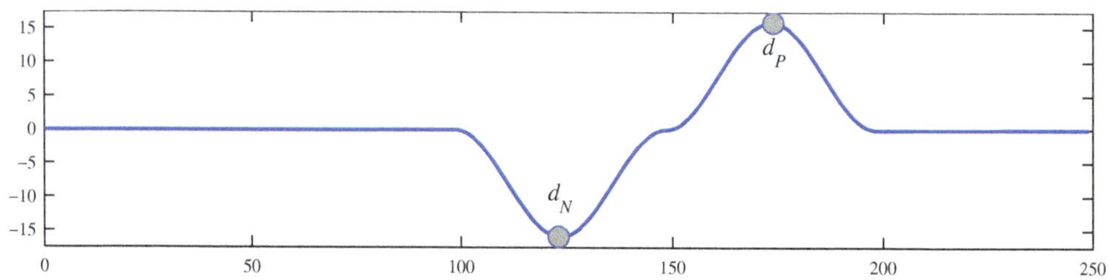

(c) Resulting sequence $d_k = s * \mathbf{w}$.

Fig. 6. Measuring process carried out as a convolution of s with an unit window of M samples.

The contact point moves a distance $\Delta x_{Q(i)}$ between two consecutive pulses. The $x_{Q(i)}$ position is measured at a time:

$$t_{(i)} = i\, T_{PRF} + \frac{T_{Q(i)}}{2} \tag{28}$$

in the following interrogation pulse, the Q position is sampled at:

$$t_{(i+1)} = (i+1)\, T_{PRF} + \frac{T_{Q(i+1)}}{2} \tag{29}$$

The elapsed time between measures (28) and (29) is:

$$\Delta t_{(i)} = T_{PRF} + \frac{\Delta T_{Q(i)}}{2} \tag{30}$$

The Q movement is:

$$\Delta x_{Q(i)} = v_{Q(i)}\, \Delta t_{(i)} \tag{31}$$

At the same time, the ultrasound pulse has also to cover the distance advanced by the contact point, thus:

$$\Delta T_{Q(i)} = 2\,\frac{\Delta x_{Q(i)}}{c} \tag{32}$$

where the constant c is the propagation velocity for Rayleigh waves on the rail. By replacing (31) on (32) and then combining on (30), it yields:

$$\Delta T_{Q(i)} \left(\frac{c - v_{Q(i)}}{c}\right) = 2\,\frac{v_{Q(i)}}{c}\,T_{PRF} \tag{33}$$

where the factor $(c - v_{Q(i)})/c$ means a Doppler shift between the frequency at which the interrogation pulses are emitted $(1/T_{PRF})$ and at which they are received. As $v_{Q(i)} << c$ it can be assumed that $(c - v_{Q(i)})/c \approx 1$. Therefore, the contact point velocity $v_{Q(i)}$ is obtained from (33) as:

$$v_{Q(i)} \approx \frac{\Delta T_{Q(i)}}{2\,T_{PRF}}\,c \tag{34}$$

where all values in (34) are known and $\Delta T_{Q(i)}$ results from the difference of two consecutive RTOF measures. The $v_{Q(i)}$ value represents the instantaneous velocity estimation of the contact point by sending a pulse i. Afterwards, the train speed v can be found out by averaging N measurements of $v_{Q(i)}$:

$$v \approx \hat{v}_{(j)} = \frac{1}{N}\left(\sum_{i=j}^{i=j+N-1} v_{Q(i)}\right) \tag{35}$$

Finally, the measuring process indicated on (27) only requires the knowledge of the wheel radius R, whose value may be smaller than the nominal one due to the wheel wear. On the other hand, the product $v\,T_{PRF} = \Delta x$ represents the spatial sampling interval which determines the length resolution of the irregularity.

4. Measuring process simulation

In practical applications, signals are interfered by electrical and structural noise leading to variations in the measured position $x_{Q(i)}$. Nevertheless the measuring method is very robust against noise due to the integration performed on (25).

Fig. 7(a) shows a simulated sequence $s(i)$ corresponding to a degenerated wheel-flat with d = 0.4 mm in a wheel of R = 500 mm. It has been acquired at intervals $\Delta x = v\,T_{PRF} = 0.6$ mm in a rail with c = 3000 m/s. In order to test the technique performance, this sequence has been deeply contaminated with white Gaussian noise and standard deviation increasing with distance. The signal represents a typical acquisition when the Time-Gain-Compensation (TGC) function, on an ultrasound equipment, has been turned on to receive similar echo amplitudes from different distances . Interferences signify a considerable uncertainty in finding the actual echo position after every pulse i.

The resulting sequence d_k obtained from the noisy signal $s(i)$ by application of (25) is shown in Fig. 7(b). It has been used a window size of $M = 267$ samples, or following (26), this corresponds to a maximum wheel-flat length of $L_{max} = 320$ mm. It can be seen the filtering effect of the sum, as well as the agreement of the negative and positive peaks with the correct $d = 0.4$ mm value ($d_N = 0.3898$ mm, $d_P = 0.4824$ mm). The averaged estimation is $d_E =$ 0.4361 mm.

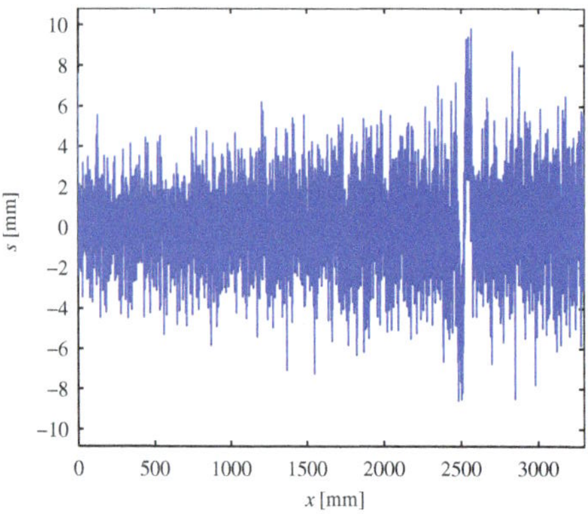

(a) Simulation of a noisy sequence s.

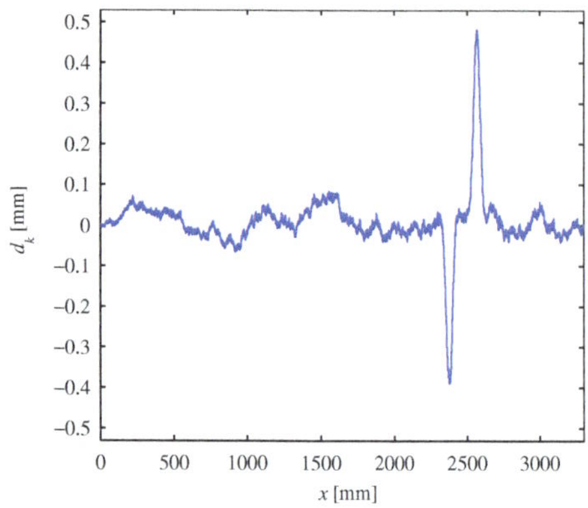

(b) Resultant d_k sequence.

Fig. 7. Simulation parameters: $R = 500$ mm, $c = 3000$ m/s, $d = 0.4$ mm, and $\Delta x = v\,T_{PRF} = 0.6$ mm.

4.1 Choosing the integration window

In the case of a single wheel-flat, the M value must be strictly chosen greater than the number of samples obtained from a cycle of s in order to provide a robust estimation of d. However, for multiple wheel-flats which are frequently formed on modern high speed trains due to the failure of disc brakes and wheel slid prevention systems (WSP) (Grosse et al., 2002; Kawaguchi, 2006). In such case, a too large integration window may include information belonging to different flats giving wrong results.

Fig. 8(a) shows a simulated s sequence acquired at $\Delta x = 0.6$ mm. The signal corresponds to three degenerated wheel-flats of $d = 0.5, 0.6$, and 0.4 mm without overlapping (separated at 157 mm) in a wheel of $R = 500$ mm. The resulting d_k sequence using an integration window $M_x = \Delta x\,M = 150$ mm (distance less than wheel-flats separation) provides a wrong information, since no indication is given about the largest defect of $d = 0.6$ mm (Fig. 8(b)). These effects have been originated by the size chosen for the window M_x. So when the integration is performed, s values corresponding to consecutive wheel-flats are combined (note that first and last cycles, corresponding to the extreme flaws, have been correctly evaluated).

Repeating the same procedure but using a smaller window $M_x = 75$ mm (half of the previous case), the resulting sequence d_k contains information according to the loss of material d of each wheel-flat: $d_{N_1} = 0.5107$ mm, $d_{P_1} = 0.4859$ mm, $d_{N_2} = 0.5876$ mm, $d_{P_2} = 0.5485$, $d_{N_3} = 0.4688$ mm, $d_{P_3} = 0.3678$ mm (Fig. 8(c)).

Therefore, the value of M should be determined by a heuristic process trying out different integration window sizes for each obtained signal s. From this process, it is interesting to know the greatest estimation which is linked to the deeper wheel-flat. Moreover, from the standpoint of railway maintenance, this value indicates whether the wheel should be reprofiled or removed from service. Fig. 8(d) shows the estimations resulting of trying out different window sizes ($1 \leq M_x \leq 150$ mm) on the signal shown in Fig. 8(a). The maximum value provides an estimation of the greater wheel-flat characteristic: $\max\left[d_E(M)\right] = 0.6326$ mm which is a value acceptably close to the real one ($d_3 = 0.6$ mm).

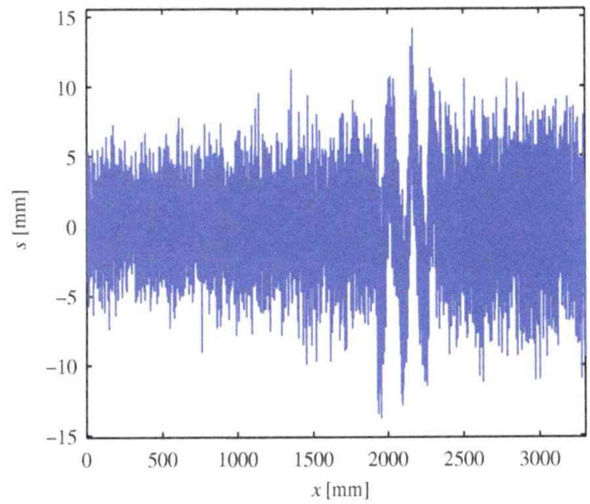

(a) Simulation of a noisy sequence s.

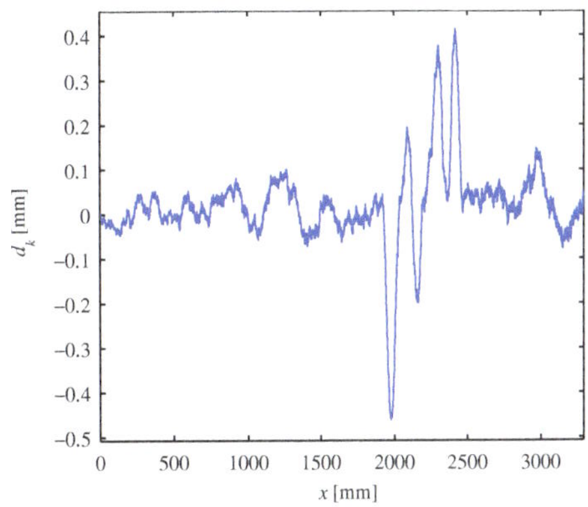

(b) Sequence d_k using $M_x = 157$ mm.

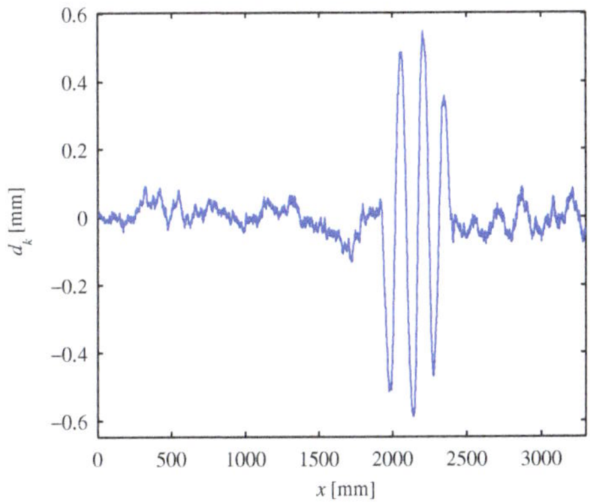

(c) Sequence d_k corresponding to $M_x = 75$ mm

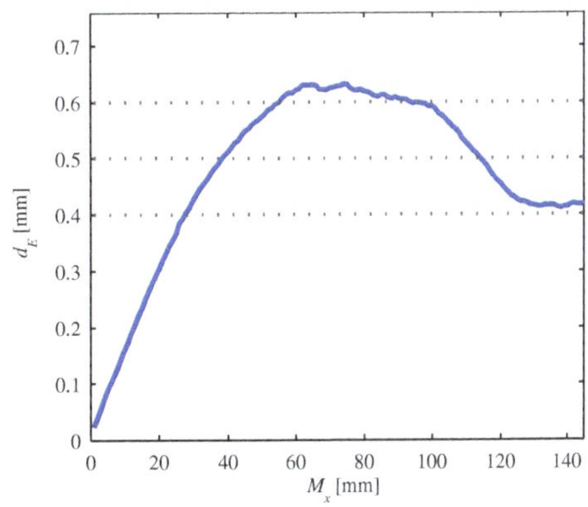

(d) $d_E(M)$ estimated by a multiple windows process.

Fig. 8. Simulation parameters: 3 wheel-flats without overlapping; loss of material for each wheel-flat $d = 0.4, 0.6, 0.5$ mm; wheel radius $R = 500$ mm and $\Delta x = \nu\, T_{PRF} = 0.6$ mm.

5. Prototype inspection system

An UltraSCOPE®, Dasel Sistemas (2010), ultrasonic testing instrument was modified to support this methodology. In order to optimize bandwidth and storage requirements, the acquisition time has to be kept small. Thus, a tracking algorithm was developed and hardware implemented to make a narrow acquisition window around the rail-wheel contact point follows the contact-echo displacement (Brizuela, Ibáñez & Fritsch, 2010).

On the other hand, the acquired signals are interfered by grain noise and many other propagation modes in the measuring rail, which cannot be removed by conventional filtering. Fortunately, this kind of noise can be partially removed because it is mostly static. Therefore, a noise cancellation procedure was also included with the tracking algorithm to avoid losing the contact echo. To this purpose, a vector formed by the difference of the absolute values between two consecutive acquisitions is obtained. Electrical noise is also reduced by applying a programmable narrow-bandpass 63-coefficients FIR filter.

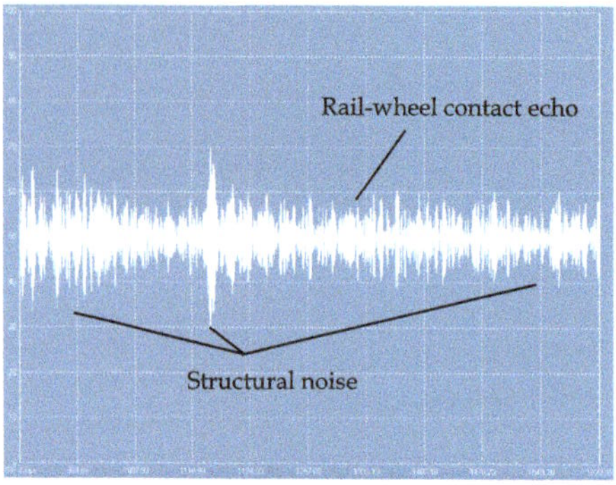

(a) Original A-scan.

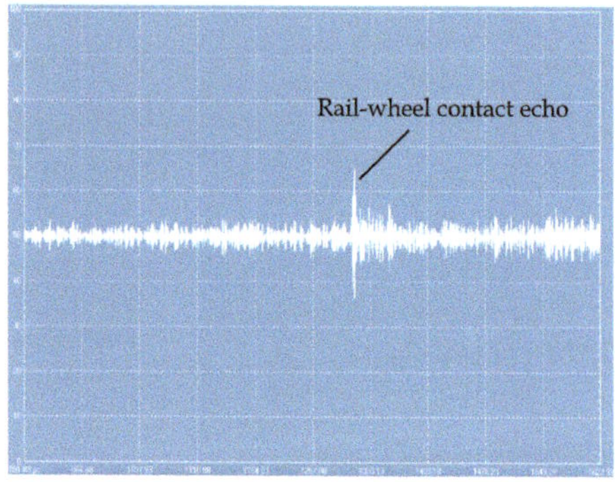

(b) Differential trace.

Fig. 9. Structural noise cancellation. In both cases the wheel is in motion near the same rail region.

Since the acquisition system operates over a differential trace, if the wheel is static or non-present, A-scans will be null excepting the electrical noise. While the wheel is in motion, A-scans will contain the information about the actual rail-wheel contact point position as a positive indication and the precedent one with negative sign. While the acquisition window is moving, the noise cancellation algorithm is performed using the samples which correspond to the same spatial point on the rail in consecutive acquisitions. Fig. 9(a) shows a high rail structural noise which masks the wheel contact-echo in the acquired signal. After structural noise removal, the wheel echo is clearly visible (Fig. 9(b)).

In addition other functions have been included, such as, Time-Gain-Compensation (TGC) to receive similar contact-point echo amplitudes from different distances; a programmable burst generator to drive transducers through a power stage (pulser). Finally, measurements are launched automatically when a wheel is detected over the rail.

The recorded echo signals are sent through a USB 2.0 interface to an evaluation computer, which looks at the position of the signal value maximum in each capture to recover the RTOF $T_{Q(i)}$ required to compute $s(i)$ and $v_{Q(i)}$ on eqs. (24) and (16).

5.1 Test bench

Finally, to evaluate the proposed technique performance, an experimental test bench was arranged (Fig. 10(a)). A 1 MHz piezoelectric transducer generates Rayleigh wave pulses in a 2000 mm long measuring rail. A pair of wheel treads with $R = 420$ mm were used to perform the experimental work.

Two artificial wheel-flats were mechanized, one on each wheel tread. Figs. 10(b) and 10(d) show the test profiles measured with a mechanical comparator (10(c)). The maximum loss of material for the *wheel-flat* #1 is $d = 0.46$ mm. By application of (1) this corresponds to a new wheel-flat length of $L = 39.3$ mm. A simple comparative between measured and new profiles can be also done in Fig. 10(b). Note that profile #1 corresponds to an asymmetric partially rounded wheel-flat.

(a) Experimental test bench.

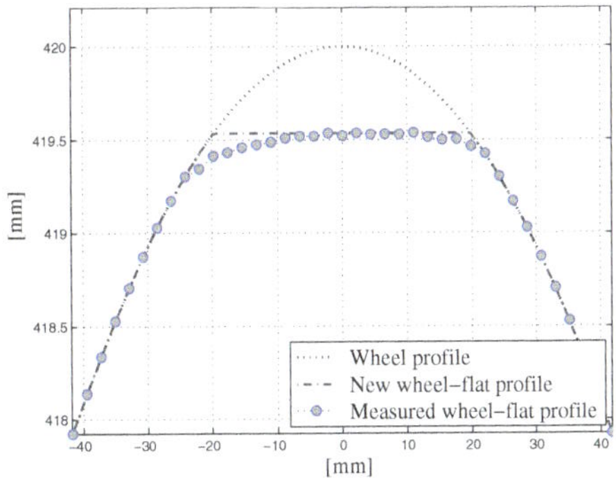

(b) *Wheel-flat* #1

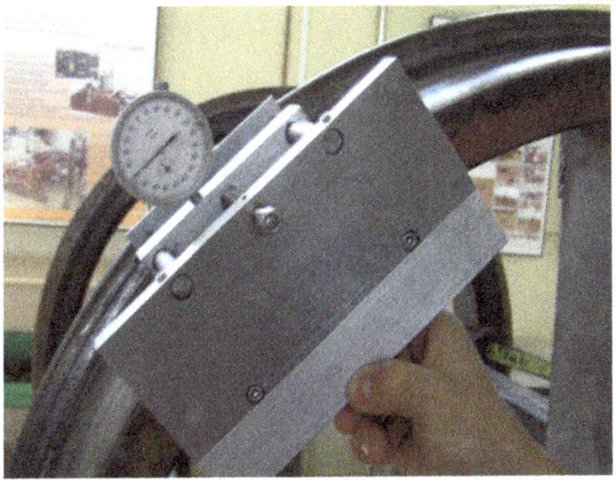

(c) Mechanical measuring instrument.

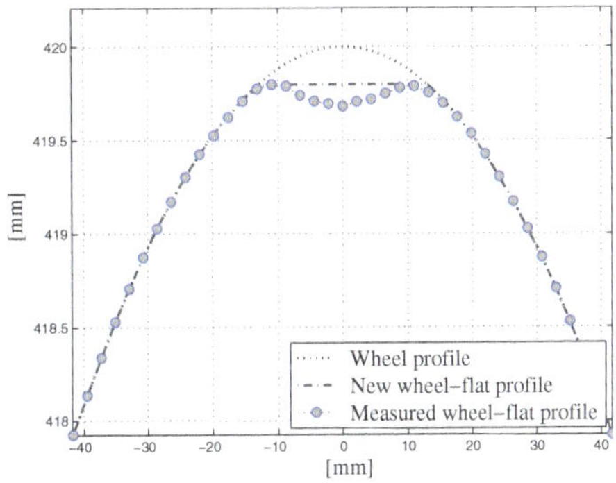

(d) *Wheel-flat* #2.

Fig. 10. Prototype inspection system used for laboratory testing and artificial wheel-flat profiles.

On the other hand, the *wheel-flat* #2 profile shown in Fig. 10(d), has a shape slightly concave (a cavity). This artificial flaw emulates a new wheel-flat, since it is not possible to roll over a cavity. The corresponding new wheel-flat is defined by $d = 0.20$ mm which yields to a length of $L = 25.9$ mm.

6. Experimental results

The wheelset was moved by hand over the measuring rail, so that the speed was not quite constant. However, the rail-wheel contact position $x_{Q(i)}$ is obtained by (22) as a function of the measured RTOF $T_{Q(i)}$ and the ultrasound propagation velocity $c = 2970$ m/s. Consequently, the mean speed v can be also estimated for every position in the rail by applying (35).

The *wheel-flat* #1 was tried out under this procedure. The estimated mean speed near the flat region was $v \approx 0.45$ m/s and the spatial sampling interval was $\Delta x = v\ T_{PRF} = 0.90$ mm. Figure 11(a) shows the contact point distance $x_{Q(i)}$ as a function of the trigger number. Note the jump in x_Q when the wheel rolled over the wheel-flat, which is shown with more detail in Fig. 11(c). The current wheel-flat length could be obtained from this graph by observing the slope of x_Q which changes between triggers #820 and #860, that is, an interval $\Delta t = 80$ ms at

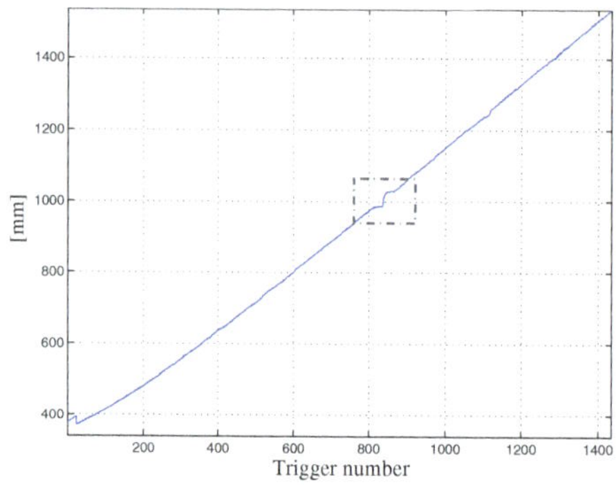

(a) Rail-wheel contact point position x_Q.

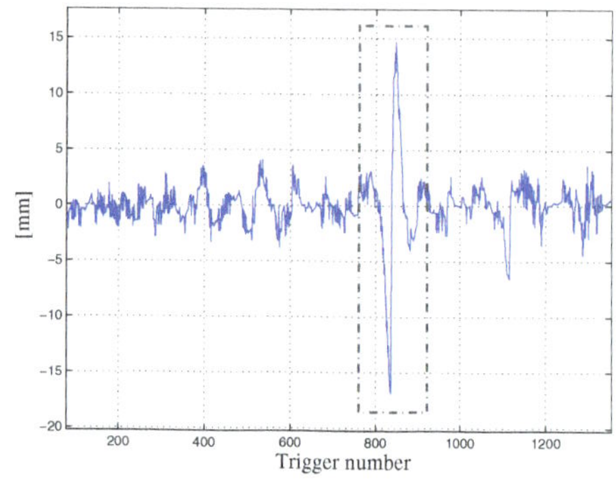

(b) Displacement signal s.

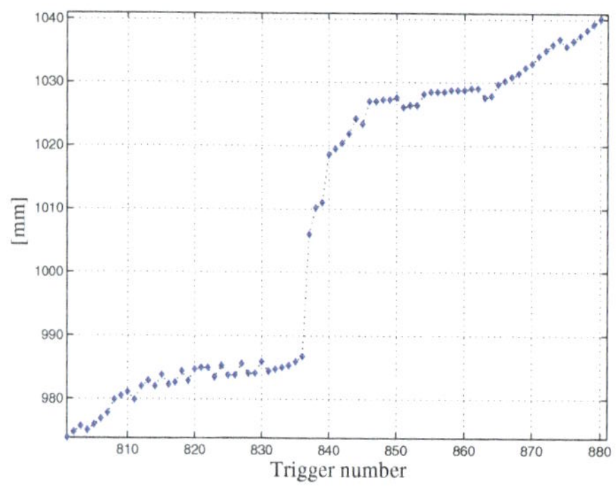

(c) Detailed view of Fig. 11(a).

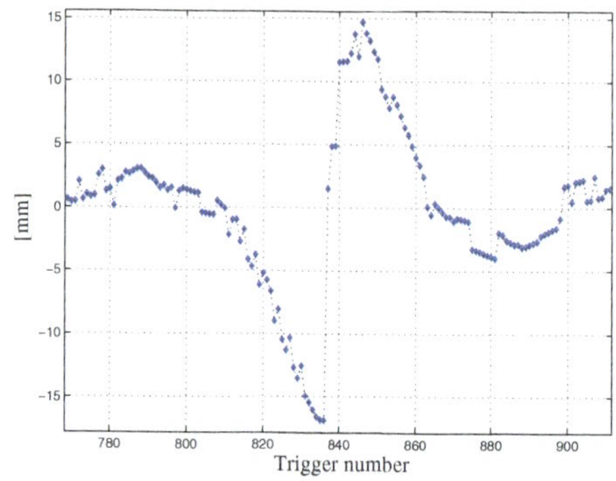

(d) Detailed view of Fig. 11(b).

Fig. 11. Wheel position obtained by measuring the RTOF ($T_{Q(i)}$) and the relative displacement between contact point and the wheel center projection as a function of the trigger number.

$T_{PRF} = 2$ ms. The wheel-flat length can be easily found out by multiplying the train speed v and Δt, which yields to $L \approx 36.4$ mm. Nevertheless, this measurement method is rather imprecise, since it depends on finding the points where x_Q changes its slope. Moreover, it gives no information about the dimensions of the original new wheel-flat (length and loss of material). Therefore, it is much better to estimate the original loss of material d by applying the area calculation on $|s(i)|$.

Figure 11(b) shows the displacement s around the flat region computed by (24). This sequence is contaminated by the uncertainty of locating the exact position of the echo signal due to residual noise (Fig. 11(d)). Then, (25) was applied with different window widths ($2 \leq M \leq$ 250 samples), obtaining $d_E(M)$ from (27).

Figure 12(a) shows the resulting estimation of d_E as a function of M. It can be seen that, for M values below those indicated by (26), that is, $M < 16$, the estimated d_E value shows errors. Above this figure, the estimation remains steady near the true value (0.46 mm), with an average $d_{E_{mean}} = 0.40$ mm and a standard deviation $\sigma_{d_E} = 0.03$ mm, in agreement with theory. The maximum estimation is obtained when $M = 27$ (or $M_x = 24.30$ mm), where $d_E(27)$

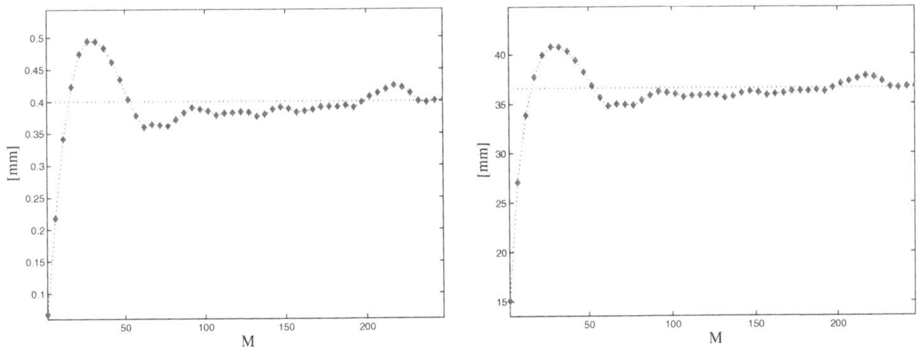

(a) Estimated loss of material d_E.   (b) Estimated new wheel-flat length L_E.

Fig. 12. Loss of material d and new wheel-flat length L_E estimations as a function of the integration window size.

Parameter		*Wheel-flat#1*			*Wheel-flat#2*		
Loss of material	d [mm]		0.46			0.20	
Equivalent new wheel-flat length	L [mm]		39.3			25.9	
Distance to transducer	x [mm]	500	1000	2000	700	900	1300
Mean speed	v [m/s]	0.33	0.45	0.54	0.21	0.25	0.34
Sampling interval	Δ_x [mm]	0.66	0.90	1.62	0.42	0.50	0.68
Highest estimation	$d_{E_{max}}$ [mm]	0.45	0.48	0.56	0.21	0.20	0.16
Mean estimation	$d_{E_{mean}}$ [mm]	0.40	0.40	0.40	0.17	0.16	0.15
Estimated maximum length	$L_{E_{max}}$ [mm]	38.90	40.75	43.72	27.07	26.16	23.57
Relative error	$\varepsilon(L_{E_{max}})$%	-1.01	3.68	11.32	4.51	0.10	-8.99
Estimated average length	$L_{E_{mean}}$[mm]	35.40	36.61	40.30	24.14	23.02	21.06

Table 1. Wheel-flats evaluation at different distances from the transducer.

= 0.49 mm a value slightly higher than the measured value. In fact, it can be seen that using a large value for M has little impact in the measurement of isolated wheel-flats.

Figure 12(b) shows the corresponding $L_E(M)$ values, using the estimations $d_E(M)$ and (1). The average value $L_{E_{mean}}$ is 36.61 mm with a standard deviation $\sigma_{L_E} = 1.39$ mm. Note that the equivalent new wheel-flat length is 39.3 mm. These results show a small error by defect, which are due to approaching the integral (15) by a sum (25). The maximum length estimated is $L_{E_{max}} = 40.75$ mm and it is found at $M = 27$ as well.

Following this methodology, both artificial wheel-flats were evaluated in several positions over the measuring rail in order to put them under different conditions of structural noise interference. On Table 1 the obtained results of these experiments have been summarized. Note that as the wheel speed has not been controlled, so the spatial sampling interval is different for each test.

For all cases, the lower relative error with regard to the true value is reached when the integration window extent is close to the defect length, so the estimation is maximum. The error increases when measurements are made beyond 1300 mm of the transducer, where

Parameter		*Wheel-flat #1*									
Decimation factor	n	1	2	3	4	5	6	7	8	9	10
Mean speed	v [m/s]	0.41	0.82	1.24	1.65	2.06	2.48	2.89	3.30	3.72	4.13
Spatial sampling interval	Δx [mm]	0.90	1.82	2.72	3.63	4.54	5.45	6.36	7.27	8.18	9.09
Window size	M [samples]	28	14	12	9	7	6	5	5	4	4
Estimated maximum length	$L_{E_{max}}$ [mm]	40.75	40.09	40.60	39.47	39.29	37.72	37.64	41.43	42.32	42.93
Relative error	$\varepsilon(L_{E_{max}})$%	3.69	2.02	3.32	0.45	-0.01	-4.01	-4.22	5.42	7.68	9.25
Estimated average length	$L_{E_{mean}}$ [mm]	36.42	35.94	35.90	34.77	38.38	36.90	37.13	34.42	34.37	34.05

Table 2. New wheel-flat estimated at different train speed using the recorded signal on Fig. 11(b).

interrogation pulse attenuation is important and structural noise interferences on the echo signal are higher, increasing the error on determining the contact point position. Nevertheless, the estimated lengths corresponding to *wheel-flat #1* have been obtained with a relative error below 12%. On the other hand, detecting the *wheel-flat #2* is more difficult because it is smaller. In this case, the amplitude of the displacement signal s is comparable to the residual noise. However, lengths have been estimated with an error below 9%, which confirms experimentally the method robustness.

6.1 Measures under different speed inspection conditions

The technique behaviour has been also tested by making the measurements at multiple speeds (more than 3m/s). The recorded displacement signals were decimated in order to increase the equivalent wheel speed. Table 2 contains the wheel-flat length estimations corresponding to the displacement signal shown in Fig. 11(b). In this case, the integration window size was bounded to $M_x = 150$ mm. The maximum length estimated is obtained when the window size is close to the equivalent new wheel-flat length (39.9 mm), however fewer samples are contained on M as the speed increases.

The estimated maximum lengths of the new wheel-flat at speeds up to 3 m/s remain close to the true value with relative errors that do not exceed 5 %, which means an inaccuracy of 1.9 mm. For higher train speeds the relative error increases because the spatial sampling interval will increase as well. Note also that, as the inspection speed increases the average wheel-flat lengths tend to decrease as a consequence of capturing fewer samples.

Consequently, inspection speeds that exceed a maximum value of 3 m/s should be avoided to keep enough resolution.

7. Conclusion

Over this chapter, a new concept supported on a theoretical background has been disclosed to detect and measure wheel-flats in the rolling surface of railway wheels. Wheel-flats are quantified by the sum of relative displacements between the wheel-rail contact point and the wheel center projection on the rail, which yields the loss of material produced by abrasion in the original flat formation. These displacements can be obtained from ultrasonic techniques by analysing the contact point echo. Two methods based on the same measuring arrangement but different principles have been described and discussed.

In the first proposed method based on the Doppler effect, wheel-flats can be easily distinguished. However, this technique does not allow measuring their length due to the uncertainty in the time-frequency analysis. Nevertheless, it provides a reliable wheel-flat indication, which can be very useful to give warning of failure.

On the other hand, the RTOF measuring alternative provides a high resolution which allows sizing wheel-flats. In this technique, the displacement signal s is obtained from the measured RTOF and the wheel mean speed can be also estimated.

This methodology has been tested by simulation using high noise levels in signals and wheel-flats with different stages of roundness. The simulation results have proven that this technique is robust against noise and the measurement is independent of wheels wear degree and wheel-flats roundness.

An experimental test bench was built to evaluate the technique performance. The specific hardware design provides a robust support for a quantitative measurement of wheel-flats, independently of the hostile railway environment, weather conditions and wheel wear. Two artificial wheel-flats of 40 and 26 mm length with different wear degree were mechanized in a railway wheelset for testing. The artificial wheel-flats were placed at different positions over the measuring rail and the estimated lengths remained close to the true value with a low relative error. Thus, simulation and experimental results agree with the theoretically expected.

The inspection speed was also taken into account for the RTOF measuring method. Recorded signals were decimated in order to increase the equivalent wheel speed. The system resolution decreases as the speed increases; nevertheless estimated lengths remained close to the real value with relative errors below 5% at maximum speed (3 m/s).

The proposed methodology, as being a dynamic technique without moving parts but with a well-characterized and stable measuring arrangement, is suitable for the railway industry. The system allows a periodical wheel inspection, improving reliability, availability and effective operations of the railway system, guaranteeing high safety standards. Moreover, it allows reducing the maintenance costs. The inspection procedure can be performed while the train is getting into a repair shop reducing the time spent in maintenance scheduled tasks. As a result, frequently inspections can allow to follow-up the wheel wear history by uploading the information into a database and optimizing wheels service.

8. References

Baeza, L., Roda, A., Carballeira, J. & Giner, E. (2006). Railway train-track dynamics for wheelflats with improved contact models, *Nonlinear Dynamics* 45: 385 – 397.

Belotti, V., Crenna, F., Michelini, R. & Rossi, G. (2006). Wheel-flat diagnostic tool via wavelet transform, *Mechanical Systems and Signal Processing* 20(8): 1953–1966.

Bray, D., Dalvi, N. & Finch, R. (1973). Ultrasonic flaw detection in model railway wheels, *Ultrasonics* 11(2): 66–72.

Brizuela, J., Ibáñez, A. & Fritsch, C. (2010). NDE system for railway wheel inspection in a standard FPGA, *Journal of Systems Architecture* 56: 616 – 622.

Brizuela, J., Ibáñez, A., Nevado, P. & Fritsch, C. (2010). Flaw detector for railways wheels by doppler effect, *Physics Procedia* 3(1): 811 – 817.

Dasel Sistemas (2010). Ultrasound technology. viewed March 22, 2012, http://www.daselsistemas.com.

Fan, H. & Jia, H. (2008). Study on automatic testing of treads of running railroad wheels, *Proceedings of the 17th World Conference on Nondestructive Testing*, 17th WCNDT 2008, Shangai, China.

Feng, Q., Cui, J., Zhao, Y., Pi, Y. & Teng, Y. (2000). A dynamic and quantitative method for measuring wheel flats and abrasion of trains, *15th World Congress on NDT*, Rome, Italy.

Grosse, M., Ceretti, M. & Ottlinger, P. (2002). Distribution of radial strain in a disc-braked railway wheel measured by neutron diffraction, *Applied Physics A: Materials Science and Processing* 74: 1400 ? 1402.

Gutauskas, P. (1992). Railroad flat wheel detectors. US Patent No. 5 133 521.

He, P., You, Z. & Teng, S. (2005). Fast algorithm of flat sliding detection in flat wheel detecting system, *Instrumentation and Measurement Technology Conference*, IMTC 2005, Ottawa, Canada.

Ibáñez, A., Parrilla, M., Fritsch, C. & Giacchetta, R. (2005). Inspección mediante ultrasonidos de ruedas de tren en operaciones de mantenimiento, *V CORENDE Proceedings*, Neuquén, Argentina.

Kappes, W., Kröning, M., Rockstroh, B., Salzburger, H.-J. & Walte, F. (2000). Non-destructive testing of wheel-sets as a contribution to safety of rail traffic, *CORENDE 2000 Proceedings*, Mar del Plata, Argentina.

Kawaguchi, K. (2006). Development of a wsp system for freight trains, *7th World Congress on Railway Research (WCRR2006 Proceedings).*, Montreal, Canada.

Kumagai, N., Ishikawa, H., Haga, K., Kigawa, T. & Nagase, K. (1991). Factors of wheels flats occurrence and preventive measures, *Wear* 144: 277 – 287.

Madejski, J. (2006). Automatic detection of flats on the rolling stock wheels, *Journal of Achievements in Materials and Manufacturing Engineering* 16(1-2): 160–163.

Pohl, R., Erhard, A., Montag, H.-J., Thomas, H.-M. & Wüstenberg, H. (2004). NDT techniques for railroad wheel and gauge corner inspection, *NDT&E International* 37: 89 – 94.

Salient Systems, Inc. (2010). Intelligent track solutions. viewed March 22, 2012, http://www.salientsystems.com.

Salzburger, H. J., Wang, L. & Gao, X. (2008). In-motion ultrasonic testing of the tread of high-speed railway wheels using the inspection system AUROPA III, *17th World Conference on NDT*, Shangai, China.

Snyder, T. & Stone, D. H. (2003). Wheel flat and out-of-round formation and growth, *Proc. 2003 IEEE/ASME Joint Rail Conf.*, Chicago, Illinois, pp. 143 – 148.

Stratman, B., Liu, Y. & Mahadevan, S. (2007). Structural health monitoring of rail road wheels using impact load detectors, *Journal of Failure Analysis and Prevention* 7(3): 218 – 225.

Vyas, N. S. & Gupta, A. K. (2006). Modeling rail wheel-flat dynamics, *Engineering Asset Management*, Springer London, pp. 1222 – 1231.

Wu, T. X. & Thompson, D. J. (2002). A hybrid model for the noise generation due to railway wheel flats, *Journal of Sound and Vibration* 251(1): 115 – 139.

Zakharov, S. M. & Goryacheva, I. G. (2005). Rolling contact fatigue defects in freight car wheels, *Wear* 258: 1142 – 1147.

Zerbst, U., Mädler, K. & Hintze, H. (2005). Fracture mechanics in railway applications – an overview, *Engineering Fracture Mechanics* 72(2): 163–194.

15

Speed Sensorless Control of Motor for Railway Vehicles

Ding Rongjun
Csr Zhuzhou Institute Co., Ltd
China

1. Introduction

In the past twenty years, AC speed-driving system has been greatly improved and gradually replaces DC speed-driving system. It has been widely used in all aspects of national economy, promotes the fast development and revolution in industrial automation. In the driving system with AC speed control, in order to realize high-precision speed or torque close-loop control, it is required to install a speed sensor on the axle of motor for detecting the rotor speed.

Due to the particularities in railway transportation application, while realizing high performance close loop control, the speed sensor also results some disadvantages:

1. Because of the severe vibration and shock in rail transportation, the speed sensor must have good vibration resistance. It requires a special manufacturing technology, so the resolution and accuracy will be lower, which will limit the control performance in low speed and become a bottleneck problem for improvement.
2. To ensure the measurement accuracy, the verticality and coaxiality of the speed sensor should be guaranteed when installing it. In the railway traction system, the diameter of axle for high power traction motor is big, and a special switchover treatment is needed to connect the speed sensor. It is hard to meet the requirement on installation accuracy. The service life of the speed sensor will be reduced to a certain extent. It is also a potential fault for the system reliability.
3. The speed sensor is a measuring element so it requires a certain working environment. In rail transportation, the speed sensor is installed on the cover of motor, exposed directly to air. The work environment is very severe, such as large temperature variance, higher relative air humidity etc. All of these factors will influence the reliability and accuracy of the sensor.
4. The feedback signal of the speed sensor would be influenced by the transmission distance. The larger the transmission distance is, the more vulnerable is the signal to be disturbed by the surrounding equipments. It is obvious for power-distributed EMUs, subway and light rail vehicles.
5. In some applications where the mechanical dimensional requirement of the motor is very rigorous, it is required to sacrifice other performances for the space to install the speed sensor.

The speed sensorless control can make up shortcomings described above and reduce the cost of system. Therefore, the research of speed sensorless control becomes a hotspot in the research of railway driving system. Along with the fast development of microelectronics, good hardware and software conditions are available for the application of the speed sensorless control on the induction motor. The utilization of speed sensorless control technology on high precision application becomes possible.

2. Basis knowledge concerned

2.1 Mathematical model of induction motor

The Mathematical model of induction motor is a high order, nonlinear, strong-coupled multivariable system. When establishing the mathematical model, some assumptions are made as follows:

1. Symmetrical difference of three phase windings is 120° in space and the magnetic motive force is sinusoidal distributed;
2. The iron loss is ignored;
3. Magnetic saturation is ignored, i.e. it is assumed that the mutual induction and self induction between the windings are linear;
4. The influence of temperature and frequency on the parameters of motor is ignored.

When analyzing and controlling the induction motor, Clarke translation is used to translate three phase coordinate system to two phase coordinate system, the relation between 3-phase and 2-phase coordinate system is shown in Fig.1.

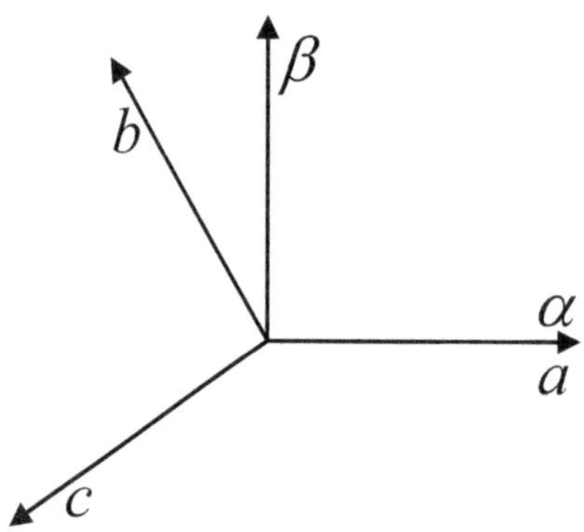

Fig. 1. The relation between the 3-2 coordinate system

The translation matrix C for 3>2 is:

$$C = \frac{2}{3}\begin{bmatrix} 1 & -\frac{1}{2} & -\frac{1}{2} \\ 0 & \frac{\sqrt{3}}{2} & -\frac{\sqrt{3}}{2} \end{bmatrix} \quad (1)$$

By deduction, the voltage equation in $\alpha - \beta$ is as follows:

$$\begin{bmatrix} u_{s\alpha} \\ u_{s\beta} \\ u_{r\alpha} \\ u_{r\beta} \end{bmatrix} = \begin{bmatrix} R_s + L_s p & 0 & L_m p & 0 \\ 0 & R_s + L_s p & 0 & L_m p \\ L_m p & \omega L_m & R_r + L_r p & \omega L_r \\ -\omega L_m & L_m p & -\omega L_r & R_r + L_r p \end{bmatrix} \begin{bmatrix} i_{s\alpha} \\ i_{s\beta} \\ i_{r\alpha} \\ i_{r\beta} \end{bmatrix} \tag{2}$$

Where $u_{s\alpha}$, $u_{s\beta}$, $i_{s\alpha}$, $i_{s\beta}$ are α and β components of stator voltage and current respectively; $u_{r\alpha}$, $u_{r\beta}$, $i_{r\alpha}$, $i_{r\beta}$ are α and β components of rotor voltage and current respectively; R_s and R_r are stator and rotor resistances; L_m is mutual inductance; L_s and L_r are stator and rotor self inductance; ω is rotor angular frequency.

The flux equation of the motor is as follows:

$$\begin{bmatrix} \psi_{s\alpha} \\ \psi_{s\beta} \\ \psi_{r\alpha} \\ \psi_{r\beta} \end{bmatrix} = \begin{bmatrix} L_s & 0 & L_m & 0 \\ 0 & L_s & 0 & L_m \\ L_m & 0 & L_r & 0 \\ 0 & L_m & 0 & L_r \end{bmatrix} \begin{bmatrix} i_{s\alpha} \\ i_{s\beta} \\ i_{r\alpha} \\ i_{r\beta} \end{bmatrix} \tag{3}$$

The electromagnetic torque equation of the motor is as follows:

$$T_e = \frac{3}{2} P_n (\psi_{s\alpha} i_{s\beta} - \psi_{s\beta} i_{s\alpha}) \tag{4}$$

where P_n is the number of pole pairs.

The electromechanical motion equation of the motor is as follows:

$$T_e - T_L = \frac{J}{P_n} \frac{d\omega}{dt} \tag{5}$$

where T_L is load torque.

After equivalent transformation deduction, Γ type equivalent circuit model of the induction motor is shown in Fig.2.

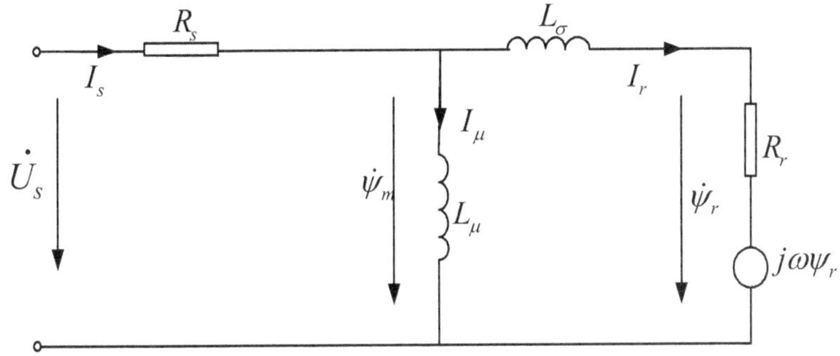

Fig. 2. Γ-type equivalent circuit model of induction motor

Generally, when designing the state observer of AC driving system, the current and flux are taken as the state variables. Now, the stator flux and rotor flux are expressed as $\vec{\psi}$, the stator voltage as the input variable expressed as $\vec{u_s}$, the stator current as the output variable expressed as $\vec{i_s}$, the state equation in α - β is expressed as:

$$\begin{cases} \dot{\vec{\psi}} = A\vec{\psi} + B\vec{u}_s \\ \vec{i}_s = C\vec{\psi} \end{cases} \tag{6}$$

$$A = \begin{bmatrix} -R_s k_1 & 0 & R_s k_2 & 0 \\ 0 & -R_s k_1 & 0 & R_s k_2 \\ R_r k_2 & 0 & -R_r k_2 & -\omega \\ 0 & R_r k_2 & \omega & -R_r k_2 \end{bmatrix},$$

$$B = \begin{bmatrix} 1 & 0 \\ 0 & 1 \\ 0 & 0 \\ 0 & 0 \end{bmatrix}, \quad C = \begin{bmatrix} k_1 & 0 & -k_2 & 0 \\ 0 & k_1 & 0 & -k_2 \end{bmatrix},$$

where $k_1 = 1/L_\mu + 1/L_\sigma$, $k_2 = 1/L_\sigma$.

As the observability of the system decides the pole assignment of the state observer directly, so the observability must be proved. Consider that the mechanical time constant of the induction motor is far larger than the electromagnetic time constant, and that the varying of the stator and rotor resistance and inductance is relatively slower, the observability can be proved directly by linear time invariant system. The same method can be used to proof the observability of the time-varying system.

The sufficient and necessary condition for the complete observability of the system [A, B, C] state is controllable matrix:

$$N = \begin{bmatrix} C^T & A^T C^T & \dots & \left(A^T\right)^{n-1} C^T \end{bmatrix} \tag{7}$$

Nonsingular, i.e. $rank\, N = n$.

A, C matrix in the (6) are substituted to the first two items of the controllable matrix N, the result after calculation is as follows:

$$C^T = \begin{bmatrix} k_1 & 0 & -k_2 & 0 \\ 0 & k_1 & 0 & -k_2 \end{bmatrix}^T$$

$$A^T C^T = \begin{bmatrix} -R_s k_1^2 - R_r k_2^2 & 0 & R_s k_1 k_2 + R_r k_2^2 & \omega_r k_2 \\ 0 & -R_s k_1^2 - R_r k_2^2 & -\omega_r k_2 & R_s k_1 k_2 + R_r k_2^2 \end{bmatrix}^T$$

so

$$\det\left[C^T, A^T C^T\right] = R_r^2 k_2^4 (k_1 - k_2)^2 + \omega_r{}^2 k_1^2 k_2^2 > 0 \tag{8}$$

i.e. $rank\ N = 4$, so the system can be observed completely.

2.2 Mathematical model of inverter

Three phase two-level voltage source inverter is used to provide power to AC induction motor. Each leg of the inverter is equal with two on-off switches. The main circuit topology of the control system is shown in Fig. 3.

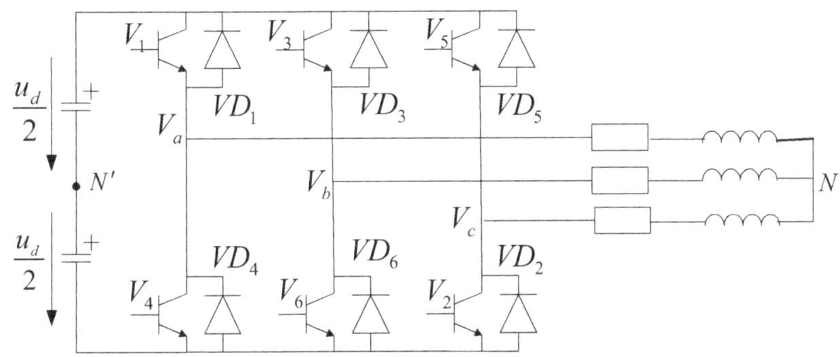

Fig. 3. Main circuit topological for inverter

Because of the on/off state of each switch, the inverter has 8 states as shown in Table-1. Sa=1 indicates that the upper switch of leg-A is on and the lower switch is off; if Sa=0, it is reverse. The other two legs are the same. the 8 on-off states of inverter are corresponded to 8 voltage space vectors U0~U7, where U0 and U7 are zero voltage space vectors. The distribution of voltage space vector is shown in Fig. 4.

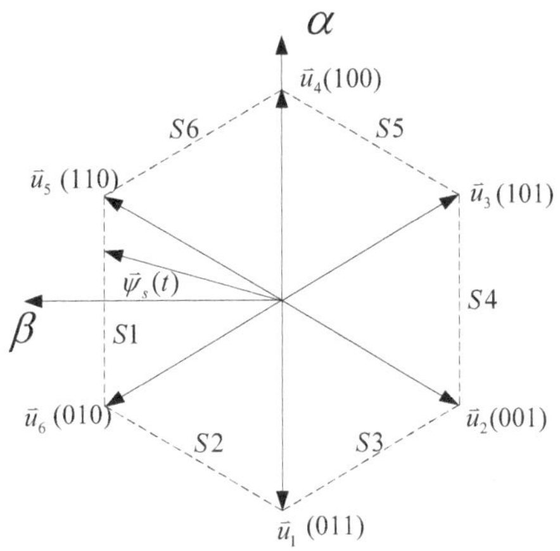

Fig. 4. Distribution of 8-Space Voltage Vectors

state	0	1	2	3	4	5	6	7
Sa	0	0	0	1	1	1	0	1
Sb	0	1	0	0	0	1	1	1
Sc	0	1	1	1	0	0	0	1

Table 1. Switch states of the inverter

According to the on-off state of the inverter, the output three phase voltages can be obtained from following equation:

$$\begin{bmatrix} u_a \\ u_b \\ u_c \end{bmatrix} = \frac{U_d}{3} \begin{bmatrix} 2 & -1 & -1 \\ -1 & 2 & -1 \\ -1 & -1 & 2 \end{bmatrix} \begin{bmatrix} S_a \\ S_b \\ S_c \end{bmatrix} \tag{9}$$

The voltage space vector u_s is defined as follows:

$$u_s = \frac{2}{3}\left(u_a + \alpha u_b + \alpha^2 u_c\right) \tag{10}$$

where $\alpha = e^{j2\pi/3}$. If $S_{abc} = 011$, the on-off state of the inverter is shown in Fig.5, then:

$u_a = -\frac{2U_d}{3}$, $u_b = u_c = \frac{U_d}{3}$. The voltage space vector at this moment is obtained as follows:

$$u_s = \frac{2}{3}\left(-\frac{2U_d}{3} + \frac{U_d}{3}e^{j\frac{2\pi}{3}} + \frac{U_d}{3}e^{j\frac{4\pi}{3}}\right) = -\frac{2U_d}{3} = \frac{2}{3}U_d e^{j\pi}$$

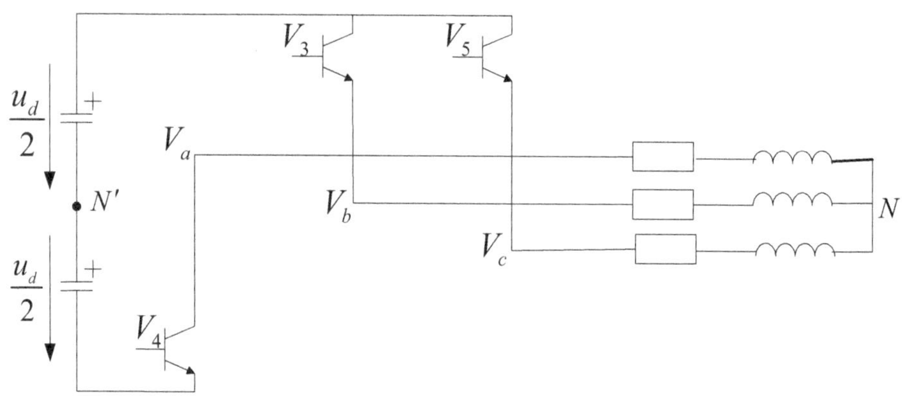

Fig. 5. The working state of the inverter at the on / off state of "011"

The voltage vector space distribution of the inverter is shown in Fig. 4. Under respectively action of 6 non-zero space voltage vectors, the flux trajectory will be hexagon. The voltage space vector to be selected decides the variance of the magnetic field. The magnetic field is changed as follows: At the beginning, the flux and torque are far smaller than the references, it is required to increase the flux and torque. Proper voltage space vector is selected for switching and can accumulate the flux, promote magnetic field rotation and increase the torque at the same time. When the flux is increased, the flux quickly comes to the hysteretic loop control bandwidth, then the hysteretic loop control unit takes action. When it is required to increase the flux, by looking up table, select the voltage space vectors which can sustain the torque while increasing the flux; when it is required to increase the torque, select the voltage space vectors which can quickly increase the rotation angle and the torque. Due to the action of the voltage space vector switching table, in the process of switching, the three phase stator magnetic field can be rotated still and the motor rotor is driven continually. Therefore, proper control on flux and torque is a prerequisite for effective speed control of three phase induction motor.

2.3 Voltage space vector modulation

The relation between stator flux and stator voltage is:

$$\vec{\psi}_s(t) = \int [\vec{u}_s(t) - R_s\vec{i}_s(t)]dt \tag{11}$$

If the speed is high, the influence of voltage on stator resistance can be ignored, then:

$$\vec{\psi}_s(t) \approx \int \vec{u}_s(t)dt \tag{12}$$

The equation (12) shows the relation between the stator flux space vector and the stator voltage space vector is an integral relation. If the action duration is short, (12) can be changed as follows:

$$\Delta\vec{\psi}_s = \vec{u}_s \cdot \Delta t \tag{13}$$

It indicates the trajectory of stator flux is the same as direction of stator voltage vector. Therefore, the trajectory of the stator flux can be controlled by controlling the stator voltage vector. And the zero voltage vector can not control the motion of the stator flux, but can stop the stator flux in motion.

If the stator-flux $\vec{\psi}_s(t)$ is as shown in the in Fig. 4, the voltage space vector $\vec{u}_1$ is output by inverter, $\vec{\psi}_s(t)$ will move along the trajectory of S1 which is in the direction of $\vec{u}_1$. When $\vec{\psi}_s(t)$ moves to the crossing point of S1 and S2, the voltage space vector $\vec{u}_2$ will be output. So $\vec{\psi}_s(t)$ will move along S2. Six effective on/off states of the inverter can be used directly to get simply the hexagon flux trajectory for controlling the motor.

3. Research on speed estimation algorithm

The state observer is used to observe the state and the parameter of the nonlinear dynamic system at real time. When the state observer is used in the speed sensorless control system, the mathematical model of the motor is used to estimate the state of the motor. This estimated state should be corrected with feedback compensation. For the convenience of direct torque control, the stator and rotor flux is selected to be observed by the state observer. The model of motor Luenberger state observer is shown in Fig. 6.

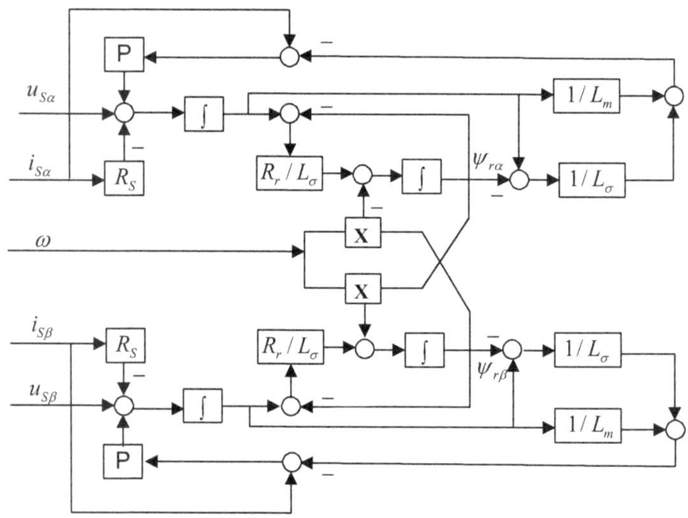

Fig. 6. Luenberger state observer for induction motor

The relation between the real motor and observer model is shown in Fig. 7. The equation of Luenberger state observer with the stator and rotor flux as the variable of state is as follows:

$$\begin{cases} \dot{\hat{\psi}} = (A + \omega J)\hat{\psi} + B\hat{u}_s + G(i_s - \hat{i}_s) \\ \hat{i}_s = C\hat{\psi} \end{cases} \tag{14}$$

where "^" reprensents estimated value, G is a calculative matrix gain to stabilize the errors equation (15).

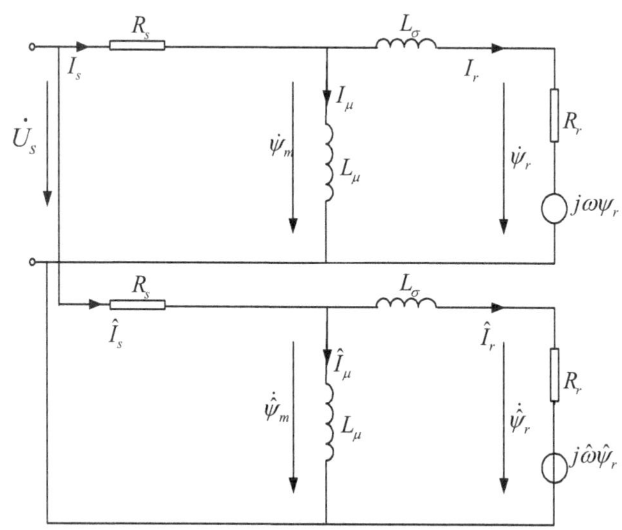

Fig. 7. Relation between actual motor model and observer model

The self-adaption law is derived on the basis of Lyapunov theory. The error equation of each state from the formula (6) and (14) is as follows:

$$\frac{d}{dt}e=\left[A-GC+\omega J\right]e+\Delta\omega J\hat{\psi} \tag{15}$$

where $e=\psi-\hat{\psi}$, $\Delta\omega=\omega-\hat{\omega}$.

when deriving the selfadaption law with Lyapunov stability theory, Lyapunov function is defined as follows:

$$V\left(e,\hat{\omega}-\omega\right)=e^{T}e+(\hat{\omega}-\omega)^{2}/\lambda_{\omega} \tag{16}$$

In the defined Lyapunov functions, $e^{T}e$ is only a special unit matrix, so we can not gurantee the stability of selfadaption flux observer in theory. When using LMI toolbox to derive the inequality matrix to gurantee the stability of the observer, sometimes,we can't get the solution of observer gain matrix G, so that, we propose a new Lyapunov function:

$$V\left(e,\hat{\omega}-\omega\right)=e^{T}Pe+(\hat{\omega}-\omega)^{2}/\lambda_{\omega} \tag{17}$$

where P is 4×4 symmetric positive definite matrix, in more general, which can also proof fully the stability of the observer in theory. The derivation of formula (17) is as follows:

$$\begin{aligned}\frac{d}{dt}V&=e^{T}[(A-GC)^{T}P+P(A-GC)+\omega(J^{T}P+PJ)]e\\&+\Delta\omega\left[\hat{\psi}^{T}J^{T}Pe+e^{T}PJ\hat{\psi}\right]-\frac{2}{\lambda_{\omega}}\Delta\omega\frac{d\Delta\omega}{dt}\end{aligned} \tag{18}$$

defin e $\frac{2}{\lambda_{\omega}}\Delta\omega\frac{d\Delta\omega}{dt}=\Delta\omega\left[\hat{\psi}^{T}J^{T}Pe+e^{T}PJ\hat{\psi}\right]$

Proper gain matrix G and P can be selected, in order to make following inequality become true, the state errors equation (15) is gradual stable, so the gradual stability of the state observer can be guaranteed.

$$(A-GC)^{T}P+P(A-GC)+\omega(J^{T}P+PJ)<0 \tag{19}$$

The self-adaption law is obtained by following formula:

$$\frac{d}{dt}\hat{\omega}=\frac{\lambda_{\omega}}{2}\left[\hat{\psi}^{T}J^{T}Pe+e^{T}PJ\hat{\psi}\right]=e^{T}PJ\hat{\psi} \tag{20}$$

So that, the speed self-adaption estimation algorithm is as follows:

$$\hat{\omega}=K_{p\omega}\left[e^{T}PJ\hat{\psi}\right]+K_{i\omega}\int_{0}^{t}\left[e^{T}PJ\hat{\psi}\right]dt \tag{21}$$

As known from Lyapunov stability theory, if there are symmetric positive definite matrixes P and G and the (19) formula can be true, then the state observer is gradually stable. $\bar{\omega}$ is the speed upper limit and can be obtained from test or other parameters. In order to ensure that the system has a certain reliability and robustness, $\bar{\omega}$ is larger than the rated speed generally.

When the speed varies in the range $[-\bar{\omega},\bar{\omega}]$, the robust stability of errors equation (15) can be decided by following two matrix inequalities:

$$B_1(P,G)=(A-GC)^T P+P(A-GC)+\bar{\omega}(J^T P+PJ)<0 \tag{22}$$

$$B_2(P,G)=(A-GC)^T P+P(A-GC)-\bar{\omega}(J^T P+PJ)<0 \tag{23}$$

The inequalities (22) and (23) are the bilinear inequalities of P and G matrix variables. If setting matrix G, then the bilinear inequality becomes the linear inequalities of P. As the same, if P is set, then the bilinear inequality becomes a linear inequalities of G. Therefore, use following iterative algorithm to get the feasible solution of bilinear inequalities (22) and (23) with LMI toolbox in MATLAB.

Order $B(P,G)=Diag(B_1(P,G),B_2(P,G))$, where $\Lambda(A)$ is the maximum eigen-value of matrix A. The following shows the algorithm for computing matrix P and G.

1. initializing: set matrix $P_0>0$ randomly, $i=0$.
2. Iterative: Order $i=i+1$, solving $\min\limits_{G}{}^{\Lambda(B(P,G_i))}$ optimizing to get matrix G_i; Solving $\min\limits_{P}{}^{\Lambda(B(P,G_i))}$ optimizing to get positive definite matrix P_i;
3. End: Matrix inequality $\Lambda(B(P_i,G_i))<0$ can be guaranteed.

4. Research on magnetizing of induction motor at unknown speed

When the speed sensorless control system is used for railway application, motor must be restarted after turning off inverter due to instantaneous over-current/over-voltage, etc. Generally, the induction motor is rotating at a high speed at that moment. In this case, the control system should be put into using again even the motor is at high initial speed. If the control system is with speed sensors, magnetizing can be guaranteed through adjusting the rotor/stator flux-linkage after obtaining the speed. However, without speed sensors, if the difference between real speed and estimated speed is small, magnetizing can be realized. Otherwise, magnetizing would be failed. Thus, a novel control algorithm proposed in this section will be used at the initial moment of motor restarting.

4.1 The theoretical principle of magnetizing of induction motor at unknown speed

Fig. 8 shows Heyland circle of stator current space vector under the same voltage space vector. As shown in Fig. 8, point b_0 is the no-load point, point b_∞ is the working point with infinite slip frequency and point b is the estimated speed point. b_1, b_2 are randomly selected as two actual speed points, distributed on both sides of point b. According to the principal of stator current Heyland circle, if the actual speed is at b_1, the estimated slip frequency is less than the actual slip frequency, i.e. the estimated speed is larger than the actual speed. If the actual speed is at b_2, the estimated slip frequency is larger than the actual slip frequency, i.e. the estimated speed is less than the real speed.

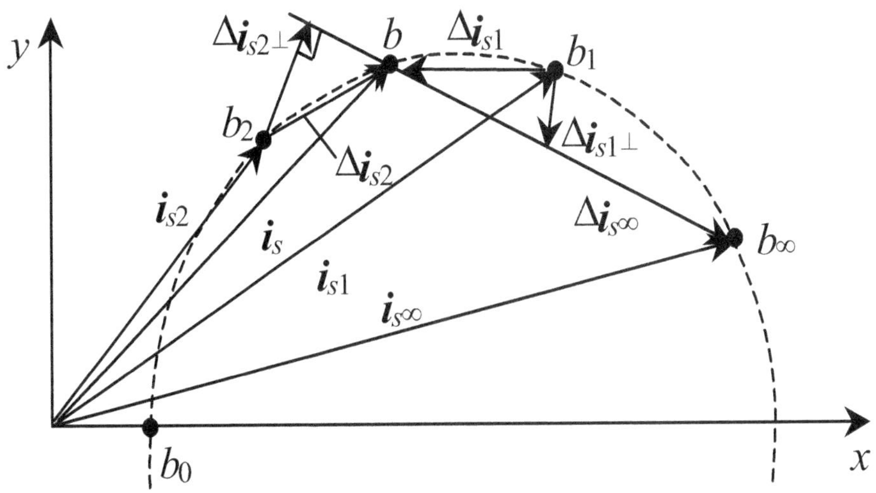

Fig. 8. Heyland circle of stator current space vector under the action of same voltage space vector

Defining $\Delta i_{s1} = i_s - i_{s1}$, $\Delta i_{s2} = i_s - i_{s2}$, $\Delta i_{s\infty} = i_s - i_{s\infty}$, if Δi_{s1}, Δi_{s2} are projected on $\Delta i_{s\infty}$, the direction is opposite. This shows exactly the relation between the estimated speed and the real speed, and becomes basis for the self optimizing fuzzy search of the initial speed.

Heyland circle of stator space current vector is based on steady-state equation of induction motor and that the parameter is sine. The conclusion above is true only at the steady state operation point of the induction motor. The rotor flux equation in the stator field rotating coordinates can be derived as:

$$\dot{\boldsymbol{\psi}}_r + \boldsymbol{\psi}_r \left[\frac{1}{T_\sigma} + \mathrm{j}(\omega_s - \omega) \right] = \frac{\boldsymbol{\psi}_\mu}{T_\sigma} \tag{24}$$

where $T_\sigma = L_\sigma / R_r$ is the rotor flux leakage time constant.

The equation (25) can be derived when the initial value of the rotor flux is zero. And you can reach that the rotor flux is the stepping response of stator flux. The time constant is T_σ and the rotor flux will reach steady state value after 5 T_σ.

$$\boldsymbol{\psi}_r = \frac{1}{1 + \mathrm{j}\omega_r T_\sigma} \boldsymbol{\psi}_\mu [1 - \mathrm{e}^{-t/T_\sigma} \mathrm{e}^{-\mathrm{j}\omega_r t}] \tag{25}$$

Through (24) and (25), the time limit of fuzzy search can be got.

4.2 The influence of instant power off on the speed sensorless control system

In the contact line of electrified railway, power supply of commutation in different sections is used, for balancing the three phases of power system as much as possible. In order to prevent short-circuit between phases, the phases shall be separated from each other by air or insulating material. As the contact line between the glass insulation parts is not conductive, so it is called neutral section. In order to prevent the trailing arc from burning the insulation parts and contact line or causing other incidents when the pantograph of locomotive passes through the neutral section, it is required the pantograph of locomotive can go into and out the neutral section insulator at current-less state, i.e. so called as passing neutral section, so the DC side of motor inverter must be power off. Whereas in actual condition, the distance of one section is only about 20 kilometers in general. If taking account of 120km/hour for electric locomotive, the locomotive must be power-off in every 10 minutes. It is known on this fact that it is very meaningful to study the influence of instant power off on the speed sensorless control system of the asynchronous motor.

Assume that the DC side of the motor inverter is power off at time t_0. After power off, the voltage on stator side is zero immediately, but the rotor rotates still due to inertia force. The synchronous speed is smaller than rotor speed at this moment, i.e. slip ratio s<0. The output electromagnetic torque will be negative at this moment. Under the action of negative torque, the rotor speed will be slower. When the rotor speed is slow down to a certain degree, the slip will be positive again. The torque is positive at this moment. The rotor speed increases and makes the slip be reduced to zero again. In the other aspect, at the moment of instant power off, due to the inductance effect of the motor, the stator and rotor current will not disappear immediately, even larger than the value at the moment of power off, and will create very big negative torque at this moment, cause an impact on the motor. The current will tend towards zero then. When the stator current decreases gradually, the electromagnetic power is smaller and smaller. The electromagnetic torque is also smaller or even zero. Presume the time at this moment is t_1, if the negative torque is constant, the motor will be slow down. Actually, the period of time from t_0 to t_1 is very short, after the electromagnetic torque tends towards zero, the motor torque equation can be expressed as follows:

$$0 = T_L + R_w \omega + J\frac{d\omega}{dt}, \quad t \ge t_1 \tag{26}$$

Where R_w is the rotational damping coefficient directly proportional to the speed. Assume the relation between the load torque T_L and rotor mechanical angular speed is linear, i.e. $T_L = T_{L0} + R_L\omega$, which is substituted to the torque equation, the result is as follows:

$$T_{L0} + (R_r + R_w)\omega + J\frac{d\omega}{dt} = 0, \quad t \ge t_1 \tag{27}$$

Solution:

$$\omega = \omega_1 e^{-\frac{t-t_1}{T_M}} - \frac{T_{L0}}{R_L + R_w}(1 - e^{-\frac{t-t_1}{T_M}}), \quad t \ge t_1 \tag{28}$$

where ω_1 is the rotor mechanical angular speed at time t_1, and can be calculated approximately from formula $T_e - T_L = R_w\omega + J\frac{d\omega}{dt}$ under the load torque -800Nm by 0.1 seconds. $T_M = \frac{J}{R_L + R_w}$ is the mechanical time constant of the motor.

It is known that after power off, the torque, stator current and rotor current, flux in the motor equation will vary in short period of time, but tend towards zero soon. The rotational speed will not be zero immediately, but decreases gradually under the action of the load torque. After analyzing the speed estimation solution of the speed sensor-less control system based on the model reference self adaptive control aforementioned, it is known that the general varying direction of the state parameters used in the speed estimation tends towards zero. Under such a case, the actual speed of the motor decreases under the action of the load torque according to the rule shown in (28). The speed estimation solution at this moment can not estimate the actual speed rightly.

4.3 Initial speed estimation method based on self optimizing fuzzy search

For self optimizing fuzzy control, if the iterative step of the self optimizing search is too small and speed convergence will be slow, it is difficult to be adaptive to some uncontrollable disturbance response. If the step is too large, the search error will be bigger and often cause vibrations. Therefore, it is proper to change the step length. At the point farther the pole, the step can be larger. At the point near the pole, the step should be smaller. Use fuzzy logic decision to change the step length.

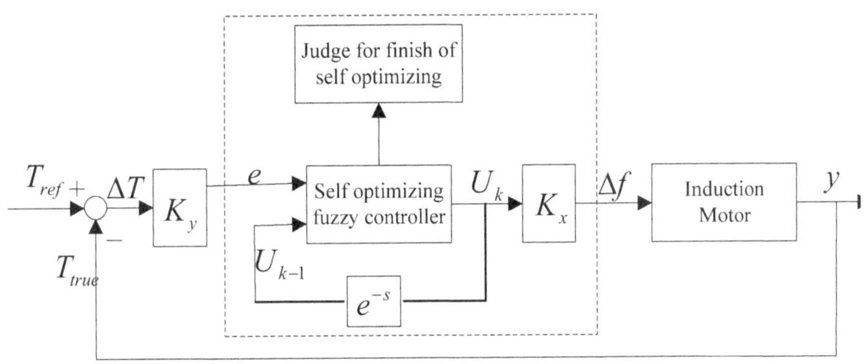

Fig. 9. Self optimizing fuzzy controller of initial rotational speed

In Fig. 9, take $k_y = k_1 I_{ref} / T_b$, where I_{ref} is current reference, k_1 is a constant relative to I_{ref}, T_b is motor break-down torque. Take $k_x = k_2 f_{max} / 24$, f_{max} is the maximum operation frequency of the motor. k_2 is a constant relevant to the motor's maximum operation frequency range, and is used to adjust the searching accuracy. The value of $u(k)$ is obtained by looking-up table 2 and table 3 according to the value of $u(k-1)$ and e.

Uk-1 / Uk / e	NB	NM	NS	PS	PM	PB
NB	PS	PS	PS	PS	PS	PS
NM	PS	PS	PS	PM	PM	PM
NS	PM	PM	PM	PB	PB	PB
N0	PM	PM	PM	PB	PB	PB
P0	NB	NB	NB	NM	NM	NM
PS	NB	NB	NB	NM	NM	NM
PM	NM	NM	NM	NS	NS	NS
PB	NS	NS	NS	NS	NS	NS

Table 2. Table of initial speed self optimizing fuzzy control rules

Δfk / Δf(k-1) / Δt	-6	-5	-4	-3	-2	-1	1	2	3	4	5	6
-6	1	1	2	2	2	2	2	2	2	2	1	1
-5	1	1	2	2	2	2	2	2	2	2	1	1
-4	2	2	2	2	3	3	3	3	4	4	4	4
-3	2	2	2	2	3	3	3	3	4	4	4	4
-2	4	4	4	4	4	4	6	6	6	6	6	6
-1	4	4	4	4	5	5	5	5	6	6	6	6
-0	4	4	4	4	5	5	5	5	6	6	6	6
+0	-6	-6	-6	-6	-5	-5	-5	-5	-4	-4	-4	-4
+1	-6	-6	-6	-6	-5	-5	-5	-5	-4	-4	-4	-4
+2	-6	-6	-6	-6	-6	-6	-4	-4	-4	-4	-4	-4
+3	-4	-4	-4	-4	-3	-3	-3	-3	-2	-2	-2	-2
+4	-4	-4	-4	-4	-3	-3	-3	-3	-2	-2	-2	-2
+5	-1	-1	-2	-2	-2	-2	-2	-2	-2	-2	-1	-1
+6	-1	-1	-2	-2	-2	-2	-2	-2	-2	-2	-1	-1

Table 3. Table of initial speed self optimizing fuzzy control

4.4 Solution on magnetizing of induction motor at unknown speed

Magnetizing of induction motor at unknown speed based on self optimizing fuzzy control is shown in Fig. 10. The difference between the real torque and the reference torque was put into fuzzy controller. The output is the stator frequency increment of next step fuzzy search. When the real current is compared with the current reference, under the control of PI controller, the modulation ratio of the voltage is output then to keep the motor current constant.

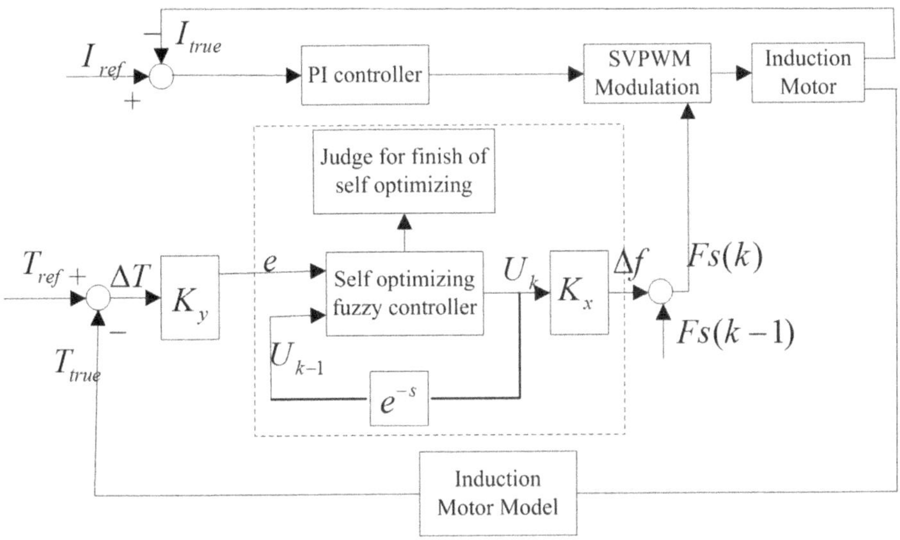

Fig. 10. Magnetizing at unknown speed control using self optimizing fuzzy control

Specific process of self optimizing fuzzy search is as follows:

1. Applied a fixed impulse signal, i.e. a DC voltage on induction motor. The sign of torque shows the rotation direction of the motor. The judgment of the rotation direction is very important. Further testing under wrong rotation direction will cause overcurrent on the motor.
2. In the direction detected above, apply a voltage space vector with the angular frequency of f_{ref}. When the actual rotation speed is unknown, it is optimal to take half of the maximum stator angular frequency $f_{\max}$ as the initial value of f_{ref}. The value of the voltage vector is set by current regulator.
3. Start the self optimizing fuzzy search on the initial speed.
4. Decide whether the self optimizing find is finished according to the condition after the self optimizing, which is the modulation index by the current regulator and actual torque.
5. Take the speed obtained in the self optimizing fuzzy search as the initial speed of the speed sensorless control for magnetizing at unknown speed.
6. Judge whether the magnetizing is successful according to the starting current. If the self optimizing result last time is different to the actual speed, you can search it again.

5. Design and test of the system

The DTC proposed by Professor M. Depenbrock has been widely used in railway driving system because of its good dynamic response performance. This section present the design and experiment result of control system based on DTC. The ISC is used at low speed and DSC with eighteen-corner and six-corner flux trajectory at high speed.

The main circuit of the system designed is shown as Fig.11. TMS320C31 and TMS320F240 are used as the MCU. It is a two-level inverter and the voltage of DC-link is 1500V. The switching parts used for three-phase inverter are IGBT. The parts labeled as VH1, VH2 are voltage sensors. VH1 detects line voltage while VH2 detects the DC-link voltage real time. LH1~LH4 are current sensors. LH1 detects DC circuit current. LH3~LH4 detect the phase-current of motor. LH2 detects the current of chopper-resistor R2.

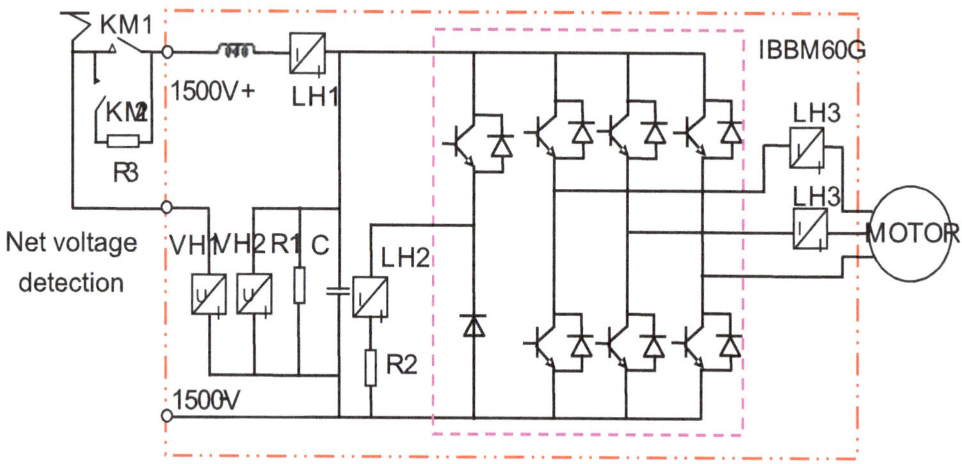

Fig. 11. Diagram of control system main circuit

The basic parameters of motor are shown in table-4. The rated power of system is 1000kVA and the switching frequency is 450Hz. The resolution of speed-sensor used for test is 90 p/r.

R_s	R_r	Lμ	Lσ	P
0.0483Ω	0.0435Ω	0.00107H	0.0196H	3

Table 4. Parameters of motor

In order to test the accuracy of estimated speed for the sensorless driving system in the condition of motoring / braking in full speed range, the speed in the motoring and braking operation is estimated. The estimation result is shown in Fig. 12 and 13. The result shows that the estimated speed can meet the real speed well.

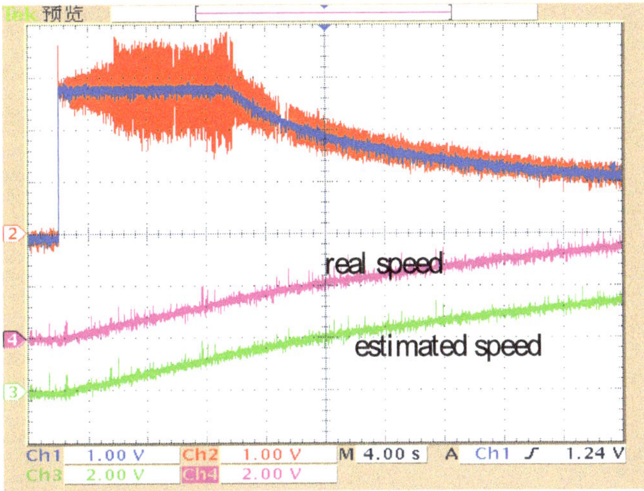

Fig. 12. Estimated speed in motoring

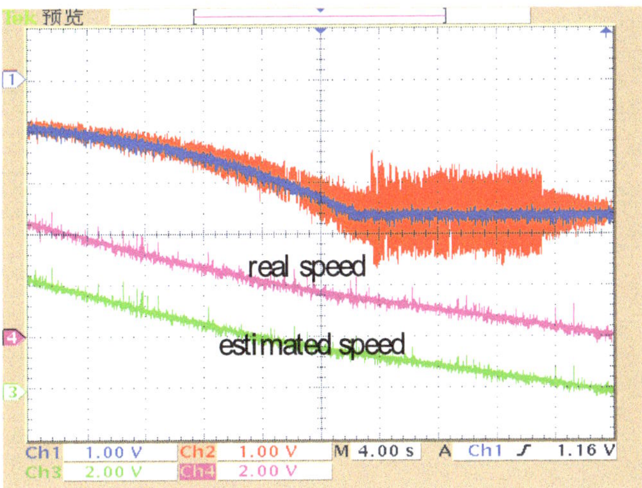

Fig. 13. Estimated speed in braking

The speed-close loop test at low speed was carried out. Fig.14 is the result. It is observed that currents of motor are sine wave. The measured speed can also follow the real speed well. The real speed set in the test is 6r/min. It can be observed that the motor has no evident stepping phenomenon.

The torque step response was carried and the results are shown in Fig. 15 and 16. The result shows that in the motoring and braking state the torque dynamic response performance is good.

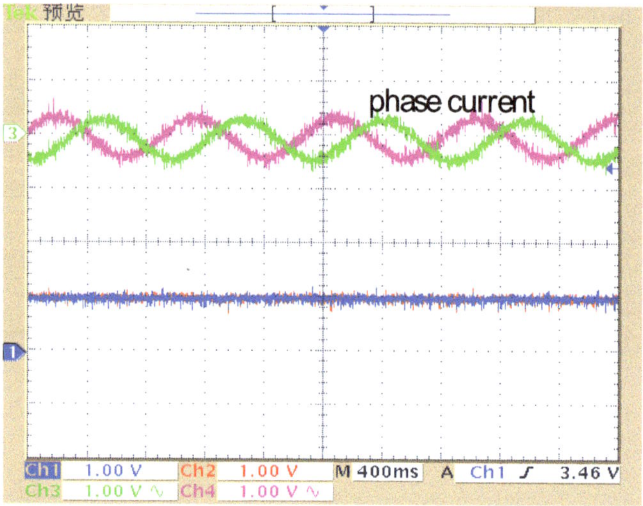

Fig. 14. Result of sensorless control at low speed

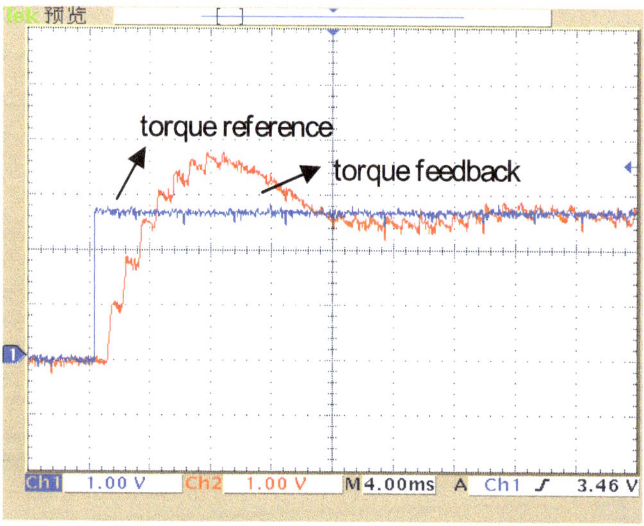

Fig. 15. Torque response in motoring

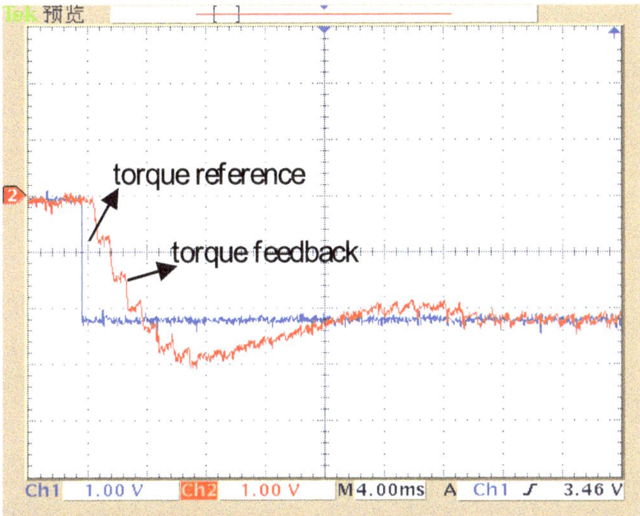

Fig. 16. Torque response in braking

Fig.17 is the stator flux under DTC at low speed. The amplitude of flux is gradually established. After the stator flux is established, it can be seen that the stator flux trajectory in low speed is approximate a circle. Fig.18 is the current of motor with the circular flux trajectory under DTC at low speed.

Fig.19 is the stator flux with eighteen-corner flux trajectory under DTC at high speed. The result shows that the stator flux trajectory is an eighteen-corner. Fig.20 is the stator current of motor. It is shown in Fig.21 that the stator flux waveform with six-corner flux trajectory under direct torque control in constant power flux-weakening zone is in a good six-corner shape. Fig.22 is the stator current.

The waveform of magnetizing sensor-less control system is shown in Fig.23 and 24. The test result shows that the initial speed self optimizing algorithm can estimate correctly the initial speed in about a few hundred milliseconds. In process of estimating the initial speed, the current impact in the restarting magnetic excitation stage can be controlled in an alloweable range, meeting the requirement on actual engineering application.

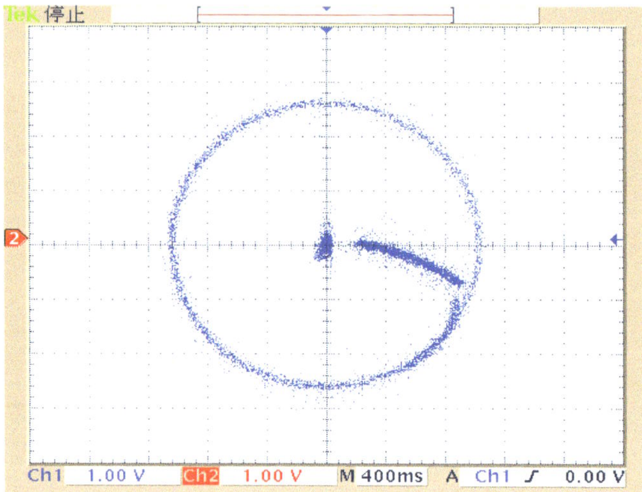

Fig. 17. Flux waveform with circular flux trajectory under DTC at low speed

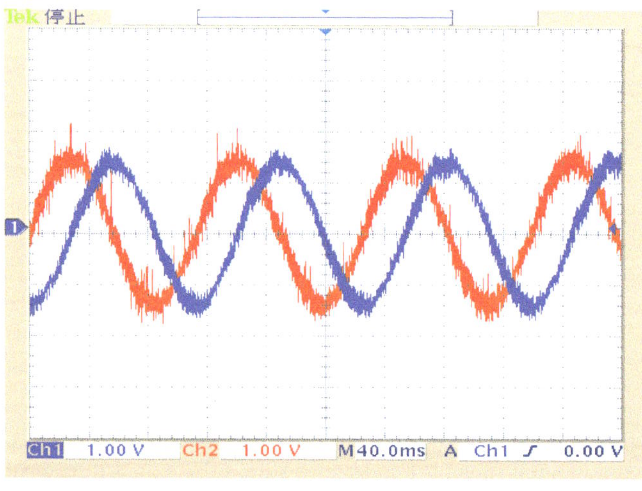

Fig. 18. Current waveform with circular flux trajectory under DTC at low speed

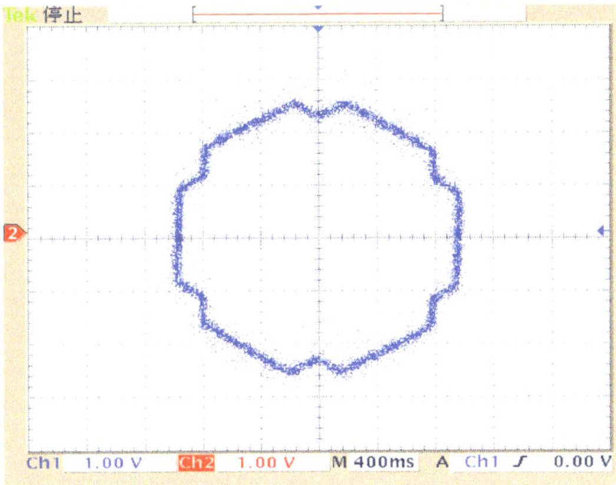

Fig. 19. The flux waveform under eighteen-flux trajectory control at high speedcorner

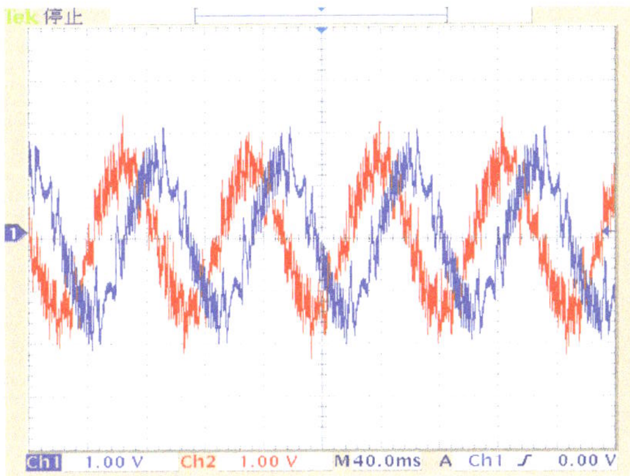

Fig. 20. Current waveform under the eighteen-corner flux trajectory control at high speed

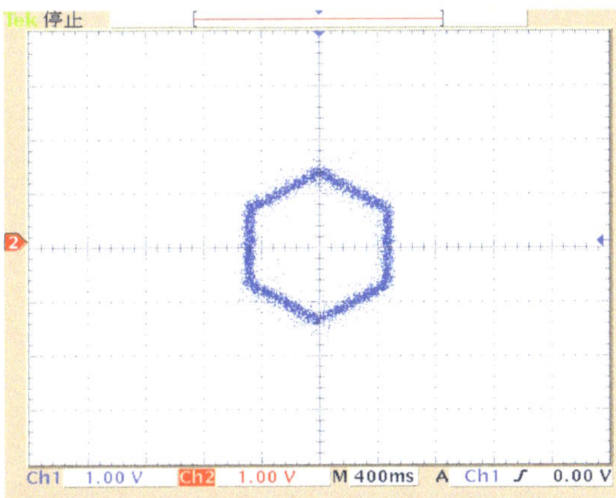

Fig. 21. Flux in flux-weakening

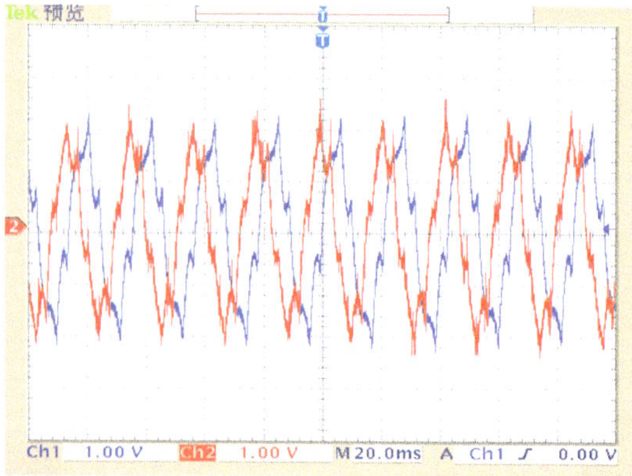

Fig. 22. Current in flux-weakening

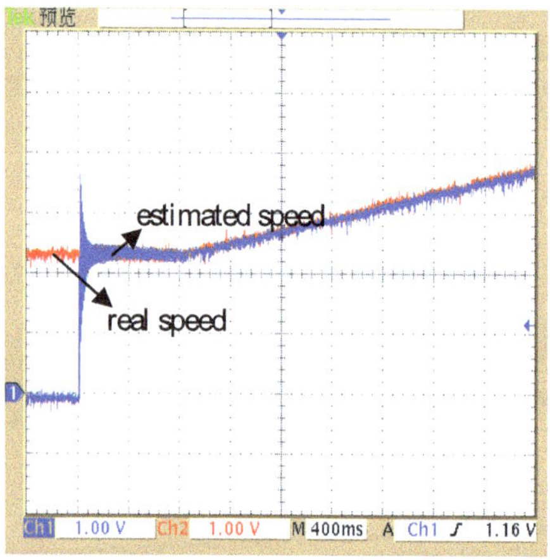

Fig. 23. Speed estimation in magnetizing process

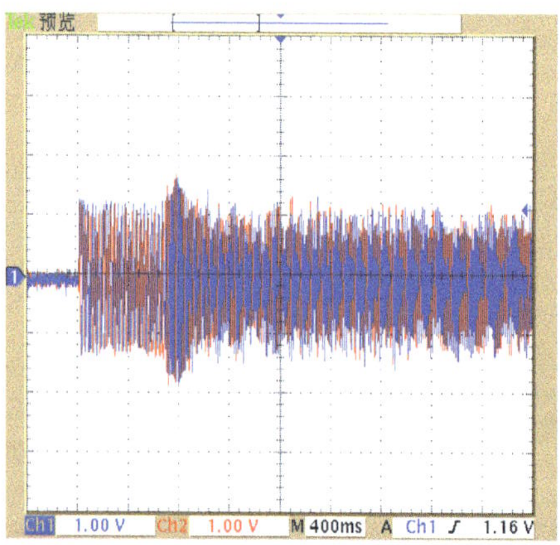

Fig. 24. Phase current in magnetizing process

6. Conclusion

Speed sensorless control of induction motor is presented in this chapter. A novel Lyapunov function is proposed to estimate the speed, especially the magnetizing of induction motor at unknown speed necessary in railway is researched deeply. The experiment result shows it is feasible and can be applied in railway vehicles.

7. References

[1] M.Depenbrock. Direct Self-Control(DSC) of inverter-fed induction machine, IEEE Transactions on Power Electronics, 1988, 3(4): 420-429.

[2] K.Yuki, K.Kondo,et. al, Development of high performance traction drive system without speed sensors, IPEC-Tokyo, 2000.

[3] K. Kondo, K. Yuki, An application of the induction motor speed sensorless control to railway vehicle traction system, 37th IAS Annual Meeting Industry Applications Conference, 2002, 3: 2022-2027

[4] TH.Frenzke, F.Hoffmann, Speed Sensorless control of traction drives-experiences on vehicles,8th EPE, Lausanne,1999

[5] G.Amler,F.Sperr,F.Hoffmann. Highly dynamic and speed sensorless control of traction drives,EPE,Toulouse,2003.

[6] A. Steimel, Stator-Flux-Oriented high performance control in traction,35th Annual Meeting IEEE IAS Industry Application Society, Rome, 2000

[7] M. Nemeth-Csoka, Open-Loop speed calculation in stator-fixed reference frame, European Transactions on Electrical Power, 1996, 6(2): 125-130.

[8] C.Schauder, Adaptive speed identification for vector control of induction motor without rotational transducers, IEEE Transactions on Industry Applications, 1992, 28(5): 1054-1061.

[9] H.Kubota, K.Matsuse, T.Nakano, DSP-Based speed adaptive flux observer of induction motor, IEEE Transactions on Industry Applications, 1993, 29(2): 344-348.

[10] H.Kubota, K.Matsuse, T.Nakano, New adaptive flux observer of induction motor for wide speed range motor drive, 16th Annual Conference of IEEE Industrial Electronics society, 1990, 2: 921-926.

[11] H.Kubota, K.Matsuse, Speed sensorless field oriented control of induction motor with rotor resistance adaption, IEEE Transactions on Industry Applications, 1994, 30(5): 1219-1224.

[12] F.Z.Peng, T. Fukao, Robust speed identification for speed sensorless vector control of induction motors, IEEE Transactions on Industry Applications, 1994, 30(5): 1234-1240.

[13] L. Zhen,X.Longya, A mutual MRAS identification scheme for position sensorless field orientation control of induction machines, IEEE Transactions on Industrial Electronics, 1998, 45(5): 824-831.

[14] Y. R. Kim, S. K. Sul, M.H. Park, Speed sensorless vector control of induction motor using extended Kalman filter,IEEE Transactions on Industry Applications, 1994, 30(5): 1225-1233.

[15] Texas Instrument, Sensorless control with kalman filter on TMS320 fixed-point DSP.

[16] S. Stasi, L.Salvatore,F.Cupertino,Comparison between adaptive flux observer and extended kalman filter-based algorithms for field oriented control of induction motor drives, EPE Conference,Lausanne, 1999.

[17] A.Steimel, J.Wiesemann, Further development of direct self control for application in electric traction, Proceedings of the IEEE International Symposium on Industrial Electronics,1996.

[18] A.Steimel, Direct self-control and synchronous pulse techniques for high-power traction inverters in comparison, IEEE Transactions on Industrial Electronics, 2004, 51(4): 810-820.

[19] Ch.Evers, K.Worner, F.Hoffmann, Flux-Guided control strategy for pulse pattern changes without transients of torque and current for high power IGBT-Inverter drives, EPE Conference,Graz, 2001.

[20] K.Worner, A.Steimel, F.Hoffmann,Highly Dynamic Stator Flux Track Length Control for High Power IGBT Converter in Traction Drives, EPE Conference, Lausanne, 1999.

[21] F. Hoffmann, S. Koch, Steady State analysis of speed sensorless control, IECON, Aachen, 1998.

[22] M.Depenbrock, C. Foerth, S. Koch, Speed sensorless control of induction motors at very low stator frequencies, EPE Conference, Lausanne, 1999.

[23] M. Depenbrock, C. Evers, Model-based speed identification for induction machines in the whole operating range, IEEE Transactions on Industrial Electronics, 2005,53(1):31-40.

[24] H. Kubota, I. Sato, Y. Tamura, et al. Regenerating-mode low-speed operation of sensorless induction motor drive with adaptive observer,IEEE Transactions on Industry Applications, 2002, 38(4): 1081-1086.

[25] I.Takahashi ,Y. Ohmori, High-performance direct torque control of an induction motor, IEEE Transactions on Industry Applications, 1989, 25(2): 257-264.

Permissions

The contributors of this book come from diverse backgrounds, making this book a truly international effort. This book will bring forth new frontiers with its revolutionizing research information and detailed analysis of the nascent developments around the world.

We would like to thank Xavier Perpinya, for lending his expertise to make the book truly unique. He has played a crucial role in the development of this book. Without his invaluable contribution this book wouldn't have been possible. He has made vital efforts to compile up to date information on the varied aspects of this subject to make this book a valuable addition to the collection of many professionals and students.

This book was conceptualized with the vision of imparting up-to-date information and advanced data in this field. To ensure the same, a matchless editorial board was set up. Every individual on the board went through rigorous rounds of assessment to prove their worth. After which they invested a large part of their time researching and compiling the most relevant data for our readers. Conferences and sessions were held from time to time between the editorial board and the contributing authors to present the data in the most comprehensible form. The editorial team has worked tirelessly to provide valuable and valid information to help people across the globe.

Every chapter published in this book has been scrutinized by our experts. Their significance has been extensively debated. The topics covered herein carry significant findings which will fuel the growth of the discipline. They may even be implemented as practical applications or may be referred to as a beginning point for another development.

The editorial board has been involved in producing this book since its inception. They have spent rigorous hours researching and exploring the diverse topics which have resulted in the successful publishing of this book. They have passed on their knowledge of decades through this book. To expedite this challenging task, the publisher supported the team at every step. A small team of assistant editors was also appointed to further simplify the editing procedure and attain best results for the readers.

Our editorial team has been hand-picked from every corner of the world. Their multi-ethnicity adds dynamic inputs to the discussions which result in innovative outcomes. These outcomes are then further discussed with the researchers and contributors who give their valuable feedback and opinion regarding the same. The feedback is then collaborated with the researches and they are edited in a comprehensive manner to aid the understanding of the subject.

Apart from the editorial board, the designing team has also invested a significant amount of their time in understanding the subject and creating the most relevant covers. They scrutinized every image to scout for the most suitable representation of the subject and create an appropriate cover for the book.

The publishing team has been involved in this book since its early stages. They were actively engaged in every process, be it collecting the data, connecting with the contributors or procuring relevant information. The team has been an ardent support to the editorial, designing and production team. Their endless efforts to recruit the best for this project, has resulted in the accomplishment of this book. They are a veteran in the field of academics and their pool of knowledge is as vast as their experience in printing. Their expertise and guidance has proved useful at every step. Their uncompromising quality standards have made this book an exceptional effort. Their encouragement from time to time has been an inspiration for everyone.

The publisher and the editorial board hope that this book will prove to be a valuable piece of knowledge for researchers, students, practitioners and scholars across the globe.

List of Contributors

J. Sadeghi
Iran University of Science and Technology, Iran

Huang Yuanliang
Jinan University, China

José A. Lozano, Jesús Félez, Juan de Dios Sanz and José M. Mera
Universidad Politécnica de Madrid, Spain

Cătălin Cruceanu
University POLITEHNICA of Bucharest, Romania

Mohammad Ali Sandidzadeh and Babak Shamszadeh
School of Railway Engineering, Iran University of Science and Technology, Iran

A.G. Hessami
Innovation Director, Vega Systems, UK

Liu Heyun and Gu Xiaosong
1School of Communication and Control Engineering, Hunan University of Humanity, Science and Technology, Hunan Province, China

Tang Wenbin
School of Energy and Power Engineering, Changsha University of Science and Technology, Hunan Province, China

Sanjeev Kumar Appicharla
Institution of Engineering and Technology, UK
The International Council on Systems Engineering, UK

Fazhou Wang and Yunpeng Liu
Wuhan University of Technology, China

X. Perpiñà, L. Navarro, X. Jordà and M. Vellvehi
IMB CNM CSIC, Barcelona, Spain

Jean-François Serviere and M. Mermet-Guyennet
Alstom Transport, Tarbes, France

Jabbar Ali Zakeri
Iran University of Science and Technology, School of Railway Engineering, Iran

Luca Fumagalli, Paolo Tomassini, Marco Zanatta, Giorgio Libretti and Marco Trebeschi
Q-Tech Srl, Rezzato (BS), Italy

Giovanna Sansoni and Franco Docchio
Laboratory of Optoelectronics, University of Brescia, Brescia, Italy

Akira Matsumoto, Takeshi Mizuma and Hirotaka Mori
National Traffic Safety and Environment Laboratory, Japan

Yasukuni Naganuma
Central Japan Railway Company, Japan

Hitoshi Tsunashima
Nihon University, Japan

Jose Brizuela and Carlos Fritsch
Consejo Superior de Investigaciones Científicas (CSIC), Spain

Alberto Ibáñez
Centro de Acústica Aplicada y Evaluación No Destructiva - CAEND (CSIC-UPM), Spain

Ding Rongjun
Csr Zhuzhou Institute Co., Ltd, China